On-Road Intelligent Vehicles

On-Road Intelligent Vehicles

Motion Planning for Intelligent Transportation Systems

Rahul Kala

Robotics and Artificial Intelligence Laboratory, Indian Institute of Information Technology, Allahabad, India

AMSTERDAM • BOSTON • HEIDELBERG • LONDON
NEW YORK • OXFORD • PARIS • SAN DIEGO
SAN FRANCISCO • SINGAPORE • SYDNEY • TOKYO

Butterworth-Heinemann is an imprint of Elsevier

Butterworth-Heinemann is an imprint of Elsevier
The Boulevard, Langford Lane, Kidlington, Oxford OX5 1GB, UK
50 Hampshire Street, 5th Floor, Cambridge, MA 02139, USA

Notices
Knowledge and best practice in this field are constantly changing. As new research and experience
broaden our understanding, changes in research methods, professional practices, or medical
treatment may become necessary.

Practitioners and researchers must always rely on their own experience and knowledge in
evaluating and using any information, methods, compounds, or experiments described herein.
In using such information or methods they should be mindful of their own safety and the safety
of others, including parties for whom they have a professional responsibility.

To the fullest extent of the law, neither the Publisher nor the authors, contributors, or editors,
assume any liability for any injury and/or damage to persons or property as a matter of products
liability, negligence or otherwise, or from any use or operation of any methods, products,
instructions, or ideas contained in the material herein.

British Library Cataloguing-in-Publication Data
A catalogue record for this book is available from the British Library

Library of Congress Cataloging-in-Publication Data
A catalog record for this book is available from the Library of Congress

ISBN: 978-0-12-803729-4

For information on all Butterworth-Heinemann publications
visit our website at https://www.elsevier.com/

  Working together
to grow libraries in
developing countries

www.elsevier.com • www.bookaid.org

Publisher: Joe Hayton
Acquisition Editor: Sonnini R. Yura
Editorial Project Manager Intern: Ana Claudia Abad Garcia
Editorial Project Manager: Mariana Kühl Leme
Production Project Manager: Sruthi Satheesh
Designer: Mark Roger

Typeset by TNQ Books and Journals

Contents

Acknowledgement

The book talks about vehicles exhibiting erratic behavior on the road and feasible strategies and tactics to avoid accidents with such vehicles. The main inspiration was Indian traffic, the dynamics of which are partly discussed in the book. Hence, all the drivers that I have met — decent or insane, motorbike or car divers, who drive with hands or hands free, who caused accidents for me or saved me from accidents, who drive at triple the speed limit or under the speed limit, who shout more and drive less or drive more and shout less, with or without a license, drunk or sober — they are all thanked 'unalike'.

A good part of the book is drawn from my PhD work. I thank my PhD supervisor, Prof. Kevin Warwick, currently serving as the Deputy Vice Chancellor (Research) at the Coventry University, for his continuous support and guidance which resulted in an excellent thesis, thereby motivating this book. Nothing felt better than the sight of a line not corrected by him for (grammatical) errors. I further thank the academic and administrative staff of the University of Reading in general and School of Systems Engineering in particular for their support in providing the necessary infrastructure and support for the research which grew up in the form of this book. Although fire alarms and the loss of internet at the lab were the most cherished moments, strong administrative and infrastructural backing ensured quality output. In the same vein, I thank the Commonwealth Scholarship Commission in the United Kingdom and the British Council for their financial assistance to the research. I also thank all the past and current members of the Call Centre Lab for all their advice, help and support from time to time. Without them, there might have been some less than serious research in a typical peace-loving lab.

In particular, I extend my thanks to Shoureya Kant for his help in the sub-problem of traffic merging, Archit Kumar for his help in the problem of traffic light operation, Prakhar Mohan Agarwal and Safeer Afaque for their help in the sub-problem of road detection, Christopher James Shackleton for enriching discussions concerning Advanced Driver Assistance Systems and the technology otherwise.

The book would have never been possible without active support from everyone at the host institution, the Indian Institute of Information Technology, Allahabad. Although the gang of students made sure to disturb me even at the odd hours to delay the delivery of the book, the academic discussions went a long way to improving the quality of the book and enriching the content of the specific chapters. I also thank Prof. G.C. Nandi and Dr Pavan Chakraborty from the Robotics and Artificial Intelligence Laboratory, whose works in robotics have been instrumental in the formulation of the book. I also thank the other academic colleagues of the institute who have been very supportive and have helped throughout the process. I also extend my thanks to the administrative staff for the uninterrupted support towards development of the book. I extend my thanks to Prof. Anupam Shukla and Dr Ritu Tiwari from the Indian Institute of Information Technology and Management, Gwalior, whose inspiration and support constantly reflects in my works.

Rahul Kala

Introduction

1.1 Introduction

Standing at the cliff of a remarkable technology, and to see it change the very fundamental picture of the most important aspect of life is similar to experiencing the majestic history being created in front of you. We have already travelled the era of discovery of automobiles, public transport, railways, ships and airways, and witnessed the impact these technologies have had on the life of one and all. We have already witnessed the automobiles scale up to the most economical scales, causing less pollution, driving at the highest speeds and levels of comforts and at the same time displaying a very high level of luxury. As a result, it is common to see people passionate about the vehicles they possess and love driving or travelling with these vehicles.

The era of the human-driven vehicles being eliminated and being replaced by the self-driving cars is almost here, with some connotations attached. These vehicles have a lot to offer including driver ease, no need of hiring a driver, ability to sleep and enjoy while the machine drives, ability to move efficiently, not worrying about the parking etc. Therefore, the technology is irresistible. This also gives the ability to frame the next generation and extremely efficient transportation networks, which will enable the quickest and safest transportation of a large number of people, operating using futuristic technologies, the most sophisticated Artificial Intelligence tools and techniques and using hardware and software solutions like never before.

It is intriguing to open the wrapper of this interesting technology to get an understanding of the complex concepts which make it happen, while presented in manner simple to understand and easy to master. This book is devoted to the same. Fundamentally, the autonomous vehicles can be seen as mobile robots, which can look around, understand the traffic scenario, make intelligent decisions and act upon them, exactly similar to the way in which humans drive the vehicles. So one needs the basic hands, legs and body to be found in the autonomous vehicles; mechanisms to understand what the eyes see and mechanisms to instruct the legs to move as desired; and some memory to remember the happenings. Once we get these building blocks, the machines can be taught how to drive, like a driving instructor does for the humans, which is the role that we will assume while designing all the planning algorithms. The ability to instruct is largely facilitated by the available tools and techniques of Artificial Intelligence, which though need to be tweaked to suit the specific requirements.

Once the vehicles are taught how to drive, their role can then be extended to traffic inspectors to decide how the traffic should operate and to regulate the traffic accordingly, thereafter designing algorithms which automatically do so. Ultimately, the role is extended to the overall transportation manager who knows the states, sources, destinations, cause of travel etc. for all vehicles and has the potential to regulate traffic so that the overall goodness is maximized. The purpose of the book is to design efficient algorithms which do all of this. It is important to realize the potential of having all vehicles talk to each other, additionally talk to the intelligent agents located at the roadside and the central transportation management authority.

The current demonstrations by academia, research labs and other companies showcase the promising future of the technology. The visitors to these demonstrations normally find it hard to believe their eyes that a vehicle is indeed driving on its own, and that the demonstration is not a scary movie or a nightmare. The people who actually get a chance to sit inside are even more excited, although a bit scared, not immediately trusting the technology. These demonstrations ultimately suggest a very soon and a very effective widespread adoption of these vehicles, with some caution and concern. With such a background, it is natural to extrapolate the developments in the related domains of robotics, vision, intelligent vehicles and intelligent transportation systems. The future will see extraordinary vehicles, engineered by unique designs suited to the specific task that they will be expected to perform, operating in the next-generation transportation systems, in which every imagination is possible. It further goes to eliminate the gap between the on-road and off-road vehicles, mobile robots and autonomous vehicles and, more surprisingly, the land, air, water and underwater robotics. In the future one does expect robots to have extraordinary capabilities, similar to the ones being widely projected in different science fiction movies. The book presents the in-depth technology behind the intelligent vehicles and intelligent transportation systems, in pursuit of designing the futuristic transportation networks.

1.2   Why Autonomous Vehicles?

Before delving deeply into the technology, it is always better to assess the end returns. Autonomous vehicles have a lot to offer in the future and clearly fascinate young researchers and technocrats, which is motivating enough to pursue research and development in the area. It is easy to portrait oneself as an architect of the future transportation system, to sculpt each and every aspect of the vehicle and the overall system. The technology behind autonomous vehicles and intelligent transportation systems provides the enabling tools to showcase such a future. Although the pros may be highly benefiting and rewarding, one must never underestimate the cons and the costs incurred or likely to be incurred in the long run. This section puts the pros and cons of the technology into perspective, to clarify the context before the book takes the readers deeper into the insight about the technology.

1.2.1   Advantages

Autonomous vehicles are largely motivated by passenger safety. Every year numerous people die or are seriously injured in road accidents. The accidents are largely caused by human errors in perception and decision-making, or by lack of attention and sometimes by late reactions. The autonomous vehicles can use the active percept of a number of redundant sensors to timely make correct driving decisions and hence eliminate accidents. Autonomous vehicles further result in more efficient travel. This means one has to spend much less on insurance, while the insurances will naturally become less expensive. The vehicles benefit from efficient onboard algorithms, communication with the vehicles around, the transportation management centres and information from different sources. This facilitates making better informed and optimal decisions. Further, these vehicles can park themselves close to each other and operate in high-speed platoons, which significantly increase the operational efficiency. The implications of a single vehicle may seem limited, but a network of such intelligent vehicles has the potential to make the overall transportation system operate efficiently. As an example, the vehicles can collectively operate to avoid congestions.

Driving is not always a pleasing and joyful experience and many people have to force themselves to drive. The autonomous vehicle relieves the humans from this mundane task, and the person can instead let the vehicle drive itself and do tasks that are more important, take rest or simply entertain one. This is especially important from the point of view of social justice, as these vehicles will enable the disabled and elderly citizens to commute who currently cannot on their own. It is also possible for vehicles to deliver goods and services without any human onboard. This facilitates the use of autonomous vehicles for delivery of courier, goods, automated taxis, continuous long-distance travel, as a means of public transportation, as guide vehicles, inspection vehicles etc. The vehicles can also be used to negotiate difficult-to-drive roads, which is not very easy for the humans to negotiate. Many times the human drivers need to be hired, which increases the cost of transportation, which means that both time and operating costs are reduced by the use of autonomous vehicles, resulting in less expensive and better services. Not only will the manpower be reduced, the number of vehicles will be reduced as well. The vehicle will be able to leave the passenger, park it and come to pick up the passenger on being asked remotely. This will result in fewer parking spaces needed around the core areas of markets and offices.

1.2.2 Concerns

The picture is not entirely beautiful; there are some black spots as well, although the limitations and concerns are much less compared to the advantages. The greatest concern in the use of autonomous vehicles is the legal and legislative aspects. These vehicles are already giving a hard time to the legislators to frame policies about their operation. An accident by a human-driven car has well-defined laws about the defaulter and the punishment, but in case of an accident by an autonomous vehicle, naturally the vehicle itself cannot be punished for poor driving. The issue raises serious ethical concerns, because a robot is not allowed to harm or kill a human, which may be possible in case the algorithms of the autonomous vehicle falter. It is still unclear how the vehicle should react when an accident is imminent, and a choice has to be made whether to save the passenger, an infant or a grown-up pedestrian. Humans are able to make such decisions, whereas vehicles may not have the same ethical judgement ability.

Costs are a matter of great concern. Autonomous driving capability adds costs to a traditional vehicle in terms of both hardware and software. Because a lot of money is invested in research and development, the initial software costs are expected to be very high. Lack of acceptance by the initial market, added to high costs, may diminish the acceptability of these vehicles. The hardware costs are as well currently very high, which need to be scaled down to be accepted. These vehicles will create sources of automation, and, as a result, numerous drivers of trucks, buses, taxis etc. will become jobless. This will create unemployment unless an alternative source of employment is generated. The technology already faces the criticisms of trade unions and workers.

There are some implementation issues as well. These vehicles always obey laws, whereas human drivers often break laws by some magnitude. One may not always make optimal separations with the vehicles in front, make close cut-ins and overtakes, get inside a roundabout even though it may not be very safe etc. This causes the autonomous vehicles to under-perform in comparison to the human drivers. Although teaching the vehicles to break traffic laws is not suggested, operating in such scenarios questions the acceptability of these vehicles. They may wait for a prolonged time at roundabouts or cause trouble to the drivers around who expect the vehicle to drive a little more aggressively. Further, the autonomous vehicle owners will constantly want to reduce the number of human-driven vehicles due to the added benefits; the people will continue to demand their right to drive on their own. This will cause debates in the future, especially considering that some people have a passion for driving and are not willing to let the vehicle drive itself. It is and will always be uncertain if these vehicles have been trained against all kinds of traffic and all kinds of traffic situations. This creates a constant fear of accident due to unseen scenarios not considered in design. Because everything is networked, privacy is another concern; so is the case of the hacking of networks.

1.3 A Mobile Robot on the Road

Intelligent vehicles operating autonomously on the road may be easily seen as mobile robots which navigate from one place to another. The fundamental technology behind autonomous vehicles is the same as that of *mobile robotics* (Holland, 2004; Siegwart et al., 2011). Both of the techniques have the problem of navigating a robot from one place to another, while avoiding collisions with other robots and entities which may otherwise be treated as static or dynamic obstacles. The output is a trajectory to be followed, which must be the shortest, fastest or the safest to be called optimal. Autonomous vehicles are a special example of the general class of mobile robots. Like mobile robots, these vehicles perceive the world, make a map representation of the percept which is constantly updated, plan the motion in multiple layers of hierarchy and finally move as per the plan. Hence, the techniques discussed in the book will be, first, from the perspective of general mobile robotics and then noting the differences that arise when the same techniques are applied for the specific problem of autonomous-vehicle driving. Mobile robots largely operate on widely bound or unbound home and office environments, whereas vehicles have a well-defined road structure and need to obey the traffic laws and best driving practices. This can be both a boon and a curse to the problem of vehicle navigation, in contrast to the general problem of the navigation of mobile robots. Throughout the book, this aspect of the difference will be highlighted.

1.4 Artificial Intelligence and Planning

Many breakthroughs in the domain of robotics have happened due to the development of *Artificial Intelligence* (AI) tools and techniques (Konar, 1999; Russell and Norvig, 2009). AI deals with the techniques of enabling machines to make decisions displaying some intelligence normally associated with humans. AI techniques work to gather information about the environment, interpret the information, model the information and use the same to make effective decisions which are acted upon. The actions may further change the environment which enables making subsequent decisions. The AI agent does the entire task of modelling and decision-making, and is programmed to act rationally or should make decisions which have the best-expected returns given the limited knowledge of the environment.

The problem of *planning* in AI specifically deals with, given some information about the environment and knowledge of the rules of operation of the environment and the agent, deciding the sequence of moves to make such that the agent gets the desired work done in the minimum expected cost and with the maximum expected returns. Planning is responsible for deciding the sequence of decisions of the agent to be made to accomplish the goal in the best possible way. If the environment changes unlike expectations, one may have to *re-plan* to still reach the goal in the best possible way. One may compute a *policy* instead, which states the actions to be taken by the agent for every possible situation. The policy is hard to initially compute, but once computed can be referred to make the best decision for any possible situation, accounting for the uncertainties in the environment.

AI and robotics are closely related to each other. Most tasks of robots are performed using AI techniques (Choset et al., 2005; Tiwari et al., 2013), whereas robotics is probably the best example to illustrate the practical utilities of AI techniques. Accordingly, throughout this book, AI forms the theoretical basis of the book with an application in robotics, specifically intelligent vehicles. Although the reference to classical and modern AI will not be explicitly made in the book, the different methodologies discussed will be deeply rooted in AI.

1.5 Fully Autonomous and Semi-Autonomous Vehicles

The vehicles operating autonomously in the transportation network is the big dream which will eventually turn out to be reality, wherein the number of autonomous vehicles on the road will slowly start rising, ultimately occupying most or all of the transportation infrastructure. There are a number of hurdles in the way of autonomous vehicles, which are slowly getting cleared. Till the time the dream becomes a reality, one may expect the intelligence of vehicles to continuously rise, while autonomous vehicles display a limited capability. As an example, vehicles could communicate with other vehicles and nearby transportation management centres, make some specific driving manoeuvres like parking and vehicle following or give out information about traffic conditions. Semi-autonomous vehicles will have limited capabilities as compared to autonomous vehicles, while requiring a human to be present in the vehicle. Semi-autonomous vehicles may vary largely in their capabilities and utility and will pave the way for the fully autonomous vehicles.

1.6   A Network of Autonomous Vehicles

A single vehicle has a limited perception, limited knowledge about the intent information of the other vehicles and knowledge about the surrounding transportation system. The Intelligent Transportation Systems allow the vehicles to communicate with each other and share information about vision, diving intents and otherwise. The communication makes the vehicles a wireless ad hoc network known as the *vehicular ad hoc network* (Gordon, 2009; Gozalvez et al., 2012; Ma et al., 2009). The vehicles may further communicate with the roadside units and ultimately to the transportation management centres. This creates a picture of the transportation system wherein every vehicle can communicate with each other and all the entities of the transportation system. The transportation management centres can relay information about prospective congestions, blockages etc. to the vehicles. The vehicles can themselves discuss plans and follow the same. The additional communication between the vehicles and other transportation entities plays a major role in enhancing the efficiency of the transportation network and helping the vehicles to make decisions that are more informed.

1.7   Autonomous Vehicles in Action

Largely, from a technological point of view, self-driving cars are not the future, but the present, even though some challenges may be stopping an immediate worldwide commercial release. A very large number of field tests have been done by academia, research labs and all major automobile and information technology companies in collaboration with each other. In all these demonstrations, the vehicle is able to navigate in challenging scenarios against diverse traffic conditions and on public roads of different cities. This section summarizes the works as a motivation towards understanding of the core technology.

1.7.1   Entries From the DARPA Grand Challenge

A significant boost in the technology behind autonomous vehicles is attributed to the *DARPA Grand Challenges*, organized by the American Defence Advanced Research

Projects Agency (DARPA). The DARPA Grand Challenges were races organized between autonomous vehicles, offering a grand prize to the winners of the race. The first challenge was conducted in 2004, wherein a number of teams collaborating between industry and academia participated; however, none of them was able to complete the course. The DARPA challenges are long-distance races requiring the teams to combat extreme conditions including intersections, turns and overtakings. This created difficulties that the teams could not completely combat; however, a distance of 11.78 km could be navigated. It must be mentioned that, before this challenge, the technology had only been demonstrated on small distances.

A legendary event happened in 2005, when the next challenge was organized. Then, five teams completed the total course. The race was won by Stanford's Stanley (Thrun et al., 2006) followed by Carnegie Mellon University's Sandstorm. The next event happened in 2007 when the challenge was a little different than before, replicating realistic urban driving, and was named the DARPA Urban Challenge. The challenge was won by Massachusetts Institute of Technology's Boss (Urmson et al., 2008) followed by Stanford's Junior (Montemerlo et al., 2008). The technical details about the vehicles are beautifully compiled in Buehler et al. (2007).

Not only did the challenges result in a lot of funding towards the area, the event gained wide publicity which attracted more and more teams towards this technology, exactly the intention behind the challenge. Besides, the challenges created a lot of well-organized scholarly literature in terms of journal papers, edited books, special issues of journals etc. which has eased the study of the technology by the newer groups.

1.7.2 Autonomous Vehicles From Different Companies

Nearly all major automobile and some car-sharing/rental companies have invested heavily in this technology and are already planning release of the first commercially available autonomous vehicles by 2018−40. All these companies are in close collaboration with major academic institutions, automation companies and information technology players. The most famous and well advertised is Google with their Google cars, which has already passed many miles of on-road tests and a few challenges to cater to. The project benefits from the works of Sebastian Thrun from team Stanford at the DARPA Grand Challenges.

Similarly, the automobile giant Mercedes-Benz also has a strong presence with its luxury self-driving cars and driverless trucks. In a more economical range is Nissan, which again has experimented widely on roads with its academic partners including major universities of the world. Similarly, Bavarian Motor Works (BMW), with a number of corporate collaborators, is to launch its own self-driving car rather soon. The list includes the likes of Volkswagen, Tesla, General Motors (GM), Honda, Jaguar, Toyota etc. Uber is another major player for its driverless taxis. Besides, nearly all major universities of the world, in collaboration with such companies, are doing intensive research and development in the area. The technology is dynamically changing and is best covered by different science news agencies and bloggers.

1.8 Other Types of Robots

The overall technology used in autonomous vehicles is closely related to the technology of mobile robots. Mobile robots have the capability to move from one place to the other. The most common mobile robots have wheels at their base. The motion is most commonly by using either a differential wheel drive or sometimes caster wheel or Ackermann steering. The other types of robots move by using two or more legs or in a snake-like manner. The different locomotion strategies differ in terms of speed, stability, dexterity, ability to handle uneven terrain etc.

Mobile robots may be indoor or outdoor. Indoor mobile robots operate typically in home, office or factory environments. The outdoor mobile robots may either be on-road or off-road. The on-road mobile robots are the autonomous vehicles discussed in the book. The off-road mobile robotic vehicles are largely used to go for exploration in unknown or hazardous environments, operate in adverse terrains or navigate across areas where roads do not exist. These are used for explorations, disaster management, rescue operations, locating and handling hazardous situations etc.

The mobility may not be limited to the ground. The robots operate over water as boats and ships. These robots need to consider the additional factor of water currents which plays a major role in navigation. The robots may further operate underwater, wherein again the role of water currents is of prime importance. These robots are needed to do underwater repairs, inspection of underwater debris etc.

Robots further can go airborne. The use of quadcopters or drones is increasingly common. Aerial robots are very useful for survey, explorations etc. However, it is difficult to operate them with high payload or for prolonged hours as their battery dies off. The robotic technology is increasingly being applied on satellites and space vehicles which may now be considered as robotic in nature with all characteristics of a robot from sensing to actuation. Even though the focus of the book is on-road intelligent vehicles, it is wise to consider the diminishing gaps between the different types of robots and the added capabilities of each of these types, motivating hybrid robots operating in the transportation system. Further, the technologies of the different types of robots are highly inter-related.

1.9 Into the Future

It is fascinating to forget all the limitations of the current technology, which are rapidly diminishing, and to imagine the transportation system of the future. It is motivating to paint one's own image of the most perfect transportation system that one can imagine. The demand in transportation systems is constantly increasing. With the developments in the different areas of life, one constantly needs to travel from one place to the other. The under-privileged sections of the society have started to enjoy the fruits of development, needing to travel around, while people have increasingly started to travel for leisure, recreation, holiday, meeting friends and relatives or otherwise. Along with the increasing traffic demand, the need for safety, luxury and efficiency in travel is steeply rising. People cannot invest a lot of time in travelling, although they would like a pleasant travelling experience.

Certainly, the transportation system of the future should facilitate high speeds, high efficiency, safety and comfort, while performing at alarmingly high demands. How the transportation system of the future will meet these requirements is a very big and open question in front of the researchers, technocrats and philosophers. The main break-through in the transportation system will be brought by the technology of autonomous vehicles, which will be capable of driving at very high speeds, while being on high alert to avoid accidents even at such high speeds. The increasing rise in computation and memory of the computers, added to the advances in AI showcase the same capa-bility. On top, the vehicles of the future can talk to all the other vehicles to make collaborative plans which are the best for performance. The vehicles can further talk to the different roadside units, traffic lights, intersection managers and overall transpor-tation management centres. Not only would the vehicles be able to know the current traffic ahead, they will be able to predict the future traffic to make plans in anticipation and to further coordinate with each other to eliminate congestions, routing the traffic in advance to eliminate any building up of congestions. In such conditions, it may be viable to assume that the manually driven vehicles may ultimately get eliminated from the traffic, baring a few special roads offering a racetrack for the humans to drive purely for joy.

It is evident that the future can offer much more. Increasing autonomy makes it possible to engineer vehicles with unconventional designs and capabilities, which operate in unconventional manners. The platoon of vehicles is a simple example, wherein vehicles closely stick to each other and navigate at high speeds. Vehicles could even attach and detach to each other in the future, saving or sharing fuel costs. All this enables moving a large number of vehicles at high speeds over a limited trans-portation network. Further, a vehicle following the normal mode of operation is not the only possibility. The future may see micro unmanned vehicles to big transportation ve-hicles. The small unmanned vehicles could drive under the larger ones, vehicles could board and de-board public transportation systems, vehicles may be able to re-adjust their size as per the situation, vehicles may even become compact to additionally use pedestrian and cyclist lanes, or if the situation gets much worse, some vehicles may have the capability to go airborne. This calls for unification of on-road vehicles, off-road vehicles, mobile robotics, aerial robotics and modular robotics; the research in all areas is progressing very quickly in shaping the future of the transportation network. Although every person may have their own picture of the future, it is clear that the future is fascinating, dynamic and will benefit from technologies unimaginable in the current point in time.

1.10 Summary

The chapter was devoted to motivating readers towards this exciting technology, at the same time presenting the saleable and salient features. The foremost thing to consider before adopting any technology, or while devoting oneself to the field, is the pros and cons of the technology. Although efficiency and comfort happen to be the biggest pros, the reliability, legal considerations and loss of employment will remain big challenges hindering growth. The technology itself is derived from the technology of mobile robot motion planning which uses the components of AI to facilitate intelligent decision-making. The classical mobile robotics theory gives a generic framework to work upon, whereas the AI tools and techniques are responsible for providing an algorithmic framework to transform the understood percept into informed decisions. The options increase when the vehicles are networked with one another, to the roadside units and ultimately to the transportation management centres. The vehicles have passed numerous field tests and are on the verge of widespread adoption, changing the traffic dynamics of modern day transportation systems. The future implications are equally interesting, wherein teams of autonomous vehicles will operate in the most advanced transportation infrastructure, maximizing the efficiency and safety at every level.

References

Buehler, M., Iagnemma, K., Singh, S., 2007. The 2005 DARPA Grand Challenge: The Great Robot Race. Springer, Berlin, Heidelberg.

Choset, H., Lynch, K.M., Hutchinson, S., Kantor, G.A., Burgard, W., Kavraki, L.E., Thrun, S., 2005. Principles of Robot Motion: Theory, Algorithms, and Implementations. MIT Press, Cambridge, MA.

Gordon, R., 2009. Intelligent Freeway Transportation Systems. Springer, New York.

Gozalvez, J., Sepulcre, M., Bauza, R., 2012. IEEE 802.11p vehicle to infrastructure communications in urban environments. IEEE Communications Magazine 50 (5), 176–183.

Holland, J., 2004. Designing Autonomous Mobile Robots, first ed. Elsevier, Burlington, MA.

Konar, A., 1999. Artificial Intelligence and Soft Computing: Behavioral and Cognitive Modeling of the Human Brain. CRC Press, Boca Raton, FL.

Ma, Y., Chowdhury, M., Sadek, A., Jeihani, M., 2009. Real-time highway traffic condition assessment framework using vehicle–infrastructure integration (VII) with artificial intelligence (AI). IEEE Transactions on Intelligent Transportation Systems 10 (4), 615–627.

Montemerlo, M., Becker, J., Bhat, S., Dahlkamp, H., et al., 2008. Junior: the Stanford entry in the Urban Challenge. Journal of Field Robotics 25 (9), 569–597.

Russell, R., Norvig, P., 2009. Artificial Intelligence: A Modern Approach, third ed. Pearson, Harlow, Essex, England.

Siegwart, R., Nourbakhsh, I.R., Scaramuzza, D., 2011. Introduction to Autonomous Mobile Robots, second ed. MIT Press, MA.

Thrun, S., Montemerlo, M., Dahlkamp, H., Stavens, D., et al., 2006. Stanley: the robot that won the DARPA Grand Challenge. Journal of Field Robotics 23 (9), 661–692.

Tiwari, R., Shukla, A., Kala, R., 2013. Intelligent Planning for Mobile Robotics: Algorithmic Approaches. IGI Global Publishers, Hershey, PA.

Urmson, C., Anhalt, J., Bagnell, D., Baker, C., et al., 2008. Autonomous driving in urban environments: Boss and the Urban Challenge. Journal of Field Robotics 25 (8), 425–466.

Basics of Autonomous Vehicles 2

2.1 Introduction

The technology behind autonomous vehicles may appear as technical jargon of challenging concepts and methodologies. This chapter is devoted to dissecting the attractive-looking autonomous vehicles to reveal the internal technology involved. The process of making an autonomous vehicle involves procurement of a commercial vehicle which serves as a platform for research and development. The vehicle is fitted with sophisticated sensors through which a computer programme may look at the world, much like human drivers look at the driving scenario with their eyes and subsequently perceive the environment by listening to the sounds through their ears.

Many vehicles nowadays come with a drive-by-wire compatibility, which means that the different driving mechanisms including the steering can already be controlled by a joystick or a computer programme. This provides the interface between the hardware and the software. The sensors are connected to a cluster of computing systems, which are networked to share data. The computing systems compute the output which is passed to the motors of the steering, brake and throttle. This drives the vehicles.

The intelligence of the vehicles, giving it the capability to autonomously navigate in the traffic scenario, is hidden in the software of the vehicle. The software is responsible for mapping the sensory percept of the sensors to the actuation of the actuators. The mapping is, however, extremely complex, considering that a large number of sensors generate a lot of data, which can be mapped to complex navigation manoeuvres. Hence, a step-by-step and hierarchical approach is adopted, wherein understanding of the sensor percept happens in different steps, which are used to make a map representation of the world. The location of the vehicle is also estimated with the help of sensors. Motion planning of the vehicle requires a map representation of the world and the current location of the vehicle, both of which can be facilitated by the said modules. The motion planning happens in multiple levels of hierarchies. The trajectory generated by motion planning is used by the control algorithms to control the vehicle and send relevant signals to the actuation units.

This chapter is organized as follows. First, the hardware aspects of the technology are discussed in Section 2.2. This includes different types of sensors, use of drive-by-wire for actuation, and the processing, power and networking units. Section 2.3 presents the software perspectives. Section 2.3.1 presents the vision systems. Section 2.3.2 discusses the mechanism for making a map of the world. Section 2.3.3 presents the technique to localize the vehicle in the map. Section 2.3.4 presents the motion-planning perspectives to make a trajectory, which is followed using control algorithms presented in Section 2.3.5. The Human–Machine Interface issues are given in Section 2.3.6. Section 2.4 specifically details the localization algorithms. These include Kalman

Filters presented in Section 2.4.1, Extended Kalman Filters presented in Section 2.4.2 and Particle Filters presented in Section 2.4.3. The different control algorithms are given in Section 2.5.

2.2 Hardware

The autonomous vehicle, from a hardware perspective, obviously consists of a modern automobile with all the standard automobile hardware including brake, throttle, steering, wheels, etc. The commercially available autonomous vehicles include additional hardware to act as a platform wherein they can be made autonomous by the use of relevant software. The additional hardware primarily consists of the sensors, actuators and processing units. Each of these is described in the next subsections. The basic concepts are highlighted in Box 2.1 and illustrated in Fig. 2.1.

2.2.1 External Sensors

The autonomous vehicle uses sensors to look at the physical world, so as to facilitate decision-making. The sensors may be of different types, while operating in different domains and with different operational frequencies. The intention is to redundantly

---

Box 2.1 Summary of Vehicle Hardware

- Sensor Redundancy
    - Uses many sensors to record the same thing
    - Combines outputs of different sensors
    - Reduces error, reduces the effect of noise, increases field of view, gets more detailed information, gets different information about the same region
    - Works on different frequencies and levels of details enabling hierarchical decision-making
- Sensors
    - *Vision*: Video Camera, Infrared camera
    - *Proximity*: Ultrasonic, SONAR, RADARs and LIDARs
    - *Motion and Internal*: Inertial Measurement Unit (accelerometer, gyroscope and magnetometer), GPS, encoders
    - *Three-dimensional sensing*: Stereovision and 3D LIDARs
- *Drive by wire*, to enable driving using a computer programme
- Computing and Networking
    - *Cluster* of computers for parallel and fast computing
    - *Networking* of the sensors, computing devices, vehicles and road infrastructure
    - *Watchdog* for error recovery
    - *Process Management* for scheduling very important to less important processes
    - Logging and visualization for debugging
- *Power* for all devices, sensors and actuators

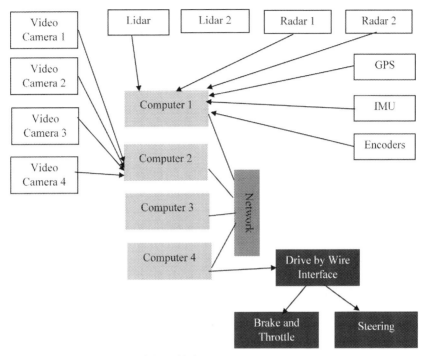

Figure 2.1 General hardware of the vehicle.

look at the world with a large array of sensors. The sensors can be highly noise prone, and relying on individual sensor readings can be dangerous, considering that the sensory percepts are used to make important decisions in driving and a wrong percept can lead to a wrong decision, ultimately potentially fatal in driving. The redundancy enables making the most likely estimate based on the percepts of multiple sensors. Further, the Field of View of a single sensor may be highly limited, which can be extended by using more sensors with overlapping Fields of View. The sensors best perform within a certain range and Field of View. Use of multiple sensors enables every sensor to focus on the area best suited according to that sensor's operation. Many times different sensors may give different information about the same region. For example, a *vision sensor* can say whether the obstacle ahead is a vehicle or a pedestrian, whereas the *proximity sensor* can be used to give the distance to the obstacle in front. In addition, different sensors work at different frequencies and have different computational requirements. Therefore, high-frequency sensors may be associated with emergency services, whereas low-frequency and rich-information sensors may be used to make coarser driving decisions.

The vehicles are thus equipped with multiple copies of different types of sensors, located at different regions of the vehicles. The most common type of sensor to use is a *vision camera*. The camera takes a snapshot of the situation and stores it as an image. Image processing and computer-vision techniques can be used to get information about the situation from the image. Wide-angle lenses can be used for extended views. Alternatively, many cameras can be used with overlapping fields of view to get extended information about the scene. This requires solving the problem of correspondence between images, to fuse two camera images to get one extended view of the world. Infrared cameras may also be used along with the normal vision cameras.

The problem with a monocular camera is that it cannot be used to project the image into the three-dimensional (3D) world, unless one uses stereovision. Hence, one cannot gettrue distance information. In navigation, the distance to obstacles ahead is information of paramount importance. *Proximity sensors* are widely used to get accurate distance information. They work by sending a wave that hits the obstacle ahead and is reflected back. By analyzing the reflected wave, an accurate estimate of the distance can be made. Two of the most common examples of proximity-sensing technology are *ultrasonics* and SONAR (SOund Navigation And Ranging), which send a sound signal and wait to hear the same back to estimate the distance. Many modern-day cars are already fitted with ultrasonics to detect obstacles in front used for Adaptive Cruise Control, collision-warning alert, automated parking, etc. Ultrasonic sensors send a signal at a particular angle at which the sensor is pointed and measure the distance from the obstacle in the same direction. One can employ an array or a ring of ultrasonic sensors to get information about all obstacles at different angles. Alternatively, one could mount the ultrasonic sensor on a pan—tilt unit which rotates along all directions to get the distance estimates. The former technique increases the cost of employing multiple sensors, but the sensors can work in parallel to give a quick response and faster updates. However, the latter method will update the map once the pan—tilt unit completes rotation along all combinations, which will slow the updates, while a high-resolution scan can be achieved.

Similarly, *RADARs* (RAdio Detection And Rangings) are widely used to get estimated distance to the obstacles. The RADARs are mounted on the vehicle and operate at high frequencies to give the distance from all obstacles ahead. The RADARs rotate to give a 360-degrees scan of the world. The sensor operates by using radio waves. Multiple RADARs can be mounted at different parts of the vehicle, front and rear, to get the complete view of objects all round. More-expensive equipment is LIDAR (Light Detection And Ranging), which uses light to scan the entire world and give the depth to all nearby obstacles in 360 degrees in the mounting plane with a small angular rotation. The LIDAR may be mounted horizontally or vertically.

2.2.2   Stereovision and 3D Sensing

Stereovision techniques use two cameras to see the same object. The two cameras are separated by a baseline, the distance for which is assumed to be known accurately. The two cameras simultaneously capture two images. The two images are analyzed to note the differences between the images. Essentially, one needs to accurately identify the same pixel in both images, known as the problem of correspondence between the two cameras. Features like corners can be easily found in one image, and the same can be searched in the other image. Alternatively, the disparity between the images can be found to get the indicative regions in the other image, corresponding to the same regions in the first image, for which a small search can be used. The disparity helps to get the depth of the point which enables projecting it in a 3D world used for navigation.

It is also common to use 3D vision sensors like a *3D LIDAR* which scan the entire world for all angles and make a *point cloud*. Every point in the point cloud corresponds to one solid angle of operation and the distance from the obstacle in that direction. In this way, the sensor can scan for all possible solid angles, and return a point cloud. The point cloud reports the distance corresponding to every angle, and is a 2.5-dimensional (2.5D) data structure. The same can be projected into a 3D world and given to the mapping server after the point cloud has been transformed into the global world reference frame. A way of having the sensor is to mount the laser on a pan—tilt unit and to operate it for all possible pan and tilt angles.

2.2.3 Motion and Internal Sensors

The vehicle also makes use of sensors to track its own motion and thus estimate the location based on the last estimates of its position. An *Inertial Measurement Unit* (IMU) is used, which consists of an accelerometer, a gyroscope and a magnetometer. The *accelerometer* measures the acceleration along all three axes of its own coordinate frame, or the coordinate frame in which it is mounted. The accelerometer works on the principle of suspending a tiny mass from a spring. The displacement of the mass along the three axes will be proportional to the force applied, and knowing the mass the accelerations along the three directions can be determined. Similarly, the *gyroscope* measures the orientation with respect to its own frame of reference. The gyroscope consists of a rotor which rotates a small wheel. The gyroscope uses the law of conservation of momentum to make the rotor resist any change in alignment due to external forces, and keep itself aligned in the same direction, even if the mounting frame changes its orientation. The angle between the mounting frame and the rotor can thus be measured, which gives the orientation of the vehicle along the principle axis. *Magnetometers* are also used. These metres measure the magnetic field and orient themselves in the direction of the magnetic field using the magnetic forces. The change in orientation of the magnetic pointer can be measured to give the orientation change of the vehicle.

Similarly, a *Global Positioning System* (GPS) can give the location of a vehicle in terms of the global coordinate frame (or latitude and longitude). GPS uses the information from satellites to compute the position. The satellites constantly broadcast their location information and the global time which is maintained by a globally synchronized atomic clock. Geostationary satellites do not change their position with respect to the earth, whereas for the other satellites the position at any time can be computed. The position of satellite and time information is used from multiple satellites to compute the position of the GPS receiver with respect to the satellite and ultimately in terms of the global coordinate system. Although GPS seems a promising technology to localize the vehicle, or to accurately know the position of the vehicle, the GPS error can be very large in areas of tall buildings and other areas where GPS reception is low. It is suggestive from the literature not to heavily rely on GPS for accurate positioning. Instead, vision systems should also be used.

The wheels of the vehicle can be fitted with *encoders*. The encoders are fitted on the rotating wheel and measure the rotation of the wheel. The system consists of a sensor fitted on the shaft along with an encoder wheel which rotates along with the wheel. A reflector is used to give pulses which are constantly detected by the sensor. The encoder wheel consists of fine markings which stop the reflector to send signals to the sensor when a marking blocks the way. Hence, tics are generated and sensed by the sensor, which correspond to the motion of the wheel. The encoders can be made to have a high resolution. The magnitude of motion of the wheels can be used to help estimate the position, orientation and velocity of the vehicle. Potentiometers work on the same principle and convert angular rotation into a potential value which can be detected and used for measurement of the rotation. Neither technique, however, accounts for slippage and is thus not a good option when the vehicle may lose traction on the way. Every sensor has some or the other limitations because of which the individual sensors alone cannot be used to get the position of the vehicle. Hence different internal and external sensors are used together to overcome each other's limitations.

2.2.4   Actuators and Drive by Wire

The other important aspect of autonomous vehicle hardware is the actuators, which physically move the vehicle as desired. Autonomy needs to be provided to all hardware units that the human operates while driving. This includes the brake, throttle, steering, gearshift (in case the vehicle does not have an automatic transmission), indicators and horns. The most important of these are the brake and throttle for maintaining a speed profile and the steering for tracing a desired trajectory. The others may be naively handled. Modern vehicles come with a *drive-by-wire* technology which enables driving the vehicle using a keyboard, joystick, or any automated programme. The technology provides a programming interface to move the vehicle using computer programmes. The computer programmes can be made to send reference signals to the steering, brake and throttle mechanisms, which have inbuilt microcontrollers to physically operate these devices based on the reference signals given. Many automobile manufacturers now support drive-by-wire enabling a ready-to-programme vehicle from an actuation perspective.

2.2.5   Processing and Networking

The sensors produce high frequency and high-volume data, which are processed using advanced data processing techniques. The large variety of sensors require a lot of processing capabilities. Similarly, the mapping algorithms responsible for integrating the outputs of the different sensors have high-processing and memory requirements. The planning and decision-making algorithms may themselves be very computationally expensive. Hence, autonomous vehicles are equipped with high-end computing systems which are continuously running. It may not be possible for a single system to do all the processing in small computational times, and, hence, different types of computations are off-loaded to different servers loaded on the vehicle.

Process Management becomes an important aspect of such a system. Every system runs multiple services in parallel, each responsible for some sensing action, data processing, decision-making, or actuation. The multiple services can communicate to each other. Some of these services may be responsible for taking emergency actions, which need to be processed in real time; whereas some other services may be responsible for taking long-term decisions and may be operated with a lower priority. The process manager deals with all these issues, ensuring that the processes are executed as per requirement, while no process starves thereby making the particular process useless.

The systems have a *watchdog* running to quickly detect software and hardware failures and to take precautionary actions accordingly. A typical action can be to off-load computation to a different system if one of the systems fails. Accordingly, if a sensor fails, it can be deactivated and any sensor from the redundant set of sensors may be used. If the failure is fatal, or if operations are not reliable, these systems can make the vehicle come to a safe position and then stop. Sometimes the vehicle may be asked to stop immediately by invocation of an emergency stop signal. Then a human can take over and manually start driving the vehicle. Automated trouble-shooting may be invoked in parallel to isolate the problem and take preventive measures. In such a case, the vehicle can indicate when it is ready to resume autonomous driving. This can cater to situations like a software error (in which case restarting the software is enough) or system failure (in which case the particular system may be rebooted). If the vehicle cannot automatically recover, service by a technician will be required.

All the sensors, computation and processed information is continuously logged in the autonomous vehicles. These logs can be very useful in isolating the problem and detecting errors. A hardware failure is clearly indicated in the log. The system logs can be used to recreate the operational scenario in which the failure occurred and replay the scenario with the ground truth and system information multiple times. This is used by programmers to debug the error.

All the systems are connected together by a *networking* framework. Further, a network exists between the different sensors and actuators which connect them to these computing devices. The protocol of communication is important. The protocol should be free from packet loss and thus must ensure reliability, and should not have high-networking overheads making the protocol very slow. Hence, Transmission Control Protocol is a poor choice for communication due to its slow operations. Similarly, the communication between the sensors and the computing devices needs to allow fast propagation of information. Multiple vehicles themselves are connected by a Vehicle Ad hoc Network (VANET), which is also connected to the central transportation management authority to manage the overall traffic.

2.2.6 Power

The *power system* is responsible for powering up all the devices which are attached to the vehicle. The devices include the processing units and the power requirement of any sensors, actuators and the networking devices. Power cords may be drawn from the vehicle to power up all the devices. It is easy to give uninterrupted power supply to all the devices to ensure that the devices keep performing for a continuous operation of the vehicle as it drives from the source to the goal.

2.3 Software

Software gives the autonomous vehicles the capability to make intelligent decisions and thus operate autonomously in the road traffic. The overall architecture is similar to the problem of mobile robot navigation (Holland, 2004; Siegwart et al., 2011; Tiwari et al., 2013). The architecture follows a typical *sense—plan—act* cycle, wherein the data are taken from multiple sensors, the processed data are used to plan the move, and the actions are made so as to physically move the vehicle, which changes the environment detected by the sensors based on which a revised plan and a revised action are taken. The vehicle continuously operates in cycles of sense—plan—act to navigate. The complete architecture may not necessarily operate in a single layer of hierarchy. The decisions are always made at multiple layers. Each decision requires its own sensing and gives its own actions. The higher layers are responsible for long-term decisions like deciding the route, which are made at the least frequency. The lower-level decisions are made with high frequency, are based on high-frequency sensors and include tasks such as monitoring the control error to generate control signals.

The different sensors work at different frequencies. Likewise, the different planning and control algorithms working at different layers also work at different frequencies and have different computational expense. The overall architecture is thus a set of nodes which work in parallel and do some job needed for the overall task of navigation. The nodes can communicate with each other and share data. So the nodes associated with each sensor can pass the sensor readings to the nodes responsible for data processing of the sensors, which can further propagate the information to the mapping nodes. The planning algorithms working on different layers of the hierarchy can further take the map or sensor information and send the decision to master or slave control units for physical motion of the vehicle. The overall architecture of the system is summarized in Fig. 2.2. The concepts are also summarized in Box 2.2. Each of the components is described in the following sections. The architecture is simplified for better understanding. The architecture adopted in many working prototypes can be found in Leonard et al. (2008), Montemerlo et al. (2008), and Urmson et al. (2008).

2.3.1 Vision

The *vision* systems are responsible for seeing the robotic world with the help of different sensors like camera, LIDARs, RADARs, etc. The sensors are redundantly placed to give an overall view of the world, whereas the same region may be covered by multiple sensors to combat noise. Each sensor records the data in its own domain. Processing techniques are needed to analyze the raw data recorded by the sensors into something meaningful which can be used by the planning algorithms. The processing of data may be different for all sensors. The vision camera requires computer vision techniques to identify the different things round. The proximity sensors may be used to compute distance to the obstacles round. Each of these distances needs to be analyzed to get information about all the obstacles. The sensor readings may be tracked with time to get the speed information. Chapter "Perception in Autonomous Vehicles" gives an in-depth treatment to the topic of vision in intelligent vehicles.

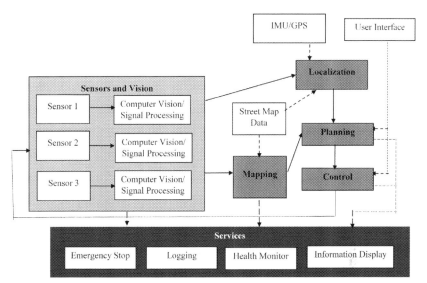

Figure 2.2 Software architecture.

Box 2.2 Summary of Vehicle Software Architecture

- **Basic architecture**
  - *Sense−Plan−Act* cycles
  - Hierarchical data processing and decision-making
- **Vision**
  - Convert raw sensor data into information
  - Get information about obstacles, roads, lanes, traffic signs, etc.
- **Mapping**
  - Map information from different sensors into one compact environment representing data structure
  - Take the most reliable reading if multiple sensors report different facts
  - *Occupancy grid* for 2D, in which each cell may be navigable or nonnavigable, decided based on integration of outputs of sensors operating in the region of the cell
- **Localization**
  - Gets the position and orientation of the vehicle in the map
  - Tracks the vehicle and other obstacles for pose estimates
  - Can be used to estimate speeds and accelerations
  - *Simultaneous Localization and Mapping* (SLAM), simultaneously computes the pose and makes the map
  - Pose represented as a Probability Density Function (PDF)
  - *Prediction Step* applies the motion to compute the updated PDF
  - *Update step* accounts for the sensor readings to update the PDF
  - *Kalman Filter* assumes a multivariate Gaussian Function as a PDF
  - *Extended Kalman Filter* extends the working of the Kalman Filter to nonlinear systems by locally linearizing the system
  - *Particle Filters* use a collection of particles with state and weight to imitate the PDF, and can represent multimodal PDF for global localization

Continued

Box 2.2 Summary of Vehicle Software Architecture—cont'd

- **Motion Planning**
  - Given a map, the current state or the source and the desired state or the goal, computes a collision-free trajectory connecting the source and goal, adhering to the vehicle and problem constraints
  - *Optimal planners* guarantee generation of an optimal trajectory
  - As the vehicle moves and the map changes, the trajectory is continuously reworked by the *replanning* algorithms
- **Control**
  - Generates signals to be given to the motors to adjust the traced trajectory very close to the reference trajectory
  - Notes error and feeds it back in *closed-loop control*
  - Controllers may work in a master—slave or hierarchical mode
  - *Lateral controller* responsible for steering
  - *Longitudinal controller* responsible for speed
  - Formulates a control strategy requiring formulation of vehicle kinematics incorporating speeds, and dynamics incorporating forces and torques
  - *Linear Control*: Proportional—Integral—Derivative (PID)
  - *Nonlinear Control*: fuzzy controller, neural controller
  - *Specialized Techniques*: Optimal Controllers, Predictive Control, Robust Controller
- **Human—Machine Interface Issues**
  - Allows humans to set and change destination, give route specifications, speed preference, etc.
  - Allows humans to give and take control
  - Asks humans to take control in cases of deadlock, the vehicle takes control in case of emergency
  - *Semiautonomous vehicles* may give information about the route, warnings and suggestions

---

2.3.2 Mapping

Multiple sensors give different information about the different areas. An area may as well be covered by multiple sensors which serve that area, and the readings of the different sensors may not agree, resulting in ambiguous information about the environment. In such a case, the *mapping* module needs to decide how much of which sensor to believe as per the conditions. The purpose of mapping is to make a single data structure which stores all information about the environment. This data structure is called the map. The map is queried by all planning and other decision-making algorithms. The information about everything in the environment including obstacles, vehicles, roads, boundaries, etc. is summarized and stored on the map. The map may be built iteratively as the vehicle moves and gets information about the different areas.

A two-dimensional (2D) map indicating the road and the drivable and nondrivable regions inside the road can be used to take most driving decisions. This 2D map is made from the environment by assessing whether the vehicle can pass through a region or not. A popular method to make and store this map is in the form of an *occupancy grid*, which divides the entire region into small grids and asks every sensor to vote about the presence or absence of the obstacle in that grid, if the region was being sensed by the sensor and the sensor did record evidence regarding the presence or absence of the obstacle. The decision which gets the majority of the votes stands. A weight addition may be done to incorporate the confidence of the classifier in its vote. Similarly, advanced fuzzy and statistical integration techniques may also be used. The actual map may be 2.5D indicating the height of the obstacles or the slope in addition, thus enabling a vehicle to decide whether to go over the obstacle or to avoid it. Because height alone is only a partial dimension, it may be referred as a 2.5D map.

For navigation in a traffic scenario, the road-network graph of the area of operation is already known with good accuracy. The graph is a good source of information regarding the roads, intersections, lanes, diversions and the distance to intersection. The local map, on the contrary, gives information about the navigability and obstacles in the area. The two maps can be integrated to get a detailed view in the vicinity and a broad global view, to be used for decision-making.

2.3.3 Localization

Localization deals with the mechanism to find out the position and the orientation of the vehicle. It locates the vehicle in the map developed by the mapping technique. Localization can be used to constantly *track* the vehicle as it moves with the help of the sensors. The motion sensors can be used to give the maintained accelerations, speeds and orientations at every moment, which can be integrated to give the position and orientation at every time. The problem is that the sensors are noisy, and the readings of the sensors will not be accurate, which will lead to wrong estimates. The vision sensors can be used to look at the world and give the position estimates with respect to the neighbouring landmarks, similar to how humans estimate their position in a known environment. These sensors, and thus the information as well, are noisy, and the landmarks may not always be within sight or reliable. The localization techniques use all this information to compute the most likely position and orientation of the vehicle from all the sensor recordings. Using similar reasoning, the localization algorithms can be used to track and estimate the pose of all the obstacles, other vehicles, and, in general, the things of interest.

GPS is a good sensor which gives the position of the vehicle (whereas the IMU can be used to further give the orientation).However, the GPS errors are very large and the sensor can only be used for global localization for higher-order decisions like route selection, selection of the exit to take at an intersection, etc. For most local decision-making regarding obstacle avoidance, lane change, lane keeping, etc., the vehicles heavily rely on the vision sensors.

The map of the environment is needed so that the localization algorithms can be used to compute the location based on the vision percept. The mapping algorithms as well make the map based on the current position of the sensors or the current position of the vehicle. All the obstacles and regions of interest are plotted with respect to the vehicle's position. Therefore, mapping requires localization, whereas localization requires mapping. Together, the problem is known as *Simultaneous Localization and Mapping* (SLAM), wherein the mapping and localization happen simultaneously. This is assuming that no known landmark exists which is always visible and the position of which is accurately determined to act as a reference point for localization and mapping.

2.3.4 Motion Planning

Motion planning is responsible for all decision-making of the vehicle. Motion planning may be done at different levels of hierarchy like route planning, complex obstacle avoidance, simply reacting to a nearby obstacle, etc. Typically, the vehicle needs to navigate in a road scenario, and motion planning is responsible for computing a feasible and safe trajectory for the vehicle to follow. The motion-planning algorithms need the map and the location as input, which is facilitated by the mapping and the localization modules. The trajectory gives a general notion about the desired motion of the vehicle in the operational scenario. As the vehicle moves, the situation may change, which is handled by continuously replanning the trajectory as per the changed environment. Chapters "Perception in Autonomous Vehicles, Advanced Driver Assistance Systems, Introduction to Planning, Optimization Based Planning, Sampling Based Planning, Graph Search Based Hierarchical Planning, Using Heuristics in Graph Search Based Planning, Fuzzy Based Planning, Potential Based Planning, and Logic Based Planning" talk about the different planning techniques used in the navigation of autonomous vehicles.

2.3.5 Control

The planning algorithms generate a reference trajectory. The *control algorithms* are necessary to control the vehicle by sending the relevant signals to all the actuation units, such that the trajectory traced by the vehicle is close to the reference trajectory. The control of an autonomous vehicle may be very hard and many times the control algorithms operate in a master—slave mode. The master-control algorithm interprets the reference trajectory to give the desired speed and steering necessary, which is given to the low-level controller which computes the signals to send to each motor to get the required speed and steering profile. The control algorithms must facilitate very small error between the reference trajectory and the trajectory traced by the vehicle, otherwise the vehicle may collide with the nearby vehicles. The vehicles maintain very small lateral separations with the nearby vehicles, and thus the margin of error is very small.

2.3.6 Human—Machine Interface Issues

The vehicle needs to interact with humans in many ways. All these issues are handled by the *human—machine interface* module. A fully autonomous vehicle drives completely independently. Nevertheless, it requires a human to interact with the system and give the destination information. Further, the human is given the power, at any time, to either leave the autonomous driving mode and drive the vehicle him- or herself, or stop driving him- or herself and enter into an autonomous driving mode. The human may as well change the destination or give route specifications, or ask the vehicle to drive at a particular speed, etc. The software caters to all these needs of the user. If the vehicle is in a deadlock, or if the vehicle seems to be making wrong manoeuvres, a human must act instantly and take the control of the vehicle, which the vehicle must facilitate.

A semiautonomous vehicle does a part of the driving itself, and the rest is performed by the human. Sometimes the software may only be used to give information to the driver. The interface is required to let the human control the vehicle and make use of all its functionalities. The vehicle may give information about the route, or give warnings from an audio, text message, image, or a tactile feedback for better understanding of the user.

2.4 Localization

Localization deals with the problem to determine the pose of the vehicle (position and orientation) at every instant of time. It is assumed that the vehicle has motion sensors which can sense the motion of the vehicle and give pose estimates,, and further that there are external sensors which can determine the pose of the vehicle with respect to some external landmarks, and thus give an estimate of the vehicle's pose. If the sensors were noise-free, any one of the sensors could be taken to estimate the vehicle's pose.However, all sensors are noisy, and all redundantly provide information about the vehicle's pose. The use of localization techniques is thus to estimate the vehicle's pose given the noisy sensor readings. The localization techniques do so by taking the most-expected pose of the vehicle per the sensors and their uncertainty models.

The pose of the vehicle may never be accurately known due to the noisy sensors. Hence, the localization techniques return a probability distribution of the pose of the vehicle. It is common to assume the vehicle's pose as the pose which has the maximum probability in the given probability density function for any decision-making. However, many algorithms can additionally incorporate the localization probability density function to make decisions that are more effective.

The localization algorithms typically try to take the pose estimates of the previous time step along with the current readings of the motion of the vehicle and the current external sensor readings. The aim is to make the new pose estimates based on all this information. The algorithms typically have two steps, prediction and update. In the *prediction step*, the vehicle is assumed to move and the motion is recorded by the suitable motion sensors, each of which has a noise model. Alternatively, the vehicle moves using some control signals which can be measured or are known before sending them to the vehicle. The dynamics of the vehicle are assumed to be known. In such a case, knowing the previous pose estimate of the vehicle, new estimates may be made. A concern is, however, that the sensors are noisy and the noise gets added into the estimates made. The second step, or the *update step*, is when the vehicle uses its external sensors to look at different landmarks and get pose estimates from them. Again, these sensors are noisy and the noise factor needs to be accounted for. These readings are used to correct the pose estimates.

Fig. 2.3 shows the general methodology of the approach. Suppose that a robot can only move in the X-axis, so the state is just described by the X-axis position value. The initial position is known with some uncertainty. The initial Probability Density Function is given in Fig. 2.3A. Now the robot is asked to move along the X-axis by three units, whereas it may not be able to move exactly three units due to noise. The uncertainty increases as an effect. The predicted Probability Density Function is given in Fig. 2.3B. Suppose the robot has a sensor which can sense its location with some uncertainty. The updated Probability Density Function is given in Fig. 2.3C. Here, the robot's sensor gave a strong indication about the position and hence the uncertainty reduced dramatically. These steps happen repeatedly as the robot moves.

A major assumption made here is that the system is assumed discrete, wherein the motion takes place in some discrete steps and, further, the observations are made from the sensors in some discrete steps. The observations and motion may, hence, be sampled at some frequency and the computations made on the sampled time steps. This is natural in robotics as the control signals are applied at discrete time steps and the motion planning algorithms plan and replan with some frequency at discrete time steps. The frequency of computation can be kept high or low depending upon the accuracy and the computational resources available.

Different algorithms make different assumptions about the system, sensors and the way to model probability density function. Three dominant algorithms are discussed: Kalman Filters, Extended Kalman Filters and Particle Filters. Even though the interest is to find the pose of the vehicle, the localization algorithms can be used to track or filter any source of information. Hence, the more general term, *state* of the system, is used for discussions.

2.4.1   Kalman Filter

The *Kalman Filter* (Choset et al., 2005; Evensen, 2003; Julier and Uhlmann, 1997; Wan and Van Der Merwe, 2000) attempts to estimate the state of the system based on the previous state, the recorded control inputs and the sensor percept. The Kalman filter stores the probability density function in the form of a Gaussian distribution, which means that the probability density function is unimodal and can be characterized by a mean and variance along with all the associated state variables. It is a multivariate Gaussian distribution that takes care of all the correlations between all the variables. The Gaussian assumption may initially seem too restrictive, however, it is extremely useful to simplify the computations of the Kalman Filter. Let the system be given by Eqs [2.1] and [2.2]:

$$x(k+1) = F(k)\,x(k) + G(k)\,u(k) + v(k) \qquad [2.1]$$

$$y(k) = H(k)\,x(k) + w(k) \qquad [2.2]$$

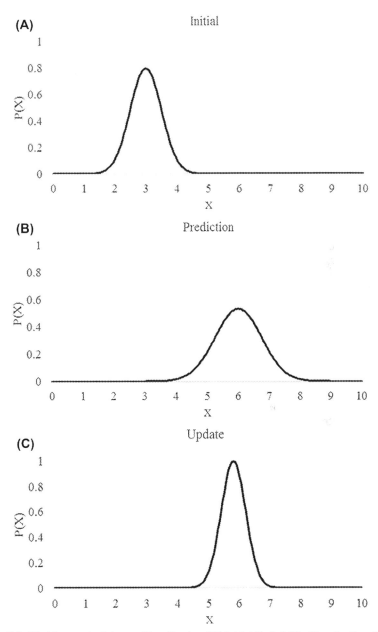

Figure 2.3 Working methodology of localization (A) Initial Probability Density Function, (B) Probability Density Function at prediction when the system moves three units on X-axis, (C) Probability Density Function at update when the sensors make a strong observation indicating the current position.

Here, $x(k)$ is the state of the system at the time instance k. $F(k)$ is related to the dynamics of the system. $u(k)$ is the control inputs given to the system at time k. The matrix $G(k)$ defines how the control affects the system. The term $v(k)$ is the *process noise*, which is assumed to be Gaussian white noise with mean 0 and covariance matrix $V(k)$. Overall, once the applied inputs and the system dynamics are known, it should be possible to compute the state of the system as the vehicle moves. As an example, the pose and speed of a vehicle may be determined based on the previous pose and speeds and the magnitude of brake, throttle and steering applied. An additional factor of noise needs to be incorporated, as the main reason of using Kalman filters is the presence of noise. The main assumption here is that the equation is assumed linear.

The term $y(k)$ is the observations made by the sensors at time k. Not all the state variables may be observable, and further the state variables themselves may not be observable, whereas the quantity that is actually observed may be some function of the state variables. The matrix $H(k)$ denotes how the state variables map to the observed variables at time k. $H(k)$ is assumed to have a full-row rank. The sensors are again noisy, and therefore a noise term $w(k)$ needs to be added. The term $w(k)$ is the *measurement noise*, which is assumed to be Gaussian white noise with mean 0 and covariance matrix $W(k)$. In a driving scenario, distance from landmarks like the lane boundaries can be a good indicator of position, which is an observed variable dependent upon the current pose and speed of the vehicle. Similarly, other observation variables can be used, all related to the current pose. Given the pose, it is possible to compute the values of all these variables, assuming that all information about the landmarks is clearly known. The matrix $H(k)$ converts the pose into the values of all the observed variables, which accounts for the presence of noise.

Assuming that some estimate is available for the time instance k, the task is to compute the estimate at time instance $k + 1$. Knowing the initial estimate (even though with a lot of uncertainty), the same principle can be used to get the estimates as the vehicle moves and the observations are made from the sensors. The problem is not only to compute the estimate at every instance of time, but also the covariance, as the estimate is a Gaussian distribution. The mean is denoted by $\hat{x}$, whereas the covariance is denoted by P. Let the term $\hat{x}(k + 1|k)$ denote the estimate of time $k + 1$, given all previous observations and state estimates till the time k with the associated covariance as $P(k + 1|k)$. Similarly, $\hat{x}(k + 1|k + 1)$ is defined as the estimate at time $k + 1$, given all observations till time $k + 1$, with covariance $P(k + 1|k + 1)$. Let all estimates at time k be computed. The task is to compute the estimate for time $k + 1$.

The first step is prediction. Here, the vehicle moves by application of a control sequence $u(k)$, which is known. This first step is to predict the state at the next time instance $\hat{x}(k + 1|k)$ based on this information alone. The prediction is given by Eqs [2.3] and [2.4]. This can be done by knowing the dynamics of the vehicle and the assumed noise model. Then, the second step of update is applied. Here, the vehicle makes some observations, whereas the mapping of state to observation is already known. These observations are used to correct the errors and make better estimates, or compute $\hat{x}(k + 1|k + 1)$. The update step is given by Eqs [2.5]–[2.9].

All the equations have been used here without proofs to make the concepts simple. The derivations are made on the principle to maximize the expectation of the observations from the state estimates.

Prediction

$$\widehat{x}(k+1|k) = F(k)\,\widehat{x}(k|k) + G(k)\,u(k) \qquad [2.3]$$

$$P(k+1|k) = F(k)\,P(k|k)\,F(k)^T + V(k) \qquad [2.4]$$

Update

$$z = y(k+1) - H(k+1)\,x(k+1|k) \qquad [2.5]$$

$$S = H(k+1)\,P(k+1|k)\,H(k+1)^T + W(k+1) \qquad [2.6]$$

$$R = P(k+1|k)\,H(k+1)^T\,S^{-1} \qquad [2.7]$$

$$\widehat{x}(k+1|k+1) = \widehat{x}(k+1|k) + R\,z \qquad [2.8]$$

$$P(k+1|k+1) = P(k+1|k) - R\,H(k+1)\,P(k+1|k) \qquad [2.9]$$

Eqs [2.3] and [2.4] simply use the previous estimate and apply the dynamics equation to get the new estimates, accounting for the known-motion noise covariance. The estimate and the covariance of this step is also called as the *prior*. The update step is harder to understand. Eqs [2.5]−[2.7] define the basic terms used to write the update equations. The first term z defined in Eq. [2.5] is the innovation error measured as the difference between the observations made and the expected observation based on the last computed estimates. Similarly, the innovation covariance is denoted by S and given by Eq. [2.6]. The term R is the Kalman gain given by Eq. [2.7]. Based on these terms, Eq. [2.8] is simply the expected estimate based on the observed values, with the covariance given by Eq. [2.9]. The estimate and the covariance of this step is also called as the *posterior*.

2.4.2 Extended Kalman Filter

The greatest problem of the Kalman filter is the linear assumption of the system dynamics, which may not always be true. The *Extended Kalman Filter* (Choset et al., 2005; Evensen, 2003; Julier and Uhlmann, 1997; Wan and Van Der Merwe, 2000) works for nonlinear systems when the dynamics and observability of the system may be nonlinear, but still known. The Extended Kalman Filter works by locally linearizing the system around the current point of operation to compute the prediction and update, whereas the actual system works on the actual nonlinear model. By continuously linearizing the system around the current state of operation, one can compute the corrections for the next time step to make the estimates. The system may now be given in a more general form. Eqs [2.1] and [2.2] are extended to account for nonlinearity, given by Eqs [2.10] and [2.11].

$$x(k+1) = f(x(k),\, u(k),\, k) + v(k) \qquad\qquad [2.10]$$

$$y(k) = h(x(k),\, k) + w(k) \qquad\qquad [2.11]$$

Here, the function f corresponds to the dynamics of the system based on the current state and the current input sequence, whereas the function h accounts for the nonlinear observability function based on the current state of the system. Both these functions have the noise terms.

The prediction and update step based on the local linearity of the system is given by Eqs [2.12]–[2.20], which are similar to Eqs [2.3]–[2.9], with the assumption of absence of the matrices $F(k)$ and $H(k)$. Both functions are nondetermined due to the nonlinear functions f and h. However, with the assumption of local linearity, the two matrices may be computed by taking a gradient of the corresponding function at the point of operation. The matrix $F(k)$ is computed in Eq. [2.13] and the matrix $H(k)$ is computed in Eq. [2.15].

Prediction

$$\hat{x}(k+1|k) = f\big(\hat{x}(k|k),\, u(k),\, k\big) \qquad\qquad [2.12]$$

$$F(k) = \left.\frac{\partial f}{\partial x}\right|_{x=\hat{x}(k|k)} \qquad\qquad [2.13]$$

$$P(k+1|k) = F(k)\, P(k|k)\, F(k)^T + V(k) \qquad\qquad [2.14]$$

Update

$$H(k) = \left.\frac{\partial h}{\partial x}\right|_{x=\hat{x}(k+1|k)} \qquad\qquad [2.15]$$

$$z = y(k+1) - H(k+1)\, x(k+1|k) \qquad\qquad [2.16]$$

$$S = H(k+1)\, P(k+1|k)\, H(k+1)^T + W(k+1) \qquad\qquad [2.17]$$

$$R = P(k+1|k)\, H(k+1)^T\, S^{-1} \qquad\qquad [2.18]$$

$$\hat{x}(k+1|k+1) = \hat{x}(k+1|k) + R\, z \qquad\qquad [2.19]$$

$$P(k+1|k+1) = P(k+1|k) - R\, H(k+1)\, P(k+1|k) \qquad\qquad [2.20]$$

2.4.3 Particle Filtering

The greatest problem with the Kalman Filters is the assumption of a Gaussian probability distribution function which is unimodal. The choice of a unimodal function may seem apt for situations wherein the sensor gives a fair enough indication of the position, and one makes a Gaussian distribution around it to account for uncertainties. However, many times the problem may be *Global localization*, wherein the robot knows about the map but does not have any indication of its position. The initial sensory percepts give some indication of the location, but numerous places in the map may have the same sensory percepts. In such a case, the probability distribution function of the robot is multimodal in nature. As the robot moves and gets more percepts, the ambiguity in terms of the multitude of locations will get resolved, and the problem henceforth will be of local localization. Similarly, as the robot moves, to which of the two locations the sensory percept belongs is ambiguous, and the distribution becomes bimodal. It may even be possible to get a series of false percepts which give a false indication regarding the position of the robot, which is hard to correct for unimodal distribution methods when the true percepts are later made.

The greatest difference between the *Particle Filters* (Arulampalam et al., 2002; Chen, 2003) and the Kalman Filters is the way to represent the probability distribution of the state. The Particle Filters represent the probability density function by making use of a set of *particles*, which sample out the ideal probability distribution function to represent the same based on a few samples only. Each of the particles in the set represents a *state* at whatever step is the algorithm. The particles are further attached with a *weight* denoting the likelihood of the particle in agreeing to the control sequence and sensor percepts. The state of the particle along with the weight represent the probability density function at any time. As the number of particles tends to infinity, the probability function approaches the ideal continuous probability density function as applicable for the system with the given dynamics, noise models, observations and control sequences. The method is a special case of the *Sequential Importance Sampling* methods, which keep sampling out particles based on their importance, while the importance is proportional to the weights. The particular technique discussed is *Sampling Importance Resampling*. Here, the probability density function at each iteration is represented and modified to get the probability density function after incorporating the motion and observation. Monte Carlo simulations are made to get the updated state of the particles after application of some control, whereas Markov Chain assumptions are made in the computation of the probabilities, in which case the historical probabilities and actions are not necessary to be stored.

Let at time instance k the particle i be $m_i(k)$ constituting the pair $<x_i(k), w_i(k)>$, in which $x_i(k)$ is the state that the particle represents and $w_i(k)$ is the associated weight. Let $M(k)$ be the set of particles at time step k. First, the vehicle moves using some control sequence $u(k)$. This gives the updated state of the particles as given by Eq. [2.21].

$$x_i(k + 1) = f(x_i(k), u(k), k) + v(k) \qquad [2.21]$$

Here, $v(k)$ is the process noise with known variance $V(k)$. The technique, hence, generates a Gaussian random number with mean 0 and variance $V(k)$, imitating the natural noise, to give the prediction of the particle. Because the factor is random, different particles get displaced by different factors, thus overall representing the effect due to the white noise.

Now the sensors are invoked, and an observation $y(k + 1)$ is made. This observation is used to update the particles. The first step is to compute the new weights associated with the particles. For any particle, the expected outcome is given by Eq. [2.22] in the absence of noise, while the observed output is $y(k + 1)$.

$$y'(k + 1) = h(x(k + 1), k + 1) \qquad [2.22]$$

This is used to compute the probability of seeing the output for the particular state represented by the particle, $P(y(k + 1)|x_i(k + 1))$. The weight x_i of the particle i is proportional to $P(y(k + 1)|x_i(k + 1))$, and is set as $\alpha P(y(k + 1)|x_i(k + 1))$, in which the factor α is the normalization constant and is used to ensure that the weights of all particles add up to 1. The weight may hence be given by Eq. [2.23].

$$w_i = \frac{P(y(k + 1)|x_i(k + 1))}{\sum_i P(y(k + 1)|x_i(k + 1))} \qquad [2.23]$$

The samples then need to be resampled. *Resampling* deletes the samples which have a very small weight in comparison with the others and are hence not very useful. This solves the problem of degeneracy, as in a particle filter; starting from an initial estimate, most of the particles become useless with very high error and hence very small weights. It is preferable to delete those particles and thereby increase particles in the areas in which the weights and thus the probabilities are high. The resampling is done by selecting particles from the pool of particles, with repetition, such that the chance of selection of a particle is proportional to its weight. Therefore, the particles with high weight get selected multiple times, whereas those with nearly zero weight are extremely unlikely to be selected and are hence deleted. This increases the samples near the areas which are more likely to represent the system state, rather than areas which are extremely less likely to represent the system state. The resampled set of particles is given the same weight $1/N$, in which N is the number of particles.

If the initial state of the system is precisely known, all the particles are located at the same state with the same weight. Otherwise, the initial particle distribution is made per the known uncertainty in the state. In the extreme condition, when no information about the state is available, the particles are randomly distributed in the state space with equal weights. The particle state and weights are updated as the controls are applied and the observations are made. Every control sequence changes the state of the system, and the weights are updated to account for the new observations, after which the particles are resampled. The resampled particles go to the next iteration of control and observation.

2.5   Control

The planning modules in autonomous vehicles give a reference trajectory for the vehicle to follow. The localization module is responsible to continuously give the position of the vehicle. The vehicle operates using the steering mechanism to orient the vehicle, and the brake and throttle mechanism to make the vehicle reach the desired speed. The *steering mechanism* is also called the lateral control of the vehicle. The lateral control helps the vehicle to manoeuvre to change lanes, perform overtaking, maintain one's lane, avoid obstacles, etc. The lateral control is the most complicated to handle. The steering is fitted with motors which are sent control signals to operate. The control algorithms are responsible for the generation of these control sequences. The *speed control* is also the longitudinal control of the vehicle. The vehicle can increase its speed by using the throttle, and reduce its speed by using brakes.

Although it is not possible to give an in-depth treatment of the problem of motion control of autonomous vehicles, which can be a book in itself, the basics are covered here for the sake of completeness. For more details, please refer to some interesting references (Kato et al., 2002; Keviczky et al., 2006; Kumarawadu and Lee, 2006; Taylor et al., 1999; Rajamani et al., 2000). The control algorithms decide the signals to be sent to each of these controllable units which maintain the trajectory profile of the vehicle. The general methodology is to compute an *error* term at any instance of time. The trajectory is already computed, while the location of the vehicle is known. This can be used to compute the error. The control algorithms attempt to constantly compute a control sequence such that this error keeps reducing as the vehicle operates.

The first step in approaching the problem from a control perspective is to design the *kinematics* of the vehicle. Autonomous vehicles are rectangular with four tires, two in the front, and two at the rear. The front tires can be steered, whereas the rear tires maintain the same orientation. This is called the double bicycle model. Knowing the angle of orientation of the steering wheels, the position and orientation of the vehicle, and the constants depending upon the geometry of the vehicle, the relation between the linear and angular speed of the vehicle to these parameters can be computed. This constitutes the kinematic model of the vehicle.

Using the kinematics, the next step is to compute the *dynamics* of the vehicle. The dynamics finds a relation between the forces acting on the vehicle, and their effect on the linear and angular acceleration, speed and position. This helps to compute the required torques and forces, which is used to generate the control sequence of the individual motors. The dynamics gives a mathematical model which can be used to propose and test the working of different control methodologies and provide convergence guarantees. It also helps to set the parameters of the control algorithms to give the best performance. It is possible that the modelled dynamics are not the same as the actual dynamics due to assumptions and unmodelled forces and effects. The errors induced as a result are subsequently removed by the working of most of the control methodologies.

Different control methodologies exist to control the vehicle. The open-loop control systems generate a control sequence which is executed on the vehicle. On the contrary,

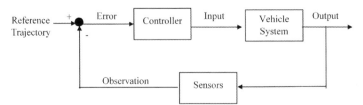

Figure 2.4 Working methodology of feedback control.

closed-loop systems take feedback in the form of error and use the same to compute the next control input given to the vehicle. The methodology is given in Fig. 2.4. The reference trajectory is applied to compute the error. The error is the difference between the reference trajectory and the current system state as observed by the use of sensors. The control methodology converts this error term into the input control signal, which is given to the vehicle which moves per the vehicle dynamics and the vehicle state is changed. The changed state is observed by the sensors to compute the new error.

The control of the autonomous vehicle may be extremely complicated, for which it is common to use a high-level controller to generate basic commands to exhibit some common vehicle behaviours, and the basic commands are given to slave controllers which generate the control sequence used to move the vehicle. Similarly, the speed and steering mechanisms may be handled separately. Sometimes the master controller may interpret the needed trajectory and generate the low-level information which may be used by the slave controllers.

The most basic controller is the *Proportional−Integral−Derivative* (PID) controller. The controller works by giving the control input as the weighted addition of the error function, integral of the error function and the derivative of the error function. The proportional term is from the human intuition that the applied control sequence is smaller when the error is smaller and vice versa. The derivative term avoids oscillations and converges the system to a small steady-state error, whereas the integral term is used to account for constant external errors in the system. Overall, the three factors help in reaching a small steady-state error, wherein the system may showcase very small oscillations in the steady state.

PID is a linear controller which causes its performance to deteriorate when applied on nonlinear systems. The nonlinear controllers can model nonlinearity. The dominant controller used is the *fuzzy controller*. The fuzzy controller takes the error and the change in error as the input, and uses fuzzy-based inference to compute the control sequence. The fuzzy systems operate by rules which can be tuned to get the desired system performance. The different rules model different parts of the input space and together model the nonlinearity of the overall space. Similarly, *neural controllers* can be used which use neural network modelling to compute the control input for the applied input error functions. The neural networks also have nonlinearity, while the system can be learnt using machine-learning algorithms to give the best control sequence when tested over a variety of scenarios.

Optimal Controllers can be used to give the optimal control sequence for any input reference trajectory. The optimal controllers generally model the problem as an optimization problem wherein the control sequence needs to be computed such that the error between the reference trajectory and the trajectory simulated using the vehicle dynamics is the least. Optimization algorithms can be used to compute such a sequence. The algorithms commonly compute the control sequence in discrete time steps. *Predictive Control* algorithms can be used to predict the performance of the controller by looking at the trajectory in the near future, and compute a control sequence which may not be the best considering only the immediate next move, but will prove to be good for the near future trajectory. Similarly, *robust controllers* can be used to get controllers which perform well under large uncertainties and errors.

2.6 Summary

This chapter explored the basic technology behind autonomous vehicles. Even though it can be very challenging to engineer and understand the complete technology behind these vehicles, the basic principles are easy to understand and appreciate. Autonomous vehicles can be easily seen as mobile robots on the road, which navigate using the basic principles of robot navigation. These vehicles look at the world with the help of sensors, make decisions by planning their actions, and implement the actions with the help of actuators. The hardware largely consists of the sensing devices, processing units and the actuation units, besides power and networking infrastructure.

A large number of sensors are fitted into the vehicle to let the vehicles redundantly look at the world and report the inferences. Common sensors include the vision camera and proximity sensors including RADARs, ultrasonics and LIDARs. Stereovision and 3D LIDARs can be used to give 3D vision. The vehicles also consist of an Inertial Measurement Unit and encoders to measure position. The actuators are used in steering, brake and throttle, besides other entities that humans operate while driving. The drive-by-wire technology gives the ability to control vehicles using a computer programme.

The vehicles are operated using sophisticated software which provide the intelligence to the vehicles to navigate autonomously. The vision algorithms take raw sensor values and make inferences based on the same. The inferences from multiple redundant sensors are given to the mapping server to represent the environment as a map. The localization module tracks the vehicle and other obstacles to estimate the position at any time. The common algorithms include Kalman Filters which make a Gaussian assumption of the probability density function assuming a linear system, extended Kalman Filters which approximate the linear system by the use of derivatives, and Particle Filters which represent the probability density function by using a set of particles representing the state with an associated weight.

The motion-planning algorithms are responsible for making the vehicle's trajectory, and broadly take care of all decision-making while the vehicle is in motion. The trajectory is physically navigated using the control algorithms. Different control algorithms like PID controllers, fuzzy controllers, neural controllers, optimal controllers, predictive controllers, robust controllers, etc. may be used.

References

Arulampalam, M.S., Maskell, S., Gordon, N., Clapp, T., 2002. A tutorial on particle filters for online nonlinear/non-Gaussian Bayesian tracking. IEEE Transactions on Signal Processing 50 (2), 174−188.

Chen, Z., 2003. Bayesian filtering: from Kalman filters to particle filters, and beyond. Statistics 182 (1), 1−69.

Choset, H., Lynch, K.M., Hutchinson, S., Kantor, G.A., Burgard, W., Kavraki, L.E., Thrun, S., 2005. Principles of Robot Motion: Theory, Algorithms, and Implementations. MIT Press, Cambridge, MA.

Evensen, G., 2003. The ensemble Kalman filter: theoretical formulation and practical implementation. Ocean Dynamics 53 (4), 343−367.

Holland, J., 2004. Designing Autonomous Mobile Robots, first ed. Elsevier, Burlington, MA.

Julier, S.K., Uhlmann, J.K., 1997. New extension of the Kalman filter to nonlinear systems. In: Proceedings of the Signal Processing, Sensor Fusion, and Target Recognition VI.

Kato, S., Tsugawa, S., Tokuda, K., Matsui, T., Fujii, H., 2002. Vehicle control algorithms for cooperative driving with automated vehicles and intervehicle communications. IEEE Transactions on Intelligent Transportation Systems 3 (3), 155−161.

Keviczky, T., Falcone, P., Borrelli, F., Asgari, J., Hrovat, D., 2006. Predictive control approach to autonomous vehicle steering. In: Proceedings of the 2006 American Control Conference, pp. 4670−4675.

Kumarawadu, S., Lee, T.T., 2006. Neuroadaptive combined lateral and longitudinal control of highway vehicles using RBF networks. IEEE Transactions on Intelligent Transportation Systems 7 (4), 500−512.

Leonard, J., How, J., Teller, S., Berger, M., Campbell, S., Fiore, G., Fletcher, L., Frazzoli, E., Huang, A., Karaman, S., Koch, O., Kuwata, Y., Moore, D., Olson, E., Peters, S., Teo, J., Truax, R., Walter, M., Barrett, D., Epstein, A., Maheloni, K., Moyer, K., Jones, T., Buckley, R., Antone, M., Galejs, R., Krishnamurthy, S., Williams, J., 2008. A perception-driven autonomous urban vehicle. Journal of Field Robotics 25 (10), 727−774.

Montemerlo, M., Becker, J., Bhat, S., Dahlkamp, H., Dolgov, D., Ettinger, S., Haehne, D., Hilden, T., Hoffmann, G., Huhnke, B., Johnston, D., Klumpp, S., Langer, D., Levandowski, A., Levinson, J., Marcil, J., Orenstein, D., Paefgen, J., Penny, I., Petrovskaya, A., Pflueger, M., Stanek, G., Stavens, D., Vogt, A., Thrun, S., 2008. Junior: the Stanford entry in the urban challenge. Journal of Field Robotics 25 (9), 569−597.

Rajamani, R., Tan, H.S., Law, B.K., Zhang, W.B., 2000. Demonstration of integrated longitudinal and lateral control for the operation of automated vehicles in platoons. IEEE Transactions on Control Systems Technology 8 (4), 695−708.

Siegwart, R., Nourbakhsh, I.R., Scaramuzza, D., 2011. Introduction to Autonomous Mobile Robots, second ed. MIT Press, MA.

Taylor, C.J., Košecká, J., Blasi, R., Malik, J., 1999. A comparative study of vision-based lateral control strategies for autonomous highway driving. The International Journal of Robotics Research 18 (5), 442−453.

Tiwari, R., Shukla, A., Kala, R., 2013. Intelligent Planning for Mobile Robotics: Algorithmic Approaches. IGI Global Publishers, Hershey, PA.

Urmson, C., Anhalt, J., Bagnell, D., Baker, C., Bittner, R., Clark, M.N., Dolan, J., Duggins, D., Galatali, T., Geyer, C., Gittleman, M., Harbaugh, S., Hebert, M., Howard, T.M., Kolski, S., Kelly, A., Likhachev, M., McNaughton, M., Miller, N., Peterson, K., Pilnick, B.,

Rajkumar, R., Rybski, P., Salesky, B., Seo, Y.W., Singh, S., Snider, J., Stentz, A., Whittaker, W.R., Wolkowicki, Z., Ziglar, J., Bae, H., Brown, T., Demitrish, D., Litkouhi, B., Nickolaou, J., Sadekar, V., Zhang, W., Struble, J., Taylor, M., Darms, M., Ferguson, D., 2008. Autonomous driving in urban environments: boss and the urban challenge. Journal of Field Robotics 25 (8), 425−466.

Wan, E.A., Van Der Merwe, R., 2000. The unscented Kalman filter for nonlinear estimation. In: Proceedings of the IEEE 2000 Adaptive Systems for Signal Processing, Communications, and Control Symposium Conference, pp. 153−158.

Perception in Autonomous Vehicles

3

3.1 Introduction

Humans have good driving skills which helps them to navigate for prolonged durations of journey without collision, while coordinating well with other drivers. Humans have good situation assessment which helps them to make the best decisions while driving. Humans are able to easily interpret the situation and compute the best possible driving decision. A pool of vehicles operating in the transportation network relies on efficient and cooperative driving by all the users of the road, thus making the transportation network manageable.

The driving capability of humans is largely attributed to their expertise in vision or to see and understand the world around. Humans use their eyes to quickly interpret the information about the roads, lanes, obstacles and other vehicles around. Further, they are alert to the blowing of horns by the other vehicles which gives them an immediate feedback on driving, thus suggesting corrective actions. Further, humans are alert and can quickly react to the changes in the environment. So suddenly spotting a pedestrian in front, a sudden change in traffic light etc. are situations reacted to by slowing down and waiting. It takes significantly less time for an attentive driver to perceive, comprehend and act upon a situation, compared to an inattentive one. One would like the same capability to be built into autonomous vehicles, before they can be put to any good use.

Perception or vision systems comprise onboard sensors and associated processing units in autonomous vehicles. An autonomous vehicle relies on these sensors to get all kinds of information about the driving environment which is used for decision making or planning the motion of the vehicle. The information needed includes road lane markings, road geometry and road orientation. Further, it is important to know about the position, speed and orientation of the other vehicles and obstacles on the road. The vehicle needs, as well, to track its own position, speed and pose. Traffic lights, traffic signs, warning messages, signs on the road including road markings that warn of pedestrians, overtaking allowed, overtaking restricted, one-way roads etc. are important things to look at while driving.

First, Section 3.2 talks about the different sensors and the way to work with each of these. The section discusses the way to fuse the outputs of different sensors working in different domains with different frequencies to make a map of the environment. One of the dominant types of sensors is the video camera which takes images of the world around and requires computer vision techniques for making inference from the images. The computer vision aspects are discussed in Section 3.3. Section 3.4 specifically talks about classification, which is one of the dominant steps for recognition and detection of objects of interest in images. Section 3.5 extends the discussion to tracking of objects using filtering and the use of optical flow. Section 3.6 specifically narrates the amalgamation of vision techniques for the specific problem of vehicle navigation, presenting about the use of vision for detection and tracking of roads, lanes, obstacles, pedestrians and traffic signs.

3.2   Perception

In this section the basic terminologies and concepts of perception are presented, which form the basis for the complete treatment of the topic from the next section onwards. The discussion starts with the basic principles of vision and delves further into more interesting topics. The concepts are nicely summarized in Box 3.1.

3.2.1   Sensor Choices and Placement

One of the most important questions that arises in any problem in robotics is the choice of sensors. A variety of sensors are available which give different levels of information, have different operational frequency, require simple to complex data processing, have different levels of noises and ultimately have different monetary costs which contribute to the cost of the overall vehicle. The most common sensor is the vision camera, which takes an image of the scenario. Further, proximity sensors are used to get the distance to the obstacle. Commonly used proximity sensors include an array of ultrasonic sensors and radio detection and ranging (radar). Light detection and ranging (lidar) is expensive but a very useful sensor. Two-dimensional (2D) lidars can give distance about all obstacles in the plane. Three-dimensional (3D) lidars can

---

Box 3.1 Summary of Perception Techniques

- Sensor Calibration
  - Convert between coordinate systems of every sensor, image, vehicle and global co-ordinate axis system. So data from a sensor can be mapped globally.
  - Calibration of a camera helps to convert a pixel in an image to that in the camera coordinate axis system, and subsequently into the global coordinate axis system.
- Computer Vision
  - *Preprocessing*: Change colour space, reduce resolution and remove noise
  - *Feature Extraction*: SIFT, SURF, HOG, PCA, LDA, ICA, Wavelet etc.
  - *Segmentation*: Edge detection (Canny, Sobel), grow-cut, dilation and erosion, classification (train a classifier to differentiate between object and non-object and pass a floating window of different scales and orientations), segmentation by depth in case of 3D vision etc.
  - *Recognition*: ANN, SVM, Decision Tree, Random Forest, Ensemble, Adaboost etc.
  - *Tracking and Optical Flow*: Track the object of interest to estimate speed, reduce computation time and errors

---

detect and virtually reconstruct all the obstacles within visibility. The sensors are redundantly used, with multiple sensors sensing about a region, to combat uncertainties and noise.

The vehicles and obstacles at the front are very important and, therefore, it is common to employ vision cameras to detect road, road boundaries and signs at the front. Similarly, lidars and radars are actively used to get distance and speed estimates of the vehicles directly in front. Ultrasonics may be used to quickly detect a hazard and immediately take safety precautions. One camera or proximity sensor may not give a complete wide-angle vision of the obstacles at front, and, hence, multiple sensors may be used and *fused*. Similarly, it is important to look at the vehicles at back, which again requires an array of sensors. Obstacles at the two sides are reasonably less important; however, having a 360 degrees view all around the vehicles significantly helps in reconstructing the operational scenario to understand the algorithm while working in the design phase and to diagnose the problems in the testing phase. The sensor outputs can be integrated to reconstruct the operation of the vehicle from the sensor logs.

Similarly, motion sensors like an Inertial Measurement Unit (consisting of an accelerometer, a gyroscope and a magnetometer) can be used to measure the motion of the vehicle. A global positioning system (GPS) is used to get the approximate location of the vehicle in a map of a city. The GPS itself cannot be used for localization alone as the errors are very high, whereas for navigation one normally needs errors of just a few centimetres. Hence, vision techniques are used to make location estimates. The internal encoders present in the vehicle also measure the motion of the wheels, which can be used to indicate the motion of the vehicle.

3.2.2   Sensor Calibration

Each sensor is mounted on different parts of the vehicle. Each sensor hence has a *local sensor coordinate axis system*. Because the vehicle is rigid, and it is assumed that the sensors are tightly placed and they do not get disturbed during the motion of the vehicle, the conversion from one sensor coordinate axis system to another is a constant transformation matrix. The vehicle is given its own coordinate axis system, called the *vehicle coordinate axis system*, which again has a constant transformation to all the sensor coordinate axis systems. Calibration techniques are used to compute all transformation matrices, made possible by measuring the distance and pose of each of the sensors with respect to the vehicle's coordinate axis system. Alternatively, the relations may be obtained by using calibration points. The sensor readings to different known obstacle positions are used to estimate the relations between the different readings which enables estimating the transformation matrix between every pair of sensors. Further, the vehicle itself is non-stationary and is moving and changing its orientation. The localization techniques, which also depend upon vision, can be used to compute the vehicle's pose with respect to the *global coordinate axis system*, thus enabling reporting the data given by the sensors with respect to the global map external to the vehicle.

The proximity sensors suggest the presence or absence of obstacles and the outputs can be transformed into the coordinate axis system of the vehicle to draw a small map around the vehicles. However, the vision cameras result in an output image. The pixels of the output image constitute the data in the *image axis system*, which may be transformed into the real world coordinate axis system with the help of a *calibration matrix*. Hence, the vision cameras may be first calibrated by using a chessboard calibration technique. The technique requires a chessboard to be placed in front of the camera the image of which is taken by the camera. All the corners of the chessboard are detected by image processing techniques and used to estimate the intrinsic and extrinsic parameters of the camera. The intrinsic parameters include the focal length and skew of the camera. The extrinsic parameters include the position and orientation (or pose) of the camera.

The different sensors operate at different frequencies and ultimately contribute to the knowledge of the world or the map of the environment. The different sensors return different types of data. Image processing and signal processing techniques enable conversion between data types. Essentially, one is interested in what obstacle or road information is present, along with the location (and orientation) of the obstacle or road. All this information is understood for every sensor, converted with respect to the vehicle's coordinate frame, ultimately to the global coordinate frame, and then given to the *mapping* server to reflect it in the knowledge of the world. The map is updated by different sensors at different frequencies. The high-frequency sensors may be used to handle critical situations which require immediate action, whereas the lesser-frequency ones may be used to make longer-term plans and actions.

3.2.3   Stereovision and 3D Techniques

The vehicles typically operate in a 3D world, and, therefore, it is important to make 3D maps of the environment and to capture the 3D information. Technically, the vehicles require 2.5-dimensional (2.5D) information. General mobile robots operate in a grid, in which every region may be obstacle or a non-obstacle, which is 2Din nature. Vehicles additionally may be interested in the height of the objects to decide whether it is possible to drive over the obstacle, manoeuvre is necessary to avoid the obstacle or to know the slope of the road. Because, apart from the 2D map, only the height is in question, the maps are called 2.5D maps.

The vehicles can sense in 3D by the use of a stereovision system. The system uses two cameras to simultaneously sense the world and estimate the depth of the point, similar to the vision system in the humans. The 3D vision by using stereovision can be computationally very expensive and a slow process, and therefore it is common to use 3D lidars to produce the 3D map of the world. The 3D lidars scan using a laser ray for all possible solid angles (across all pan and tilt angles). The distance corresponding to every angle is measured. This generates a point cloud telling about the location of an obstacle at a certain depth and angle, and it is a 2.5D data structure. The data can be projected on a 3D map.

3.2.4   Multisensors and Information Fusion

The preceding sections suggest different types of sensors, giving different outputs which are interpreted, transformed and given to a mapping server which creates a map of the entire environment. The mapping server is responsible for fusion of the different outputs. All sensors may redundantly report the facts about the environment. Ideally, if the sensors are noise-free, the multiple sensors report the same facts if a region is redundantly serviced by multiple sensors. However, noise results in conflicting facts reported by the different sensors and, hence, the *integration unit* of the mapping server needs to decide which sensor to believe in case of any anomaly. A simple majority voting of sensors may work many times; however, the technique assumes all sensors to be equally informative about all regions at all times. A weighted majority voting may be used to prefer some sensors for some kinds of data and facts. Better integration techniques make use of fuzzy arithmetic and other uncertainty reasoning methods for integration.

3.3 Computer Vision

One of the dominant and richest sensors used in the vehicles is the vision camera, which outputs an image. The image is a rich source of information, not only about the presence or absence of obstacles, but also about the type of obstacles. The image is high-resolution data. The intention is to process the image to get an indication of the road and non-road regions, information about the lanes and lane markings, information about the presence and absence of obstacles around and detection of the obstacles. The detection of obstacles is particularly interesting, as it enables the vehicle to spot the traffic lights, traffic signs, other vehicles around, pedestrians, road blockages etc. on the road. All of these have immense importance in making driving decisions. When used in a stereovision framework or when used in combination with proximity sensors like lidars, the vision camera can be used to know the type of obstacle, whereas the proximity sensor or stereovision system accurately reports the distance to the obstacle. The intention is to first identify the different types of obstacles, lane markings and road information, and to further localize them in the world map.

The image given as an input is, hence, processed using computer vision techniques to identify the different types of obstacles and information about the road. The typical process used is shown in Fig. 3.1. The different steps are illustrated in the subsequent sections.

3.3.1 Image Preprocessing

The original image is high resolution. Although the high-resolution image is instrumental in giving accurate details about the environment, the same may be computationally very expensive to operate, especially when the results need to be given in near-real time. Hence, the image may be *reduced in resolution* before operating any further. The Red—Green—Blue (RGB) colour space further is prone to high variance due to changing lightening conditions. The same colour, as captured by a camera, may look very different at different lightening conditions which change at different times of a day even for a well-illuminated room. Hence, the image may be changed to a different *colour space*, typically Hue Saturation Value (HSV) in which the last

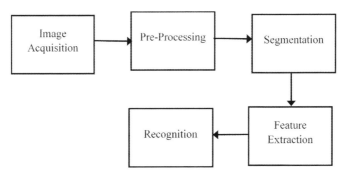

Figure 3.1 General methodology for computer vision.

component corresponds to light intensity, lightness and a—b colour components (LAB) in which the first component corresponds to lightness and Luma, Blue difference and Red difference colour components (YCbCr) in which the first component corresponds to light.

The image obtained from the vision camera may be prone to noise and, therefore, it is common to use *noise-removal techniques* to eliminate noise. Salt-and-pepper noise is common in images, whereas the noise may also be Gaussian in nature or sometimes generated from complex noise models. The use of median filters, mean filters, Gaussian filters and some specialized filters are common approaches to remove noise. Wavelet decomposition is another way to remove noise. Similarly, some statistical filters can be used for noise removal. The noise-free image is used for further processing. Although noise removal removes the noise content of the images, it may even lead to some loss of data as an effect.

3.3.2 Feature Extraction

The image is used to detect and identify the objects in the image. The overall image is of high dimension over which a classifier cannot normally be built. Hence, *feature extraction* technique is used to reduce the dimensionality of the data while leading to better classification. The features are attributes of the data which maximize the inter-class distance while minimizing the intra-class distance. The features are strong indicators of the class of the object. Features exploit the redundancy in the original data to best represent the same data in a lower dimension. While designing features for a problem, it is not necessary that the individual features be the best for the particular problem; however, the combination of features should be the best for classification. Two good features may not perform well if they convey nearly the same information, whereas two mediocre features can be a very good combination if one feature complements the other feature for classes not easily separable by one feature alone.

Numerous approaches are used to get features out of the image. The discussions here are largely for 2D images; however, many of the features can be generalized for 3D maps and depth images, if stereovision or 3D lidar scans are possible. The *Scale Invariant Feature Transform* (SIFT; Lowe, 1999) is a powerful feature vector which selects characteristic feature points in an image as feature descriptors which are largely invariant to illumination, scaling and rotation. Difference of Gaussians is applied at multiple levels. For this a Gaussian filter with the stated variance value is convolved with the original image. The SIFT features are the points which lie at the extrema of the difference of Gaussians. The SIFT features of an image can be mapped with the template of the object available in the data set. The matching involves finding the corresponding points between the pair of images. Based on this a matching score is found which describes the matching of the set of images. The matching also informs the translation, scaling and orientation between the two images, which can be used to localize the object.

Another related technique is the *Speeded Up Robust Features* (SURF; Bay et al., 2006) which is faster to compute than the SIFT descriptors. SURF uses an integral image, which stores the sum of all intensity values from the corner of the image. The integral image enables easy computation of the Hessian matrix. A Hessian matrix is computed for all points, which is formed by the derivatives of the greyscale image. The matrix is computed for multiple scales to get the image in multiple scales or the scale space. A higher value of the determinant measures a stronger change in derivatives and, hence, is marked as a point of interest. A Haar wavelet response is taken around the region of interest to get the feature descriptor.

Similarly, the *Histogram of Oriented Gradients* (HOG; Dalal and Triggs, 2005) is a very similar and another widely used feature descriptor. The approach first takes a derivative of the image. One set of derivatives is taken in the vertical axis (similar to finding horizontal edges). Another set of derivatives is taken in the horizontal axis (similar to finding vertical edges). The derivatives can be obtained by taking derivative-masking windows, or by subtracting the consecutive vertical or horizontal pixels. The vertical and horizontal derivatives can be used to get the local orientation of the region by application of the tan inverse function. Like this all orientations are taken. A histogram is made out of all orientations, which is then normalized. About 10 bins are made, each with some range of angles between 0 and 360 degrees. All orientations are kept in the bins and the totals per bin makes the histogram. The normalized histogram is used as a feature descriptor.

Sometimes the image may have far too many features. Alternatively, all the pixels of the image may be taken as features, resulting in a very big feature vector. The task is then to select the best features, exploiting the redundancy in features as visible from the historical datasets. The dominantly used technique is the *Principal Component Analysis* (PCA, Jolliffe, 2014). The PCA assumes the data to be represented in a different coordinate axis system, such that the dominant axis which most nearly represents the data can be computed. The algorithm assumes linearity of the data to first get the dominant axis which maximizes the separation between the data. The axis is a weighted combination of the different feature axes. This is done by computing the correlation matrix between all features and thereafter computing the eigenvalues and eigenvectors. In this manner, the PCA transforms the data to a different axis system such that the initial axis is dominant in which the data are most separated, whereas the utility of the later axis is smaller. Only the top few features may be considered for the task of classification, whereas the latter ones may be simply ignored. The assumption of linearity can be dropped by the use of nonlinear PCA which assumes a nonlinear function to estimate the axis of the transformed data.

A similar technique is the *Linear Discriminant Analysis* (LDA; Mika et al., 1999). Similar to the PCA, new features are made by using the linear combination of the original features, to transform the data into a separate feature space. However, LDA tries to transform the data such that the interclass separations are increased, whereas the intraclass separations are small. This makes the distinction more class conscious and is therefore widely used in feature selection. Similarly, *Independent Component Analysis* (ICA; Hyvärinen et al., 2001) does the same thing by decomposing the original data into new independent features which have the minimum redundancy or the minimal mutual information, and the features together most nearly represent the complete data. The reduced set of features can be used for classification.

Wavelet-decomposition techniques are also applied extensively and used as features. Essentially, one keeps decomposing the image at different levels. The informative part of the decomposition is used and, by repeated application of different levels of decomposition, a much smaller image may be obtained. A variety of wavelets may be used like Haar, Daubechies, Coiflets etc. Similarly, the corners can also be used as features of the images. Harris corner detection is a common technique for the same. The most prominent corners may be considered. Similarly, most prominent edges may also serve as features.

3.3.3   Segmentation and Localization

Segmentation is responsible for separating out different things of interest in the given image. Segmentation becomes a very simple problem to solve if a high-resolution 3D map of the world is available by lidars or stereovision. The different vehicles and obstacles are present at different depths and, hence, the problem is only to detect changes in depths around the object of interest.

Segmentation is needed for detecting road regions, lanes, background which is not necessary for processing, different vehicles ahead, other obstacles or different traffic signs. All these regions are separated from their surroundings by an edge. Hence, *edge-detection* techniques may be used to separate out the different regions of interest. An edge is determined by a sharp change in the intensity values of the pixel. A vertical edge is a sharp change in intensity of a pixel with the one at the left or right of the pixel, whereas a horizontal is a sharp change in the pixel value with the one above or below. The edges may be taken as the derivative of the image. A vertical and a horizontal derivative is taken, whereas the edge is defined as the norm of the horizontal and vertical counterparts. Correspondingly, the direction of the edge may be defined by using the tan inverse function over the horizontal and vertical edge components. Unfortunately, the primitive method of putting a threshold on the magnitude of gradient results in too many edges and a very adverse effect of noise. Canny edge detection (Canny, 1986) and Sobel edge detection (Gao et al., 2010) are popular methods to use for edge detection.

A *grow cut method* can be easily applied to separate out the road region from the non-road region, or to separate out the lane markings from the surroundings. In this method, regions which are clearly known to be road or lanes are first detected and some seed points are placed. These points grow in all directions, recursively and stop when the difference in colour levels is more than a threshold. The region grown by the seed points is segmented out. Any segmentation may result in too many noisy decisions. The application of dilation and erosion operators (Haralick et al., 1987; Ji et al., 1989) are used to remove small regions of noise. Postprocessing techniques like elimination of small edges and computing connectivity may be applied on the edges.

Characteristic sign boards, vehicles and obstacles can also be detected using *classification techniques*. Templates of the object to detect are given to the system, reduced in size by feature extraction. A floating window of different sizes is passed through the image and the level of matching with the template is checked. If a window has a very strong matching with the template, the local neighbourhood is searched to get the window that has the best matching, and correspondingly the detected object is returned.

The segmentation techniques also return the *location* and orientation of the object of interest. For roads and lanes, the complete boundary may be of interest which is returned. For a monocular camera system, the calibration matrix may be used to get the perspective image of the road and lane boundaries. For a stereovision system, the same object and the same points of interest in the object may be detected by both the cameras to get the 3D information. Alternatively, the same object may be detected by a monocular camera and a proximity sensor to localize it in the 3D world. The same information, along with the class label as discussed later, is given to the mapping module to make a 3D map of the world.

3.4   Recognition

Recognition is the task to map every input, represented by a set of features, into the corresponding class. It is a decision-making process, wherein like an expert, a machine needs to decide to which class the input belongs. Specifically, given a block of image, the role of the classifier is to decide whether it represents a road, lane marking, traffic signal and its state, a traffic sign and the specific traffic sign out of the allowed signs, a text warning, a vehicle, an obstacle, a pedestrian etc. The classifiers rely on a *machine-learning* approach (Shukla et al., 2010a,b), wherein historical data are available. If the historical data are labelled, consisting of the inputs and the corresponding class labels, the learning is called supervised learning, whereas if the historic data are unlabelled, the learning is called unsupervised learning. If only a small part of the historic data are unlabelled, the training so performed is called semi-supervised learning. Here, the focus is primarily on supervised learning. The learning algorithm tries to exploit the redundancy in the data by finding out patterns, trends and relations amongst the input attributes and the class labels. Essentially, a small change in feature value rarely changes the class label and therefore inputs with similar set of features generally have the same class labels.

The historical dataset can be represented by a few features as detailed in the feature extraction technique. The features are normally normalized to generally lie between

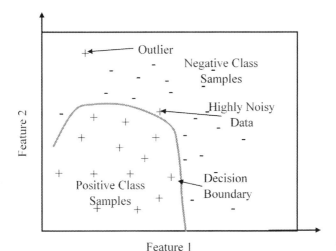

Figure 3.2 Feature space for a classificatory problem.

0 (or -1) and 1 (or any similar bound). This forms a *feature space* for the classificatory problems, wherein each feature is an axis and every input from the historical data set is a point in this higher dimensional feature space, with the point labelled by the class label to which it belongs. The feature space for a dummy problem is given in Fig. 3.2 for understanding. If the choice of features is good, the different class labels form clusters in this feature space. Characteristics of a good set of features is that the clusters must be easily separable with high interclass separation and a low intraclass separation. There may still be outliers which are labels far away from the expected cluster centre. Further, noise in the data may make the clusters mingle with each other and therefore the decision boundary separating the classes may be hard to construct. Different classifiers adopt different methodologies to construct the decision boundary and thereby separate the different classes in this feature space.

3.4.1 Neural Networks

Neural networks are one of the most dominant classifiers used for classification. The neural networks imitate the thinking process of humans. The human brain is composed of billions of neurons which are all connected to make a network. The sensory organs sense the real world which triggers the receptive neurons which do a little processing of the input percept and pass the processed data as information to the connected neurons, which do a little further processing. The complex decision making from the native data happens through layers of processing by these neurons.

Similarly, *artificial neural networks* (ANNs) (Haykin, 2009) are a collection of neurons stored in a layered manner. The most common architecture is the Multilayer Perceptron. Every neuron of a layer is connected to the neurons of the preceding and next layer through connections which have weights. Each artificial neuron takes the inputs from the preceding layer and multiplies the inputs by the associated connection weights and adds all the inputs so received. The weighted addition passes through a nonlinear activation function, which is the output of the neuron and is passed as input to the next layer. The nonlinear activation function enables the decision boundaries to be nonlinear. Typically, an additional input of unity value, called bias, is also added at each layer.

The network comprises one input layer which receives the inputs, one or more hidden layers, and one output layer which gives the system output. A typical configuration is shown in Fig. 3.3. The number of input neurons is same as the number of features. The number of output neurons is the same as the number of output classes, wherein each output neuron attempts to give the posterior probability of the associated class. Rescaling and normalization may be required to make the probabilities lie between 0 and 1, and to sum all the posteriors to 1. Typically, one hidden layer is taken. Each neuron in the hidden layer dominantly models a part of the decision boundary in the input space, whereas the output neurons integrate the different part boundaries to make the total decision boundary.

This gives the representation capability to model any type of decision boundary. However, out of all infinitely possible decision boundaries, the best one needs to be considered, which is done by optimal setting of the parameters, consisting of the architectural parameters (number of hidden layers, connections and number of neurons), weights and biases. The weights and biases are learnt by a gradient descent approach called the *Back Propagation Algorithm*. The algorithm sets the weights and biases by computing the correction as the gradient of the error function. The labelled outputs and, hence, the error function for the output neurons is known, for which corrections

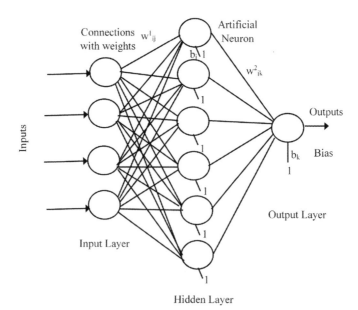

Figure 3.3 Artificial neural network.

can be applied. The gradients are back-propagated to the hidden and then the input layer to compute the corrections. Training happens in multiple iterations, typically every piece of data used in every iteration. The training may be done in small batches of data in case the data are too large in size.

Too many hidden neurons can lead to too many weights to be learnt, which can slow the learning process and also require more training data. Further, there is a very large chance of *over-fitting* the data as more neurons mean more representational capability or capability to model more complex decision boundaries. The outliers and noisy data may hence make the decision boundary bend largely to fit such data. This will make the system perform very well on the training data; however, the performance on the validation data can be very poor due to very high sensitivity of the decision boundary. Similarly, too few neurons in the hidden layer can *underfit* the data, wherein the neurons cannot collectively model all the turns and modalities of the decision boundary and, hence, the performance on the training as well as validation data is very poor. The cases are shown in Fig. 3.4.

The complexity of the data is unknown and, hence, the number of neurons required cannot be statistically computed. One, hence, needs to experiment against different numbers of neurons and different hyper-parameters to maximize the performance on the validation dataset, to expect good performance on the testing dataset. Of course, once testing has been performed, it is not advisable to change the hyper-parameters. This can make the hyper-parameters humanly 'trained' on the testing data and thus the hyper-parameters become specialized to the testing data. The testing data hence no longer becomes representative of the live data obtained from the application.

3.4.2 Support Vector Machines

Support Vector Machines (SVMs) is another dominant classifier widely used for machine learning (Steinwart and Christmann, 2008). The SVMs assume the data linearly separable in the feature space and constructs a linear hyperplane which separates the positive and the negative classes by the largest margin. The margin is defined as the distance of the hyperplane from the nearest positive and negative class sample present in the historic data. Obviously, the actual decision boundary may not be linearly separable, for which the SVM uses the kernel trick. The data are projected to a higher-dimension feature space by the use of kernels. The trick is that data nonlinearly separable in the feature space may be linearly separable in a higher-dimensional space by a good choice of kernel and by the application of multiple kernels. This is shown in Fig. 3.5. The outliers and noisy data may still have a severe effect and, hence, a term is added to make the decision hyperplane resistant to some data by a small margin.

Multiclass classification problems can be converted into multiple binary classification problems to be used by binary classifiers. The first methodology is *one-vs-all*, wherein the number of classifiers is the number of classes, and each classifier separates its own class against all the other classes. This reduces the number of classifiers and, hence, the training time. However, it creates a highly imbalanced data set for every classifier wherein the number of positive examples (of the class that the classifier is

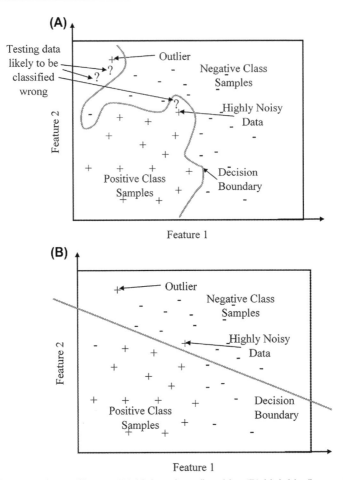

Figure 3.4 Bias variance dilemma (A) high variance/low bias (B) high bias/low variance.

trying to separate from the data set) are far more than the number of negative examples (samples of all the other classes), in which conditions the classifiers do not perform well. Alternatively, *one-vs-one* methodology makes separate classifiers separating every pair of class, from which the decision may be made by using voting. This creates far too many classifiers, however, to keep the data balanced. Hybrid approaches are also used.

3.4.3 Decision Trees

Decision Trees are also extensively used, the common algorithm used being the C4.5 decision tree (Quinlan, 1993). The decision tree works on the principle to recursively divide the feature space based on some feature at every level. The division is such that the information gain is maximized after the division of the space. Every division

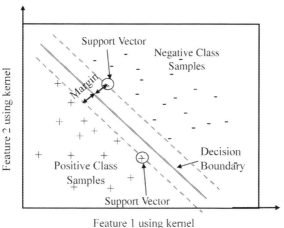

Figure 3.5 Support vector machine.

becomes a node of the tree with branches pointing out to the different subdivisions created. Each division is a range of some input feature, which is annotated on the tree branches and become the unit conditions of the resultant rule that discriminates between the classes. The division happens till all data items on the divided search space lie in the same class. Such a node becomes a leaf node which is labelled with the class label of the data items in that part of the feature space. A typical decision tree is shown in Fig. 3.6.

On the testing data, the tree is iterated from the root till the leaves based on the decisions at each node. All leaf nodes are associated with a class label, which is the

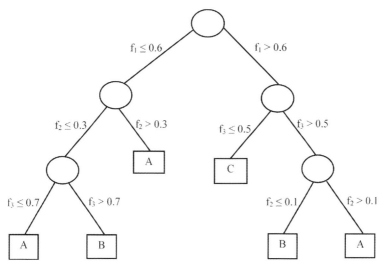

Figure 3.6 Decision tree.

output of the classifier. Alternatively, one may traverse through all branches, noting the division criterion, and convert all traversals from root to source as a rule. The set of all non-overlapping rules makes the resultant system. Many decision trees may be used in an ensemble architecture, called as *Random Forest*, implying a collection of trees.

3.4.4   Adaptive Boosting

Many times it may be preferable to make a collection of classifiers to classify a particular problem, rather than relying on one classifier alone. The individual classifiers may have modelling limitations restricting them to imitate the optimal decision boundary, which can be overcome by the use of multiple classifiers. Hence, problems may opt for an *ensemble* of classifiers (Kuncheva, 2014). Here, the same problem is redundantly solved by an array of classifiers. The decisions of the classifiers is reported to an *integrator*, which integrates the individual decisions to make the final decision of the system. The individual classifiers constituting the ensemble need to individually have high accuracy to make a valuable addition to the ensemble. At the same time, the individual classifiers must differ to make the overall approach better than the individual classifiers. The two aims contradict each other. A typical architecture is shown in Fig. 3.7.

The contradictory aims can be obtained by making classifiers with different hyper-parameters, by using different types of classifiers, by using different set of features or by using different data to train different classifiers. The last approach is the best suited for increasing the diversity and is used in a technique called bagging. *Adaptive boosting* algorithms probabilistically assign data to different classifiers. Each classifier is trained on a small set of data given to the classifier. The classifiers are weak in the sense that their accuracy is marginally above the accuracy of a random classifier. Each classifier is given data which are normally misclassified by the other classifiers, which leads to very high diversity amongst the classifiers. The trained classifiers are used in testing, wherein the input is given to all the classifiers which make their decisions and the decisions are used by the integrator to make the final decision.

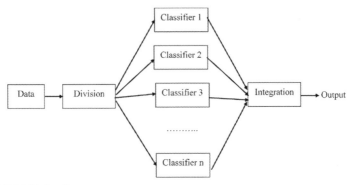

Figure 3.7 Multiclassifier systems.

3.5   Tracking and Optical Flow

It is possible that computer vision techniques result in misclassified information due to noise, which can lead to wrong decisions. Mostly, the recordings are taken as a continuous stream of video, the video consisting of a set of frames. There is a very small change in going from one frame to the other, and therefore the decisions of classification and localization must also not change dramatically. Consider a vehicle in front which is detected by the front camera. One can continuously detect the vehicle and estimate a track of its speed and position. The speed is also needed for motion planning as shall be presented in the subsequent chapters. Now if the vehicle suddenly disappears or dramatically changes its position, it obviously is an error which can be corrected based on speed estimates. Tracking is hence useful to correct errors and filter out the data to more realistic estimates. Further, knowing the approximate positions of the areas of interest, one may as well narrow down the search to a smaller than expected region to speed up the computations. Extended Kalman Filters and Particle Filters are widely used for tracking (Choset et al., 2005).

A related technology is *optical flow* (Horn and Schunck, 1981; Barron et al., 1994) which measures the change in the pixel's position with time. A point in the current frame is also recorded after a unit span of time. The difference between the two positions, or the derivative of the position of the point with respect to time, constitutes the optical flow. The optical flow can be used to estimate the relative motion between the camera (or the vehicle) and the obstacle or the other vehicles. This helps in estimating the relative speed and prospective position of the obstacles that is useful in planning decision making. The proximity sensors, if available, of course give better speed estimates by constantly tracking the change in proximity with the vehicle.

3.6   Vision for General Navigation

The vehicle operating on the road needs vision for a variety of things, the most dominant of them being the detection and tracking of roads and lanes. Further, the vehicle needs to detect and track all the other vehicles and other obstacles around it. This section highlights all these aspects of vision in navigation. The problems are summarized in Fig. 3.8 and in Box 3.2.

3.6.1   Vehicle and Obstacle Detection and Tracking

The vehicles and other obstacles can be easily detected using computer vision techniques (Bertozzi and Broggi, 1998; Labayrade et al., 2005; Leonard et al., 2008). The obstacles are normally the outliers in the road region, which are neither roads nor lane markings. The vision cameras can be used to detect obstacles ahead, behind and otherwise around the vehicle, whereas the proximity sensors will be needed to give information about the distance of the vehicle from the obstacles in different directions.

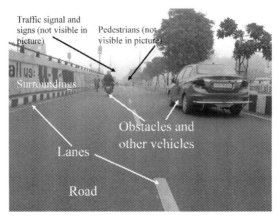

Figure 3.8 Perception in autonomous vehicles.

One of the most convenient methods to detect and thereafter track the obstacles is by the use of 3D vision, primarily using lidars. Although vision sensors are very good in detecting the type of obstacle for decision making, the 3D sensors can very quickly give a detailed 3D map of the world produced as a point cloud, in which every point tells about the distance to the obstacle in the direction. The point clouds can be easily converted into 3D maps. One of the problems with such maps is the small blind spots corresponding to which no data were collected. This may be due to the resolution of the operation or maybe due to noise. Hence, the 3D maps so produced need to be postprocessed to remove small blind spots by associating them to the nearest data obtained. The maps may further be postprocessed to filter out irregular data which appears highly distinctive from the neighbours by using filtering. Some smoothing may also be applied. The different vehicles and obstacles have different depths and show sharp changes in depths. The 3D map so produced may be clustered to get different obstacles. Every obstacle is thus registered in the map and in case of anomalies fusion from different sensors is done to produce the map.

The problem of *tracking* can be similarly handled. While tracking, it is important to register new obstacles and delete the old obstacles which are no longer in the vicinity. This can be done by checking the new obstacle positions with the estimates from the prior percept. If no obstacle in the list of obstacles around matches a particular sensed obstacle with a strong confidence, it is considered as a new obstacle and registered at the current location, from which it is tracked. If an obstacle is not sensed again for long, it may be deleted from the list of obstacles being tracked.

3.6.2 Road and Lane Detection and Tracking

The foremost problem is that of the detection of roads (Bertozzi et al., 2000; Leonard et al., 2008; Wang et al., 2004). Roads normally have a black colour, which is very distinct from the white lane markings. The other vehicles and obstacles can also be black, which can be problematic for a road detection system. However, the other

---

Box 3.2 Perception in Autonomous Vehicles

- **Vehicle and Obstacle Detection**
  - Identified as outliers, which are neither road regions nor lane markings
  - Vision camera detects which vehicle/obstacle, proximity sensors give distance to the vehicle/obstacle
  - 3D vision, primarily lidars, convert the raw point cloud to 3D maps straightaway, from which obstacles and vehicles can be segmented by *depth*
  - Vehicles/obstacles can be *tracked*. This requires registering new vehicles/obstacles, updating the old ones and deleting the ones not in sight for long
- **Road Detection**
  - Distinctive black colour and, hence, *colour filtering* or a simple classifier taking pixel colour as an input can be used
  - Postprocessing using *dilation* and *erosion*
  - In 3D vision, road regions are the set of points with the *lowest z coordinates*
- **Lane Detection**
  - Distinctive white colour and, hence, detected by colour filtering or colour-based classifier
  - *Thinning* by using thinning algorithms or edge detection
  - Lanes very different from *common widths* are noise
  - *Connectivity* checking algorithms combine edges if one starts very close to the end point of another.
  - Local *smoothing* and curve-fitting algorithms estimate the boundary
  - *Tracking* done on the coefficients/control points of the curve-fitting tools
  - Road and lane markings are fused to get the road and lane information
- **Pedestrian Detection**
  - Make a classifier to separate humans and non-humans, scan all regions with all scales and orientations.
  - Posture of humans is very complicated and changes drastically while walking. Hence, make different classifiers for *different parts* (like face, legs and hands) and integrate the results.
  - 3D vision techniques can use depth of a *narrow obstacle* as an indicator
  - Tracking as normal
- **Understanding Traffic Signs**
  - Very distinctive *colour* (bright red, bright blue or bright yellow) on which colour filtering can be used to indicate approximate presence.
  - Very distinctive *shape* (circular, triangular, rectangular or diamond shaped) which can be searched by classification or edge-detection techniques.
  - Approximate location and presence can be used to read the sign like a standard computer vision problem.

vehicles and obstacles can be easily spotted by the 3D sensing technologies. Hence, a simple *colour filtering* may be applied. A classifier is made which takes the coloured pixel as input and predicts whether it belongs to the road class or the non-road class. The colour filters are, however, very sensitive to the time of the day, and, hence, transformation to a relatively illumination-invariant colour space may be helpful. Samples are taken from different roads and at different times of the day to carry out the classification. Post-processing using dilation and erosion is always suggestive to fill in the small gaps created by the misclassified road regions.

Similarly, the lanes are marked by a thick white paint. Colour filtering is useful to detect the lane regions. The problem is now to apply thinning on the lanes so as to reduce the thick road paint into thin regions, not affecting their connectivity. Thinning algorithms can be used for the same. The only criterion is that the thinning algorithm must give the centreline of the white lane, rather than either of the edges. Alternatively, edge-detection techniques may be used to first get the edges, and then algorithms may be used to merge the edges corresponding to one block of road paint into a single edge. The lane detection techniques rely on a very strong prior that the lanes have thick edges of approximately known sizes. Hence, widths much smaller or larger than the expected width can be directly discarded and taken as noise.

The problem with the lane markings so obtained is that they do not normally maintain the *connectivity* property. The lack of connectivity may be due to misclassification, noise, presence of obstacles etc. Hence, the task is then to connect the edges so obtained into continuous lanes. A road may have any number of lane markings, all of which need to be detected and marked. Only some lanes may be visible at any point of time. The different edges representing the lane markings so produced are then processed to induce connectivity. Different methods can be used for the same. End points and orientations at the end points can be noted for each edge. If the ending point and orientation of one edge matches the starting point and orientation of another edge, then the edges are said to represent the same lane marking and are joined. This gives a reduced set of lanes, each lane as a set of edge components. The edges may first be locally smoothed knowing that the shape of the lane will never be irregular. Curve fitting may be done by regressing a parabolic or a cubic curve against the edge segment using optimization techniques. The entire boundary can also be modelled using cubic polynomials, splines or similar curve-fitting techniques.

The road detected and the lane data need to agree on the extremity of the road and lanes. Hence, the information about the extremity is redundantly obtained by the road-detection and the lane-detection techniques. Both techniques may be used to uniquely identify all the lanes and thereafter to regress the lane boundary. The lane boundary at the vicinity especially needs to be known with high precision as that affects the immediate motion of the vehicle. Further, it is suggestive to first convert the image from the current view into the perspective view which provides an overhead view of the road. Camera calibration techniques are useful for the same. In any case, the road and lane information is mapped to the real world coordinates.

The 3D sensing can also be used to detect the roads. The point cloud produced by 3D lidars can be projected on a 3D map, changing the depth at every angle into a point with x, y and z coordinates. Here, the z coordinate denotes the height of the detected obstacle. The road regions have the least possible height. Because of irregular road surfaces and noise, the bottom few points may be taken and regarded as road regions. The sensors can also be used to detect irregularities in the road surface including uneven road surface, an upwards slope, some small obstacle on the road surface which can be easily surpassed etc. The information from all sources is fused to make an informed decision about the roads and lanes.

Tracking of lanes is not entirely similar to tracking of landmarks by using conventional filtering techniques. The lane is characterized by the control points of the curve used to regress the lane. The changes in lane changes these control points and these points need to be adapted as the vehicle moves and the lane shape changes. The new estimates of the lane from the new frames are collected and are matched against the old lane boundary represented by the old control points. The filtering technique modifies these control points based on the old control points, motion profile of the vehicle and the new frames.

The information about the road, lane markings, vehicles and other obstacles is represented in 3D maps as discussed. The information is, however, needed as a 2D map representing the drivable and non-drivable region. This conversion is simple. Irregularities of road and small obstacles can be regarded as the passable obstacles. Multiple sensors can vote regarding the presence or absence of obstacle at each region. In case of anomaly, the confidence of the individual sensors may be taken. A rather pessimistic approach is suggestive, meaning that if anything is uncertain about a region, or in case of moderate anomalies, it is regarded as an obstacle. Similarly, even if there is an indication that the region is obstacle-free from multiple sensors, however, the confidence of that for any sensor is not high, the vehicle may better stay away from the region.

3.6.3 Pedestrian Detection

Pedestrian detection (Gavrila and Munder, 2007; Shashua et al., 2004; Suard et al., 2006) is a very interesting problem from a safety perspective. On numerous occasions, the vehicles are very vigilant in driving; however, accidents happen due to sudden pedestrian crossing of the road. Pedestrian detection is hence a very important topic. The problem can be solved by using the same techniques as discussed earlier. A classifier can be made to distinguish between images of pedestrians and non-pedestrians. Floating windows of different positions, sizes and orientations may be passed to this classifier which gives the probability of the window containing a pedestrian. Standard image-based features may be used. The technique is capable of efficiently detecting pedestrians and giving the location. The detection can be enhanced with 3D vision techniques which can efficiently segment out the pedestrians from the rest of the surroundings based on the depth value. Pedestrians, unlike other vehicles, have a very thin occupancy and, hence, cannot be detected by a small array of ultrasonic sensors. Further, due to variations in clothing and posture, edge detection alone cannot be used for segmentation. Further, tracking algorithms, typically Kalman filtering or particle filters, may be used for better prediction.

However, the naïve technique of pedestrian detection faces problems due to the highly complicated posture of the pedestrians. As humans walk, the hands and legs swing giving a very different appearance. The posture of walking also significantly differs between different humans. Humans may as well bend while walking, walk by looking around or may engage themselves in activities like talking which severely affects the walking posture. This makes it very difficult to make a good classifier for pedestrian detection. The problem can be solved by detecting each of the simple features of the human body like face, legs and hands separately. Each of these features is detected by a separate classifier. The output and confidence score of each classifier is then used by a decision-making tool for evaluation. The tool may be a set of simple and/or rules, may be a simple integrator or may be a classifier in itself.

3.6.4   Understanding Traffic Signs

The problem of *traffic signs* can also be naively solved using the approaches of computer vision discussed earlier (de la Escalera et al., 1997; Bahlmann et al., 2005; Chen et al., 2011). In fact, most of the discussions on computer vision dominantly apply to traffic sign detection as they are primitive and simple to detect. The major difference in the detection of a traffic sign is the prior information about the colour and shape of these signs. Most of the traffic signs are pained with a bright red, bright blue or a bright yellow colour depending upon the type of information communicated by the traffic sign. The brightness may get dull with time and may vary at different times of the day; however, the chroma information of the traffic sign is very distinctive. Hence, colour-based features work very well in traffic sign recognition. Simple colour filtering may be used to get some initial indication regarding the presence or absence of a traffic sign and its prospective location. Similarly, colour-based features like a histogram of colours are powerful indicators of traffic signs along with the edge and other features.

The other dominant prior is the shape of the traffic sign. The shape of the traffic sign is usually circular, triangular, rectangular or diamond shaped, which makes it very different from the other objects in the surroundings. The shape and colour of traffic sign is local to every region and known in advance. So a simple primitive shape search using relevant filters is a good indicator of the location and the type of traffic sign. For example, a triangular traffic sign can be judged by characteristic placement of the three corners with a red (or similar) colour background. Alternatively, a classifier may be trained on triangular regions and used for detection. Edge-detection techniques also segment out the shape edges, which can be checked for the presence of a particular shape. Once the complete traffic sign is known, any simple classifier can be used to recognize the information contained in the traffic sign, knowing that there are very few types of traffic signs.

3.7   Summary

The chapter explored the use of vision as a facilitating technology behind intelligent vehicles. Vision relies on a large sensor array to redundantly look at the world with the help of multiple sensors. The sensors may be a vision camera, ultrasonics, lidars or a stereovision setup. All the sensors capture different types of information and operate at different frequencies. The interpreted information is transformed into the world coordinate axis system and is given to the mapping server to make a map representation of the world, which is used by the planning and decision-making algorithms.

A number of technologies rely on the vision camera as a source of information. The camera takes input in the form of an image and processes the same using computer vision techniques. Initially, noise-reduction techniques are applied to get a better image, which is then processed to detect and recognize the objects of interest. Feature-extraction techniques extract out relevant features in the image useful for processing. The segmentation algorithms aim to segment out the different objects of interest. The segmented-out objects are then used for the task of recognition by a classifier. The typical choices of classifiers include neural networks, support vector machines, decision trees and adaptive boosting. The objects may be tracked in real time using filtering algorithms like Particle Filters and Extended Kalman Filters.

The vision algorithms are used to understand the traffic environment as the vehicle moves in the traffic scenario. The vision algorithms are used to detect the road and the lane marking, to indicate the drivable regions and, therefore, the general driving trajectory. Further, the algorithms are used to detect the obstacles and other vehicles, which are also tracked and must be avoided while driving. The vision systems are also useful to detect pedestrians and take precautionary measures like stopping on suddenly sighting a pedestrian and similarly looking for traffic signs for adherence or information.

References

Bahlmann, C., Zhu, Y., Ramesh, V., Pellkofer, M., Koehler, T., 2005. A system for traffic sign detection, tracking, and recognition using color, shape, and motion information. In: Proceedings of the 2005 IEEE Intelligent Vehicles Symposium, pp. 255−260.

Barron, J.L., Fleet, D.J., Beauchemin, S.S., 1994. Performance of optical flow techniques. Systems and experiment. International Journal of Computer Vision 12 (1), 43−77.

Bay, H., Tuytelaars, T., van Gool, L., 2006. SURF: speeded up robust features. In: ECCV 2006, Lecture Notes in Computer Science, vol. 3951, pp. 404−417.

Bertozzi, M., Broggi, A., 1998. GOLD: a parallel real-time stereo vision system for generic obstacle and lane detection. IEEE Transactions on Image Processing 7 (1), 62−81.

Bertozzi, M., Broggi, A., Fascioli, A., 2000. Vision-based intelligent vehicles: state of the art and perspectives. Robotics and Autonomous Systems 32 (1), 1−16.

Canny, J., 1986. A computational approach to edge detection. IEEE Transactions on Pattern Analysis and Machine Intelligence 8 (6), 679−698.

Chen, L., Li, Q., Li, M., Mao, Q., 2011. Traffic sign detection and recognition for intelligent vehicle. In: Proceedings of the 2011 IEEE Intelligent Vehicles Symposium, pp. 908−913.

Choset, H., Lynch, K.M., Hutchinson, S., Kantor, G.A., Burgard, W., Kavraki, L.E., Thrun, S., 2005. Principles of Robot Motion: Theory, Algorithms, and Implementations. MIT Press, Cambridge, MA.

Dalal, N., Triggs, B., 2005. Histograms of oriented gradients for human detection. In: Proceedings of the IEEE Computer Society Conference on Computer Vision and Pattern Recognition, pp. 886−893.

de la Escalera, A., Moreno, L.E., Salichs, M.A., Armingol, J.M., 1997. Road traffic sign detection and classification. IEEE Transactions on Industrial Electronics 44 (6), 848−859.

Gao, W., Zhang, X., Yang, L., Liu, H., 2010. An improved Sobel edge detection. In: Proceedings of the 2010 3rd IEEE International Conference on Computer Science and Information Technology, vol. 5, pp. 67−71.

Gavrila, D.M., Munder, S., 2007. Multi-cue pedestrian detection and tracking from a moving vehicle. International Journal of Computer Vision 73 (1), 41−59.

Haralick, R.M., Sternberg, S.R., Zhuang, X., 1987. Image analysis using mathematical morphology. IEEE Transactions on Pattern Analysis and Machine Intelligence 9 (4), 532−550.

Haykin, S., 2009. Neural Networks and Learning Machines, third ed. Pearson, Upper Sadle River, NJ.

Horn, B.K., Schunck, B.G., 1981. Determining optical flow. In: Proceedings of the SPIE 0281 Techniques and Applications of Image Understanding Conference, vol. 319. http://dx.doi.org/10.1117/12.965761.

Hyvärinen, A., Karhunen, J., Oja, E., 2001. Independent Component Analysis. John Wiley & Sons.

Ji, L., Piper, J., Tang, J.Y., 1989. Erosion and dilation of binary images by arbitrary structuring elements using interval coding. Pattern Recognition Letters 9 (3), 201−209.

Jolliffe, I., 2014. Principal Component Analysis. Wiley Statistics Reference. Wiley.

Kuncheva, L.I., 2014. Combining Pattern Classifiers: Methods and Algorithms, second ed. John Wiley & Sons, Hoboken, NJ.

Labayrade, R., Royere, C., Gruyer, D., Aubert, D., 2005. Cooperative fusion for multi-obstacles detection with use of stereovision and laser scanner. Autonomous Robots 19 (2), 117−140.

Leonard, J., How, J., Teller, S., Berger, M., Campbell, S., Fiore, G., Fletcher, L., Frazzoli, E., Huang, A., Karaman, S., Koch, O., Kuwata, Y., Moore, D., Olson, E., Peters, S., Teo, J., Truax, R., Walter, M., Barrett, D., Epstein, A., Maheloni, K., Moyer, K., Jones, T., Buckley, R., Antone, M., Galejs, R., Krishnamurthy, S., Williams, J., 2008. A perception-driven autonomous urban vehicle. Journal of Field Robotics 25 (10), 727−774.

Lowe, D.G., 1999. Object recognition from local scale-invariant features. In: Proceedings of the International Conference on Computer Vision, pp. 1150−1157.

Mika, S., Fitsch, G., Weston, J., 1999. Fisher discriminant analysis with kernels. Neural Networks for Signal Processing IX (1), 41−48.

Quinlan, J.R., 1993. C4.5: Programs for Machine Learning. Morgan Kaufmann Publishers.

Shashua, A., Gdalyahu, Y., Hayun, G., 2004. Pedestrian detection for driving assistance systems: single-frame classification and system level performance. In: Proceedings of the 2004 IEEE Intelligent Vehicles Symposium, pp. 1−6.

Shukla, A., Tiwari, R., Kala, R., 2010a. Real Life Applications of Soft Computing. CRC Press, Boca Raton, FL.

Shukla, A., Tiwari, R., Kala, R., 2010b. Towards Hybrid and Adaptive Computing: A Perspective, Studies in Computational Intelligence. Springer-Verlag, Berlin, Heidelberg.

Steinwart, I., Christmann, A., 2008. Support Vector Machines. Springer, NY.

Suard, F., Rakotomamonjy, A., Bensrhair, A., Broggi, A., 2006. Pedestrian detection using infrared images and histograms of oriented gradients. In: Proceedings of the 2006 IEEE Intelligent Vehicles Symposium, pp. 206−212.

Wang, Y., Teoh, E.K., Shen, D., 2004. Lane detection and tracking using B-Snake. Image and Vision Computing 22 (4), 269−280.

Advanced Driver Assistance Systems

4

4.1 Introduction

The deployment of autonomous vehicles in the transportation systems may be a slow process owing to technological, social, monetary and legal factors. On the contrary, serious limitations of the traditional vehicles make their use very inconvenient. Automobile is a highly competitive and fast-growing sector which pushes industry to ship the most sophisticated vehicles with as many intelligent features as possible. The *Advanced Driver Assistance System* (ADAS) is a collection of numerous intelligent units integrated in the vehicle itself. All these units perform different tasks and assist the human driver in driving. One may choose as many features as possible, whereas the features may themselves display a range of intelligence, autonomy and monetary costs. Driving is a complicated task requiring a significant level of expertise and experience, which is clear from the magnitude of research put into making a fully autonomous vehicle navigate in complex and challenging driving scenarios. Even though the humans may be able to learn driving relatively easily, it can take a prolonged time to really master the art of driving, especially in challenging and unseen scenarios.

Modern-day vehicles are seen as a perfect integration of numerous systems, each of which varies in intelligence, which all interact and are overall controlled by the human. Each of the systems takes care of some critical aspect of driving. The human driver may hence have to do very little, being assisted by the sophisticated systems onboard. Care must though be given in the design of these systems. Although good designs may aid the driver in efficient driving, a poorly designed system may unnecessarily distract the human driver in interactions, whereas the outputs of these systems may not be worthy of the distraction made. Similarly, wrong outputs can be fatal to the human driver.

ADAS has already made it to the vehicles that we drive these days. The vehicles have SatNav or similar systems to suggest routes to the user, some of which also give warnings of heavy traffic, road closures and suggest alternative routes. Automatic parking is becoming increasingly common. Adaptive Cruise Control (ACC)-based systems automatically maintain the speed of the vehicle so that you do not have to maintain a perfect balance of brake and accelerator while driving on a long straight road. Modern vehicles can already sense collision-prone situations based on speeds and distances maintained with the vehicle in front.

Imagine driving in a futuristic transportation system. One would expect a vehicle to do a lot more, if not completely drive the vehicle. Already, many people love driving

On-Road Intelligent Vehicles.

expensive cars on scenic or otherwise lovely roads to drive upon and therefore are un-likely to prefer vehicles driving all on their own. One expects the vehicle to be routed by the best route possible avoiding any congestion that may be likely to occur in the future. One may as well expect all information regarding the route be given in advance to the driver including suggestions to fill fuel or maybe buy something on the way. Further, one may expect the vehicle to do all the dirty manoeuvres of parking, getting out of parking, overtaking, driving in a queue of vehicles, driving in congestion-prone situations etc. The vehicle should detect and avoid any fatal accident even if the human driver makes a serious mistake, and the vehicle should warn about any likely accident well beforehand so that the human driver can adjust. Vehicles with better and more active sensory percepts, along with faster rational decision-making abilities, are more capable of assessing any likely hazardous situation and suggesting or taking pre-ventive measures.

The fascinating aspect of the technology is that the vehicles can communicate with each other. The humans can rarely talk to the other drivers around to communicate in-tents and hence make a unanimous travel plan which is benefitting to all the other drivers. The only communication which is possible is talking to the drivers around in case of a very unusual situation like a deadlock. The use of horns and indicators is another form of communication which helps in decision making. Humans can judge driving gestures very well, which acts as an indirect communication between the hu-man drivers. Intelligent vehicles are blessed with the power to communicate by creating ad hoc networks within themselves in close proximity, communicate to the transportation infrastructure present at the roadside and thus form a global network or communicate with the transportation authority to get information or to request spe-cial services.

Communication may help the vehicles to see through areas beyond their percep-tion range and to rectify each other's perceptions. Further, it may help the vehicles to correct the localization errors, which can be critical in robotics. More impor-tantly, communication empowers the vehicles to unanimously make travel plans and rectify any anomalies well on time. They may as well coordinate with each other to predict and avoid congestion at some areas, avoid a likely deadlock at an intersection, foresee two vehicles steering towards each other or making the same cut in.

To ease the discussions, the ADAS systems are first classified into information-based assistance systems and manipulation-based assistance systems. The information-based systems only provide suitable information and warnings to the human driver. They neither implement any driving decisions, nor physically operate the vehicle even by a small amount. On the contrary, manipulation-based assistance systems may also operate the vehicle in case of emergency or when asked by the human driver. Section 4.2 describes the different information-based assistance systems. Section 4.3 describes the manipulation-based assistance systems. The modal-ity of feedback being audio, display or tactile is a key issue. These issues are discussed in Section 4.4. The multi-vehicle systems that rely on communication between the vehicles are discussed in Section 4.5, whereas the communication mechanism is also illustrated in Section 4.6. Section 4.7 summarizes the work.

4.2 Information-Based Assistance Systems

While driving one may be concerned about a variety of things including the route to take, whether one is going on the correct route, the likely congestion levels to expect, whether it is possible to reach the destination on time, whether the current state is safe, and if not what needs to be done etc. All these questions can pre-occupy the mind of the driver which have immense importance in driving. The *information-based assistance systems* take care of all these questions, while enabling the human driver to control the vehicle based on the information provided. These systems do not themselves control the vehicle in any way. The systems are briefly summarized in Box 4.1.

Box 4.1 Information-based Assistance Systems

- **Advanced Traveler Information Systems (ATIS)**
  - Dynamic re-routing
  - Anticipate traffic and congestion
  - Information about vehicles around, roads ahead, places of interest around, diversions available, planned closures etc.
- **Inattention Alert Systems**
  - Distraction and fatigue
  - *Indicators*: eye movements (including percentage closure and blink indicators), head movement (including yawning), gaze and biological signals (including EEG, ECG and sEMG)
  - *Modality Detection of Face, Eyes (including blink), Lips (including yawning), Gaze and Head Pose*: Machine Learning Approaches, Colour Filters, Histogram, Clustering, Grow-cut, Corner Detection, Edge Detection, difference of consecutive frames, modality tracking, optical flow, 3D extraction techniques and distances between the modalities. First face is detected, which gives the approximate locations of mouth, lips and eyes.
  - *Features for image and signal modality data*: Time Domain features (average/median/maximum/minimum value, number of zero crossings, standard deviation, variance), Frequency Domain (FFT), Wavelets, Energy-based, PCA, LDA and ICA
  - *Classification*: neural networks, support vector machines, decision trees, fuzzy inference systems and adaptive boosting algorithms.
  - *Hybrid Techniques*: Fuse two or more modalities at data, feature, score or decision level.
- **Measuring Driver Performance**
  - Also an indicator of inattention.
  - Way of handling the steering wheel, reversal speed of the steering and position on the seat indicates fatigue
  - *Simple Indicators*: Distances with the vehicle in front, fluctuations in speed, ability to make smooth lane changes and ability to maintain the vehicle strictly within the lanes
  - *Complex Indicators*: Whether the driver is following the traffic signs and signals, safety of overtaking, rash driving, poor cut-ins, response time to emergency situations and illegal entry at the intersections.

4.2.1 Advanced Traveller Information Systems

The *Advanced Traveller Information Systems* (ATISs) give information about the transportation system to the driver to make key decisions (Chorus et al., 2006; Levinson, 2003; Shekhar et al., 1993). These systems themselves also suggest the routes to take to avoid congestions, while taking the shortest and the fastest route as far as possible. Routing is a key problem in ATIS. The dynamic and historic information about traffic may be available from the cameras at different roads and sometimes even from the other cars. The information can be used to *anticipate* the future traffic. The anticipated traffic is used for deciding the route of a vehicle. A good routing strategy must be able to minimize *congestion* from happening in the future, while making the vehicle take the quickest, shortest, and safest route. The quickest route may not always be the shortest and similarly may not be the fastest, due to the dynamic nature of the traffic and varying speed limits on different roads. Fuel economy also becomes an important criterion while making a trade-off between the objectives and hence selecting a route. Safety of a route measures the likeliness of congestion and hence getting struck for a prolonged time. People may also like to avoid toll roads, roads which may be damaged or roads which are harder to drive upon and require more caution.

Besides, it is important to display the information to the user. The live traffic and the anticipated traffic need to be suitably represented in the route map, over which the route may be plotted, for information to the driver. The driver may prefer to take the suggested route, or make the route himself/herself based on the presented information. These systems can detect road closures, lane blockages, accidents, congestions etc. at different parts of the transportation network (or the same may be fed in by the transportation authority) and suggest alternative routes to take. *Dynamic re-routing* is very important to optimally utilize the transportation infrastructure. Similarly, these systems are helpful to decide whether to take main roads or the bypass to balance traffic in the transportation system.

ATIS can be used to display any kind of information of use to the driver, including information about the route, vehicles around, roads ahead, places of interest around, expected congestion levels, diversions available, planned closures, construction works etc. All these affect some or the other driving perspectives and are of use to the driver. The systems may even indicate if it is preferable to lane change, overtake or take a diversion instead.

4.2.2 Inattention Alert Systems

Driving can be highly uncertain wherein any pedestrian, vehicle or obstacle may suddenly come in front of the vehicle. Similarly, many times the roads may have sharp and unanticipated turns and speed breakers which suddenly come in front while driving. The reaction time of humans needs to be very small to react to all of these in time. If the driver is inattentive, this reaction time may increase to a few 100 milliseconds or even a few seconds. Within this time there may be situations requiring sudden reaction of the human driver, which he/she may not be able to offer. This is a grave cause of accidents on the road. Many road accidents are attributed to inattention, primarily drowsiness, fatigue and distraction, of the drivers.

Driving for prolonged hours can lead to *fatigue* and drowsiness in the human driver. Many people need to drive between cities for prolonged hours. Many others may need to drive after prolonged hours of work and a rather hectic schedule. The drivers do not take recommended rests to reach their destination early. This causes a sense of fatigue wherein the body becomes tired and is late in responding to sudden reflexes. Under more stress, the body becomes drowsy wherein the person is nearly on the verge of sleeping. The extreme case is when the person has micro-sleeps during driving, which is a big risk. During fatigue, a person is often seen yawning, whereas the eyes may close quite often. The driver feels lazy, irritated and depressed. The driver may show delayed responses to stimuli, which includes a lazy handing of the steering, often resulting in the vehicle not sticking to the lanes and not making smooth manoeuvres.

Similarly, *distraction* is very common in human drivers. The drivers may get distracted from anything happening in the surroundings, presence of other people around, from the other people talking or playing of music, doing some physical activity like adjusting the stereo system or eating or sometimes just in their own thoughts. This causes a lack of concentration on the physical driving task and reacting to any changes in the environment. Being a boring and routine job makes it increasingly likely for drivers to lose their concentration. Unlike fatigue, distraction may not be caused by a lack of rest and may not be easily controllable for the driver.

The *inattention alert systems* constantly monitor the human driver while he/she is driving. The level of inattention, either fatigue or distraction, is closely monitored. On increasing to a threshold, the system may issue warnings and further still may start sounding alarms to alert the driver about the inattentive status. The driver, hence, constantly knows about the inattention and on being advised by such systems can opt to take rest for some time or allow a co-passenger to drive. Subsequently, the system may also offer to drive on its own or take safety precautions, if the system has the same capability.

4.2.2.1 Indicators of Inattention

A prominent and natural way of judging for distraction is by the *eye movements* of the drivers. Attentive drives tend to concentrate on the road and the vehicles in front. One is also cautious to look at the vehicles around in cases of mergers, intersections, traffic lights and traffic signs around. In distraction, one may spend less time looking ahead and may instead concentrate on the road surroundings, vehicle interiors, mirror etc. (Rantanen and Goldberg, 1999). A sudden change in eye movement may be observed during distraction. For detecting fatigue and sleepiness, eye closure is a valuable indicator. One of the best features used to detect drowsiness is the percentage of time the eye is more than 80% closed or PERCLOS (Dinges and Grace, 1998). Eye closure duration, blink frequency, blink amplitude, blink duration and microsleep events are other prominent indicators which are widely used (Friedrichs and Yang, 2010). Similarly, eyebrow movements can be used as the indicators of drowsiness.

Head movements are similar indicators for distraction, wherein distracted drivers may seldom show frequent changes in head movements whereas the concentration may not be focussed on the road ahead. Similarly, less frequent head movements indicated a condition of fatigue; the head movements themselves are taken to relieve the stress and get the driver in a more comforting posture. Facial expressions are other indicators of distraction and fatigue. Yawning is a very prominent indicator for drowsiness. Fatigued drivers tend to yawn frequently. Similarly, talking is an indicator for distraction.

The distraction can also be prominently measured based on signals obtained from the driver's body, for which the drivers must wear specialized caps with electrodes to measure the biological signal; but it is a fair indicator of distraction, fatigue and sleepiness (Jap et al., 2009; Liu et al., 2010). *Electroencephalographic (EEG)* signals carry a lot of information about the brain waves and are widely used to indicate brain activities which cause a drastic change in some part of the brain. Distraction and fatigue can both be conveniently measured using the EEG signals. An *Electrocardiogram (ECG)* reports the changes in the heart signals which can also be interpreted to measure the change in the thought process of the human driver. *Surface electromyogram (sEMG)* places nodes on the surface of the skin to measure the muscular signals and hence indicate the activity of the human driver.

4.2.2.2   General Procedure for Detecting Inattention

The data used to detect distraction and fatigue may come from any source, but need to be used to make a decision regarding the presence or absence of inattention and the magnitude of inattention. This can be done using machine-learning tools. The decision follows a typical procedure of a pattern recognition system, shown in Fig. 4.1. The first step is to collect the data. The visual data like face and eye can be collected from infrared cameras or video cameras. The biological signals require specialized electrodes placed on the body for fetching the data. The raw data collected from any source may be too noisy to work with. *Noise removal* can be done using image-processing and signal-processing techniques, typically passing the signal through a band-pass filter to remove the unnecessary frequencies corresponding to noise. The noise-free signal is then taken. The signals may require *segmentation*. In the case of image data of a

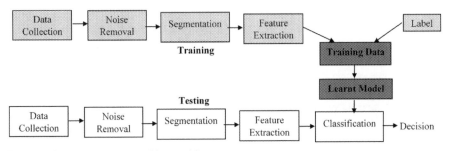

Figure 4.1 The pattern recognition problem.

person, one may need to extract the eyes and mouth for fatigue analysis, which is called the problem of detection of individual modalities. The segmented image or signal is generally too large for the machine-learning techniques to work. Hence, *feature extraction* is used to convert the raw data into some useful features. The features are then passed to a machine-learning tool which does the task of *classification*. The classifier produces the final output, which is the state of the driver and the level of inattention.

4.2.2.3 Feature Detection

The problem of *detection of face* may be solved as a machine-learning problem, wherein a large number of images of faces are shown to the machine-learning tool along with the images of non-faces. The images are reduced in dimensionality using *features* like *Scale Invariant Feature Transform* (*SIFT*), *Speeded Up Robust Transform* (*SURF*), *Wavelets* and other features discussed in Section 4.2.2.4. The machine learns to distinguish between the faces and non-faces. Using this data, the newly acquired image from the camera may be traversed by a floating window of different sizes, throughout the image. The machine is fed with these images, and outputs the possibility of the image window being a face. The face captured by the highest probability window is taken. The problem of face detection in this scenario is reasonably simpler as the camera and the sitting position of the driver are both fixed. Hence, there is a small region within which the face must lie.

Alternatively, using *colour filters* is another natural approach to face recognition. The range of values that the skin colour may take can be computed against a variety of test subjects. A colour filter allowing colours of this band filters out most of the skin areas along with some other noises in the background. The image so formed may be divided into connected components, only taking the most dominant connected component while filtering out all the other smaller connected components corresponding to noise or other body areas. A bounding box may then be drawn around the face, bounding this connected component. Sometimes, the face region as detected by the colour filter may be grown out by using image-processing operators.

In this problem, because the camera is located such that it nearly aims at the face and the small regions around, a *histogram* of colours shows a peak at the region of the skin colour, stating that the skin colour is the colour found in abundance, which can be used as an indicator to extract the face. *K-means clustering* can also be applied to separate the skin colour from the background colour. Another way to intuitively use the colour information for extracting the face is to select some seed points which have a high confidence of belonging to the face. The points are allowed to grow like a cellular automata till the colour is nearly the same. This method is called the *grow-cut method*. The growth stops on reaching the face boundary, as a sudden change in colour is observed. This method also segments out the face from the eyes and lips.

Another key possibility to extract the face is by the knowledge that, mostly, the background is constant in an image taken of a person driving, whereas the face may show some movement as the person drives. Taking *differences of consecutive frames* can give an indicative boundary of the moving objects in the image. The small regions or regions with small boundaries may be deleted, as they correspond to noise, thus getting the face boundary. A *fusion* of all the methods above is usually suggestive for extracting the face as different approaches may face problems in different scenarios.

It is preferable to use *tracking systems*, which can continuously keep track of the face which helps significantly in face detection and removing the noise. The location of the face at every instant of time will be around the area where it was seen previously. The speed of motion of the face can itself be estimated, to estimate the prospective positions of the face in the future. Tracking-based systems use this information to avoid outliers and have a focused detection of the face around the expected areas. Kalman filtering and particle filtering are common approaches to track face.

Three-dimensional (3D) detection and recognition techniques have increasingly become popular. These techniques use a stereovision camera to estimate the depth of the object. The depth is added information as compared to 2D vision techniques. It becomes increasingly easy to segment out the face using 3D techniques as the regions around the face are marked by a sharp increase in the depth. Depth alone has the potential to conveniently extract the face. The areas with low depth are generally the face, whereas the areas with a high value of depth correspond to the background. Smaller detected regions can be taken as noise, whereas the detection can be supplemented by the methods discussed previously. The features used can also be extended from taking Red—Green—Blue (RGB) (or equivalent) colour channel as an input, to Red—Green—Blue—Depth (RGBD) (added depth) channel as an input.

Apart from the face, inattention-detection systems also require location of the eyes and lips which are strong indicators of distraction and fatigue. The lips are particularly useful for detection of *yawning*. Lips can also be identified by machine-learning techniques, given a dataset of images of lips and non-lips. Alternatively, knowing the position of the face, the approximate position of the lips can be determined. The lips have a different colour than the skin, and are separated in the grow cut method used in face detection. Another way of detecting lips is by using a grow-cut method starting with the seed points as known regions in the lips. Alternatively, corners are valuable indicators of lips and *corner detection* techniques can be applied over the lower part of the image of the face. *Edge detection* is also useful to detect the lip boundaries.

The *eyes* can be conveniently detected as darkest regions in the segmented-out face. Alternatively, a machine learning-based detection system may be designed using images of eyes and non-eyes. Knowing the approximate location of the eye, which is possible after face detection, edge-detection techniques become very helpful. The eyes have a distinctive colour as opposed to the rest of the face and hence leave an edge which can be detected by edge-detection techniques like Sobel and Canny.

Blink is also modelled in inattention-detection systems, especially for fatigue. Blink is a dynamic gesture corresponding to the eye. Taking a difference of consecutive frames of the eye can give information about the blink. Further, *optical flow* can be used to get rich features from blink. Detection systems may seldom differentiate between pupils and eye-lids, which is also a good indicator of blink. *Gaze* is also indicated by using the eyes as a feature. Gaze can be detected by using pupil as a reference. Tracking of the pupil may be done using filtering techniques to indicate the change in gaze.

Head pose was also used to detect inattention. The head pose can be indicated by the relative distances between different features of the face including eyes, lips and nose. The head pose may be constantly tracked by using optical flow and filtering techniques to give the dynamic features corresponding to changes in the head pose.

4.2.2.4 Feature Extraction and Classification

The next important task is feature extraction, wherein the complete image or signal is broken down into a lower dimension, representing good *features* that can be used for classification of the input. The features must be such that they facilitate a small intra-class distance and a large inter-class distance. The number of features needs to be limited as it significantly affects the classifier performance and necessitates the availability of more data. For inputs representing signals, the popular features to use include features in the time domain and features in the frequency domain. The *time domain features* extract features in the temporal representation of the signal. The features include average value, median value, maximum value, minimum value, number of zero crossings, standard deviation, variance etc. In the *frequency domain*, the signal is first converted into the frequency domain and the frequencies are selected. The use of Fast Fourier Transform (FFT) is a historic and popular technique. *Wavelet decomposition* techniques are widely used, wherein the signal is decomposed using wavelets. Energy features are also used, which take the features as the energy content of the input signal.

Feature-selection techniques aim in reducing the dimensionality by a careful selection of features for the problem. Approaches include filter and wrapper. *Filter approaches* assess the data statistically and generate new features which more clearly separate the data. The top features may be selected for the actual classification. These techniques do not use any classifier to assess performance and are generally faster, whereas they do not model any dependence on the classifier. The techniques include Principal Component Analysis (PCA), Linear Discriminant Analysis (LDA), Independent Component Analysis (ICA) etc. *Wrapper approaches*, on the other hand, use a classifier to judge the performance of the selected features and aim to select the best set of reduced features, which give the least error on the training data. *Hybrid approaches*, using both filter and wrapper methodologies, are widely used.

The last important stage is *classification*, wherein the actual decision is made regarding the input indicating inattention or not. The classifiers take a labelled set of data as input, which correspond to the cases of inattentive driving and attentive driving. This is normally marked by a human expert based on his/her judgement. The inputs are reduced in dimension by extraction of features. The feature to class mapping is learnt by the classifier. Based on the learning, the classifier is able to map any new input to its class, and hence decide whether (as per the current inputs) the driver appears attentive or not. Popular classifiers are neural networks, support vector machines, decision trees, fuzzy inference systems and adaptive boosting algorithms (Shukla et al., 2010a).

4.2.2.5 Hybrid Techniques for Inattention Detection

Overall, there are different techniques to assess inattention and to indicate the level of inattention. Some techniques are primarily for distraction whereas others are primarily for fatigue. The two inattentions may interfere with each other, making detection difficult. *Hybrid techniques*, making use of multiple techniques that have been discussed, suggest better implementations. Here, the performance is not restricted to the limitations of one modality alone. Therefore, EEG-based and eye-based systems may be used simultaneously to detect fatigue. Similarly, more modalities may be used. The general approach is shown in Fig. 4.2.

The *fusion* or *integration* (Kuncheva, 2004; Shukla et al., 2010b) of decisions from each modality becomes an important criterion, which decides the overall response to the system. Typically, fusion may be performed at different levels. *Data level fusion* suggests that the raw data of different modalities is combined, and the combined data is then passed through feature extraction and classification. *Feature level fusion*

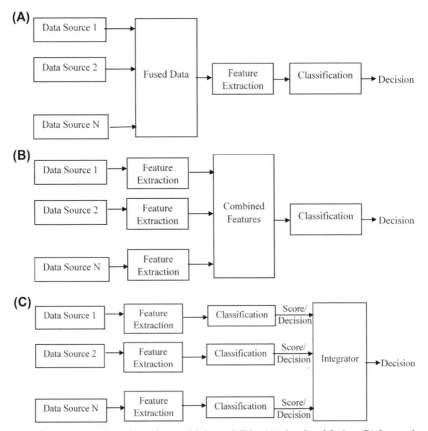

Figure 4.2 Inattention detection using multiple modalities (A) data level fusion (B) feature level fusion (C) score and decision level fusion.

systems, on the contrary, first independently perform feature extraction on every modality, and then combine the features, which go to the classifier.

The problem with these methods is the multiplication of the features given to the classifier which deteriorates as the classifiers generally face the problem of the *curse of dimensionality*, or that the classifier performance exponentially reduces on increasing the number of features. *Score level fusion* systems classify every modality separately, based on their own features, and produce a probability density function or a score function for every modality. The scores for each modality are then fused to produce the overall scores by an *integrator*, based on which decision is taken. Taking the average is a natural technique. Alternatively, maximum, minimum or median scores may also be considered. Fuzzy integrators are also common which use fuzzy arithmetic for integration of the results. *Decision level fusion* systems, instead, make every classifier take a decision and then fuse the decisions. Poling or voting is a popular mechanism. The different techniques are also summarized in Fig. 4.2.

4.2.3 Measuring Driver Performance

One is normally cautions about the driving performance of oneself and the others recruited to drive. Performance can be used by the transportation authority to continuously assess the driver and accordingly extend or cancel one's licence to drive, suggest driving lessons or issue periodic warnings. *Driver performance systems* sit inside the vehicle and continuously rate the driver's performance.

One of the main application areas of driver performance is to detect inattention, in which sense this sub-section is a continuation of the last sub-section. Because of distraction, the driving performance of the person is affected. Systems can also monitor the irregularities in driving patterns to indicate an inattentive state of the driver. However, the inattention should be prominent before a warning sign may be issued. The time it takes for the system to judge an prominent level of distraction is very risky for the driver.

One of the best ways to indicate the driver's performance is from the distances from the vehicle in front, fluctuations in speed, ability to make smooth lane changes and the ability to maintain the vehicle strictly within the lanes. Causing lane drifts and uncertain manoeuvres are examples of poor driving. Similarly, keeping too small a distance from the vehicle in front and constant fluctuation in speeds is not regarded as good driving. The deviation of the vehicle from the lane centre is a simple and good indicator of driving performance and hence also inattention. Different indicators may need different sensors and environment percept to decide the driver's performance.

Some of the sophisticated systems can also track the traffic signs and signals and judge whether the driver is following the traffic signs and signals or not. Further, some systems can track the vehicles around to judge safety of overtaking, rash driving, poor cut-ins, response time to emergencies and illegal entry at intersections. The driver can be given a warning both before and after the poor driving manoeuvres, which acts as a continuous feedback for the driver to be cautious in future driving manoeuvres.

The way of handing the steering wheel, force applied at the wheel and reversal speeds of steering also indicate a fatigued and inattentive driver. Drivers tend to make more and large steering corrections within the wakeful period followed by no steering correction in the sleepy periods. The sharp change is easily observable and can be used for detecting fatigue. Similarly, pressure distribution on the seat is another indicator of the drowsiness of the driver. Drivers may tend to continuously adjust their position in the seat in the wakeful periods, while generally adopting positions that are more relaxing rather than the positions offering greater attention.

4.3   Manipulation-Based Assistance Systems

Although information is invaluable for the human driver, the information-based assistance systems load the driver with a lot of facts, leaving it to the human driver to interpret those facts and take actions. Hence, many times these systems may be a burden for the human driver, providing information that the human driver may not need or may not readily trust. The *manipulation-based assistance system* goes one level higher and manipulates the vehicle on behalf of the human driver. It may be used to take emergency actions as there may not be enough time to warn the human driver who then takes the action. In addition, these systems may be used to perform some regular or simple manoeuvres which do not require much of human driver's wisdom. They may also be used to carry some specialized and precise manoeuvres like overtaking and parking, which may be hard for human drivers.

4.3.1   Safety Alert and Emergency Stopping

Safety alert systems predict prospective collisions and risky situations and warn the driver. The driver may look at the warnings and take corresponding safe actions. Sometimes the situation may be too adverse for the driver to be warned and the situation may require an immediate action. In such a situation, these systems also let the vehicle take preventive actions like *emergency braking* (Coelingh et al., 2010; Miller and Huang, 2002; Sengupta et al., 2007; Vahidi and Eskandarian, 2003).

The most common and the easiest to understand system is the *forward collision avoidance system*. This system constantly monitors the vehicle in front, the distance maintained and the speed of the vehicles. Accordingly, the system computes the *Time To Collision* (*TTC*), which is the time after which the vehicles may collide if they keep travelling with the same profiles. Correspondingly, they compute the *Time To collision Avoidance* (*TTA*), which is the time in which a reaction must be generated, otherwise a collision is bound to happen. If the TTC is much higher than the TTA, the situation is safe and no warning is issued. It is assumed that the driver is aware about the situation and can take precautionary measures. When the difference between the two times starts getting closer, the system keeps generating warnings. The warnings are smaller to start with, which get more intense if the distance reduces, unless it is apparent that the driver has started taking the precautionary measures. If the times get very close, the system takes the vehicle control and starts applying emergency braking. Similarly, these systems may also be used in cases when the vehicle at the back is very likely to collide, in which case the system tries to speed up the vehicle to create a safe gap.

Similarly, many times a vehicle may opt to make a close cut in, requiring immediate speed adjustment from the vehicle at the back. Many accidents are caused by such close cut-ins. The close cut in normally happens due to improper assessment of the human driver and conflicting intents of the two vehicles. Such cases may force a driver to brake and control the speed immediately. *Collision avoidance systems* can be used to track all vehicles and quickly detect a close cut in, based on which an avoidance action can be taken. Correspondingly, these systems can also be used to display how safe a lane change is. A driver can hence be assured of making safe lane changes so as not to cause trouble to any other vehicle. These systems can also stop a vehicle from making a lane change when it is not safe to do so, thereby avoiding accidents.

A more difficult mechanism is the *intersection collision avoidance system*, wherein the system avoids collisions with the vehicle operating in an intersection. Uncontrolled intersections leave it for the vehicles to decide when it is safe to cross an intersection. Often different vehicles make opposing plans which results in the vehicles nearly colliding with each other. Because the collisions may not always be from the vehicle directly ahead or directly behind, these collisions are hard to infer. These systems track all the vehicles in the surroundings and their intents. When a vehicle appears to be in a collision-prone state, the systems suggest precautionary actions and sometimes even take the precautionary actions.

A very challenging problem is detection of *pedestrians* and making corresponding collision avoidance systems. Pedestrians may sometimes not be very cautions of crossing the road, especially children. Vehicles coming at a moderate speed can be of a great threat to these pedestrians. Not all drivers are careful in sighting pedestrian crossings and slowing down to be cautions. Unless the pedestrians are detected on time, the driver has negligible time to stop the vehicle to avoid accidents. These systems hence sight pedestrians and stop a vehicle well on time to avoid collision with the pedestrians, and resume the vehicle later on. Fig. 4.3 summarizes the safety alert systems.

4.3.2 Adaptive Cruise Control

Adaptive Cruise Control (ACC) systems control the speed of the vehicle and free the human driver from constantly monitoring and regulating the speed by the use of brake and throttle. In an empty road, ACC helps the human driver by taking

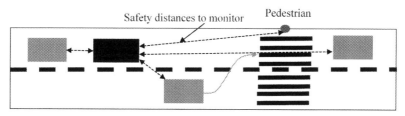

Figure 4.3 Safety alert systems.

the speed control, whereas the human is still required to control the steering. These systems facilitate driving at a constant speed using simple control systems. In a traffic-prone environment, these systems can assess the distance from the vehicle in front and maintain a fixed distance from the vehicle in front, while driving at a safe and appropriate speed. The distance from the vehicle in front may be measured by radio detection and ranging (radar), ultrasonic sensors or light detection and ranging (lidar). Of course, the human driver can override the actions of the ACC by pressing the brake or the throttle. For detailed discussions, see Marsden et al. (2001); Martinez and Canudas-de-Wit (2007); Rajamani and Zhu (2002); and van Arem et al. (2006).

Especially in lane-oriented organized traffic, for most parts of the journey a vehicle is simply driving straight in one's own lane and/or following a vehicle driving in front. *Lane keeping* and *vehicle following* are two dominant behaviours of driving. In the case of straight roads, a simple ACC is enough as barely anything needs to be done on the steering. More practically, and for roads which have small to large curves, the speed and steering need to be both monitored and corrected for the vehicle to follow another vehicle. The lane-keeping behaviour simply keeps track of the roads and the lane boundaries of travel. The trajectory to be followed is simply the middle point of the lane, which can be easily computed based on the knowledge of the lane boundaries. Most roads are smooth, meaning the trajectories are easily traceable. A vehicle-following behaviour is nearly the same, with additional constraints on maintaining sufficient safety distance from the vehicle in front. The safety distance depends upon the speeds of both the vehicles.

Here, a slightly more complicated vehicle-following behaviour is discussed, as demonstrated by Gehrig and Stein (2007). In this behaviour, a vehicle is asked to follow a vehicle while escaping obstacles on the way. The vehicle being followed may at times become completely occluded by the obstacles. The approach is meant to assist human drivers by taking most of the driving from them, whereas the human may be asked to take over the control in case of safety concerns, obstacles close by, no reasonably safe trajectory possible, high lateral acceleration, uncertainties with dynamic obstacles etc. The environment is sensed with calibrated stereovision. The authors attached an elastic band to the vehicle being followed. One end point is attached to the vehicle being followed while the other end is at the vehicle itself. The band moves as the vehicle being followed moves.

The band is repelled by all the obstacles that lay around. The repulsion considers the position, orientation and speed of the vehicle. The repulsion by obstacles is modelled separately from the repulsion from the road boundaries. Different types of obstacles like pedestrians, vehicles etc. can also be modelled separately based on their behaviour. These repulsions disallow the vehicle to collide with any obstacle or drive outside its lane. The elastic band also has an internal force which forces it to attain the same trajectory as the leader vehicle being followed. The band is maintained to be always collision-free and smooth and denotes the trajectory of the following vehicle. If the band becomes collision-prone due to sudden changes in the obstacle, waypoints are used to construct a feasible plan which attains the best shape as per the elastic band forces. A controller is used to make the

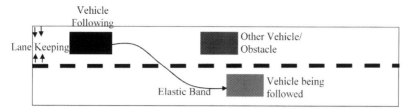

Figure 4.4 Vehicle following.

vehicle travel the trajectory represented by the band, while the trajectory keeps adapting itself to the changing environment. Fig. 4.4 gives a small summary of the work.

4.3.3 Overtaking Assessment and Assist

Overtaking is a very interesting traffic behaviour wherein a faster moving vehicle or a vehicle capable/desirous of attaining higher speeds sees a slower moving vehicle at the front and decides to change lanes to surpass it (Hegeman et al., 2009; Jin-ying et al., 2008; Wang et al., 2009). This calls for a lane change to the right lane (to-left traffic rule assumed), also called the *overtaking lane*, travelling on the right lane till the slower vehicle is surpassed with an additional safety distance, and returning back on the left lane or the normal driving lane. The process is shown in Fig. 4.5. Different vehicles on the road may have different preferences in terms of speeds. Overtaking gives an opportunity for the faster vehicles not to be penalized by the lower speed of the slower vehicles in front. While overtaking, however, it is preferred that none of the other vehicles are punished by a significant amount to host the overtaking. So the overtaking procedure should complete with no or small change of speed or driving profile of the vehicle being overtaken or any other vehicle in the vicinity.

Overtaking has two important tasks, deciding feasibility of the overtake attempt and actually overtaking. The *feasibility of the overtaking attempt* is primarily based on the distance between the overtaking vehicle and the vehicle being overtaken, vacancy at the overtaking lane for the overtaking duration in the overtaking zone and the prospective plans of the other vehicles in the overtaking lane or desirous to enter the

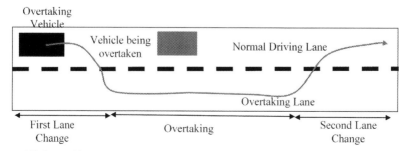

Figure 4.5 Overtaking.

overtaking lane. Physically *performing the overtaking* manoeuvre chiefly calls for a lane change to the right overtaking lane, a lane-keeping at the overtaking lane and a lane change to the normal driving lane. The behaviour may be decomposed into simple lane change and lane keeping behaviours.

Planning an overtaking refers to mechanisms to judge the feasibility of overtaking, mechanisms to carry the individual behaviours of lane keeping and lane changing, and mechanisms to decide when to initiate which behaviour and in what duration to aim completion of the behaviours. Various overtaking planning systems exist in the literature for organized traffic. One such system is discussed, as demonstrated by Naranjo et al. (2008). The authors used fuzzy control for overtaking. Feasibility of overtaking was based on the preconditions that the overtaking vehicle must be moving on a straight road and on the same lane as the vehicle being overtaken, the speed of the overtaking vehicle must be higher than the vehicle being overtaken and the overtaking lane must be free and long enough for the overtaking. Distance between the overtaking vehicle and the vehicle being overtaken was considered. The longitudinal distance of the overtaking procedure was assumed largely dependent upon the target speed of the overtaking vehicle and the function was regressed based on the actual data.

The physical tracking of the trajectory was done using fuzzy controllers. One controller was responsible for the lane change behaviour and the other for lane-keeping behaviour. Both controllers used lateral error and angular error as the input variables. The variables were divided into separate membership functions and rules were written separately for the two behaviours. Using these fuzzy controllers, the authors demonstrated successful overtakings on straight roads.

4.3.4 Automated Parking Systems

Parking is a common problem in general driving. Placing a vehicle in an empty parking slot is a difficult job and often requires multiple sharp steering attempts. Many modern vehicles already come with a parking assist, wherein the vehicle can park itself with some assistance from the driver. This relieves the driver from constant assessment and manoeuvring the vehicle. Parking may be parallel, perpendicular or en echelon. Each of these has its own difficulties. Parking is a very important task that an autonomous vehicle must be able to carry, which goes a very long way to make the autonomous vehicles drive from parking to parking via roads, intersections, mergers etc.

The problem of parking is very close to the general problem of robot motion planning. Existence of a well-structured lane and vehicle-following and -overtaking behaviours make autonomous vehicle planning different from mobile robotics. Both these factors are absent in parking. Hence, practically any robot motion-planning algorithm can be used. The problem is shown in Fig. 4.6. A vehicle normally goes back and forth a small number of times, before succeeding in placing itself perfectly in the empty parking slot, which is one feature that the motion-planning algorithm should facilitate. Common approaches include Rapidly-exploring Random Trees, Graph Search and some reactive techniques (Han et al., 2011; Paromtchik and Laugier, 1996a,b; Hsieh and Ozguner, 2008).

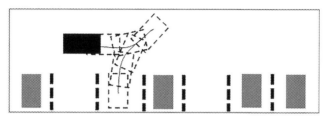

Figure 4.6 Parking.

4.4 Feedback Modalities to Driver

Sections 4.2 and 4.3 have discussed giving some information or warning to the user. The information or warning given may be either *visual* displayed on a screen in front of the human driver, *acoustic* played in the vehicle for the driver to listen or *tactile* which can be felt by the driver. These constitute three basic modalities in which all information is given to the human driver in the driving process. Different modalities are suitable for different kinds of information. This section very briefly discusses the pros and cons of different modalities for different kinds of information. For detailed discussions, please refer to Adella et al. (2011), Cao et al. (2010), Kaufmann et al. (2008), Kim et al. (2010) and Mulder et al. (2008).

The most natural choice of information is visual, which clearly allows a user to see and decide the course of action. The pro of this modality is the level of detail that can be input. One can display complicated maps, routes, congestion levels, detailed warning messages etc. with good details. The problem with this modality is that it distracts the driver. A driver spending much time on the display is hazardous to driving as the concentration should always be on the road. Hence, the information displayed must always be simple and easy to interpret. Displaying complicated information also results in the drivers tending not to look at the screens and driving on their own. The visual display is only used for non-emergency situations due to the time it takes to glance and interpret the situation.

The audio modality may be either in the form of *detailed sound instructions* given to the driver or in the form of *warning tones*. The pro of using detailed sound instructions is that they do not draw the attention away from the driver like the visual counterparts, while being restricted to audio instructions alone. It still takes a long time to interpret the information and can be used in non-emergency situations or situations requiring a moderate level of reaction time. The warning tones can though be used for emergencies. Warning tones are easy and fast to interpret and do not require much of a driver's attention. The details that can be given are low. In other words, only a few types of warning tones can be used without confusing the driver. The length and volume of the tone may correspond to the level of emergency. Too many tones or a prolonged used of warning tones can annoy the driver, thus distracting him/her.

Tactile modality is the most effective for suggesting emergency actions. A vibration may be produced in the seat, steering or sometimes pedal, used to indicate the action to be taken. It is very necessary to produce tactile feedback at the place where the correction needs to be applied. The vibrations produced must convert to natural reactions of the driver, without requiring intensive interpretation. A tactile feedback should be sparingly used as excessive tactile feedback can seriously annoy the driver, leading to de-installation. Further, tactile feedback can only be used to give a very small set of warnings without much detail.

Hybrid feedback modalities are also used, wherein the information may be given to more than one modality. As an example, in moderate situations a visual warning may be displayed; when the situation becomes more intense, an audio warning may be used; and subsequently a tactile feedback may be given. Alternatively, a warning may be sounded, and for details the driver may be referred to the display console. It is important to have good designs of hybrid modality systems so as not to distract the user, while giving all the information in a comforting manner.

4.5   Multi-Vehicle Systems

Increasing autonomy in vehicles is making it possible for most, if not all, vehicles to talk to each other and with the transportation infrastructure at the roadside, or with the central transportation authority. The vehicles in the vicinity can form small ad hoc networks to talk to each other and communicate information. This results in a significant advantage to autonomous vehicles.

One of the best examples in the use of multi-vehicle technology is in *multi-vehicle perception*. The view of one vehicle is always occluded by the other vehicles. Many times sensor uncertainties in one vehicle may be large, disallowing effective decision making. Multi-vehicle perception allows the vehicles to share their perceptions of the world. Each vehicle makes a small local map of the surroundings which is communicated to all the other vehicles, especially the vehicles nearby. The integration of all the local maps of different vehicles produces an accurate overall map of the region, which can be used to make effective decisions in navigation. Each vehicle best perceives the nearby surroundings, whereas for the areas far away, the uncertainties are bound to be large. Hence, it is better to rely on the vehicles ahead for information of the further areas.

A good example is the *see-through technology*. Vehicles need to know the occupancy of the vehicles in the lane ahead to best decide the feasibility and utility of overtaking. The slow vehicle just ahead may make it impossible to assess the scenario of the vehicles ahead. Multi-vehicle perception allows the vehicle to see through the eyes of the vehicle ahead. The situation ahead may either be displayed on the screen of the vehicle ahead, or may be communicated by an ad hoc network. Similarly, knowledge of the vehicles ahead affects decisions of lane changes and slowing down in anticipation.

An extended advantage of the technology is *multi-vehicle localization*. The vehicles largely use a Global Positioning System (GPS) to give an indicative position, while using a combination of vision, proximity and motion sensors to give the precise location. GPS uncertainties may be very large in areas highly occluded by buildings or otherwise. The localization by vision and proximity sensing also depends upon the locations of the other vehicles. Sharing percepts allows the vehicles to co-localize themselves. Even if a vehicle may not be able to estimate its own location, knowing the locations of the vehicles around and the distances maintained with these vehicles gives a better estimate of location.

Besides, many times vehicles may make contradictory plans like cutting in on a lane at the same time, two vehicles simultaneously trying to overtake, simultaneously trying to avoid an obstacle, simultaneously attempting to enter a roundabout etc. These situations happen because the vehicles cannot otherwise talk to each other and make unanimous plans. The plan of one vehicle is always made by making some assumptions about the navigation plan of the other vehicles. Communication allows the vehicles to talk to each other and resolve the discrepancies, as well as *coordinate* to make mutually benefitting plans.

Vehicle Platooning is a very interesting and simple technology, which allows multiple vehicles to closely follow each other, one after the other. The vehicles run by using control systems which primarily take the distance and direction from the vehicle in front as the input. The aim is to follow the vehicle as closely as possible, while maintaining safe distances. The leader of the platoon may make critical navigation decisions, whereas the rest of the vehicles simply follow each other. Communication enables broadcast of messages indicating expected drop in speed, braking or an increase in speed. Communication may further enable dissemination of distances for better estimate of the speed to operate upon. Platoons sometimes facilitate the vehicles to join the platoon, leave the platoon, split the platoon and re-merge after splitting.

4.6   Communication

Communication allows the vehicles to talk to each other and thus benefit from the advantages of multi-vehicle techniques noted in Section 4.5. It also allows the vehicles to talk to a *Road Side Unit* (*RSU*) and ultimately to the central *transportation management unit*. This section very briefly discusses the issues of communication in intelligent transportation systems. For an in-depth discussion into various issues involving communication, the readers are referred to more specialized texts (Gordon, 2009; Gozalvez et al., 2012; Ma et al., 2009; Schroth et al., 2006; Toor et al., 2008).

Vehicle to Vehicle (*V2V*) communication allows two vehicles in the vicinity to talk to each other. The network is also called a Vehicle Ad hoc Network (VANET) and is a special type of mobile ad hoc network, wherein the agents forming the network are mobile and thus the network keeps changing its size and architecture. Typically, the vehicles use WiFi technology to communicate, following the Institute of Electrical and Electronics Engineers (IEEE) 802.11 and more recently 802.11p standards or the Wireless Access in Vehicular Environments (WAVE) communication standards. Even though the vehicles can communicate in any protocol, standards allow for interoperability between different providers. It is extremely likely that standard transportation services, in the future, will be provided by multiple operators, who must be able to communicate with each other to provide the best services to the end user. The infrastructure of one operator must be able to talk to the infrastructure of the other operator, if the other operator can serve a part of the user request. The National Transportation Communications for Intelligent Transportation System Protocol (NTCIP) governs and lays down the protocols of communication. The vehicles join, disjoin and re-join the local networks as they move. The vehicles may also communicate with the RSU (*V2R communication*) or the road infrastructure (*V2I communication*). This allows them to interact with the roadside units which play a regulating role for the traffic.

The RSUs are internally connected to each other by a high-speed and reliable backbone, which allows the RSUs to talk to each other and ultimately with the vehicle. Because of this technology, an incident detected by an RSU at one road can be intimated to a very distant vehicle, requesting re-routing to avoid congestion. The transportation management unit is also ultimately connected to the same backbone, and the unit controls the RSUs and indirectly all the vehicles.

The network backbone may be made by using optical fibres, or simply using coaxial cables for communication. The coaxial cables are cheaper to use, however operate at lower speed and face a greater problem of attenuation. Telephone networks may also be used for the purpose of communication. Sometimes it may be better to have wireless communication using microwave signals. Licenced microwave signals operate at a dedicated frequency as licenced to the operator and guarantee no interference by any other operator or service. Cellular networks are convenient to use as they are already provided by the cellular operators. The reliability largely depends upon the availability of bands to broadcast the service messages. Operators operate by multiplexing of the signals from the individual subscribers. The overall communication framework is shown in Fig. 4.7.

A transportation system needs to broadcast different types of messages, each having different priorities and different magnitudes of data. The communication systems may be used to broadcast video messages as recorded by the other cars and closed-circuit

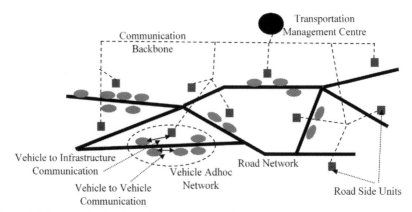

Figure 4.7 Communication in intelligent transportation system.

television (CCTV) cameras; readings of sensors like radars, lidars etc.; specialized service requests asking a vehicle to allow overtaking, to speed up, slow down or change lane; a change in display of the changeable message sign; an intimation of a risky situation and thus the possibility of an accident; suggestions regarding the route to take; or sometimes just advertisements, news items and updates about the journey. Messages corresponding to likely accidents, passage of emergency vehicles etc. have a very high priority and must be delivered as soon as possible; the lower-priority messages like information about the route can be done only so as not to cause congestion in the network traffic. Correspondingly, the needed reliability from the protocol and channel depends upon the application. Time and network overheads are normally important factors to consider.

Privacy, authentication and *security* (Hartenstein and Laberteaux, 2008; Lin et al., 2008; Raya and Hubaux, 2005) are becoming increasingly important issues in the use of VANETs. It is anticipated that future VANETs will have the users sending a lot of private information including web browsing on these networks. Further, the location of the vehicle itself and its operation can be considered as private information. The VANETs must not allow hackers to intrude the network and flood it with wrong information, which may seriously impact the traffic.

4.7 Summary

The chapter explored the domain of the ADAS, wherein the automated systems assist the human driver. Some of the assistance systems merely provide some information to the human driver, whereas some others go ahead and actually take some control of the vehicle. The discussion started with the ATIS, which gives a lot of information to the human driver to route oneself well. Many ATISs also suggest good routes to efficiently reach the destination. Inattention alert systems warn the driver on being distracted or on being in a state of fatigue. Eye movements, head pose, gaze, EEG, ECG etc. are good indicators to estimate inattention. Similarly, the automated systems can assess the driving capability of the driver based on his/her manoeuvres, which is also an indicative of inattention.

One of the most important applications of ADAS is in a safety warning system, wherein the system can assess a collision-prone situation and first give warning, and later take control of the vehicle to avoid a collision. ACC and similar systems allow the vehicle to make simple manoeuvres. Similarly, overtaking-assist systems assess the safety of overtaking and further performing the overtaking manoeuvre. Parking forms one of the dominant areas of application. In all these areas, it is important to choose between visual, auditory or tactile feedback modality.

Technology allows multiple vehicles to talk to each other and share perceptions and information. This helps the vehicles to better see the world around, localize themselves, rectify motion plans and intents and also operate in a platoon. The communication is facilitated by vehicle-to-vehicle and vehicle-to-infrastructure communication systems.

References

Adella, E., Várhelyia, A., dalla Fontanab, M., 2011. The effects of a driver assistance system for safe speed and safe distance — a real-life field study. Transportation Research Part C: Emerging Technologies 19 (1), 145−155.

van Arem, B., van Driel, C.J.G., Visser, R., 2006. The impact of cooperative adaptive cruise control on traffic-flow characteristics. IEEE Transactions on Intelligent Transportation Systems 7 (4), 429−436.

Cao, Y., Mahr, A., Castronovo, S., Theune, M., Stahl, C., Muller, C., 2010. Local danger warnings for drivers: the effect of modality and level of assistance on driver reaction. In: Proceedings of the 14th International Conference on Intelligent User Interfaces, Hong Kong, pp. 239−248.

Chorus, C.G., Molin, E.J.E., Wee, B.V., 2006. Use and effects of advanced traveller information services (ATIS): a review of the literature. Transport Reviews 26 (2), 127−149.

Coelingh, E., Eidehall, A., Bengtsson, M., 2010. Collision warning with full auto brake and pedestrian detection — a practical example of automatic emergency braking. In: 13th International IEEE Conference on Intelligent Transportation Systems, Madeira Island, Portugal, pp. 155−160.

Dinges, D.F., Grace, R., 1998. PERCLOS: A Valid Psychophysiological Measure of Alertness as Assessed by Psychomotor Vigilance. Federal Highway Administration, Office of Motor Carrier Research and Standards, Washington, DC.

Friedrichs, F., Yang, B., 2010. Camera-based drowsiness reference for driver state classification under real driving conditions. In: Proceedings of the 2010 IEEE Intelligent Vehicles Symposium, San Diego, CA, pp. 101−106.

Gehrig, S.K., Stein, F.J., 2007. Collision avoidance for vehicle-following systems. IEEE Transactions on Intelligent Transportation Systems 8 (2), 233−244.

Gordon, R., 2009. Intelligent Freeway Transportation Systems. Springer, New York.

Gozalvez, J., Sepulcre, M., Bauza, R., 2012. IEEE 802.11p vehicle to infrastructure communications in urban environments. IEEE Communications Magazine 50 (5), 176−183.

Han, L., Do, Q.H., Mita, S., 2011. Unified path planner for parking an autonomous vehicle based on RRT. In: Proceedings of the 2011 IEEE International Conference on Robotics and Automation, pp. 5622−5627.

Hartenstein, H., Laberteaux, K.P., 2008. A tutorial survey on vehicular ad hoc networks. IEEE Communications Magazine 46 (6), 164−171.

Hegeman, G., Tapani, A., Hoogendoorn, S., 2009. Overtaking assistant assessment using traffic simulation. Transportation Research Part C 17 (6), 617−630.

Hsieh, M.F., Ozguner, U., 2008. A parking algorithm for an autonomous vehicle. In: Proceedings of the 2008 IEEE Intelligent Vehicles Symposium, pp. 1155−1160.

Jap, B.T., Lal, S., Fischer, P., Bekiaris, E., 2009. Using EEG spectral components to assess algorithms for detecting fatigue. Expert Systems with Applications 36 (2), 2352−2359.

Jin-ying, H., Hong-xia, P., Xi-wang, Y., Jing-da, L., 2008. Fuzzy controller design of autonomy overtaking system. In: Proceedings of the 12th IEEE International Conference on Intelligent Engineering Systems, Miami, Florida, pp. 281−285.

Kaufmann, C., Risser, R., Geven, A., Sefelin, R., Tscheligi, M., 2008. LIVES (LenkerInnenInteraktion mit VErkehrstelematischen Systemen) — driver interaction with transport-telematic systems. IET Intelligent Transport Systems 2 (4), 294−305.

Kim, M.H., Lee, Y.T., Son, J., 2010. Age-related physical and emotional characteristics to safety warning sounds: design guidelines for intelligent vehicles. IEEE Transactions on Systems, Man, and Cybernetics, Part C: Applications and Reviews 40 (5), 592−598.

Kuncheva, L.I., 2004. Combining Pattern Classifiers: Methods and Algorithms. Wiley, Hoboken, New Jersey.

Levinson, D., 2003. The value of advanced traveler information systems for route choice. Transportation Research Part C: Emerging Technologies 11 (1), 75–87.

Lin, X., Lu, R., Zhang, C., Zhu, H., Ho, P.H., Shen, X., 2008. Security in vehicular ad hoc networks. IEEE Communications Magazine 46 (4), 88–95.

Liu, J., Zhang, C., Zheng, C., 2010. EEG-based estimation of mental fatigue by using KPCA–HMM and complexity parameters. Biomedical Signal Processing and Control 5 (2), 124–130.

Ma, Y., Chowdhury, M., Sadek, A., Jeihani, M., 2009. Real-time highway traffic condition assessment framework using vehicle–infrastructure integration (VII) with artificial intelligence (AI). IEEE Transactions on Intelligent Transportation Systems 10 (4), 615–627.

Marsden, S., McDonald, M., Brackstone, M., 2001. Towards an understanding of adaptive cruise control. Transportation Research Part C: Emerging Technologies 9 (1), 33–51.

Martinez, J.-J., Canudas-de-Wit, C., 2007. A safe longitudinal control for adaptive cruise control and stop-and-go scenarios. IEEE Transactions on Control Systems Technology 15 (2), 246–258.

Miller, R., Huang, Q., 2002. An adaptive peer-to-peer collision warning system. In: IEEE 55th Vehicular Technology Conference, vol. 1, pp. 317–321.

Mulder, M., Abbink, D.A., Boer, E.R., 2008. The effect of haptic guidance on curve negotiation behavior of young, experienced drivers. In: IEEE International Conference on Systems, Man and Cybernetics, Singapore, pp. 804–809.

Naranjo, J.E., González, C., García, R., de Pedro, T., 2008. Lane-change fuzzy control in autonomous vehicles for the overtaking maneuver. IEEE Transactions on Intelligent Transportation Systems 9 (3), 438–450.

Paromtchik, I.E., Laugier, C., 1996a. Motion generation and control for parking an autonomous vehicle. In: Proceedings of the 1996. IEEE International Conference on Robotics and Automation, pp. 3117–3122.

Paromtchik, I.E., Laugier, C., 1996b. Autonomous parallel parking of a nonholonomic vehicle. In: Proceedings of the 1996 IEEE Intelligent Vehicles Symposium, pp. 13–18.

Rajamani, R., Zhu, C., 2002. Semi-autonomous adaptive cruise control systems. IEEE Transactions on Vehicular Technology 51 (5), 1186–1192.

Rantanen, E.M., Goldberg, J.H., 1999. The effect of mental workload on the visual field size and shape. Ergonomics 42 (6), 816–834.

Raya, M., Hubaux, J.P., 2005. The security of vehicular ad hoc networks. In: Proceedings of the 3rd ACM Workshop on Security of Ad Hoc and Sensor Networks, Alexandria, Virginia, USA, pp. 11–21.

Schroth, C., Eigner, R., Eichler, S., Strassberger, M., 2006. A framework for network utility maximization in VANETs. In: Proceedings of the 3rd International Workshop on Vehicular Ad Hoc Networks, Los Angeles, California, USA, pp. 86–87.

Sengupta, R., Rezaei, S., Shladover, S.E., Cody, D., Dickey, S., Krishnan, H., 2007. Cooperative collision warning systems: concept definition and experimental implementation. Journal of Intelligent Transportation Systems: Technology, Planning, and Operations 11 (3), 143–155.

Shekhar, S., Kohli, A., Coyle, M., 1993. Path computation algorithms for advanced traveller information system (ATIS). In: Proceedings of the 1993. Ninth International Conference on Data Engineering, Vienna, pp. 31–39.

Shukla, A., Tiwari, R., Kala, R., 2010a. Real Life Applications of Soft Computing. CRC Press, Boca Raton, FL.

Shukla, A., Tiwari, R., Kala, R., 2010b. Towards Hybrid and Adaptive Computing: A Perspective, Studies in Computational Intelligence. Springer-Verlag Berlin, Heidelberg.

Toor, Y., Muhlethaler, P., Laouiti, A., 2008. Vehicle Ad Hoc networks: applications and related technical issues. IEEE Communications Surveys & Tutorials 10 (3), 74−88.

Vahidi, A., Eskandarian, A., 2003. Research advances in intelligent collision avoidance and adaptive cruise control. IEEE Transactions on Intelligent Transportation Systems 4 (3), 143−153.

Wang, F., Yang, M., Yang, R., 2009. Conflict-probability-estimation-based overtaking for intelligent vehicles. IEEE Transactions on Intelligent Transportation Systems 10 (2), 366−370.

Introduction to Planning

5.1   Introduction

Driving is associated with continuous and much *decision making*. Many decisions are made intentionally by thinking hard, whereas many others may be taken spontaneously and largely by instinct. While driving, one needs to constantly decide which route to take, whether to follow the same route or another, whether to take a diversion, whether to overtake, the speed to maintain, the distance to maintain from other vehicles and road boundaries, the lateral position or the lane to occupy while driving, whether to change a lane or lateral position, the trajectory profile to maintain while taking a turn etc. All these are very important questions, and many times different drivers tend to answer these questions differently based on preference, expertise and experience.

Artificial Intelligence (Konar, 1999; Russell and Norvig, 2009) and Soft Computing (Shukla et al., 2010) techniques are widely used for solving questions like these, for different characteristics of the environment or problem. The challenge is to solve the problem of going from place *A* to place *B*. As an example, place *A* may be one's home and place *B* may be one's office. *Planning* deals with all macro- and micro-decision making to reach a particular state called the goal, while starting from a particular state, called the source. Planning systems give a single valid move or a sequence of valid moves to take the system from source to goal subject to the validity of the assumptions made and the modelling of the problem. Optimal planners guarantee the best possible solution.

We do a lot of planning and of different types in our daily lives. We plan holidays, careers, what to do the next day, in what posture to read this book etc. The complexity of the problem and the assumptions of the environment vary from problem to problem, which governs the choice of the algorithm used to plan. The domain of Artificial Intelligence and Soft Computing already provides a variety of algorithms for different types of environments. Each of these algorithms may further be adapted based on the requirements or specific knowledge of the problem, or multiple methodologies may be mixed to reach an overall solution to the problem.

The environment in any planning problem may be static or dynamic (Russell and Norvig, 2009). A *static* environment does not change with time and hence allows the use of computationally expensive algorithms. A *deterministic* environment is one which gives the same outputs every time the same input is applied for the same state of the system, in contrast to a stochastic environment in which the outputs have some degree of uncertainty and depend upon factors which are not under control. *Discrete* problems have a discrete set of possible configurations, whereas continuous problems have a continuous range of configurations to deal with. Some problems are *fully observable*, wherein all information needed to make a decision is available, in contrast to partially observable problems wherein not all information may be available. A problem is *episodic* if the current decision does not depend upon past or future decisions, whereas if the best decision is a set of decisions which must be taken in a particular order, the problem is called sequential. A *single-agent* environment is acted upon by a single entity which is under control, whereas multiagent problems have multiple entities simultaneously acting on the environment.

Although it may be the best to take the environment as dynamic, stochastic, continuous, partially observable, sequential and multiagent, while making no assumptions at all, the algorithms suiting such a setting may not do full justice to the environment. Hence, either practical assumptions are made about the environment, thereby enabling the choice of efficient algorithms, or algorithms are adapted to cope with some of the assumptions made. This is the major challenge associated with problem solving.

This chapter first looks into the different layers of planning for an autonomous vehicle in Section 5.2. Then the chapter fundamentally studies the traffic environment into two types in Section 5.3, organized traffic wherein the vehicles drive within lanes, and unorganized traffic wherein lane discipline is not followed. The problem of planning can be taken as a robot motion-planning problem, the primitives for which are discussed in Section 5.4. Section 5.5 extends the ideas to motion planning for multiple robots, wherein coordination between the robots is the main challenge. Section 5.6 adapts the problem specifically to the problem of motion planning for multiple autonomous vehicles. Section 5.7 presents the different planning contexts for autonomous vehicle driving. Section 5.8 summarizes the chapter.

5.2 Layers of Planning

The problem of the navigation of autonomous vehicles is dynamic, due to dynamically changing traffic; stochastic, due to the uncertainty associated with the motion of any vehicle; continuous, due to the various possible positions and orientations on the road; partially observable, because the visibility distances ahead and behind are limited; sequential, as each driving decision affects future decisions; and multiagent, due to the presence of other vehicles.

It is evident that such a problem cannot be solved efficiently, because traffic is real-time (Buehler et al., 2007). A very common way of handling the problem is *abstraction*, wherein the entire problem is broken down into simpler problems and solved in a hierarchical manner. Every hierarchy may make different assumptions about the environment and may solve some part of the overall problem. A higher-level hierarchy in abstraction does not account for the finer details, for which a lower-level hierarchy is deputed. The solution for a higher-level hierarchy usually becomes one of the inputs for the lower-level hierarchy, which corrects for the wrong assumptions made by the higher level and corrects the plan or constructs a partial plan to be used by the higher level of abstraction. The solution of the finer-level hierarchy may still percolate further to the lower levels of hierarchy.

For the problem of navigation of autonomous vehicles, different layers may be identified. The layers are summarized in Fig. 5.1. The highest-level planning is used

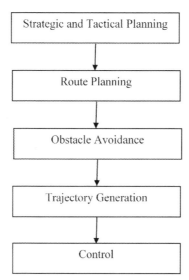

Figure 5.1 Layers of planning.

to answer the higher-level mission queries like in what order to visit the desired destinations and which destinations to visit. Route planning is responsible for deciding the route from one place to another, the route being a specification of the roads and intersections to take. The next layer takes decisions regarding the manner of avoiding obstacles, overtaking vehicles and avoiding vehicles. The trajectory-planning layer is responsible for constructing a smooth and feasible trajectory, which is traced using a control algorithm.

5.2.1   Strategic and Tactical Planning

The highest level of planning is *strategic* and *tactical* planning. Humans usually have complicated specifications of their plans. They need to go places and do some activities, whereas the places and activities may be dependent on each other and may change with time or when new information is available. Consider that you need to shop at 10 different shops at different places, whereas not getting the desired item may add a destination or cancel some other. Journeys are sometimes complicated missions that need to be planned. Traffic information is usually restricted in this layer of abstraction to make the planning easy. Alternatively, one may include traffic information to make plans that are more realistic.

5.2.2   Route Planning

Route planning is used to decide the route to take from one place to another. Consider the problem that you need to go from your office to a particular shop. Here, the office and the shop become the source and the destination, respectively. The output of the route-planning algorithm is the route to take, which enlists all the roads and intersections to take to reach the destination from the source.

The *road network graph* for most cities is well known. The transportation authority maintains a road network graph which is also maintained by numerous services. The road network graph consists of every intersection of multiple roads as the vertex and all roads as the edges. All two-way roads, in which you can travel inbound and outbound, are undirected edges, whereas all one-ways are directed edges. The number of lanes, maximum speed etc. can also be stored. The weight of the edge is taken as the length of the road (if minimizing distance) or the length of the road divided by the maximum speed (if minimizing travel time). This converts the problem into a standard graph search problem. Dijkstra's algorithm is a common search algorithm for graphs which gives the shortest path from the source to the destination when all the edges are nonnegative, an assumption which is true for this problem.

The source and the destination may not be preexisting vertices in the road-network graph and, hence, these are added as additional vertices at the query time and all intersections to which they connect are added as edges. If a road is blocked, the corresponding edges are removed.

While working in this layer of abstraction, numerous assumptions are made. The most important assumption usually made is that no other vehicle occupies the road, in which case the vehicle can travel with the maximum permissible speed. In addition, it is assumed that different lanes have the same distances which are the average length of the road. Roundabout distance may be ignored or may be taken as a constant. Naïve methods also do not consider the time wasted in traffic congestion and traffic lights. Hence, the computations are not accurate and are made from a variety of assumptions; however, the assumptions enable decision making in small computation time while operating at a higher degree of abstraction.

Many routing algorithms (discussed in chapters: Potential Based Planning and Logic Based Planning) can make assumptions about the repeated traffic trends to get a better indication of the expected traffic congestion and operating speed, and thus more reasonable assumptions of travel time and route selection. Similarly, many algorithms can simulate the motion of other vehicles to get an indicative measure of the traffic congestion and accordingly select the route. Many other algorithms navigate vehicles such that congestion never happens.

5.2.3 Obstacle Avoidance

The route-planning algorithm gives a route to take and hence divides the problem into navigation on straight roads, intersections, diversions and mergers, which essentially constitute the route. The roads in any of these scenarios will be filled with static obstacles or other vehicles which act as obstacles. Each of these needs to be avoided using some strategy. In the worst case, the different obstacles will form a labyrinth to be avoided by the vehicle by considering all possibilities in the most efficient manner. Uncertainties and lack of knowledge of the travel plans of other moving vehicles makes the problem hard. Avoiding other vehicles also calls for decisions to whether to overtake the vehicle, to follow it or to simply stay away from it. Even at the micro-level, numerous decisions need to be made for avoiding a single obstacle. These include deciding the side on which to avoid the obstacle, the distance to maintain from the obstacle, the speed to operate at, to constantly adapt the decisions if the obstacle is moving or the environment is dynamic etc. All these decisions are taken at this level of abstraction. Still, the dynamics of the vehicle, the steering mechanism, the control signals to send etc. are finer details that may be skipped.

5.2.4   Trajectory Generation

This layer deals with generating a *trajectory* for the vehicle to follow. The obstacle avoidance layer answers all decision-level questions to enable avoidance of all obstacles. This layer uses those decisions and constructs a trajectory to be followed by the vehicle. The trajectory is a specification of both the path to take and the time information associated with the path. Therefore, the linear and angular speeds and accelerations can be derived. The trajectory so constructed must be smooth so that a control algorithm can trace the trajectory with minimal errors. Many times smoothing can mean altering the plan so generated to induce smoothness, while still maintaining the feasibility of the trajectory and closely maintaining a short trajectory. This, however, may still not guarantee that a control sequence may exist to trace the trajectory generated.

5.2.5   Control

The ultimate task is to send *control* signals to the braking, throttle and steering modules to navigate the vehicle. This is the control problem. The control systems in place take a reference trajectory as input, specifying the linear and angular positions and speeds to maintain. This input is transformed into input signals to the specific motors in the braking, throttle and steering mechanisms. Any error is fed back and adapted by the control systems. Due to the complexity of the problem, systems may usually have a separate master controller and a slave controller. Here, the *master control system* looks at the desired and actual speeds and positions of the vehicle and computes the error and hence the corrections to make. The corrections are passed to a *slave control system* which generates the low-level controls to the individual motors aiming at making the necessary corrections.

5.3   Types of Traffic

Outdoor autonomous vehicles are essentially of two types. The first type of vehicles is the ones which drive on a road, and these will be the types of vehicles discussed throughout the book. Alternatively, there are off-road vehicles which are meant to navigate in rough natural terrains. Section 5.2 highlights the important assumptions made at each of the layers of planning. An important aspect to consider is that the system is multiagent and there are numerous vehicles on the roads which need to be taken care of. This necessitates important assumptions about traffic. Two kinds of traffic systems widely exist. Organized traffic systems have vehicles strictly adhering to lanes and strictly driving within lanes, unless making a legal lane change. In unorganized traffic systems, the vehicles can travel anywhere on the road without caring about the lanes. The demarcation may not be difficult as organized traffic may seldom show unorganized traits and vice versa.

5.3.1 Organized and Structured Traffic

General navigation, as organized in most countries, consists of vehicles laterally organized within *lanes* and needing to strictly drive within them. Two vehicles cannot occupy the same position in the same lane. A vehicle cannot occupy two lanes. This is with the exception of transition during a lane change in which one would have to temporarily lie partially on two lanes. This rule is even valid for two-wheelers. Even though two two-wheeler bikes may simultaneously fit into a single lane, they must occupy two lanes.

In an *organized traffic* system, vehicles only have the option to change to the left or right lane or drive in the current lane. The major task of planning in such a context includes deciding the lane of travel (Hall and Caliskan, 1999), designing trajectories for lane keeping and lane changes (Glaser et al., 2010) and deciding the speed of travel. The advantages of organized traffic are higher safety as the vehicles can drive carefree in one's own lane, clearer intentions of other vehicles as they will either drive straight or cut in, and fewer lane changes or lateral movements which signify a more comfortable driving experience and hence shorter travel distances and times. The two types of traffic are summarized in Box 5.1.

5.3.2 Unorganized and Unstructured Traffic

A number of countries follow *unorganized traffic* in which the vehicles may place themselves laterally anywhere in the road. No lane discipline needs to be followed and the lanes may be absent or for guidelines only. In such a scenario the planner can construct trajectories keeping the vehicle anywhere inside the road, while varying the speed of travel. Hence, the planning decisions are more complicated and the planner needs to select from a large set of continuous values unlike discrete lanes. Planning thus becomes overly complicated and computationally intensive.

The advantages of such a traffic system include a higher *traffic bandwidth*. Traffic bandwidth is the maximum number of vehicles which can travel at a slice of the road per unit time. For lane-based travel, the traffic bandwidth is restricted to the number of lanes. If the traffic system has vehicles significantly smaller than the width of the lane, two vehicles may co-occupy a lane or, more generally, multiple vehicles may co-occupy a lesser number of lanes, which increases the traffic bandwidth. Further, unorganized traffic can *host more overtakes* with each overtake signifying better travel for the overtaking vehicle. If the traffic system has vehicles with diverse speeds, there arises a need of constant overtaking of some slower vehicle by some faster vehicle. Diverse speeds necessitate constant overtakings at different segments of the road. In

---

Box 5.1 Characteristics of Unorganized Traffic over Organized Traffic

Pros
- Larger traffic bandwidth
- More overtakes/more efficient

Cons
- Safety
- Unclear intentions
- Large lateral movements
- Larger travel distances
- Less driving comfort

When is Unorganized Traffic Better?
- Diverse vehicle widths
- Diverse vehicle speeds
- Speed diversity necessitates overtaking
- For example, *Indian traffic!*

Migrate From Organized to Unorganized Traffic?
- Intelligent vehicles will bring diversity
- The future is diverse
- *Current defiance of lanes:* Motorists ignoring lanes and overtaking by emergency vehicles

---

unorganized traffic, a vehicle can request other vehicles to shift laterally as much as possible to give some extra space to host the overtaking. In organized traffic, this would only be possible if a spare lane was available for overtaking.

Unorganized traffic is better for systems in which the vehicles vary largely in terms of sizes and preferred travel speeds, and vice versa. Hence, it is better for countries with *diverse traffic* to opt for unorganized traffic and vice versa. To motivate the notion take the example of traffic on Indian roads. Vehicle widths vary from two-wheeled motor bikes, to three-wheelers (including auto rickshaws), to four-wheeled cars and eight-wheeled trucks, whereas the speed varies from manually ridden bicycles to fast cars. As a result, it is natural for drivers not to follow the marked lanes and overtaking becomes a common sight. The expectation of a smaller (slow-moving) vehicle is to respect a larger (faster-moving) vehicle and to allow it space to overtake, and vehicles may temporarily use the 'wrong side' of a dual carriageway to carry out an overtake.

For the same reasons it is possible for organized traffic to show unorganized trends when traffic becomes diverse. Sometimes it may be allowed for motorists to drive within lanes (Sewall et al., 2011), which increases the traffic bandwidth, at the same time making it quicker for the motor bikes to travel in a congested road. Further, it is common for a vehicle to drift on one side only to allow an emergency vehicle to overtake and pass through. The overtaking might not have been possible without the other vehicle drifting, not following the marked lanes. These examples illustrate that, as the diversity increases, unorganized traffic becomes more advantageous.

Autonomy may make vehicles vary in speed capabilities and size, depending upon their purpose, precision, control and autonomy. Lanes are constructed assuming a fixed width of vehicle and may well make the road underutilized. In addition, constant overtakes are important when the traffic is too diverse. As a result, it may be 'painful' for a fast vehicle to follow a slow vehicle. Increased autonomous vehicles may bring increased diversity to the currently organized traffic landscape which may continually pressurize diminishing of the lanes, up to the point when the entire traffic becomes unorganized.

The core focus of chapters 'Optimization Based Planning', 'Sampling Based Planning', 'Graph Search Based Hierarchical Planning' and 'Using Heuristics in Graph Search Based Planning' is unorganized traffic. The chapters specifically look into the aspect of trajectory planning of autonomous vehicles for a non-lane-based traffic. Details of the dynamics of such traffic for collision avoidance (Jain et al., 2009; Mohan and Bawa, 1985), traffic prediction (Vanajakshi et al., 2009) and traffic simulation (Paruchuri et al., 2002) can be found in the literature.

5.3.3   Off-road Vehicles

Off-road vehicles are not restricted to a structured road architecture for their travel. These vehicles travel in open spaces. The *terrains* may normally be rough for such vehicles. The vehicles themselves are designed so that they do not tumble down on encountering rough terrains. This is unlike on-road vehicles which are meant to travel on smooth well-built roads and offer maximum comfort to the human driver. The map for navigation of these vehicles may be obtained from satellite imagery, aerial robots hovering around that area, historic information from flying robots in that area, onboard camera or a combination of these. The map is not a simple specification of obstacle-prone and obstacle-free areas, but specifies different terrains which are navigable by different costs and risks of accidents.

A motion-planning algorithm at the coarser level tries to construct the safest and shortest route in such a space. Further, at the finer level, the vehicle may try to navigate through plain terrains rather than driving through uneven surfaces. In case of uneven terrains, the least slope and hence the safest areas need to be identified and navigated through. A typical example is the Mars Rover which can be guided by a human operator at the coarser level, whereas the finer-level planning for avoiding obstacle and taking less risky trajectories is done autonomously using the onboard cameras.

5.4   Motion-Planning Primitives

The generalized problem of motion planning of autonomous vehicles, wherein the lanes are not followed, closely resembles the problem of *motion planning for mobile robots* (Choset et al., 2005; Tiwari et al., 2013). The autonomous vehicle can be seen as a robot which navigates from the source to the destination. Although it must be said that algorithmically, there is a big difference between the problem of motion planning for mobile robots and autonomous vehicles, which is due to the presence of a *narrow and elongated road structure* in contrast to the largely unbounded maps of mobile robots. Hence, mobile robotics deals with behaviours of obstacle avoidance and going towards the goal; whereas autonomous vehicle navigation is largely attributed to the behaviours of *overtaking*, *vehicle following* and *driving along the road*. This section gives the primitives of motion planning over which all the motion-planning algorithms will be built in the subsequent chapters.

5.4.1 Configuration Space and Problem Description

The *configuration* of a robot is given by the minimum number of variables needed to fully describe the current state of the robot. Only variables which are of relevance and change as the robot travels from the source to the goal are considered. Variables which are not necessary for the specific problem being solved, or remain constant in all possibilities as the robot travels from the source to the goal, can be left out in configuration description. So a point robot operating in a two-dimensional (2D) plane has a configuration of (x,y), a circular robot operating in a 2D plane has a configuration of (x,y) as the rotation of a circle does not change the state, and a rectangular robot in a 2D plane has a configuration of (x,y,θ) in which (x,y) denotes the position and θ denotes the orientation of the robot. The vehicles can be easily considered as rectangular in shape with a configuration of (x,y,θ). The size of the vehicle remains constant as the vehicle travels, and hence the size is not mentioned in the configuration.

The *configuration space* (C) is a space which denotes whether every possible configuration is collision-prone and nonnavigable or collision-free and navigable. The number of axes of the configuration space is the number of variables in the configuration representation or the degrees of freedom. Every configuration is a point in this configuration space. The point may be denoted by either black (0) denoting that the particular configuration causes a collision between the robot and an obstacle or the robot is in self-collision; or white (1) denoting that the particular configuration is collision free.

This divides the configuration space (C) into two disjoint regions, a free configuration space (C^{free}) including configurations without collisions, and a collision-prone configuration space (C^{obs}), given by Eq. [5.1].

$$C^{\text{free}} = C \backslash C^{\text{obs}} \hspace{4cm} [5.1]$$

The source (S) is the current configuration of the robot which must be collision free $(S \in C^{\text{free}})$. Similarly, the goal (G) is the destination configuration in the free configurations space $(G \in C^{\text{free}})$.

The *workspace* (W) is the physical space on which the robot operates. It is the physical world with the physical robot operating in it. It is the part of the world where the

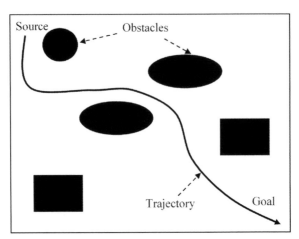

Figure 5.2 Problem of robot-motion planning.

robot operates. This is unlike configuration space in which the robot with its current configuration is just a point in the space. The robot arena in planning is usually bounded, even though the bounds may be large enough. The robot effectively operates in the region between the source and goal, and maybe some region beyond. The workspace is 2D for a robot operating in a plane, and three- dimensional (3D) for a robot not restricted on the plane. Sometimes extra depth information may be important, eg, to distinguish between obstacles that the robot can travel under and not travel under, or travel over and not travel over, which is a 2.5−dimensional (2.5D) environment. The workspace may be collision-prone W^{obs} or collision free W^{free}.

It is assumed that the source $S(\in C^{free})$ and the goal $G(\in C^{free})$ are already known. The problem is to make a trajectory $\tau{:}[0,1] \rightarrow C^{free}$ such that the robot starts from the source (or $\tau(0) = S$) and ends at the goal (or $\tau(1) = G$) and every configuration in between is collision-free ($\tau(s) \in C^{free}$, $0 \leq s \leq 1$). The trajectory could also be parametrized with time, in which case it also specifies the time and speed information of travel, with the robot at position $\tau(t)$ at time t. Fig. 5.2 shows the configuration space of a point robot, which is also the same as its workspace. The black regions denote the obstacles which the robot must avoid. The source, goal and trajectory are shown in the same figure.

5.4.2 Metrics to Judge Planning Algorithms

The chief objective of the algorithm is to find a trajectory which makes the robot reach the destination in the least possible time. Sometimes the aim is the least trajectory length. If the trajectory is to be traversed by the robot with nearly equal speed throughout, minimizing *travel time* and minimizing *travel distance* are nearly the same thing.

The other most important objective is to maximize *smoothness*. Smoothness is given by the curvature of the trajectory and a smoother curve facilitates higher travel speeds and lower control errors. Nonholonomic robots have to obey nonholonomic constraints which depend upon the positon and speed of the robot. The nonholonomic robots cannot take sharp turns, and smoothness of trajectory is an essential requirement. A related objective is to maximize the *travel speed*. Another common objective is the *clearance* to maintain or the minimum distance to maintain from the obstacles around. A high clearance means larger safety distances maintained from the obstacles, which gives ability to combat sensing and actuation errors. Many times the objective is taken as the fuel economy, wherein the aim is to take smoother paths navigable by high speeds. Further, for problems like autonomous vehicles, passenger comfort is an important objective, which asks the algorithms to minimize changes in accelerations or jerks.

Optimality is a property of the algorithm to guarantee the best results as per the chosen objective as long as the stated assumptions hold. The algorithms which guarantee optimality of the solution are called optimal algorithms. An algorithm is called *resolution-optimal* if the optimality can only be ascertained under the strict limitation of the current resolution of operation. As the resolution is increased, the resolution-optimality increases or the algorithm tends towards an optimal algorithm. Similarly, an algorithm is called *probabilistically optimal* if the probability of the algorithm being optimal tends to one as time tends to infinity. After infinite time these algorithms guarantee the optimal solution, whereas the solutions returned before this time may not be optimal. The solution tends to the optimal solution with an increase in time. A solution is called *near-optimal* if it is almost like the optimal solution with a very small difference in the solution cost as compared to the optimal solution. For most problems, near-optimal solutions are acceptable.

Completeness is the property of the algorithm to guarantee returning any solution in a finite amount or time or to graciously return no solution possible and exit for any problem. Complete algorithms may not guarantee optimality, but do guarantee any one of the numerous feasible solutions. Similarly, *resolution-complete* algorithms guarantee completeness subjected to the resolution of operation and *probabilistically complete* algorithms guarantee completeness as time tends to infinity. *Near-complete* algorithms nearly give a guarantee of completeness.

Ideally, one would like to have algorithms which are complete and optimal. However, such algorithms usually take a lot of *computational time* and a lot of *computational memory*, which is not always possible to devote. Many times the onboard computation and memory is limited, whereas, at other times, the problem may be real-time or the environment may be static for brief times only. This places a very important classification in the operation of environment which is whether the algorithm is able to work only in static environments, in partially dynamic environments, in very dynamic environments, and whether the algorithm can accommodate suddenly appearing and disappearing obstacles. Correspondingly, the level by which the algorithm can handle sensing and actuation uncertainties is of importance.

Another important aspect to judge algorithms is on their perception of the environment. Some algorithms need the environment fully observable, some algorithms can handle the environment being partially observable at the start and the environment gets sensed as the robot moves, whereas some other algorithms require minimum percept from the environment which can be easily obtained. Similarly, some algorithms work only for structured environments in which the obstacles are all assumed polygons or regularly shaped, whereas some other algorithms can handle an unstructured environment wherein no structural information about the obstacles is available and the environment is usually in the form of a grid map, aerial image, depth image or a 3D image.

5.4.3 Deliberative and Reactive Planning

There are two fundamental ways to deal with the problem of trajectory planning. The *deliberative methods* take time to compute the different possibilities and all combinations of them and compute the entire trajectory to be followed by the robot. The deliberative algorithms take a long time to think and produce a result. The results are near optimal and near complete. The algorithms, however, take a long time to compute the result and are largely valid for static environments.

On the contrary, the second fundamental way to solve the problem is to react instinctively to the current situation and output only the immediate move. These algorithms are called *reactive algorithms*. The move alters the situation which is enacted upon by the next reactive decision. In this manner by planning and taking one step at a time, the robot reaches its goal. The optimality and completeness cannot be guaranteed. Many times the reactive algorithms trap a robot wherein the robot either fails to move or starts oscillating and repeating its moves in a loop. Because the reactive algorithms disallow the robot to deliberate and plan its way out, the robot may be trapped until external human help is provided. These algorithms are, however, very fast to compute and can accommodate dynamic environments and even sudden changes in the environment.

The classification between deliberative and reactive algorithms may not always be clear. Different algorithms may have different levels of deliberation and reactiveness. Due to the contrasting natures of the deliberative and reactive algorithms, it is common to hybridize the two algorithms to make a *hybrid planner*. The deliberative algorithm may be used to guide the reactive planner. Hence, the deliberative algorithm contributes towards completeness and optimality; while operating at lower resolutions the algorithm is able to generate results faster. At the same time, the reactive algorithm takes care of environmental changes and dynamic obstacles, while being computationally fast. Similarly, multilayers of hierarchies may be used, the higher ones as deliberative and the lower ones as reactive.

A *narrow corridor* is a classic problem with different motion-planning algorithms and continues to pose a serious challenge. Here, the path between the source and the goal is separated by a very narrow corridor, compared to the size of the robot. The reactive algorithms may either fail to orient the robot to enter the narrow corridor, and instead get the robot stuck, or the robot may enter the narrow corridor and show oscillations. The iterative solutions or the probabilistically complete solutions may take a significantly long time to sight and construct a trajectory through the corridor. The corridor is only passable when operating at high resolutions, thus necessitating high-resolution settings for which the resolution-complete algorithms may take significantly long.

5.4.4 Planning and Replanning

Deliberative algorithms produce a trajectory for the robot to trace. The trajectory is only valid for a static environment. If the environment changes, the trajectory may not be optimal and may even be infeasible to follow. A common method is to make the vision system continuously *monitor* the environment and report any changes in environment. Upon detecting any change in the environment, the previous trajectory is discarded and a new one is made. This methodology can only be used when the environment is mostly static with very few and infrequent changes. Once the environment changes, the robot stops and waits for a few seconds to compute a new trajectory and starts following the new trajectory.

Another method is to rework the trajectory and adapt it to the changed environment. If the environment changes by a small amount, small adjustments to the trajectory may be easily made, rather than doing the computationally intensive planning all over again. *Local algorithms* try to search the near vicinity and within a limited search space to find a solution, rather than global algorithms which search the entire search space. Local algorithms are widely used as replanning algorithms, which search in the close vicinity of the previous trajectory. The optimality of local algorithms cannot be ascertained.

Many algorithms also try to reuse the past computations to adapt the trajectory as per changes in the environment. The information about the past computation may be available in a summarized form of a roadmap, partial solutions, processed set of nodes etc. Normally, this information is discarded and only the computed trajectory is returned. Replanning algorithms may optionally store this information permanently and use it when the environment changes to quickly compute the new trajectory.

5.4.5 Anytime Algorithms

Another method to classify the deliberative algorithms is anytime algorithms and non-anytime algorithms. The *anytime algorithms* very quickly compute a nonoptimal solution and the solution keeps improving with time. As time tends to infinity, the solution tends to the optimal solution. Optimality and completeness hence improve with time. Such algorithms are called anytime algorithms, as a solution is available anytime in the search process. One could get a suboptimal solution at small computation times and near-optimal solutions at large computation times. For motion-planning problems, anytime solutions are desirable as the available computation time depends upon a lot of factors, whereas a solution must be available within this time. For example, the planning may be done at a frequency of a few seconds, and, within this time, the algorithm should return the result.

5.5 Multirobot Motion Planning

In the future, it is believed that the homes and offices will be occupied by numerous robots, rather than just one. Each robot may be suited for a specific task. Further, multiple robots can divide the work amongst themselves thereby completing the task in less time. The robots may be homogeneous or heterogeneous depending upon the need. *Multirobotics* calls for challenges at different fronts including using multiple robots to share vision to reduce errors, multiple robots for simultaneously making a map of the real world, multiple robots making a shape or multiple robots simultaneously solving a complicated mission (Arai and Ota, 1992; Parker et al., 2005).

The problem of *multirobot motion planning* deals with planning the path of n robots, each robot starting from its own source S_i to its own goal G_i. The robots must not collide with any obstacle on the way as well as not collide with each other. Let the plan of multiple robots be denoted by $\tau = \cup_{i=1}^{n} \tau_i$. Here, τ_i denotes the trajectory of the robot i which starts at its source and ends at its goal, while ensuring that the robot does not collide with any static obstacle as well as any two robots do not collide with each other, or, $\tau_i(0) = S_i$, $\tau_i(1) = G_i$, $\tau_i(s) \in C^{\text{free}}$, $R_i(\tau_i(s)) \cap R_j(\tau_j(s)) = \phi$, $i \neq j$, $0 \leq s \leq 1$. Here, $R_i(q)$ is a function which projects a robot from its configuration space to the volume occupied by the robot in the workspace. The occupancy of no two robots in the workspace must intersect for collision-free travel. A synthetic problem is shown in Fig. 5.3.

While working with multirobotics, it is possible that not all robots reach their goals with their individual optimal plans. If five robots need to travel along a road which is one-lane only, the robots will have to go one by one, thus only one robot will have its optimal trajectory, whereas the other robots will have to travel by suboptimal trajectories. A robot sacrificing its optimal trajectory so enable some other robot attain a better trajectory is called *cooperation*. Cooperation may not harm the cooperating robot by a large amount, while it may make the trajectory of the robot seeking cooperation significantly better. Cooperation is commonly seen in human society wherein one offers ones time and help, if available, to the needy person. In multirobot motion planning, it is not important whether the individual robotic trajectories, as per the choice of metric, are optimal. However, the overall plan as per the choice of metric should be optimal.

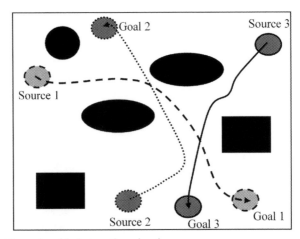

Figure 5.3 Problem of multirobot motion planning.

In multirobotics, it becomes important to decide key questions such as which robot asks for cooperation, whether cooperation should be extended to some other robot, to what extent must cooperation be extended, when cooperation must be asked for, whether cooperation would result in some benefit etc. There may not always be simple answers to the questions. *Coordination* deals with the task of coordinating between the robots. Coordination strategies decide the motion policies of the robots, especially when the optimal plans of the robots are conflicting with a mutual collision between the robots.

Coordination techniques may be centralized (Sánchez-Ante and Latombe, 2002) or decentralized (Lumelsky and Harinarayan, 1997). Being computationally expensive, centralized strategies are rarely used unless the number of robots is highly limited. Decentralized coordination is widely used. A very common technique for decentralized coordination is prioritization. These are discussed in the next subsections.

5.5.1   Centralized Solutions

Centralized techniques look at the entire problem as a whole and try to search for a solution. The configuration space of the resultant problem is hence the joint configuration space of all the robots given by Eq. [5.2].

$$C = C_1 \times C_2 \times C_3 \ldots \times C_n \qquad\qquad [5.2]$$

The free configuration space includes all the points wherein none of the robots collides with any static obstacles and no two robots collide with each other, given by Eq. [5.3].

$$C^{\text{free}} = \{ <q_1, q_2, q_3, \ldots q_n> : q_i \in C^{\text{free}}{}_i, R_i(q_i) \cap R_j(q_j) = \phi \qquad [5.3]$$

The source of the resultant problem is the joint sources of all the robots $<S_1, S_2, S_3 \ldots S_n>$ and the goal is the joint goal for all the robots $<G_1, G_2, G_3 \ldots G_n>$. All joint actions of all robots are possible. The dimensionality of the configuration spaces follows the *curse of dimensionality* principle by which the complexity of the problem becomes extremely large on every addition of dimension in the configuration space. The centralized approaches make such a complex configuration space, and, hence, the approach is computationally feasible in acceptable computation times for a few robots only. However, because all possible combinations of moves of all robots are considered, capable of generating all kinds of plans, the centralized techniques can be optimal and complete subject to the applied planning algorithm.

5.5.2   Decentralized Solutions

In *decentralized coordination*, each vehicle is planned separately. While doing so it is possible that the prospective plans of two or more robots are conflicting and a collision is likely. In such a case, a *coordination policy* is made which decides how to rectify the collision. While doing so it may be possible that collision is still there or, in pursuit of rectifying some collision between two robots, a collision between two other robots appears. Hence, the strategy may be applied repeatedly till no collision appears.

The technique initially works over all the robots separately and hence the computational time is very small. For rectifying collision-prone plans, other robots must certainly be considered. In a problem with a large number of robots, only some of the robots are potentially colliding and hence such cases are very limited. This technique shows a very small increase in computational time per robot, on increasing the number of robots. However, due to the decomposition of solving for every robot separately and then rectifying the results, the optimality and completeness cannot be guaranteed unless problem-specific assumptions are used.

The reactive algorithms can work in a highly dynamic environment and make instantaneous moves. Reactive algorithms can model the other robots as moving obstacles and normally avoid the other robots without explicitly modelling for coordination. Care must, however, be given as moving robots can act as mischievous obstacles, continuously blocking the way of the robot when the robots have opposing plans.

5.5.3 Prioritized Motion Planning

Prioritization (Bennewitz et al., 2001, 2002) is a common method to solve the problem of decentralized coordination of multiple robots. Simply said, the coordination policy states that, in case of a collision, it is the duty of the lower-priority robot to take an action to avoid collision with a higher-priority robot. The higher-priority robots need not do anything when the collision is with a lower-priority robot. The priorities may be randomly assigned, a number of probability assignments may be tried and optimized, the priorities may be based on the importance of the robot or the problem-specific heuristics may be used for priority assignments.

In this coordination, all the robots plan one after the other. First, the highest-priority robot plans, as it need not consider the plans of the other robots. Then the second highest-priority robot plans to avoid all static obstacles and avoid collision with the highest priority, the plan of which is already known. Similarly, at any time, the plans of all higher-priority robots are known, which are used to plan for a lower-priority robot. The lowest-priority robot plans at the end, avoiding all other robots already planned. In this technique, the highest-priority robot shows no cooperation and does not even move a little to benefit the other robots. The highest-priority robot travels by its optimal path. The lower-priority robots cooperate with the higher-priority robots if needed. This *uncooperative nature* affects both optimality and completeness.

5.5.4 Path–Velocity Decomposition

A common approach in single-robot motion planning for static obstacles is to optimize only the path without any indication of the speed. Once the path is available, the maximum safe speed per section of the trajectory can be set based on the curvature information. Alternatively, the robot may be asked to trace the computed path by a pre-specified constant speed. In multirobotics for robots with communication, however,

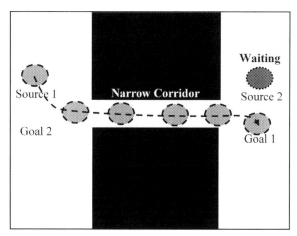

Figure 5.4 Robot 2 will have to wait for Robot 1 to cross the narrow corridor.

speed plays an important role. Consider two robots stationed just outside a narrow corridor and they need to pass the narrow corridor, whereas the narrow corridor can accommodate just one robot. This is shown in Fig. 5.4. One of the two robots will have to wait for the other robot to pass through. A solution is not possible if both robots maintain a constant speed. More generally, consider one robot plans after the second robot has planned its trajectory as per prioritization. For the lower-priority robot, the map is continuously changing. At some sections, slowing would result in a more favourable map in the future whereas at some other sections waiting for a robot to pass may be beneficial, something commonly employed by humans while walking. The planning cannot be done assuming a constant speed of navigation.

Simultaneously planning for both path and speed for all robots can be computationally very expensive. Thus, *path−velocity decomposition* (Kant and Zucker, 1986) attempts to decompose the problem. The path of the robots can be optimized first, without considering the speed of travel. Once the paths are optimized, the speeds may be assigned such that there is no collision, which is another and simpler optimization problem. Alternatively, many schemes may apply altering speed and path optimizations. Some other schemes may use heuristics to set the speeds, while searching for a path.

5.6 Motion Planning for Autonomous Vehicles

If the vehicle is assumed as a robot and the traffic operates without the notion of lanes, the problem of motion planning for autonomous vehicles can be easily taken as the general problem of motion planning of mobile robots. However, most of the near-complete and near-optimal algorithms for mobile robotics are computationally too intensive, whereas traffic is a highly real-time system. Hence, the problem modelling should be such that it facilitates the assumption of an *elongated and narrow width road structure* and the notions of *overtaking* and *vehicle following*. Accordingly, the coordinate axis system is taken and the problem is modelled.

5.6.1 Road Coordinate Axis System

The scenario is assumed available as a *map* on which the algorithm works. Two coordinate axis systems are redundantly used. The first is the Cartesian coordinate system (XY), which is the system in which the map is available as an input. The other is the *road coordinate system* (X′Y′). In this system X′ axis is taken to be the longitudinal direction of the road. In case the road is curved, this axis will also be curved. For algorithmic purposes, X′ axis is taken as the first or right road boundary. The Y′ axis, meanwhile, is taken to be the lateral direction of road. The width of the road may not be constant. Hence, Y′ axis is taken as the ratio of distance of a point from the X′ axis in the lateral direction to the road width at that point. The two axis systems are shown in Fig. 5.5. Consider point P(x,y) in the Cartesian coordinate system. The corresponding point P(x′,y′) in the road coordinate system is given by Eq. [5.4].

$$P(x, y) = P(x'y') = P\left(x', \frac{a}{w}\right) \tag{5.4}$$

Interconversion of any point between the axis systems may sometimes be required. Conversion from road to Cartesian coordinate systems involves finding the corresponding points on the two boundaries and hence computing the ratio. For conversion from road axis to Cartesian axis system the corresponding lateral point is found on the X′ axis by a small local search and the road width is measured.

5.6.2 Problem Formulation

Planning for autonomous vehicles is done at multiple levels. The strategic and tactical level deals with deciding the places to visit, order of visit of places etc. Meanwhile the

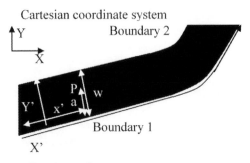

Figure 5.5 Road and Cartesian axis systems.

substrategic level deals with deciding the route or the roads to take; the decision on which may depend upon traffic prediction, expected travel speeds and/or road length. The lowest tier of planning deals with the generation of a feasible trajectory for each vehicle. A separate simple or complex control system may be required to physically drive the vehicle. The *middle tier planning* may deal with path validation, replanning (whenever required) and iteratively building up the trajectory using the lowest level of planning. This algorithm in chapters 'Optimization Based Planning', 'Sampling Based Planning', 'Graph Search Based Hierarchical Planning' and 'Using Heuristics in Graph Search Based Planning' only focusses upon the lowest level of planning which is trajectory generation. This planning may be different for merging regions, intersections, parking, blocked road and normal road scenarios, in which the focus is only on the normal road scenarios.

It is assumed that a segment of the road is available, which is the input to the algorithm. The map may be processed to extract the left and right boundaries of the *road segment*. Any number of obstacles of any size and shape may lie in the road segment. However, the obstacle framework should not block the complete road, or be such that no trajectory is possible for any vehicle. The problem of *deadlock* or *blockage avoidance* for autonomous vehicles is not taken in entirety in this book, although some preliminary ways to handle the problem have been given. The arrival times (T_i), speeds (v_i), orientations (θ_i) and sizes ($l_i \times w_i$) of vehicles are available as they enter the segment of road being planned. For simplicity, it is assumed that all vehicles are rectangles of size $l_i \times w_i$; nonrectangular vehicles may then be bounded to the smallest fitting rectangle. The purpose of the algorithm is to generate a valid trajectory τ_i from the current position (S_i) to any position at the end of the road segment (G_i), such that no two vehicles collide ($R_i(\tau_i(t)) \cap R_j(\tau_j(t)) = \phi$, $i \neq j$) and no vehicle collides with a static obstacle ($\tau_i(t) \in C^{\text{free}}$). The generated trajectories need to follow the nonholonomic constraints of the vehicle.

Without loss of generality, throughout the book it is assumed that the traffic operates on a *to-left* driving rule as operative in the United Kingdom, India and a number of other countries. Further, it is largely assumed that the inbound and outbound traffic operates on the same road without a physical barricade in between. If the inbound and outbound traffic operates with a physical separation, the algorithms presented may be independently called for both inbound and outbound parts. On not sighting any oncoming vehicle, the algorithms would specialize to the case of separate inbound and outbound traffic. Further, in case of absence of a physical barrier between the inbound and outbound traffic, it would be assumed that the inbound vehicles can partly or fully occupy the lanes meant for outbound traffic, especially to overtake a vehicle. This would, however, be done with great caution and on maintenance of enough safety distances, further not to overly penalize the outbound traffic. Similarly, the outbound traffic may come in the lanes meant for inbound traffic. Overtakings resulting from an inbound traffic vehicle skipping to the outbound traffic and vice versa are called *single-lane overtaking*. This is largely because such overtakings normally happen when a single lane is available for inbound and outbound traffic and no overtaking lane is normally available, forcing a vehicle to drive on the wrong side for overtaking.

5.6.3 Objective Function

The prime factor used to measure the algorithm's success is its ability to generate a *feasible trajectory*, if such a trajectory exists. There are a number of objectives for the planning algorithms. Considering the speed diversity, the first objective is to *enable overtaking* to happen, if there is a safe possibility for it to happen. The vehicles roughly travel the length of the road and the lateral movements rarely cause a significant loss in terms of travel distance or travel time. Overtakings can, however, make a big difference in travel time with an early overtake the most preferred.

A number of planning algorithms presented are cooperative, for which a corresponding objective is to enable a vehicle at the back to overtake, in case the same can be completed by moving aside and giving way. *Cooperation* for this book is defined as the lateral displacement that a vehicle takes, only to aid some other vehicle to overtake. This is summarized in Fig. 5.6. A higher cooperation is preferred only when it finally leads to overtaking.

The planning algorithms further attempt to maximize the *separation* between the vehicle and any other obstacle, road boundary or other vehicle as far as possible. Clearance or the safe distance is taken as a very important objective even though it causes large travel distance and travel time. This is because of the associated factors of safe distance and uncertainties. Further, the planning algorithms enable to minimize travel time, minimize trajectory length, maximize travel speed, maximize smoothness and minimize lateral movements on the road. Some objectives may be contradictory to each other, whereas some others may imply similar things depending upon the manner in which these are modelled. The different algorithms behave in different ways and hence there is no unanimous objective function. Individual algorithms discuss the manner in which they model these objectives or trade-off between the opposing objectives.

5.6.4 Continual Planning and Per-segment Planning

The planning algorithms can be classified into two categories. The first class of algorithms is the *continual planning algorithms* which plan as the vehicle moves. These algorithms either involve reactive agents that plan a unit move of the vehicle as a reaction to the immediate scenario, or algorithms that make a more deliberative plan which gets adapted or extended as the vehicle moves. This class of algorithms hence does not have a separate planning phase and another execution phase, as both these phases go hand in hand.

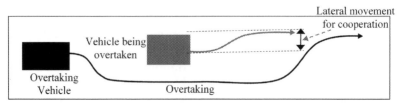

Figure 5.6 Cooperative overtaking.

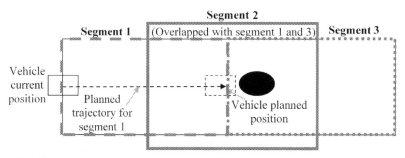

Figure 5.7 Segment overlapping in per-segment planning.

The other category involves the *per-segment planning* algorithms. These algorithms plan a vehicle's trajectory for an entire road segment in a planning phase. A control scheme may be used for making the vehicle travel as per the planned trajectory in the execution phase. The road is hence assumed to be divided into segments, whereas the planner is called for every segment when the vehicle physically enters the segment.

An important concept of this category is that the segments are to be made *overlapping* as shown in Fig. 5.7. While planning each segment, the planning algorithm's vision is restricted to that segment only. Hence, it is possible that the planning algorithm generates a trajectory such that the vehicle is placed at a position from which it cannot overcome obstacles when it reaches the end of the planned segment. Consider an obstacle-free road segment (first segment in Fig. 5.7) with only one vehicle, which may prefer to drive straight. On reaching the segment end (also the start of the next road segment), it may discover an obstacle very close. It would have been better if the vehicle had known about the obstacle earlier. For similar reasons the road is not divided into disjointed road segments. Each vehicle's trajectory is planned when it is at the start of the road segment. As the segments are overlapping, planning of a segment invalidates the last sections of the trajectories of the previous road segment plan.

5.6.5 *Planning With and Without Communication*

The most general way of dealing with the problem is not to assume any kind of *communication* between the vehicles. This allows for human-driven vehicles or autonomous vehicles moving with a different protocol to drive on the road. In such a case, the planning algorithm may attempt to *guess* the possible movements of the vehicles around and hence compute its trajectory. Human drivers tend to use driving gestures, horn blows and other indicators for driving, as well as adherence to traffic laws and best practices in driving clarify vehicle intents. Chapters 'Graph Search Based Hierarchical Planning' and 'Using Heuristics in Graph Search Based Planning' do not assume any communication between the vehicles.

Developments in autonomous vehicles motivate a futuristic traffic scenario wherein all vehicles on the road would be autonomous. Considering the autonomous vehicles are associated with safety constraints, the chances are bright that the human-driven vehicles vanish from the road very quickly. In such a case, vehicles can talk to each other over ad hoc networks via an *intervehicle communication* scheme which allows them to collaboratively formulate an optimal travel plan (Reichardt et al., 2002; Tsugawa, 2002). The advantage of the scheme is that the plans can be rectified for correctness; while planning without communication, one has to guess the motion of the other vehicles and correctness (and optimality) depends upon the guess. Further, stronger coordination strategies may be used as the vehicles can discuss the ways to collaboratively act on a situation, before acting on it. The algorithms presented in chapters 'Optimization Based Planning' and 'Sampling Based Planning' assume communication.

5.6.6   Planning and Coordination

The problem of trajectory generation of a vehicle attempts to compute an optimal trajectory according to the specified objective function. In practice, a number of vehicles may be using the road and it may not be possible for all vehicles to be simultaneously optimal without colliding with each other. The optimal plan may hence involve some vehicles to move as per near-optimal trajectories. The problem of *coordination* deals with the issues of constructing trajectories of different vehicles considering the other vehicles around. Luckily, in the case of traffic systems, a vehicle never steers left or right unless there is a strong reason to do so like overtaking, cooperating for an overtake or obstacle avoidance. Mostly the vehicles travel straight and therefore decisions may sometimes be based on the same, unless obvious from the driving posture of the vehicle.

Speed setting is an important task with multiple vehicles. The speeds are highly governed by whether the vehicle is overtaking, being overtaken, following a vehicle or allowing a vehicle to pass through. These heuristics make it easy to coordinate between vehicles; however, they pose a challenge to integrate these heuristics in common planning algorithms.

5.7   Planning for Special Scenarios

Driving on a straight road is one of the most dominant scenarios of operation. The roads themselves may be of different types and challenging due to different reasons attributed to both the structure of the road as well as the traffic. However, many special scenarios exist in driving which must be explicitly discussed. The list of special scenarios may not be exhaustive. Many places have peculiar and uncommon problems associated with them, which poses a challenge at all levels.

Overtaking (Jin-ying et al., 2008; Wang et al., 2009) is one of the most difficult manoeuvres which requires a high level of precision and control. The challenge is first to judiciously assess an overtaking opportunity. Missing out on an overtaking opportunity can reduce the efficiency of travel, while taking risky overtakings is never suggested. Once a decision regarding whether to overtake has been made, the major task is then to plan and execute the overtaking trajectory. In a lane-based system, this involves changing to the overtaking lane, driving in the overtaking lane and returning back to the original lane of travel once ahead of the vehicle being overtaken.

Parking is one of the most common applications of intelligent vehicle technology. Parking is a tedious job, which many human drivers do not prefer to do. In addition, the parking manoeuvres are normally tight requiring multiple close movements of the vehicle. Automatic parking systems (Han et al., 2011; Hsieh and Ozguner, 2008) enable a vehicle to control its own steering and thus automatically park the vehicle. It is similar to the conventional problem of motion planning of mobile robots with the difference that the operational area is small, requiring the vehicle to go into narrow parking areas and the requirement to move forth and back multiple times to get the correct alignment.

Merging is the problem wherein two roads merge into one (Papageorgiou et al., 2008; Antoniotti et al., 2007; Raravi et al., 2007). The problem is first to schedule the different vehicles and decide whether the mainline vehicle drives in next or the ramp vehicle drives in next. The task is then to move the individual vehicles such that the merger is smooth and does not impact the traffic. From the point of view of motion planning, it is important to move the vehicles such that congestion is avoided and the merging is safe while still being efficient.

Another interesting problem is driving in an *intersection*. An intersection is the point at which multiple roads meet. Many intersections have roundabouts at the intersections. The controlled intersections are controlled by traffic lights. Intersection management (Dresner and Stone, 2004, 2006, 2007) plays an important aspect in such problems, whereas the vehicle simply obeys the orders of the intersection manager. The uncontrolled intersections are navigated by the wisdom of the drivers alone. While driving in an intersection, the first decision to be made is whether to go through the intersection, or to let some other vehicle to pass through. Two vehicles simultaneously occupying the intersection with opposing plans can result in a collision. Hence, the vehicles must prioritize themselves and move in a fair and justifiable manner. Once the vehicle decides to move, the task is then to make a trajectory allowing it to move inside the intersection and to safely leave the intersection. Caution must be given as the other vehicles may come in a compromising pose which must be avoided without accidents.

Traffic may have some not-so-common scenarios that the humans can tackle primarily using their common sense and driving expertise. Driving through an accident site or a site where the road has been partially blocked is one such situation, which calls for a person to navigate through whatever road is navigable. Congestion may be common in such sites, which calls for greater caution and expertise. Many times the complete road may be blocked, in which case the person has to take a U-turn, maybe even in the middle of the road. Roads may not be wide enough to facilitate a U-turn, which forces constant back and forth manoeuvres to facilitate a complete turning. Sometimes vehicles may come into a state of deadlock, especially when the inbound and outbound traffic share a common road and some part of the road is blocked. Although humans can easily tackle such situations in the extreme, by talking, and deciding which vehicle backs up and allows the other vehicle to move first. Gestures are commonly used which are widely understood by humans. Detecting blockages and taking an alternative course of action is a very challenging problem for the autonomous vehicles.

5.8 Summary

The chapter introduced the notion of planning for autonomous vehicles. Planning a complete autonomous vehicle amidst other autonomous or human driven vehicles is a very complicated task. To ease the task the complete planning problem needs to be solved in a hierarchical manner, each hierarchy operating at some degree of abstraction with some assumptions. The most important hierarchies are route planning which tells the vehicle the roads and intersections to take, trajectory planning enabling the vehicle to avoid all static obstacles and other vehicles and control which helps the vehicle to track the trajectory. Planning scenarios may be organized and lane-oriented, or unorganized. In unorganized traffic, the vehicle need not adhere to lane discipline, which is a more challenging and generalized problem. The book is largely focussed upon unorganized traffic.

The problem of motion planning in unorganized traffic is similar to the problem of robot-motion planning. The principles of robot-motion planning were discussed for a strong foundation. A common picture of motion-planning algorithms is that there are a variety of solutions which behave differently for different scenarios. The choice of algorithm is highly dependent upon the scenario of operation and validity of the underlying assumptions. More generally, the problem is of multirobot motion planning wherein the robots need to coordinate with each other, whereas they may or may not be able to communicate via communication. The key features of planning of autonomous vehicles that differentiate it from the general problem of robot-motion planning include the use of a road coordinate axis system wherein the lateral and longitudinal axis are considered, and the behaviours of overtaking and vehicle following. The chapter presented the notion of trajectory planning for a variety of scenarios including straight roads, overtaking, vehicle following, intersections, mergers, parking and some special scenarios.

References

Antoniotti, M., Deshpande, A., Girault, A., 2007. Microsimulation analysis of multiple merge junctions under autonomous AHS operation. In: Proceedings of the IEEE Conference on Intelligent Transportation Systems, Boston, USA, pp. 147−152.

Arai, T., Ota, J., 1992. Motion planning of multiple mobile robots. In: Proceedings of the 1992 IEEE/RSJ International Conference on Intelligent Robots and Systems, pp. 1761−1768.

Bennewitz, M., Burgard, W., Thrun, S., 2001. Optimizing schedules for prioritized path planning of multi-robot systems. In: Proceedings of the 2001 IEEE International Conference on Robotics and Automation, Seoul, Korea, pp. 271−276.

Bennewitz, M., Burgard, W., Thrun, S., 2002. Finding and optimizing solvable priority schemes for decoupled path planning techniques for teams of mobile robots. Robotics and Autonomous Systems 41 (2−3), 89−99.

Buehler, M., Iagnemma, K., Singh, S., 2007. The 2005 DARPA Grand Challenge: The Great Robot Race. Springer, Berlin, Heidelberg.

Choset, H., Lynch, K.M., Hutchinson, S., Kantor, G.A., Burgard, W., Kavraki, L.E., Thrun, S., 2005. Principles of Robot Motion: Theory, Algorithms, and Implementations. MIT Press, Cambridge, MA.

Dresner, K., Stone, P., 2004. Multiagent traffic management: a reservation-based intersection control mechanism. In: Proceedings of the Third International Joint Conference on Autonomous Agents and Multiagent Systems, NY, USA, pp. 530−537.

Dresner, K., Stone, P., 2006. Multiagent traffic management: opportunities for multi-agent learning. In: Lecture Notes in Artificial Intelligence, vol. 3898. Springer Verlag, Berlin, pp. 129−138.

Dresner, K., Stone, P., 2007. Sharing the road: autonomous vehicles meet human drivers. In: Proceedings of the 20th International Joint Conference on Artificial Intelligence, pp. 1263−1268 (Hyderabad, India).

Glaserm, S., Vanholme, B., Mammar, S., Gruyer, D., Nouveliére, L., 2010. Maneuver-based trajectory planning for highly autonomous vehicles on real road with traffic and driver interaction. IEEE Transactions on Intelligent Transportation Systems 11 (3), 589−606.

Hall, R.W., Caliskan, C., 1999. Design and evaluation of an automated highway system with optimized lane assignment. Transportation Research Part C Emerging Technologies 7 (1), 1−15.

Han, L., Do, Q.H., Mita, S., 2011. Unified path planner for parking an autonomous vehicle based on RRT. In: Proceedings of the 2011 IEEE International Conference on Robotics and Automation, pp. 5622−5627.

Hsieh, M.F., Ozguner, U., 2008. A parking algorithm for an autonomous vehicle. In: Proceedings of the 2008 IEEE Intelligent Vehicles Symposium, pp. 1155−1160.

Jain, A., Menezes, R.G., Kanchan, T., Gagan, S., Jain, R., 2009. Two wheeler accidents on Indian roads - a study from Mangalore, India. Journal of Forensic and Legal Medicine 16 (3), 130−133.

Jin-ying, H., Hong-xia, P., Xi-wang, Y., Jing-da, L., 2008. Fuzzy controller design of autonomy overtaking system. In: Proceedings of the 12th IEEE International Conference on Intelligent Engineering Systems, Miami, Florida, pp. 281−285.

Kant, K., Zucker, S.W., 1986. Toward efficient trajectory planning: the path-velocity decomposition. International Journal of Robotic Research 5 (3), 72−89.

Konar, A., 1999. Artificial Intelligence and Soft Computing: Behavioral and Cognitive Modeling of the Human Brain. CRC Press, Boca Raton, FL.

Lumelsky, V.J., Harinarayan, K.R., 1997. Decentralized motion planning for multiple mobile robots: the Cocktail Party model. Autonomous Robots 4 (1), 121−135.

Mohan, D., Bawa, P.S., 1985. An analysis of road traffic fatalities in Delhi, India. Accident Analysis & Prevention 17 (1), 33−45.

Papageorgiou, M., Papamichail, I., Spiliopoulou, A.D., Lentzakis, A.F., 2008. Real-time merging traffic control with applications to toll plaza and work zone management. Transportation Research Part C 16, 535−553.

Parker, L.P., Schneider, F.E., Schultz, A.C., 2005. Multi-robot Systems: From Swarms to Intelligent Automata, vol. 3. Springer-Verlag, New York.

Paruchuri, P., Pullalarevu, A.R., Karlapalem, K., 2002. Multi agent simulation of unorganized traffic. In: Proceedings of the ACM 1st International Joint Conference on Autonomous Agents and Multiagent Systems, New York, pp. 176−183.

Raravi, G., Shingde, V., Ramamritham, K., Bharadia, J., 2007. Merge algorithms for intelligent vehicles. In: Next Generation Design and Verification Methodologies for Distributed Embedded Control Systems, pp. 51−65.

Reichardt, D., Miglietta, M., Moretti, L., Morsink, P., Schulz, W., 2002. CarTALK 2000: safe and comfortable driving based upon inter-vehicle-communication. In: Proceedings of the 2002 IEEE Intelligent Vehicle Symposium, vol. 2, pp. 545—550.

Russell, R., Norvig, P., 2009. Artificial Intelligence: A Modern Approach, third ed. Pearson, Harlow, Essex, England.

Sánchez-Ante, G., Latombe, J.C., 2002. Using a PRM planner to compare centralized and decoupled planning for multi-robot systems. In: Proceedings of the IEEE International Conference on Robotics and Automation, Washington, DC, pp. 2112—2119.

Sewall, J., van den Berg, J., Lin, M.C., Manocha, D., 2011. Virtualized traffic: reconstructing traffic flows from discrete spatio-temporal data. IEEE Transactions on Visualization and Computer Graphics 17 (1), 26—37.

Shukla, A., Tiwari, R., Kala, R., 2010. Towards Hybrid and Adaptive Computing: A Perspective. Studies in Computational Intelligence, Springer-Verlag Berlin, Heidelberg.

Tiwari, R., Shukla, A., Kala, R., 2013. Intelligent Planning for Mobile Robotics: Algorithmic Approaches. IGI Global Publishers, Hershey, PA.

Tsugawa, S., 2002. Inter-vehicle communications and their applications to intelligent vehicles: an overview. In: Proceedings of the IEEE Intelligent Vehicle Symposium, vol. 2, pp. 564—569.

Vanajakshi, L., Subramanian, S.C., Sivanandan, R., 2009. Travel time prediction under heterogeneous traffic conditions using global positioning system data from buses. IET Intelligent Transportation Systems 3 (1), 1—9.

Wang, F., Yang, M., Yang, R., 2009. Conflict-probability-estimation-based overtaking for intelligent vehicles. IEEE Transactions on Intelligent Transportation Systems 10 (2), 366—370.

Optimization-Based Planning

6.1 Introduction

The problem of motion planning for autonomous vehicles is to move a set of vehicles within a road segment. As discussed in chapter 'Introduction to Planning', the road segment may be available from higher-level planning, chiefly employing Dijkstra's algorithm (Cormen et al., 2001) for the entire road network graph. A complete road may be broken down into overlapping segments. The chapter considers a general segment of the road which may be occupied by multiple vehicles. The task is to plan and navigate all the vehicles, such that the vehicles do not collide with each other and are able to navigate the road segment in a near-optimal manner while planning in near-real time.

In this chapter, the problem is solved with the assumption of *communication* between the vehicles. Due to safety concerns, it is viable to assume that after autonomous vehicles are introduced and become common on the road, human-driven vehicles would start declining till the point when they become extinct. Considering the safety of other vehicles, human-driving in certain areas may even be declared illegal over the span of time. The benefits of a fully autonomous vehicle transportation system are further encouraging. This chapter explores the use of a Genetic Algorithm (GA) for solving the problem.

The greatest challenge associated with undertaking the task of planning in the absence of speed lanes is the need to deal with a continuous domain of lateral positions. A vehicle may lie anywhere laterally within the road, unlike the lane-oriented traffic in which the lateral decisions are largely about deciding the lane of travel. In such a case, determining the most efficient trajectory which overcomes an obstacle, overtakes a vehicle or simply moves to a better lateral position is a requirement which needs to be fulfilled in a computationally inexpensive manner. Further, because there are multiple vehicles in the road segment, choosing a noncomputationally intensive but *cooperative coordination* strategy is an important requirement. A vehicle must give some other vehicle space to overtake and facilitate some other vehicle to pass through, considering the relative importance and speeds of the vehicles.

Close overtaking is the best example to illustrate the requirement. Considering efficiency concerns over a diverse traffic landscape, every close overtake must be performed. A close-overtaking trajectory is tightly bound on all sides which makes it difficult for the algorithm. The problem is also called the *narrow corridor problem* wherein determining the trajectory is difficult if it goes through a narrow corridor. Further cooperation of the vehicle being overtaken may result in an overtaking attempt being feasible even if the overtaking is narrowly possible. Therefore, the algorithm must allow a vehicle to cooperate and make way for the overtaking vehicle, and then make the overtaking trajectory to carry out overtaking.

GA is an *optimization* algorithm which can find optimal value of an objective function by changing the function parameters. The GA is inspired from the process of natural evolution. It imitates the evolution process of humans wherein a population corresponding to multiple individuals is maintained, which undergoes breeding to generate the offspring-making individuals of the next generation. The fittest individuals survive the evolution whereas the weaker ones are eliminated. Similarly, GA maintains a pool of solutions to the problem, each individual representing the encoded solution to the problem in the form of genes. The fitness of each solution is judged by a fitness function. At every generation, specialized genetic operators are applied between the selected individuals to form the next generation of the population pool. The fittest individuals usually replicate with small mutations or deviations, whereas the weaker ones are eliminated. The fittest of the final surviving individual is adjudged as the optimal solution.

The chapter makes use of GA for the purpose. The GA is asked to optimize a *Bézier curve*. The greatest advantage of GA is that it is *probabilistically optimal*, meaning that the probability of optimality tends to 1 as optimization time tends to infinity. It is also *probabilistically complete*, meaning that the probability of finding a solution, if one exists, tends to 1 as the optimization time tends to infinity. GA fails when the optimal trajectory has too many turns or the trajectory goes through a very narrow region. Every turn is encoded as a gene in the GA and too many turns means optimization with too many variables. The GAs encounter the problem of the *curse of dimensionality*, meaning that the performance drastically reduces on an increase in the dimensionality or the number of variables optimized. Further, it is very difficult to generate samples inside narrow corridors, meaning that it is hard to sample trajectories that pass through a narrow region tightly bounded by obstacles on all sides. The sampling-based approaches generally fail in such scenarios.

Road scenarios are marked with fewer steering manoeuvres and too-narrow regions defy safety concerns and are hence undesirable. Hence, for this problem GA is a good choice of algorithm. Further, the iterative nature of GA makes it very desirable for such problems. It is an *anytime algorithm* and a near-optimal solution to the problem is available at any time in the optimization process. The solution improves with time. The near-optimal solution can be extracted at any moment for the immediate motion of the vehicle, and hence the algorithm is best suited for such problems in which a small and usually fixed optimization time is available.

The biggest disadvantage of GA is, however, its *computational cost*. Sections 6.4 and 6.5 highlight the manner in which the problem characteristics can be exploited to make it possible to run GA on such a real-time problem. The other problem is the presence of multiple vehicles on the road which requires an optimal *coordination strategy* working in near-real time. Because the general coordination strategies like priority-based planning (Bennewitz et al., 2001, 2002), co-evolutionary GAs (Potter, 1997; Stanley and Miikkulainen, 2004) or any other related techniques may either not be optimal or be computationally intensive, the requirements are difficult to keep. The work uses *prioritized-based coordination* which is noncooperative but computationally inexpensive. Every vehicle is assigned a priority and the vehicles are planned strictly in the order of priority. A lower-priority vehicle is responsible for collision avoidance with a higher-priority vehicle. *Traffic-inspired heuristics* are embedded into the strategy to make it near optimal and cooperative, at the same time being computationally inexpensive. To do so, the general operation of the traffic is observed, prevalent traffic and social rules are assessed and methods are designed by which these rules can be coded over the GA to make a coordination strategy.

Segments of different sections including text and figures have been reprinted from Kala and Warwick (2014), Applied Soft Computing, Vol 19, R. Kala, K. Warwick, Heuristic based evolution for the coordination of autonomous vehicles in the absence of speed lanes, pp. 387−402, Copyright (2014), with permission from Elsevier.

6.2 A Brief Overview of Literature

Most works in the domain of motion planning for autonomous vehicles exist for organized or a lane-oriented traffic. Some notable works, however, also exist for unorganized traffic (or extendable for unorganized traffic), which is the key focus of the chapter. Chu et al. (2012) constructed a number of candidate paths from which the best path was selected. The strategy can be used for avoiding obstacles by a single manoeuvre only. Paruchuri et al. (2002) simulated the vehicle behaviours on straight roads and crossings without traffic lights. The limitations include no cooperation between the vehicles and that the overtaking decision module does not generalize to a high number of vehicles with unorganized patterns.

The algorithms used in robot motion planning may be of use for planning autonomous vehicles in the absence of lanes, although they may require to be remodelled per the traffic scenario. Optimization-based methods are widely used for planning the trajectory of a single robot or multiple robots. Many of these approaches are, however, offline and cannot be executed in real time. The use of the road coordinate axis system instead of the Cartesian coordinate axis system can be very helpful when using these algorithms for autonomous vehicles. Many of these approaches are for a single robot only and hence noncooperative. Other approaches with multirobot coordination techniques are largely computationally intensive.

A number of good approaches using GA can be found which are all offline and hence cannot be directly used. Xiao et al. (1997) planned the path of a robot offline by using a GA. The path was also updated by an online planner. Xidias and Azariadis (2011) solved the problem of integrated routing and planning of a group of vehicles. A centralized approach was proposed and all travel information parameters were embedded in a single chromosome. Kala et al. (2011) used a multiresolution collision-checking approach to compute the trajectory of a single robot using a GA. The resolution was coarser at the start of the GA optimization, which became finer as the GA optimized the path. These approaches are largely offline and noncooperative.

Garcia et al. (2009) used the Ant Colony Optimization Algorithm along with a memory unit to restrict the magnitude of exploration in search for a path. The algorithm worked for discrete spaces only. In another related work, Lepetic et al. (2003) designed planning algorithms for robot soccer. Scenarios were precomputed in terms of relative positions of ball with respect to the robot. The intermediate solutions were generated using interpolation. In the problem of autonomous vehicle navigation, pre-computation is not possible due to the discrete decisions between overtaking and not overtaking.

Some good works also exist for multiple robots using centralized optimization, which can be computationally expensive for a large number of robots. Lian and Murray (2003) solved the problem of motion planning for a single robot, a swarm of robots and uncoordinated multiple robots using quadratic programming and spline curves. Similarly, Klancar and Skrjanc (2010) used Bernstein−Bézier curves which were optimized to produce short path length and high clearance. Kapanoglu et al. (2012) used optimization for the problem of area coverage using rectilinear moves. Szlapczynski and Szlapczynska (2012) also used optimization for planning the tra-jectory of a ship.

Some variants use cooperative co-evolution for time efficiency, which is still too computationally intensive to be applied for real-time applications. Kala (2012) used cooperative co-evolution for motion planning of multiple robots moving rectilinearly in narrow corridors. A memory unit was proposed for sharing the shortest-path infor-mation between landmarks. Wang and Wu (2005) also made use of cooperative co-evolution. Chakraborty et al. (2008) solved the problem using both centralized and decentralized methodologies using differential evolution. Only the next immediate step of the robots was optimized. The approach was noncooperative.

6.3   A Primer on Genetic Algorithm (GA)

The first and the simplest tool used to solve the problem is the GA. The GA is an *optimization*-based technique. That said, the GA can do a lot more than conventional optimization; however, to ease the discussion GA is studied as an optimization tool in this section and then modelled differently to solve the problem of motion planning for autonomous vehicles in Section 6.4. Optimization techniques are used to find the min-imum or the maximum value of an optimization *objective function*. The techniques do so by changing the parameters which affect the value of the optimization objective. The general nature of the optimization problems is given by Eq. [6.1].

$$\text{Calculate Min}(F(x_1, x_2, x_3 \ldots x_n)) \qquad\qquad\qquad [6.1]$$

$$LB_i \leq x_i \leq UB_i$$

Here, $F()$ is the optimization objective function with a total of n parameters. Each parameter x_i is bounded between a lower bound (LB_i) and an upper bound (UB_i). There may be more linear or nonlinear constraints.

The GA is inspired by the principle of *natural evolution*. In natural evolution, numerous individuals of every species exist in the ecosystem. All the individuals collectively are known as the population pool. The individuals of the population pool interact with each other during their life cycle. The interactions may be more with the individuals closer or alike, and less with the others. The individuals are essentially chromosomes which are a collection of genes. The genes together make up all the properties of the individual and completely characterize the individual. The individuals mate and produce new offsprings who constitute the future individuals of the population pool. Typically, two parents mate to generate the next generation of children. The children contain the mixed characteristics or genes from the parents, with some genes derived from the first parent and other genes derived from the second parent. While doing so, however, many times genetic errors may happen resulting in different or new genes in the children, produced as a result or error while copying the specific genes. These errors are very small as compared to the total collection of genes. Out of all the children generated, the fittest survive whereas the weaker cannot survive in the hostile environment and die in the process. This is in accordance with Darwin's theory of evolution. The surviving children constitute the next generation of the population pool. In this manner evolution happens generation by generation, and the latter generations are fitter and more suited to the changing environment.

6.3.1 General Algorithm Framework

As an imitation to the natural evolution process, the GA attempts to solve the problem by generating random solutions to the problem or by random assignment to the parameter values. Each such solution is called an *individual*. The solution in the problem-specific domain is called the phenotype. The phenotype needs to be encoded in the form of genes making the chromosome. This encoded solution is called the genotype. The *population* is a collection of individuals at any particular point of time or *generation*. The GA is hence a multiindividual method in which the search for the optimal solution happens by employing multiple individuals which work in coordination with each other. The initial population may be randomly generated. It is also possible to use heuristics to analyse the problem domain to identify some good characteristics or partial solutions which are likely to have a good fitness value. Correspondingly, the initial population may be generated biased towards some values of genes.

The individuals at any generation are subjected to a number of *genetic operators* to create the individuals of the next generation of the population pool. Essentially, the fittest individuals are selected and *crossover* is applied to get the children. The fittest individuals are likely to generate a better solution and are selected more often than the weaker solutions, which may not be selected at all in the evolution process. The crossover operation generates children with gene values intermediate between the parents. Some children are additionally subjected to *mutation* wherein small changes are applied to their gene values. This makes the population pool of the next generation.

Here, *fitness* is an assessment of the goodness of the solution for the particular problem. The fitter be an individual, the more is its chance to replicate and dominate its characteristics in the population pool. The genetic operators try to generate individuals with a better fitness value. The *fitness function* hints the evolutionary process towards which characteristics or genes to prefer, and which ones not to prefer. The overall process is shown in Fig. 6.1.

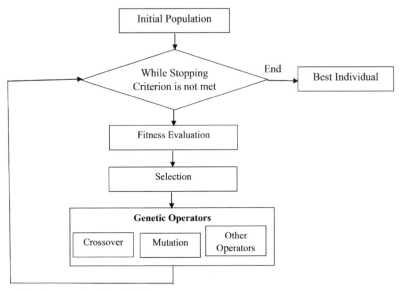

Figure 6.1 Genetic algorithm (GA).

6.3.2 *Individual Representation*

The purpose of *individual representation* is to encode the phenotype representing the real solution to the problem represented as a solution in the actual problem domain, into a genotype or the representation which the GA can handle. This representation is in the form of a collection of genes. The representation technique should facilitate the conversion of the genotype into phenotype for fitness assessment and for porting the final optimized solution into the problem domain. Similarly, conversion of phenotype into genotype may be needed to insert specific individuals in the initial population pool.

Each gene constituting the individual may be binary, only storing 0 and 1. The GA resulting from such a representation is called a *Binary* GA. For the above generic definition of the problem, each variable x_i representing a real value must be converted into a binary equivalent for the GA to work. The number of bits used for every variable must be fixed which constitutes the resolution of the algorithm. On the contrary, every gene may also be allowed to store real numbers. The GA resulting from this representation is called a *Real Coded* GA. In this work the stress is on Real Coded GAs, which are easier to understand and develop. For the generic definition of an optimization problem, the individual is simply all the variables appended one after the other shown in Fig. 6.2.

0.54	0.04	0.87	0.65	......	0.39
x_1	x_2	x_3	x_4	....	x_n

Figure 6.2 Individual representation.

A tree-based representation is another popular choice for representing individuals, wherein the genotype is in the form of a tree with each node presenting a parameter and its relation. Genetic programming is a dominant technique which uses such a tree-style representation, designed for automatic evolution of programmes which inherently have a tree-like structure. Specialized operators are made to work with tree-based representations. It is also common to give a linear encoding to a tree-based representation to suit conventional genetic operators. Grammatical evolution is a common technique. Further, many times the algorithm metaparameters may also be embedded along with the parameters representing the variables of the optimization objective. This gives a self-adaptive nature to the approach wherein the metaparameters are also optimized. Evolutionary strategies is a common technique using this principle.

For this problem the representation and conversion is simple. For the problem of trajectory planning, the phenotype represents a trajectory usable by the vehicle whereas the genotype is a small set of real numbers. This problem is addressed in Section 6.4.

6.3.3 Genetic Operators

The purpose of genetic operators is to generate the next-generation population from the current generation population. The different operators are described in the following subsections.

6.3.3.1 Scaling

First, every individual in the pool must be assigned an expectation value which denotes its likelihood of being selected in the evolutionary process. This operation is known as *scaling*. The expectation value can be easily taken proportional to the fitness value. A normalization may be required to get the range of values within the desirable range. This is called *fitness-based scaling* wherein the expected value ($E(I)$) of any individual (I) is given by Eq. [6.2].

$$E(I) = E_{\min} + \frac{F(I) - F_{\min}}{F_{\max} - F_{\min}} (E_{\max} - E_{\min}) \qquad [6.2]$$

Here, $E_{\max}$ and $E_{\min}$ are the maximum and minimum desired expected values, whereas $F_{\max}$ and $F_{\min}$ are the maximum and minimum fitness value returned by the fitness function.

The problem with fitness-based scaling is that very soon a very few individuals get very good fitness and correspondingly a very high expected value. Hence, they get selected again and again, whereas the others get eliminated not being selected for the evolution. This causes a *premature convergence* in the population pool, wherein very soon all individuals have nearly the same characteristics. Hence, an alternative method is adopted to stop a few individuals from dominating the population pool very early. Here, the expectation value is kept proportional to the rank of the individual and the technique is known as *rank-based scaling*. Even if a few individuals have significantly better fitness as compared to the rest of the individuals in the population pool, the expectation is assigned based on the ranks alone, the distribution for which does not change. The expectation value is given by Eq. [6.3].

$$E(I) = E_{min} + \frac{N - Rank(I) - 1}{N - 1}(E_{max} - E_{min}) \qquad [6.3]$$

Here, $Rank(I)$ is the rank of the individual I in the population pool with rank 1 denoting the best individual. N is the population size.

The *top scaling scheme* selects the top few individuals and gives them equal expectation values. The top few individuals hence have equal chances of being selected, whereas the others get eliminated.

6.3.3.2 Selection

The *selection* operator actually selects the individual for the evolution process. The individuals with high fitness value may be selected multiple times, whereas the weaker ones may not be selected at all. The individuals not selected are simply deleted. The selection is a stochastic process with the possibility of selection given by the expectation values. The implementation of this concept is different for different selection schemes, a few of which are briefly discussed.

The *roulette wheel selection* scheme makes a roulette wheel with different individuals as different sectors of the wheel. The selection is summarized by Fig. 6.3A. The possibility of selection of a particular sector in a roulette wheel is given by its circumference, which is kept as the expectation value of the individual. The wheel is rotated once and the individual to which the pointer points when the wheel becomes stationary is selected. For selection of multiple individuals, the wheel is rotated multiple times and each rotation gives a selected individual.

The problem with this scheme is that the fitter individuals occupy a large circumference and are likely to be selected multiple times leading to a premature convergence. Hence, a different selection technique known as *Stochastic Universal Sampling* is used. Here, the roulette wheel has as many pointers or selectors as are the number of individuals to be selected from the population pool shown in Fig. 6.3B. Each pointer selects one individual. For selection of multiple individuals, the wheel is rotated only once and the individuals corresponding to all the pointers are selected. Of course, the fittest individuals may be selected multiple times with multiple pointers pointing to that particular individual.

Another common selection scheme is *tournament selection*. Here, tournaments are organized between a few individuals. In any tournament between two individuals, the probability of winning of the fitter individual is kept as p, whereas the probability of winning of the weaker individual is kept as $1 - p$. The winner of the entire tournament is selected.

6.3.3.3 Crossover

The *crossover* operator imitates the reproduction of the natural species by making children from their parents. The children always carry the characteristics of their parents

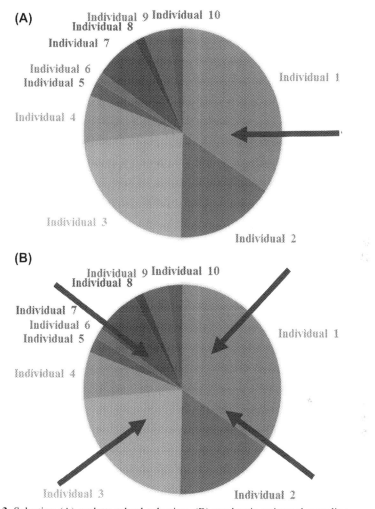

Figure 6.3 Selection (A) roulette wheel selection, (B) stochastic universal sampling.

and are formed by the exchange of genes from the parents. Any number of parents may be used to produce the children, although the most common example is a *binary cross-over* in which two parents are used to produce two children. The multiparent crossover operators extend the same idea by exchanging genes between multiple parents to produce children. The working of binary crossover is discussed here. The higher-order crossovers are a natural extension to the same principles. The ratio of the number of individuals selected for crossover to the total population size is called the *crossover rate*. In other words, crossover rate denotes the proportion of individuals which undergo crossover.

The simplest crossover is the *one-point crossover* wherein a random point is selected in the chromosome of the parents. The parents are placed on top of each other

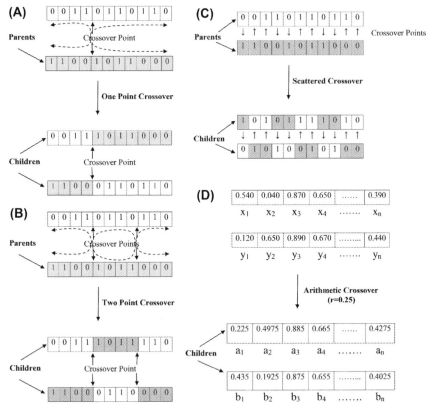

Figure 6.4 Crossover (A) one-point crossover, (B) two-point crossover, (C) scattered crossover, (D) arithmetic crossover.

and the genes lying after the crossover point are exchanged. This gives rise to two children from two parents. The technique is shown in Fig. 6.4A. Similarly, *two-point crossover* selects two random crossover points instead of one. The parents are kept on top of each other and the genes in-between the two crossover points are swapped. The technique is shown in Fig. 6.4B. Both techniques have the problem that genes located close to each other are likely to go together to the same child. Hence, a *scattered crossover* technique is used. Here, half the genes are randomly copied from the first parent to the first child, whereas the other half is copied from the second parent to the first child. The second child takes the other genes. It behaves like multipoint crossover in which the number of points are equal to the chromosome length and at each point a random decision is make whether to swap the genes or not. The technique is shown in Fig. 6.4C.

Another crossover technique used largely for the real coded GAs is the *arithmetic crossover*. Each individual represents a point in a multidimensional search space (see Section 6.3.5). The two parents are two such points in the search space. This crossover technique attempts to generate children in-between the line joining the two parents.

Using this technique the ith gene of the children A and B with parents X and Y is given by Eq. [6.4].

$$a_i = rx_i + (1 - r)y_i \qquad\qquad [6.4]$$

$$b_i = (1 - r)x_i + ry_i$$

Here, r is a random number between 0 and 1. The technique is shown by Fig. 6.4D.

6.3.3.4 Mutation

The *mutation* operator tries to bring into the population new characteristics or gene values which are nonexistent in the population pool. The addition of new characteristics to the population pool may be good, in which case the mutated individual enjoys a good fitness value and is bound to be selected multiple times, or the addition of new characteristics may be bad leading to the individual being eliminated from the population pool. This makes the individuals explore in search of the optima by constantly changing the gene values.

The *mutation rate* decides the magnitude of changes to be made in an individual to produce the mutated individual which constitutes the individual of the next generation. In a binary GA, the gene is simply flipped with 0 changed to 1 and 1 changed to 0, with a probability given by the mutation rate. The mutation is given in Fig. 6.5A. In real coded GA, the maximum percent change made in the value of any gene is given by the mutation rate, whereas the actual deviation is produced by a random number. It is advisable to keep the mutation value small so as not to cause randomness to the search process, whereas too small mutation rates may hardly make any changes to the individual resulting in very slow convergence. A popular technique is thus to use *Gaussian mutation* in which the magnitude of deviation to make in an individual is given by a Gaussian distribution which has the inherent property that the small

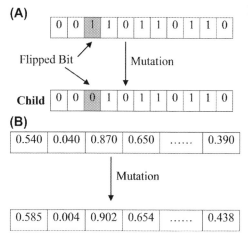

(A)

Figure 6.5 Mutation (A) bit mutation, (B) gaussian mutation.

values are most likely and are often generated, at the same time the higher values are also rarely generated. The mutation applied to a synthetic individual is shown in Fig. 6.5B. Alternatively, it is common to make two mutation operators called *soft mutation* with a small mutation rate which is called very frequently and *hard mutation* with a very large mutation rate which is called rarely.

6.3.3.5 Other Operators

Depending upon the problem and its characteristics, a number of other standard or nonstandard genetic operators may be designed. A common operator is *elite* which identifies the top few individuals of every generation and passes them straight to the next generation. It is sometimes possible that the best individual may not survive in the selection, or even if it survives, may be mutated to reduce its fitness. The elite operator ensures that the best individual is safe and hence the fitness of the best individual does not deteriorate for a constant fitness function.

Similarly, if the problem has constraints it may be possible to get individuals which do not adhere to the constraints and hence cannot be accommodated in the population pool. A common technique is to introduce a *penalty* in the fitness value of these individuals so that they are no longer selected. This also creates evolutionary pressures to reduce infeasibility and ultimately eliminate it. Alternatively, one may make a custom *repair* operator which is applied to all infeasible individuals. This operator changes an infeasible solution to a feasible one by changing its genes such that the feasibility condition is met. In the specific problem of trajectory planning, a trajectory hitting an obstacle is a constraint violation which can be handled in this manner.

Another operator commonly used is *insert* which inserts new individuals to the population pool. A common use of this operator is when the fitness function changes with time and can show trends of sudden change. In such a case, random solutions introduce an intent to create enough *diversity* in the population pool so that there are always individuals with characteristics to react to and survive any change of fitness function. The random solutions are also useful to add new characteristics to the population pool to eliminate premature convergence.

6.3.4 Stopping Criterion

The GA may go on indefinitely improving the individuals along with the generations. However, the algorithm needs to be terminated after some time so that the optimal solution can be extracted. At the later generations, *convergence* happens when all individuals have nearly the same characteristics, and it is extremely unlikely to create new characteristics that aid the optimization. It is hence better to terminate the algorithm when convergence happens.

The *stopping criterion* states the condition on meeting which the optimization terminates. Multiple criteria may be set in which case any of them must meet to terminate the optimization process. Common criteria include limiting the maximum number of generations, maximum optimization time, stall generations or number of generations in which no improvement is seen in the GA and stall time in which no improvement is seen in the GA.

6.3.5   Exploration and Exploitation

Consider the objective optimization of Eq. [6.1] which can easily be taken as the fitness function of the GA. If this fitness function is plotted, it would be a surface in a high dimensional ($n + 1$) space with n axis representing each of the variables and 1 axis representing the output. This is known as the *fitness landscape*. Because a multidimensional curve cannot be drawn in the book, let us consider a function with two variables which can be easily drawn as a 3D surface and as a contour. Let these be given by Fig. 6.6. Each individual is a point in this contour with the fitness value given by the value of the contour. The population pool is a group of points in this fitness landscape. Initially, the individuals are randomly generated and so the points are randomly distributed in the fitness landscape. As generations pass, the individuals in the high-fitness areas replicate by the selection operator and attract other individuals towards them by the crossover operator. The individuals in the poor-fitness areas are eliminated by the selection process. The mutation operator causes individuals, especially those in the high-fitness areas, to explore their surroundings by taking a small step at a random direction.

Exploitation refers to the evolutionary forces or the genetic operators which cause the individuals to come close to each other and converge at some point in the fitness landscape which is usually the global optimum. *Exploration* refers to the evolutionary forces or the genetic operators which cause the individuals to go away from each other

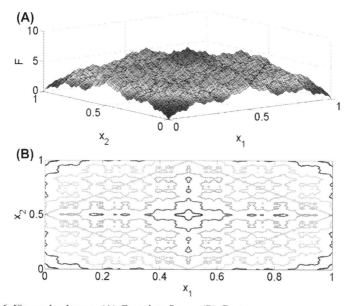

Figure 6.6 Fitness landscape. (A) Complete Space, (B) Contours.

in pursuit of, or search for, the global optimum. Normally, exploitative pressures are kept larger than the explorative pressures which cause the population pool to contract and eventually converge to a very small region or the global optimum.

Attempting to cause very early convergence by parameter settings can lead to premature convergence into a local optimum as much explorative efforts were not made to search for the global optimum. Very small exploitation can lead to no convergence or a very small convergence which necessitates the need to run the algorithm for a prolonged time. Absence of convergence makes the algorithm behave similar to a random search algorithm, which searches by continuous generation of random individuals. It is important to keep a judicious balance between the two forces, whereas the computational time is a major decider between the contributions of the two forces. Availability of long computational time means that time can be spent on explorative settings, whereas the need to give results in very short computational times will necessitate exploitative settings.

The problems are magnified when the fitness landscape has high *modality* or too many crests and troughs. Too many local optima can deceive the evolution resulting in convergence at the local minima or generating evolutionary pressures near the local minima. Large modality calls for large exploration to sight the global optimum. If possible, choosing fitness functions with limited modality is an important criterion in the design of fitness functions.

Based on these discussions, it is intriguing to study the effect of different parameters of the GA. The larger the mutation rate, the greater is the exploration, demanding more computational time at the benefit of missing a local optimum. Similarly, large crossover rates may result in exploitation and premature convergence. Keeping the same number of individuals in the population, or the population size, results in better optimization due to more exploration at the expense of computation time. If the number of generations is, however, reduced to keep the time constant, the algorithm will become more explorative. Similarly, carrying optimization for a large number of generations results in better optimization with more computational expense. Cutting down on the population size to keep the computational time constant can be an exploitative setting.

6.4   Motion Planning with Genetic Algorithm

This section makes use of GA for the problem. Before selecting any algorithm for the problem, it is always wise to consider its pros and cons. The pros and cons of GA are summarized in Box 6.1. The challenge in the work is to plan the motion of the vehicles without the assumptions of lane-based driving. Not considering lanes necessarily increases the problem of planning and coordination to a high level but provides for a more general solution. The most important problem faced here is in overtaking, due to the requirement for safety under optimal steering conditions, resulting in a small margin for error in the efficiency of driving. However, the vehicle may ask other vehicles for support to make a close overtake possible. Such a solution is inspired from

---

Box 6.1 Pros and Cons of Genetic Algorithms (GAs)

Pros
- Probabilistically optimal
- Probabilistically complete
- Iterative

Cons
- Computational cost
- Do not work well in narrow corridors
- Do not work well with too many turns (implying large dimensionality)
- How to introduce cooperative coordination?

---

observed driving in low-width high-density roads, especially where vehicles vary in speeds to a large degree.

Whilst driving along straight roads may be a relatively straightforward affair, making efficient turns for most crossings or other natural turnings of the road, requires expertise. Turning too close to the inner boundary may require reducing the speed, at the same time turning along the outer boundary may make the path too long and suboptimal. Here, the same is done by using *Bézier-based motion planning* linked with a GA for the optimization procedure. The GA is adapted to work on real-time scenarios by using a space–time-approach model of different vehicles and a globally referenced individual representation. A GA is probabilistically optimal and probabilistically complete and can work in continuous spaces. The objectives of efficiency and safety can be easily knit to a single objective function for evolutionary optimization. Road scenarios have limited possible homotopies for a vehicle, which translates to a limited modality of the fitness landscape (given the fact that a *repair operator* is designed which maps every possible trajectory to a reasonable looking, feasible and short trajectory). For a limited-modality fitness landscape, the optimal solution can be found reasonably early or, to put it another way, the probability of finding the optimal solution is high. Here, assumptions are made on a limited number of obstacles and other vehicles, and with a limited-resolution map.

The algorithm is a real-life application of *dynamic evolutionary algorithms*, which optimize an objective under a changing fitness landscape. The operational scenario continuously changes as the vehicles move. At every instant the latest instance of scenario is given to the GA for optimization. A single population pool of the GA is maintained for a vehicle, which adapts to the changing scenario. This is an example of *incremental learning* in which the learnt model (GA population) has to be incrementally updated against the changing data (scenario). The changes may be gradual requiring the GA to incrementally learn by small changes to the model (GA population), or the changes may be large effectively requiring reworking of the complete model. Enabling optimization to be carried out in real time involves a variety of challenges, which are handled by various methods in this chapter. As the solution is designed for a specific application, deterministically adaptive strategies can be framed for each of the challenges.

Traffic rules of everyday driving can in fact play a major role in *coordinating* the motion of multiple vehicles in general scenarios. Rules, such as driving on the left- (or right-hand side of the road, and overtaking on the right (or left)), play a major role in enabling drivers to plan their motion amidst multiple vehicles. Although tackling the complete problem of vehicle coordination would be extremely large, it is observed that embedding these rules as heuristics can play a major role in realizing an overall efficient strategy. No off-the-shelf solution to coordination is available from the mobile robotics literature which can guarantee optimality, cooperation and short computational times, as those solutions do not use these traffic heuristics.

The problem considered here consists of motion planning of N robotic vehicles. Each robotic vehicle R_i has a predefined source S_i and a predefined goal G_i. It is further assumed that the specification of all the roads, in terms of their start, end and connectivity with other roads in the geographical map, is known a priori. The planning must ensure that each vehicle reaches its destination, and that there is no collision with any other vehicle along the way. The road may have static obstacles which can all be sensed as and when they appear.

For simplicity, it is assumed that all vehicles are rectangular in shape with length l_i and width w_i. Every vehicle starts from its source S_i at a time T_i and reaches its goal G_i. The vehicle is assumed to disappear after its journey has been completed. At any time t during its journey, the vehicle is at a position X_i (x_i, y_i) measured in terms of global coordinates, moving with a speed of v_i in a direction of θ_i.

The approach developed here for solving the multivehicle coordination problem (see Fig. 6.7) consists of Dijkstra's algorithm for path selection; this enables the vehicle to reach its goal from the source. The algorithm works over the road map, which is known for any city with all the roads and intersections well identified. This step is common to all algorithms presented in this book and has already been explained in chapter 'Introduction to Planning'. The algorithm can also detect road blockages which lead the vehicle to replan, again using Dijkstra's algorithm. As a result, the coarser path always points to the valid path at the coarser level being tracked by the vehicle. The next level of planning consists of Bézier -curve planning, which is optimized by a GA (Section 6.4.1). The GA uses a set of genetic operators (Section 6.4.2) to compute the optimal path measured by the fitness function (Section 6.4.3). This level deals with uncertainties associated with the relative positions and speeds of the other vehicles and road boundaries. The output of the planner at any time step is the Bézier curve or a trajectory, which is used to move the vehicle. The source of the finer-level planning is always the current vehicle position, whereas the goal is the next distant crossing in the coarser path. As the vehicle approaches its current goal, so the goal is updated to the next crossing. In this manner, the coarser path constantly guides the finer-level planning by making it reach one crossing after another up to the point at which the vehicle's final destination is reached.

Planning is performed independently for all vehicles. Coordination dictates the manner in which the vehicles avoid each other, whereas each vehicle is planned using finer-level planning. The developed coordination strategy (Section 6.5) states that a

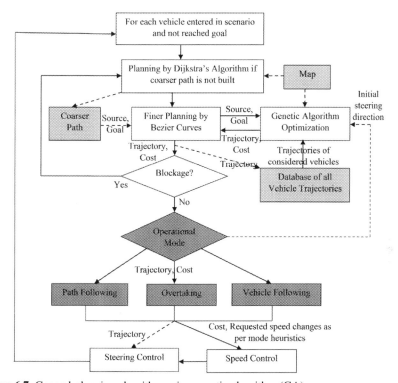

Figure 6.7 General planning algorithm using genetic algorithm (GA).

vehicle only considers the vehicles ahead in its planning which is in consensus with natural human driving wherein one largely considers the prospective motions of the vehicles in front to make one's navigation plan. The vehicle trajectories are represented as a space–time graph. While planning, a vehicle to the rear treats all vehicles ahead as dynamic obstacles the dynamics of which are known. Every vehicle attempts to move by its highest possible speed (Section 6.5.1).

The resultant coordination strategy cannot display cooperative overtaking, which is handled by plugging in an additional measure in the coordination strategy, according to which a vehicle may ask another vehicle to make some initial turn or alter its speed to better suite its plan. The initial turn is handled by the GA, whereas speed is handled by the speed control module. Two heuristic strategies are designed that make use of this measure, namely overtaking (Section 6.5.2) and vehicle following (Section 6.5.3). Both these strategies assess the current scenario and heuristically compute a speed and steering assignment for the different vehicles. The general traversal in the absence of a vehicle ahead is by the path-following strategy.

The key features of the approach are summarized in Box 6.2. The overall approach is also summered in Box 6.3.

First, Dijkstra's algorithm is used to compute the route of the vehicle. For any vehicle R_i, the output is a set of crossings (or vertices) that it must traverse in strict order. Hence, the path of R_i becomes $S_i \rightarrow P_i^1 \rightarrow P_i^2 \rightarrow P_i^3 \rightarrow \ldots \rightarrow P_i^v \rightarrow G_i$. The optimality

---

Box 6.2 Takeaways of Solution Using Genetic Algorithms

- The design of a GA which gives results within *short computational times* for traffic scenarios.
- Employment of the developed GA for *constant path adaptation* to overcome actuation uncertainties. The GA assesses the current scenario and takes the best measures for rapid trajectory generation.
- The use of *traffic rules* as heuristics to coordinate between vehicles.
- The use of heuristics for constant adaptation of the plan to *favour overtaking*, once initiated, but to cancel it whenever infeasible.
- The approach is tested for a number of diverse behaviours including obstacle avoidance, blockage, overtaking and vehicle following.

Box 6.3 Key Aspects of Solution Using Genetic Algorithms

- Use *road coordinate axis system* for better sampling
- Optimize *as the vehicle moves*
  - Tune plan
  - Overcome uncertainties
  - Compute feasibility of overtake
- Individual representation
  - Variable number of turning points with lateral and longitudinal component
  - Source, first directional maintenance point and goal are fixed
  - Points always *sorted* as per lateral axis
  - All points *longitudinally ahead* of the current position of the vehicle
  - Trajectory is the *Bézier curve* from these points
- Genetic operators
  - *Repair*, sort and delete points behind current position, delete excess points to shorten/smoothen the path.
  - *Insert* random individuals
  - *Crossover*, suited for variable-length chromosomes
  - *Mutation*, randomly deviate points
- Fitness function
  - Trajectory length
  - Length of infeasible trajectory
  - Length of trajectory without safety distance

---

of Dijkstra's algorithm is based on the assumption that different roads have similar traffic. The road network consists of a large number of vehicles in all, and jointly analysing all of them for the overall optimal plan is not possible in real time due to the amount of necessary computation.

As the environment is dynamic, the computed path may, at any instance of time, be found to be *blocked*. The algorithm must, in this case, take alternative means to act in

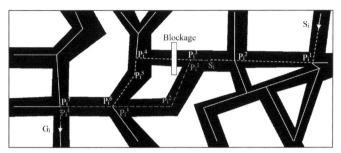

Figure 6.8 Road map with blockage. Each point of intersection of two lines (of paths) is a vertex. The shortest path from source to goal is shown as a dotted line. The list of vertices is returned by Dijkstra's algorithm (S_i, P_i^1 to P_i^7, G_i) shown in white. The path may change upon detection of a blockage. The new path is shown in grey.

response to such blockages. This process consists of blockage detection and replanning. In this algorithm, it is assumed that if the next level of planning algorithm failed to generate feasible solutions for consecutive *block* iterations, then the path is regarded blocked. In such a scenario the current position of the vehicle is added as the new source S_i as a new vertex to the road map, the goal is left unchanged. The corresponding connections and weights are updated, to account for the blockage. Dijkstra's algorithm is called again to return a new path. The movements now take place as per this path, as is shown in Fig. 6.8. In the case that no feasible path is possible, the journey of the vehicle is terminated.

The next (finer) level of the planner is carried out by the use of *Bézier curves* (Bartels et al., 1998). The purpose of the planning algorithm is to generate a path which the vehicle may follow using its own control mechanism. The generated path needs to be as smooth as possible, so that the vehicle may be controlled maintaining high speeds as per the nonholonomic constraints. The other purpose in the use of Bézier curves is to enable the vehicle to make efficient turns in all scenarios of crossing, general road turn or while overtaking. This is shown in Fig. 6.9.

The GA planning algorithm needs its own source and goal for planning. The source is always the current position of the vehicle (X_i with the vehicle oriented at an angle θ_i). At every time step, the planner generates a trajectory and the vehicle moves as per the same. At the next time step, the changed position becomes the source for the planning. The goal is fixed as the next crossing P_l^j, that is at least η units apart from the current position of the vehicle X_i. As the vehicle continues its journey using this plan, as soon as it comes close enough to the directed goal P_l^j, the goal is changed to the next position in the vehicle's path P_l^{j+1}. However, this change is not performed in the case when P_l^j is the final goal (Kala et al., 2010a). P_i^k always denotes the crossing just behind the vehicle.

6.4.1 Individual Representation

A GA is used to generate an optimal Bézier curve for motion. One of the major tasks in the use of the GA is to devise an *individual representation strategy*. The Bézier curves

Figure 6.9 Path generated by the vehicle in multiple scenarios using GA. A vehicle at every instant in time generates a feasible path as per the given map, considering the other moving vehicles (not shown in the figure). (A) Scenario 1, (B) Scenario 2.

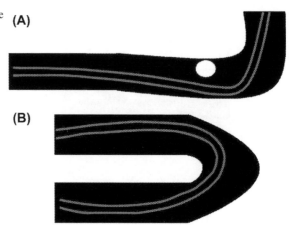

may be easily generated by means of *control points* in the map. For a better solution in the generation of feasible paths, the road axis system is taken.

The complete Bézier curve path specification (or genetic phenotype) consists of (in order) the source X_i, direction maintenance points, a variable number of additional Bézier control points (subjected to a maximum of P^{max}) and the goal. The directional maintenance points are added to assure that the Bézier curve generated is initially inclined in the direction the vehicle is currently facing or θ_i. Out of all these points constituting the Bézier curve, only the Bézier control points are nonfixed which are hence the points optimized by the GA. All the control points are placed one after the other to form the genotype. Because different paths may have a different number of control points, the genotype is of *variable length*. Each point is represented by two numbers representing values across the two axes in the road coordinate system. The bounds for each gene must be known for the optimization process. Every second gene, representing the X-axis component of the control points, can vary from a minimum of 0 to $\sum_{t=k}^{j-1} \left\| P_i^{t+1} - P_i^t \right\|$. Every other gene, representing the Y-axis component of the control points can vary from 0 to 1, in which 0 points to the right boundary and 1 points to the left boundary. This ensures that the GA produces points inside the road. The mapping between the genotype and the phenotype is given by Fig. 6.10.

It is assumed here that all the points are *sorted* as per the values corresponding to the X-axis. This is based on the fact that a vehicle always goes ahead in the direction of the road. Although it may steer by small or large amounts, it never steers large enough to face backwards on the road. Hence, in a valid curve, a latter point always lies ahead in the road from an earlier point. Further, it is assumed that no point lies *behind* the current position of the vehicle X_i, as per the road coordinate system. This is because the vehicle may have crossed the points behind and these are not needed in the curve because the vehicle would not turn back to touch these points.

The road coordinate axis system used for the generation of individuals enables sampling points that lie within road boundaries, unlike the Cartesian coordinate axis system in which the road may be a portion of the entire map, and hence only some of the

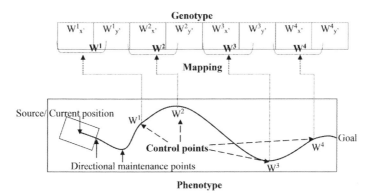

Figure 6.10 Individual representation strategy. Directional maintenance points generate a curve in the heading direction of the vehicle. Bézier control points (variable in number) constitute the genetic genotype, to be optimized by the GA.

points sampled are within road boundaries. In other words, the probability of generation of an obstacle-free point is much higher in the road coordinate axis system. Further, as a result of sorting, any path (in the road coordinate axis system) consists of small path segments between any two points which increase the probability of it being feasible; as compared to a randomly produced path in the Cartesian coordinate axis system wherein two adjacent points would be fairly far apart. By sorting, a number of genotypic representations correspond to a single phenotypic path. This makes the work of the GA easier. A higher probability of generation of a feasible path means the algorithm quickly comes up with a feasible plan which may be optimized in later iterations.

6.4.2 Genetic Operators

Evolution in the GA is performed by a variety of genetic operators. The first operator used is *repair*. This operator ensures that the points are sorted as per the values of the X-axis in road coordinate system, all points lie ahead of the current position of the vehicle, and that the individual being considered contains the smallest number of control points. For sorting, the genotype is converted into an array of points consisting of X and Y coordinates. The array is sorted as per the X component. All points in the array behind the current position are deleted, which signify the points that the vehicle has crossed and are no longer needed in the curve representation. The control points are then deleted one by one, until no further deletion results in a better path in terms of its fitness value. This is an implementation of the *shorten* operator. The deletion of points from the main path of the vehicle may well yield a better and shorter path (Xiao et al., 1997). The operation may be time-consuming for most newly generated or random individuals; however, as the algorithm proceeds, with minor or no change in the scenario, the optimized individuals are already short and hence need not result in much consumed time. The operator is shown in Fig. 6.11A.

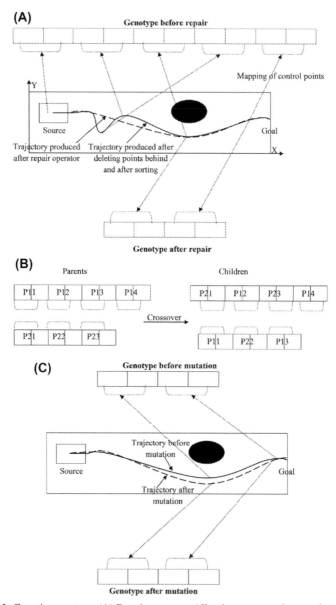

Figure 6.11 Genetic operators. (A) Repair operator. All points are sorted as per the values of the X-axis, points behind the vehicle are deleted and control points are deleted till fitness improves. (B) Crossover operator. Because parents are of different length, children are made of mean lengths with random genes exchanged. Symbols denote which gene from parents goes to which child. (C) Mutation operator. Points are moved by a maximum magnitude of 'mutation rate' picked from a Gaussian distribution.

Every deletion of a control point from the path makes the resultant path shorter and smoother. The initial set of points may contain unnecessary twists and turns. However, after this smoothing operation it is ensured that every turn results in the vehicle avoiding an obstacle or remaining within a road boundary. Hence, given a good location of control points, the resultant paths are locally shortest and smoothest. It is assumed that the map does not contain a range of closely packed obstacles such that a generated smoothened path might not be navigable physically by the vehicle. In cases in which many objects occur, the vehicle may initially fail to generate a feasible path, reducing its speed as a result until it is possible to do so.

The second operator used is *insert*, in which random individuals of the highest possible length are generated. These individuals obtain a desirable shape after the application of the repair operator. It is possible that due to scenario changes a trajectory which once looked poor may now be desirable. A converged GA population may not have enough exploratory potential to search for such new and interesting solutions. Hence, random individuals are added to the population pool. The next operator used is the standard *crossover* which generates children by exchanging the characteristics of the parents. Because the number of points (or genes) is variable, the first task is computation of the number of points in the two children (Kala et al., 2010b). The two children have $ceil((n_1 + n_2)/2)$ and $floor((n_1 + n_2)/2)$ points, in which n_1 and n_2 are the number of points in the two parents, *ceil* rounds up to the nearest integer and *floor* rounds down to the nearest integer. These points are randomly taken from the parents, as per the scattered crossover technique. For each vacant point in the first child, a random decision is made whether to take the point from the first parent or the second parent. The corresponding point from the other parent is given to the second child, if vacant positions are available. Leftover points may be used to fill in vacant positions in the children. The operator is shown in Fig. 6.11B.

The next operator used is *mutation*, which changes the value of some point per the mutation rate. In other words, the operator moves the point by some small magnitude over the road to affect the trajectory. The total percentage change that the operator may make is determined by the mutation rate, which is taken from a Gaussian distribution. The relevant operator is shown in Fig. 6.11C. The final operator used is *elite*, which transfers the best individuals of one generation to the next generation. This operator ensures that the best solution found so far is not killed by the other operators, but is retained till the end and can be used for moving the vehicle. Stochastic universal sampling is used to select the individuals for the different operators. Rank-based scaling is used in which the different individuals are assigned weights proportional to their rank in the population pool. The initial population is generated randomly.

Ideally, the optimal individual of a generation is still the optimal individual even after motion of the vehicle as per the travel plan, unless there is a change in P_i^k or P_l^j. This is because the Bézier curve remains the same with the start point shifted to the current position of the vehicle, as the vehicle moves. If the vehicle has crossed a control point, it would be naturally deleted by the repair operator. In case the vehicle physically crosses P_i^k or P_l^j (in other words, when it physically crosses a crossing or there is a change of goal), there is a change of the coordinate system across the movement. Hence, the complete population is reinitialized.

6.4.3   Path Fitness Evaluation

The last part to be considered in the implementation of the GA is the fitness function. Each vehicle cannot be allowed to approach too closely to an obstacle (which may be another vehicle). This allows scope for measurement error as well as enabling a vehicle to make an emergency stop, if necessary, to avoid a collision. Hence, each vehicle must always keep a *clear minimum distance* of d_1 from its front and d_2 from its side. Here, the minimum front distance d_1 and side distance d_2 are given by Eqs [6.5] and [6.6]

$$d_1 = c_1 v_i^2 \tag{6.5}$$

$$d_2 = c_2 v_i^2 \tag{6.6}$$

in which c_1 and c_2 are constants.

The fitness function is taken to be the *length* of the path. However, at the same time a penalty is added for infeasible paths. The penalty is proportional to the *length of the segments of the infeasible paths* which lie inside obstacles. This creates evolutionary pressures for the GA to minimize the length of the infeasible path up to the point when the infeasible path has been completely eliminated (if possible). The intent is always, of course, to avoid the obstacles, whilst maintaining the *minimum safety distance*. The resultant overall fitness function may therefore be given by Eq. [6.7].

$$\text{Fit} = l + \alpha l_1 + \beta l_2 \tag{6.7}$$

Here, l is the total length of the Bézier curve, l_1 is the length of an infeasible segment, and l_2 is the length of the path segment in which the minimum safety distance is violated. $\alpha > \beta > 1$. The GA first tries to reduce the length of the infeasible path. If and once a feasible trajectory is obtained, the GA optimizes to reduce the length of the trajectory without the needed safety distance. Then the focus is to minimize the path length.

It is evident that the fitness function of the GA needs to optimize a number of factors considering the vehicle dynamics such as journey time, vehicle speed, path smoothness, distance from obstacles, sharpest turn etc. However, unlike planning for mobile robots, roads are generally smooth enough which in itself practically ensures that any general path smoothing strategy leads to smooth curves. On top of this, vehicles are restricted to low speed as per the traffic laws in force, which means that the allowable speed considering nonholonomic constraints is much higher than the actual speed of travel of the vehicle. For the same reasons there is, in reality, no difference between minimizing time, minimizing path length or increasing smoothness. The safety distance feasibility measure is similar to the normally encountered factor ensuring a minimum distance from obstacles in mobile robotics.

The factors of safety distance (and feasibility) and path length are clearly contradictory. Shorter paths are possible by violating the safety criterion and keeping as many points in the curve as possible very close to the boundary. Unlike other research in this area, this factor is specifically added rather than checking for nonholonomic

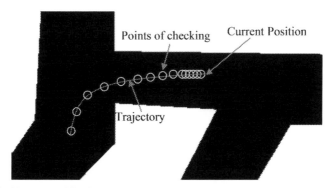

Figure 6.12 Checking of feasibility and distance computation by sampling points. From the entire trajectory generated during planning, only some points are used for feasibility checking and distance computation. These points are closely spaced near the source (current point of the vehicle) and more distance apart later.

constraints or smoothness, and allowing the vehicle to go close to the boundary or obstacles. This is done to account for sensor and actuation uncertainties.

Computation of the feasibility and path length, by point-to-point iteration of the Bézier curve, may be a very costly procedure, and the results of the GA might not be realized in real time. Hence, the computation is performed at points *interleaved by some minimum separation*, in which the separation decreases along the curve. This means that in the initial segment of the curve, each point is checked and, further still, lesser points are checked. The concept is shown in Fig. 6.12.

6.5 Coordination

In decentralized planning, every vehicle is planned separately. Two vehicles, planned separately, may thus have conflicting trajectories if they do not consider each other in their planning. Both vehicles cannot consider each other's trajectories in their planning, because the trajectory computation of the first vehicle is dependent on the availability of the trajectory of the other vehicle and vice versa. The problem is more complicated with the presence of multiple vehicles, wherein any set of vehicles may have conflicting trajectories, whereas the alternative trajectories may be further conflicting. *Coordination* deals with the manner in which such conflicts are handled, or the manner in which vehicles avoid each other; however, this only occurs while each vehicle is planned separately in a decentralized manner.

In addition to the static obstacles, the vehicle needs to ensure that it does not collide with any of the other vehicles. This is done by a space−time-graph approach. Each vehicle maintains a copy of the current plan being followed in the form of a hash map of vehicle positions, hashed with time. Every vehicle considers only the vehicles that are *ahead* of it at the current time, although they can be moving in any direction. Hence, this is an implementation of a priority-based system, in which the priorities are founded on the position of the vehicles. This is (once again) inspired from natural driving, in which a human driver mainly considers the front view while making their trajectory plan. The purpose of the algorithm is to use the *traffic heuristic* (of overtaking and vehicle following as is presented in Sections 6.5.2 and 6.5.3)-inspired coordination strategy rather than a coordination strategy which jointly optimizes the trajectories of all the vehicles. A traffic inspired heuristic is capable of quickly generating feasible and near-optimal paths.

Coordination dictates the manner in which any two vehicles avoid each other. As per the designed heuristic strategy, a rear vehicle is responsible for avoiding a collision with the vehicle in front. This strategy is, however, noncooperative, and there is no incentive for a front vehicle to aid a rear vehicle in getting a better path. Hence, an additional clause is added that a rear vehicle may request a front vehicle to veer to a particular side, if and by whatsoever amount possible. Whether the request is to be made and, if yes, in what direction depends upon the *overtaking* and *vehicle-following heuristic* discussed in Sections 6.5.2 and 6.5.3. The magnitude by which the vehicle in front veers upon being requested denotes the amount of cooperation which is an algorithm parameter. Further, a vehicle to the rear may request a vehicle in front to alter its speed by some amount, which again depends upon the overtaking and vehicle-following heuristic.

It is expected that after a short time, paths of vehicles will change due to optimization, emergence of other vehicles, obstacles etc. Hence, there is no point in making too far-sighted plans. In natural human driving the focus is on avoiding collisions with nearby vehicles and not avoiding vehicles which are a long way off. Therefore, in this system, a vehicle need not consider collisions with vehicles which are recorded after a large span of time. If a vehicle, faced by a possible collision, is already inside the region where the unsafe distance threshold is breached, slowing of the vehicle will take place.

The algorithm employed here uses a form of communication in which one vehicle can communicate with another vehicle to update its planned path. Initially, when a new vehicle comes into the scenario, it is far from any of the other vehicles and hence even if communication takes a reasonable time to be established, it is acceptable. Once communication is established however, each vehicle knows each other's plans and makes necessary corrections so that an overall feasible travel plan is generated. Plans then only need to be recommunicated when the path actually travelled by a vehicle largely deviates from the one originally communicated at the time of generation of the plan. Hence, the algorithm can handle communication errors to some extent.

The basics of coordination are summarized in Box 6.4. Box 6.5 specifically mentions the key concepts associated with the overtaking and vehicle-following traffic heuristics.

6.5.1 Determining Speed

For a single vehicle operating in a static environment, its trajectory can be computed and traced at any desirable speed without affecting the feasibility. However, operational speed is important to consider for the case of multiple vehicles. The relative

---

Box 6.4 Key Aspects of Coordination Strategy

- Priority-based coordination
- Only vehicles *ahead* considered
- Cyclic path and speed optimization
  - *Path optimization* by GA
  - *Speed optimization*, increase speed by δ if path feasible, decrease speed by δ if path infeasible
- Cooperation added by *traffic heuristics*
  - Overtake
  - Vehicle following
- Vehicle can *request* another vehicle to cooperate by
  - Slowing down
  - Turning right/left
  - Decision directed by traffic heuristics

Box 6.5 Key Aspects of Traffic Heuristics

- Assess situation for overtaking or vehicle-following behaviours
- If overtaking
  - Give initial turns to the other vehicles to best overtake
  - Alter speeds of the other vehicles to best overtake
  - Cancel overtake if it seems dangerous
- If following a vehicle
  - Give initial turns to the other vehicles to best overtake in the future
  - Alter speeds of the other vehicles to best overtake in the future
  - Initiate overtake if it seems possible

---

speed of two vehicles determines the coordination or the manner in which the vehicles avoid each other. An optimal coordination plan in the case of multiple robots deals with the optimal planning of the trajectory as well as the optimal planning of speed. The proposed method uses a *path−velocity decomposition* (Kant and Zucker, 1986) scheme wherein path components are optimized by the trajectory planner which works with the assumption that the vehicle would continue to travel with its current constant speed. Considering the needs of the online nature of the algorithm, the velocity component is optimized by a simple mechanism which attempts to assess the scenario and assign the best current speed of the vehicle as the vehicle moves. The attempt is to assign the maximum speed possible to the vehicle, so long as the increased speed does not result in a collision or loss of safety distance.

A simple heuristic rule is used to set the instantaneous speed of the vehicle. At any time the speed of the vehicle may *either be increased by a magnitude of δ or decreased by a magnitude of δ* (Sewall et al., 2011). The speed is always subject to a maximum of

v_i^{max} which is a property of the vehicle. In the case when the planning routine indicates that the path planned appears to be without any collision and safe distances can be maintained, an attempt is made to increase the vehicle's velocity. However, if the planned trajectory suggests that a collision is likely or that minimum safe distances cannot be achieved, the velocity is decreased. The resultant velocity is hence given by Eq. [6.8].

$$v_i(t + \Delta t) = \begin{cases} \min\left(v_i(t) + \delta, v_i^{\text{max}}\right) & \text{if no penalty} \\ \max(v_i(t) - \delta, 0) & \text{if penalty} \end{cases} \qquad [6.8]$$

Considering that the sampled time Δt and the speed interval δ are small, and keeping the path being traced as fixed, it should be possible for a vehicle to quickly obtain the highest possible speed to trace the path. On reaching the highest speed, any attempt to further increase the speed would be turned down as it would make a penalty-free path penalty prone. At the same time, because the current path is without penalty, no decrease in speed would take place. Hence, for a constant path, the vehicle can obtain and maintain its highest speed possible. In reality, as the vehicle speed changes, the algorithm would also modify the path which may better suit the altered speed. For unit iteration, there is a small change in speed which causes a small change in path. Hence, as the vehicle moves, altering speed and path adjustments would be applied leading to convergence which gives a near-optimal plan for navigation.

Having a collision-free near-optimal navigation plan should normally mean that the vehicles stick to it and navigate it using their control mechanisms. However, uncertainties need to be handled. The GA handles uncertainties by maintaining enough separation for the vehicle, at all times, from obstacles, other vehicles and road boundaries. However, actuation errors or uncertainties may make the vehicle lie at a position somewhat different from that expected. Such errors can grow with time. These sensing errors may, however, be corrected with time, as the vehicle approaches obstacles or other vehicles. Hence, the path and speed are constantly adapted to cope with these uncertainties, to make the overall path near optimal and collision free. The experimented results provided are in fact obtained using a lazy control mechanism which is intentionally used to produce large errors. An important reason for using the GA was in fact to overcome these errors. The decision to overtake a vehicle or to follow it instead can be altered based on these uncertainties, which is done outside the GA by a separate decision-making module.

6.5.2 Overtaking

A slower vehicle lying ahead of a faster vehicle, occupying some lateral coverage of the road in use by the faster vehicle behind, is too restrictive for the faster vehicle, forcing it to drive at a lower speed to avoid an accident. *Overtaking* deals with first making the two vehicles lie on different lateral coverage of the road such that the slower vehicle does not block the faster vehicle. It would then be expected that the faster vehicle soon lies completely ahead of the slower vehicle, post which the two vehicles may occupy more favourable lateral positions on the road.

One of the major issues associated with the algorithm is to enable faster vehicles to overtake slower vehicles. As per natural human driving, whenever feasible, a faster vehicle may well overtake a slower vehicle. This makes the travel plan of the faster vehicle more efficient. This section (and Section 6.5.3) gives details on the *heuristics* which lead to the coordination between vehicles. In the algorithm these heuristics are in the form of a vehicle requesting another vehicle to turn, or in the form of a vehicle requesting the other vehicle to slow down. The heuristics are completely derived on the basis of generally observable driving in the absence of speed lanes for comfortable to risky scenarios. The heuristic is based on the philosophy that vehicles need to cooperate, as far as possible, to allow any possible overtake. A key notion here is that not only are the plans monitored and constantly adapted to enable the completion of a successful overtake, the *decision to overtake* itself is monitored and can be altered on sensing a potential threat and reinitiated later.

Initially, the faster vehicle is situated behind the slower vehicle when the overtake procedure starts. The overtake procedure is different for the case when no other vehicle (apart from the vehicle overtaking and the vehicle being overtaken) is present, as opposed to the case when there is an additional vehicle in the overtaking zone.

The first case involves an attempt to overtake by a faster vehicle R_1 which is initially behind a slower vehicle R_2 $\left(v_2^{max} < v_1^{max}\right)$. The first task associated with the overtaking procedure is to move the vehicle R_2 leftwards (assuming a driving on the left with overtaking on the right traffic rule) if necessary, in the case when it is in front of R_1. This is important as R_2 does not consider R_1 in its trajectory planning, as R_1 initially lies at the rear of R_2. This move is issued as a *request broadcast* from R_1 to R_2. As an analogy, the process is similar to blowing a horn, indicating an attempt to overtake. The vehicle R_2 considers the request and attempts to move to the left only in the case that the generated path for it to do so is feasible, taking into account possible collisions and minimum safety distances. The opposite rightwards motion of the vehicle R_1 is as per the computation of its trajectory. As long as R_2 moves leftwards, there is more scope for R_1 to overtake, and its trajectory keeps improving with time. The leftwards indicative movements are appended as extra control points in the direction maintenance points in the Bézier-based planning. Eventually, R_1 and R_2 are reasonably far apart when abreast of each other. In a very short time R_1 is able to evolve a trajectory, considering the trajectory of R_2, such that no collision occurs as well as the overtaking procedure taking place.

Ideally, the two vehicles travel as per their own planned trajectories, and the overtaking procedure is completed successfully. However, in this course of time, there may be uncertainties in vehicle control, making the planned trajectories prone to collision. Whenever R_2 detects a possible collision or a minimum safety distance is not maintained, it slows down. This gives greater scope for R_1 to proceed with the overtaking. However, when R_1 detects the absence of a minimum safe distance, it may not slow down, but instead it asks R_2 to slow down. This ensures that R_1 does not lag behind R_2 in the overtaking process. It may again be noted that slowing of R_2 reduces the safe distance which needs to be maintained, which may again contribute towards making overtaking feasible. Many times, whilst overtaking is being carried out, it may be possible that, due to large uncertainties, which may be attributed to a poor control mechanism, overtaking is completely infeasible. Hence, while overtaking, if R_1 detects that there is a collision (excluding the safe distances) for a few consecutive time steps, it abandons the overtaking procedure. In such a case it proceeds to slow down.

The overtaking mechanism also appears to work well in the scenario of multiple vehicles. However, the procedure may involve the cooperation of other vehicles as well, ie, those not directly involved but which would be affected by the overtaking in terms of disruption to their own trajectory. Here, the first task is to decide whether the vehicle should overtake or not. Suppose that the faster vehicle behind is R_1 and needs to overtake the slower vehicle R_2, with the vehicle R_3 ahead and approaching from the opposite direction. Here, R_3 may in fact be one or many vehicles of the same kind.

Overtaking, at any time of the journey, is regarded as possible if R_1 can draw a feasible trajectory without colliding with R_2 or R_3 as well as maintaining minimum safe distances from them (assuming that R_2 and R_3 already had collision-free trajectories). In such a case, if R_2 or R_3 lie in front of R_1, it would request both of them to move to their left-hand side, respectively, to allow the overtaking procedure of R_2 to occur. Each of these would, however, move as requested only if their generated paths are feasible. The motion of R_1 would then be guided by the planned trajectory of the Bézier curve. As the two vehicles order themselves at their respective left sides, the path of R_1 starts to get both shorter and easier.

As was the case for two vehicles though, it may not ultimately be possible for all three vehicles to navigate as per their respective individual plans and complete the overtaking procedure without a collision. In such a case, a number of actions are required if the overtaking procedure is to go ahead safely. Firstly, vehicles R_2 and R_3 need to slow down when the possible collision is detected. This gives ample time for R_1 to comfortably overtake. Vehicle R_1, on seeing a safety distance not being maintained, may ask R_2 or R_3 to slow down further, dependent on which vehicle it is likely to collide with. However, vehicle R_1 may itself also need to slow down if a collision is detected. If R_2 or R_3 do not cooperate and take plans contrary to the ones desired to safely complete an overtaking procedure, R_1 would soon come into a collision-prone situation and would need to abandon the overtake. The complete overtaking mechanism is shown in Fig. 6.13.

From an analytical perspective, overtaking starts after the vehicles have found a unanimous travel plan by which no collision occurs with wide-enough separations during overtaking. The larger the separation, so the more likely it is that overtaking can be actually carried out. Overtaking can be cancelled at any time if the vehicle has a separation which is not large enough for it to fit in. Once cancelled, the reinitiation of an overtaking procedure may take place at any time when a feasible plan has been constructed. If a vehicle is moved such that sufficiently wide separation is available at all times, overtaking may go ahead with the same speed profiles of vehicles. However, when a vehicle moves, if its separation is not wide enough for it, an attempt is made to increase the separation to the desired threshold by modifying the speed profiles of vehicles. Slowing other vehicles down is usually the best alternative.

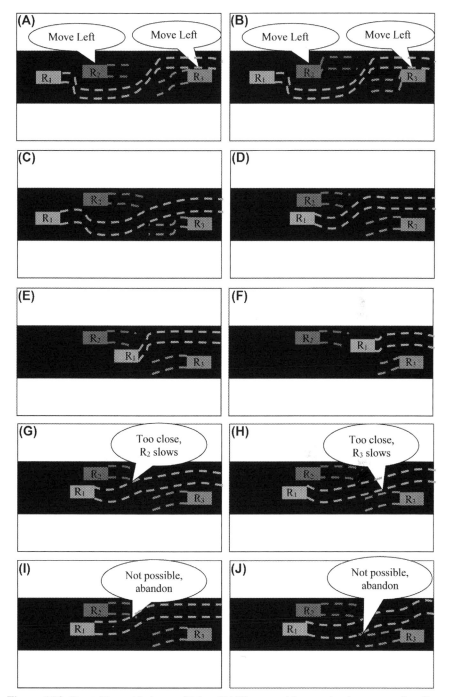

Figure 6.13 Overtaking with three vehicles. (A) The overtaking mechanism starts when R_1 foresees the possibility of overtaking. This is when it is able to construct a feasible and safe trajectory of its motion such that no collision occurs, whilst considering the feasible trajectories of other vehicles. (B) Because R_2 and R_3 are ahead of R_1, these are asked to move in their respective leftwards directions, implemented by giving a leftwards turn to their Bézier curves. (C) R_1 makes a smooth and short trajectory to overtake which is followed by (D) an overtaking manoeuvre. In (E), R_1 has completely overtaken R_2 and in (F) it avoids R_3. (G) to (J) show potential problems which can stop the overtaking procedure from occurring.

6.5.3 Vehicle Following

On many occasions, even though a vehicle to the rear of another may be both capable of and desire a higher speed than the vehicle in front of it, overtaking may not be possible. This can be due to a vehicle in the opposite direction approaching an overtaking point which must be avoided during that specific overtaking action. Conversely, in some scenarios the road may not be broad enough to accommodate vehicle overtaking. In such a case the vehicle simply needs to *slow down and follow* the vehicle ahead. *Vehicle following* is a common traffic behaviour wherein one vehicle constantly adapts its speed to always maintain a safe separation from the vehicle in front, whereas the vehicle may prefer to steer to be roughly behind the vehicle in front. As a result, the two vehicles seem to be following at roughly the same speed with roughly the same lateral positions as long as the vehicle-following behaviour is displayed.

Because the vehicle cannot generate any feasible trajectory at its current speed, it is forced to slow down (ie, if it does not, it will collide with the vehicle in front of it). It keeps slowing in this manner till it generates a feasible trajectory. This feasible trajectory may mostly be available when the vehicle is slow enough not to collide with the vehicle in front. This would happen when the vehicle to the rear is travelling at a speed which is equal to or even slower than the vehicle in front. The vehicle may accelerate at times, thereby increasing its speed; however, when this makes the distance of separation lower than the safe distance, the vehicle would again need to decelerate. Hence, the average speed of the two vehicles would be approximately the same, whilst the scenario remains. If road conditions change, then the situation may change as well.

In such a case, even whilst it is still following the vehicle in front, the vehicle to the rear must be as ready as is possible for overtaking. This means a number of things: it must attempt to push the vehicle being followed as far to its left as possible, any oncoming vehicle as far to its left as possible, to position itself somewhere in the middle, and to start to accelerate ready for the a priori known change in road scenario. This would enable the vehicle to immediately overtake, as soon as it is possible. This mode of operation is summarized in Fig. 6.14.

6.6 Results

The entire planning algorithm was simulated in a custom MATLAB design. The complete scenario was read as an image file drawn using a paint utility tool. Scenario specifications required the number of vehicles, their entry times, entry locations, maximum and initial speeds, acceleration limits etc. All these were read from the scenario specification module. The testing methodology was to individually test each of the features displayed in the algorithm. Accordingly, scenarios were set up which were challenging and exploited the capability of the module under test. First, the algorithm was tested under a number of scenarios, and then analysis of the vehicle behaviour and the algorithm parameters was carried out, which conveys a greater insight into the algorithm's capabilities.

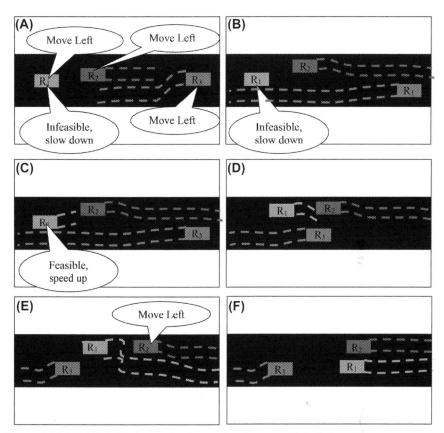

Figure 6.14 Vehicle-following mode of operation. In case vehicle R_1 is unable to compute any feasible trajectory to overtake, it must enter into vehicle-following mode. (A) R_2 and R_3 are requested to move leftwards as they are directly ahead. (B) Vehicle R_1 needs to slow until it is able to compute a feasible trajectory, which would normally mean until it is slow enough to follow R_2. (C) Generation of a feasible trajectory results in an attempt to accelerate. R_1 goes through cycles represented in (B) and (C). R_3 is soon about to pass. (D) R_1 and R_2 move as per the planned trajectories. Once R_3 has passed, R_1 gets ready to overtake R_2 as discussed in Fig. 6.13. (E) First, R_2 is asked to move as far left as possible, whereas R_1 tries to generate feasible and short trajectories. (F) Overtaking is eventually completed.

6.6.1 Higher-level Planning

The first experiment focussed on the higher-level planning of the algorithm. Although finding a route with a given road map is quite a trivial task, the experiment focussed on the ability of the vehicle to drive on the roads, which largely tested the integration between the higher-level planning and lower-level planning. To make it challenging for the lower-level planning to find safe trajectories as per the higher-level path, some obstacles were added on the road. The corresponding path traced by the vehicle is shown

(A)

(B)   Vehicle position at the
                                           time of blockage

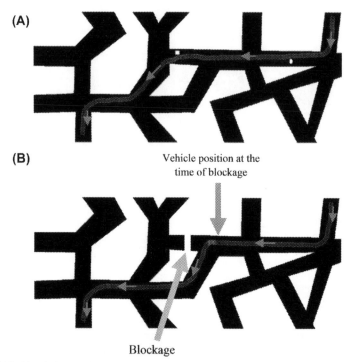

Blockage

Figure 6.15 Simulation results over a variety of scenarios using GA. A vehicle is made to move in a variety of maps. (A) Tests the ability of the vehicle to steer smoothly around turns and take optimal turns in lesser time, greater speed and smaller path length. (B) Tests the ability of the vehicle to detect blockage and replan the path.

in Fig. 6.15A. This result clearly shows that the vehicle was able to decide which strategy would be the best to avoid an obstacle. The presence of an obstacle though slowed the vehicle's speed a little, which is understandable.

The next feature of the algorithm considered was blockage detection which was tested via another similar experiment. The experiment tested the ability of the lower-level planning algorithm to detect the blockage, the higher-level planning algorithm to replan the path and the ability to make a smooth transition. The vehicle was successfully able to detect the blockage and replan accordingly in real time. A typical traced trajectory is shown in Fig. 6.15B. In this case, an alternative path was available which was used for navigation. It may be seen that there is a sharp change in the path of the vehicle after the blockage is confirmed by the algorithm. Although the algorithm generates infeasible paths, the vehicle's speed is constantly lowered which again helps in making the sharp turns required when a blockage is discovered.

6.6.2 Overtaking

One of the major features of the algorithm is overtaking, which was tested by the next set of experiments. First, some simple experiments are presented to highlight the various concepts used in the algorithm. Because overtaking is not done at road cross-ings, a straight-line road was employed as an example indicator. This may be inter-preted as a segment of the road in one of the previous scenarios. The first experiment was carried out with two vehicles approaching each other on either side of the road. It was seen that the two vehicles easily coordinated themselves on their left sides. The corresponding trajectory is shown in Fig. 6.16A. Initially, heuristics played a role, in which each vehicle attempted to shift the other vehicle more to its left — this is indicative of the ego type of behaviour exhibited by both vehicles. How-ever, the vehicles soon generated feasible trajectories, avoiding each other with a safe distance between.

The second case is one in which a slower vehicle was initially ahead of a faster vehicle. The vehicle to the rear decides to overtake. This case is shown in Fig. 6.16B—D. The trajectories follow the norms set in the design and it can be seen that the two vehicles are always separated well apart while the overtaking procedure occurs. The scenario was then reversed with the faster vehicle being ahead of the slower vehicle. In such a case an overtaking action will not take place. This was indeed observed by experimentation. The trajectories of the vehicles are shown in Fig. 6.16E. Because the vehicle to the rear had a slower speed, computation of its path was not affected by the vehicle ahead. Hence, it followed more or less exactly the same path as the vehicle in front of it.

Having only two vehicles, to some extent, eases the overtaking process. That said, the main ability of the algorithm to be tested is to be able to carry out a close overtaking

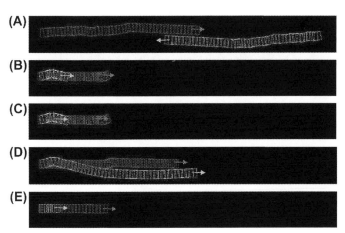

Figure 6.16 Simulation results for multiple vehicle scenarios using GA. Multiple vehicles are moved along a straight road which (A) tests the ability of two vehicles to avoid collision by moving to the left side of the road. The vehicle is further tested for its overtaking ability. In (B), a higher-speed vehicle finds itself behind a slower vehicle. It moves rightwards whereas the vehicle ahead moves leftwards. In (C), the two vehicles are fairly well apart and start moving along their computed paths. In (D), overtaking is regarded as complete, as the vehicle is fast and well ahead. In (E), a slower vehicle is behind a faster vehicle, in which both vehicles follow the same path with no collision possible.

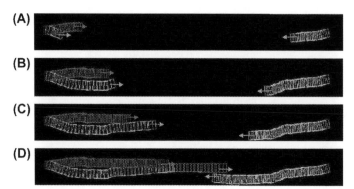

Figure 6.17 Simulation results for overtaking between three vehicles. A faster vehicle needs to overtake a slower vehicle while another vehicle is heading towards the overtaking zone. Overtaking is computed as feasible by the overtaking vehicle. In (A), the two vehicles attempt to move leftwards to give enough room for the overtaking vehicle, which moves rightwards. (B) The overtaking vehicle executes an overtaking trajectory. (C) Overtake of vehicle initially ahead is completed, and (D) finally, the oncoming vehicle is avoided.

procedure. This is investigated here by extending the situation to three vehicles, with the third vehicle approaching from the opposite direction. The oncoming vehicle therefore restricts overtaking, which must be completed before any potential collision with the oncoming vehicle. The speed of the oncoming vehicle is kept low enough to make overtaking feasible, but high enough to make overtaking close, which the algorithm must perform. The vehicles have errors associated with their movements, which must be overcome in the small overtaking gap available. The trajectories of this motion are shown in Fig. 6.17A–D. It can be seen that the feasibility was correctly computed, based on the time of overtake initiation. The initial overtaking trajectory appears similar to the case with only two vehicles; however, the challenge was to timely align the overtaking vehicle to avoid a collision with the oncoming vehicle. In this case, a sufficient margin of safety was available after the overtaking procedure.

6.6.3 Vehicle Following

Vehicle following is a complex behaviour which not only tests the ability of the algorithm to enable one vehicle to follow another vehicle, but also to be prepared for any potential overtaking procedure in the future. Balancing the two acts is a challenge which the algorithm must deal with appropriately. This is tested by an experiment similar to overtaking, with the difference that the speed of the vehicle is adjusted to make an overtaking action infeasible. The trajectories are shown in Fig. 6.18A–D. It may be seen that in this case the vehicle first followed the slower vehicle, primarily requiring adjustment (slowing) of speed and location of drive. The vehicle in this case also attempted to be as far to the right as possible. However, in case it came close to the oncoming vehicle, it also attempted to drift to the left — balancing the need to overtake

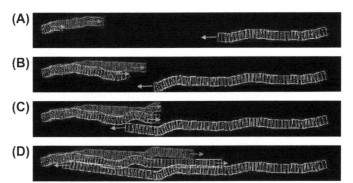

Figure 6.18 Simulation results for vehicle-following behaviour. A faster vehicle behind a slower vehicle is unable to generate a feasible overtaking trajectory due to an oncoming vehicle and hence needs to slow down and follow the vehicle in front. (A) The vehicle attempts to align itself in the middle of two vehicles, giving enough room for the oncoming vehicle to pass by. (B) The vehicle is expected to drift leftwards whereas the oncoming vehicle does the same to avoid collision. (C) The oncoming vehicle is completely avoided and overtaking may now take place, which is initiated. The vehicle moves rightwards to execute an overtaking trajectory. The other vehicle cooperates by moving leftwards. (D) Overtaking is completed.

with that of avoiding the oncoming vehicle. As soon as the oncoming vehicle had passed, the vehicle in question then proceeded to overtake the slower vehicle. It can be seen that there was an early (failed) attempt to overtake, but at all times the specified minimum safe distances had to be maintained.

6.6.4 Vehicle Behaviour Analysis

Here, the behaviour of the vehicle is metrically assessed, when placed in a variety of situations on the road. The intent is to monitor how the vehicle manages its speed and the available road width for its navigation. The factors of efficiency and safety can be contradictory, and yet these need to be balanced for optimal travel.

Efficiently making turns was one of the prime inspirations behind the use of Bézier curves. The vehicle performance is assessed in this respect. Measurements were taken of the desired trajectory for the vehicle around a curve (in the form of a map), the time of completion of the map and the distance travelled at a variety of maximum speeds. The corresponding plots are given in Fig. 6.19. At lower speeds, the vehicle travelled near the inner edge of the curve and hence the distance traversed was low; however, this distance increased as the speed increased and the vehicle started travelling towards the outer edge of the curve. Subsequent increases of speed had no effect on distance travelled. This may sound counterintuitive at first; however, it must be remembered that at very high speeds the paths generated to safely travel around a curve are infeasible and the vehicle needs to lower its speed until such a path becomes feasible. All units are arbitrary and specific to the simulation tool. These relate to the real-world units by constants.

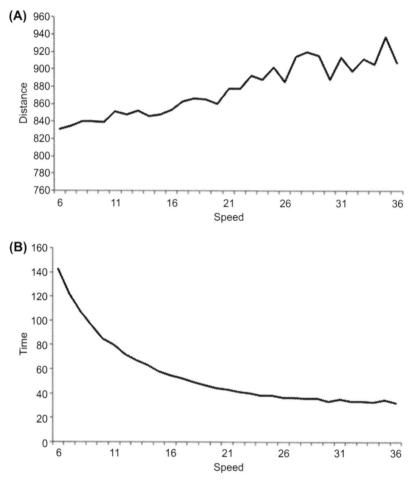

Figure 6.19 Relation between (A) distance of travel with speed (B) time of travel with speed.

In terms of time taken to travel round a curve, results are interesting. Increased speeds usually imply less travel time; however, the increased distance travelled round a curve implies a longer travel time. However, speed is dominant; the time of travel decreases with increase in speed, although this decrease is not uniform. At higher speeds, the vehicle first slows down which reduces the average speed (differentiating between average and maximum speed) to some degree.

6.6.5   Genetic Algorithm Parameter Analysis

The major parameters associated with the algorithm are within the GA which significantly affects overall algorithm performance. Because the solution needs to be returned in real time, it is not possible for the GA to have either a large number of generations for convergence or much time for its optimal solution search. For this reason, some GA operators were designed to constantly add diversity to the population pool, whilst others attempted to tune the trajectory and converge. Because both these aspects become important in different situations, the analysis cannot be done on the basis of a set of benchmark situations, in which clearly one or other strategy would be of more use. Considering the operation of the repair operator and its ability to quickly produce reasonable paths and small probabilities of invalidation of paths, it can be argued that less effort is needed to add diversity and more effort to tune the trajectory. The same strategy is used in the algorithm.

In such a context, the population size of the GA becomes a major parameter of the algorithm, which needs to be high enough to assure a feasible solution in a low number of generations and low enough to assure a small minimum execution time. In all experiments, this value was fixed at 20 because of the following: At many times the population is reinitialized for the algorithm, and hence it is necessary that the solution generated in a single generation of the algorithm (which makes the immediate move of vehicle) is feasible and near optimal (if not optimal). Hence, first, one specific curved scenario is considered and the number of individuals required to generate a decent feasible path are analysed. The total number of individuals or paths generated was increased and the best path length was noted. The corresponding graph is given in Fig. 6.20. It required at least six individuals for the generation of feasible paths, avoiding collisions and ensuring minimum safe distances. As the number of individuals was increased, a small decrease in path length was visible. The improvement obtained for a much higher number of individuals was minimal. Experimental evaluations reveal that use of a Cartesian coordinate axis system for the GA individual representation gives infeasible solutions for the part of the graph shown in Fig. 6.20. This gives an indication of the superiority of the road coordinate axis system over the Cartesian coordinate axis system.

Similarly, another experiment was performed with a different number of obstacles on the curved path. The average number of individuals required for the generation of a feasible path was noted. The graph so generated is given in Fig. 6.21. An increase in the number of obstacles means a greater difficulty in the generation of the feasible path

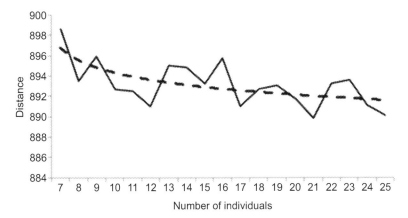

Figure 6.20 Relation between distance of travel and the number of random individuals used.

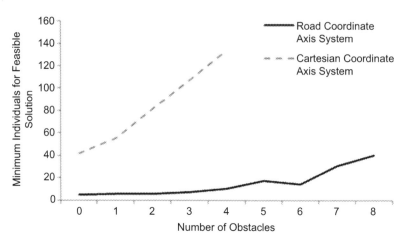

Figure 6.21 Relationship between the number of individuals needed for the generation of a feasible path and the number of obstacles in the path.

as the vehicle needs to make more turns to avoid obstacles. Further, it means fewer possibilities of feasible paths. The increase in difficulty also depends on the place where the obstacle is placed. An increase in obstacles hence increased the number of individuals required, with this increase being more at higher obstacle paths. Any further fitting of obstacles on the path was not possible due to limited road size.

Interestingly, at one instance the number of individuals required actually decreased with an increase in the number of obstacles. This appears to be due to the fact that the effect of one extra obstacle was to guide the Bézier curve in a manner that another obstacle was (automatically) avoided. From both of these experiments it can be inferred that a population size of 20 may be a good choice for the simulation as it stands. This ensures the generation of feasible paths for reasonably complex scenarios, whilst the paths are also short enough for simpler scenarios.

In Fig. 6.21, the road coordinate axis system is compared to a Cartesian coordinate axis system, in which the points used for individual representation of the GA were initially generated randomly in the road segment being planned. In this experiment, the aim was to test to what degree the road coordinate axis system was better (or worse) than the Cartesian system. Results clearly indicate a significant improvement in the number of individuals required when using the proposed coordinate axis system. Further, for more than four obstacles the Cartesian coordinate system was unable to generate a feasible solution even for a very high number of individuals. Hence, the Cartesian coordinate system is limited to planning with a reasonably small number of obstacles only.

6.7 Summary

Planning and coordination of multiple autonomous vehicles on a road network is an exciting problem. From the point of view of a single vehicle, the challenge lies in effective strategic planning at the top level to obstacle avoidance at the lower level. The real-time nature of the problem, however, eliminates the possibility of developing fancy solutions which are time costly. Unlike most mobile robotics studies, the cost of collisions in vehicular systems is likely to be high and must be avoided by all means. The presence of multiple vehicles stresses the need for coordination strategies between vehicles at all times such that no collision occurs between the vehicles and such that each vehicle has a fair path along which to move forwards at its own desired speed as much as possible. The practical high diversity in speeds of the vehicles, together with a relatively small workplace make the problem very different from existing (central processing) solutions in the research domain of mobile robotics.

In this chapter, GA was used to solve the problem of navigation of vehicles within the road segment under the assumption of communication. The greatest challenge was generating quality solutions in near-real time. The solution takes some inspiration from traffic rules, which theoretical analysis and experimental observations verify to be optimal, considering a large number of scenarios and the presence of uncertain incidents. These traffic heuristics, used with an evolutionary algorithm for the generation of Bézier curves, are able to navigate the vehicle in a variety of scenarios. The experimented scenarios range from straight roads and roads with a large number of obstacles to close overtaking and infeasible overtaking. This more or less covers the variety of scenarios that the vehicle may face in real life. Uncertainties are the key element that makes software simulations differ from real world experiments. Here, uncertainties were introduced by making the vehicle move as per its own kinematic model, and not necessarily the planned trajectory. The planner at times did generate sharp curves, which could not be followed by the vehicle. These errors were corrected with time.

References

Bartels, R.H., Beatty, J.C., Barsky, B.A., 1998. Bézier Curves. An Introduction to Splines for Use in Computer Graphics and Geometric Modelling. Morgan Kaufmann, San Francisco, CA, pp. 211−245.

Bennewitz, M., Burgard, W., Thrun, S., 2001. Optimizing schedules for prioritized path planning of multi-robot systems. In: Proceedings of the 2001 IEEE International Conference on Robotics and Automation, Seoul, Korea, pp. 271−276.

Bennewitz, M., Burgard, W., Thrun, S., 2002. Finding and optimizing solvable priority schemes for decoupled path planning techniques for teams of mobile robots. Robotics and Autonomous Systems 41 (2−3), 89−99.

Chakraborty, J., Konar, A., Chakraborty, U.K., Jainm, L.C., 2008. Distributed cooperative multi-robot path planning using differential evolution. In: Proceedings of the IEEE World Congress on Evolutionary Computation, pp. 718−725.

Chu, K., Lee, M., Sunwoo, M., 2012. Local path planning for off-road autonomous driving with avoidance of static obstacles. IEEE Transactions on Intelligent Transportation Systems 13 (4), 1599−1616.

Cormen, T.H., Leiserson, C.E., Rivest, R.L., Stein, C., 2001. An Introduction to Algorithms, second ed. MIT Press, Cambridge, MA.

Garcia, M.A.P., Montiel, O., Castillo, O., Sepulveda, R., Melin, P., 2009. Path planning for autonomous mobile robot navigation with ant colony optimization and fuzzy cost function evaluation. Applied Soft Computing 9 (3), 1102−1110.

Kala, R., 2012. Multi-robot path planning using co-evolutionary genetic programming. Expert Systems With Applications 39 (3), 3817−3831.

Kala, R., Shukla, A., Tiwari, R., 2010a. Fusion of probabilistic A* algorithm and fuzzy inference system for robotic path planning. Artificial Intelligence Review 33 (4), 275−306.

Kala, R., Shukla, A., Tiwari, R., 2010b. Dynamic environment robot path planning using hierarchical evolutionary algorithms. Cybernetics and Systems 41 (6), 435−454.

Kala, R., Shukla, A., Tiwari, R., 2011. Robotic path planning using evolutionary momentum based exploration. Journal of Experimental and Theoretical Artificial Intelligence 23 (4), 469−495.

Kala, R., Warwick, K., 2014. Heuristic based evolution for the coordination of autonomous vehicles in the absence of speed lanes. Applied Soft Computing 19, 387−402.

Kant, K., Zucker, S.W., 1986. Toward efficient trajectory planning: the path-velocity decomposition. The International Journal of Robotics Research 5 (3), 72−89.

Kapanoglu, M., Alikalfa, M., Ozkan, M., Yazıcı, A., Parlaktuna, O., 2012. A pattern-based genetic algorithm for multi-robot coverage path planning minimizing completion time. Journal of Intelligent Manufacturing 23 (4), 1035−1045.

Klancar, G., Skrjanc, I., 2010. A case study of the collision-avoidance problem based on Bernstein−Bézier path tracking for multiple robots with known constraints. Journal of Intelligent Robotic Systems 60 (2), 317−337.

Lepetic, M., Klancar, G., Skrjanc, I., Matko, D., Potocnik, B., 2003. Time optimal path planning considering acceleration limits. Robotics and Autonomous Systems 45, 199−210.

Lian, F.L., Murray, R., 2003. Cooperative task planning of multi-robot systems with temporal constraints. In: Proceedings of the 2003 IEEE International Conference on Robotics and Automation, pp. 2504−2509.

Paruchuri, P., Pullalarevu, A.R., Karlapalem, K., 2002. Multi agent simulation of unorganized traffic. In: Proceedings of the ACM 1st International Joint Conference on Autonomous Agents and Multiagent Systems, New York, pp. 176−183.

Potter, M.A., 1997. The Design and Analysis of a Computational Model of Cooperative Coevolution. George Mason University, Fairfax, Virginia (Ph.D. thesis).

Sewall, J., van den Berg, J., Lin, M.C., Manocha, D., 2011. Virtualized traffic: reconstructing traffic flows from discrete spatio-temporal data. IEEE Transactions on Visualization and Computer Graphics 17 (1), 26−37.

Stanley, K.O., Miikkulainen, R., 2004. Competitive coevolution through evolutionary complexification. Journal of Artificial Intelligence Research 21 (1), 63−100.

Szlapczynski, R., Szlapczynska, J., 2012. On evolutionary computing in multi-ship trajectory planning. Applied Intelligence 37 (2), 155−174.

Wang, M., Wu, T., 2005. Cooperative co-evolution based distributed path planning of multiple mobile robots. Journal of Zhejiang University - Science A 6 (7), 697−706.

Xiao, J., Michalewicz, Z., Zhang, L., Trojanowski, K., 1997. Adaptive evolutionary planner/navigator for mobile robots. IEEE Transactions on Evolutionary Computing 1 (1), 18−28.

Xidias, E.K., Azariadis, P.N., 2011. Mission design for a group of autonomous guided vehicles. Robotics and Autonomous Systems 59 (1), 34−43.

Sampling-Based Planning 7

7.1 Introduction

It may not be possible to generate all possible trajectories in a continuous domain of space and time for solving the problem. *Sampling-based approaches* attempt to sample out the search space in pursuit of optimal trajectories. *Probabilistic Road Map (PRM)* and *Rapidly Exploring Random Trees (RRTs)* are two widely used algorithms in sampling-based motion planning. PRM maintains a sampled road map of the space by generating random samples in the entire space, which become the vertices of the road map. It is usually better to take samples near obstacle boundaries and inside narrow corridors rather than in open spaces. Connections between vertices are tested by a small local search and collision-detection algorithm. The feasible connections are added as edges of the road map. PRM is very useful for multiquery motion planning, wherein the road map needs to be produced in an offline manner once, and the road map can then be used multiple times to compute a path between any source and destination as desired.

RRTs on the contrary, maintain a single sampled version of the space in a tree structure. As the algorithm proceeds, the tree is extended and hence the algorithm becomes clearer in identification of clear areas and the obstacle-prone areas. Initially, the source position is kept as the only node and root of the tree. At each iteration, this tree is grown by extending a node of the tree by some step size in a direction which is obtained by the generation of a random sample in the search space. The tree tries to grow towards this random sample. A general strategy is to bias the extension of the tree towards the goal which generally means to attempt to find out the obstacle-free and obstacle-prone areas near the path from the current position to goal, as opposed to the far-off areas. Once a tree has been sampled, the trajectory is simply the path from source to goal through the intermediate tree nodes. The approach is suited for single-source -oal problems as the tree is rooted at the source and the algorithm terminates on reaching the goal.

The chapter deals with the use of RRT for the problem. The chapter still assumes communication between the vehicles. Even though the problem characteristics can be exploited to make GA computationally inexpensive, sometimes the computational requirements may be very hard making the use of GA impossible or there may be too many obstacles and vehicles to avoid. RRT is, in general, *much faster* than GA with the loss that the algorithm is *neither globally nor locally optimal*. It does not guarantee an optimal way to avoid the obstacles and vehicles, nor does it decide whether to avoid/overtake from the left or right. It also does not specify optimal distances to maintain from obstacle/vehicle boundaries while avoiding/overtaking them. However, as opposed to most reactive techniques, the algorithm is *probabilistically complete* in nature.

Practically, RRTs show local near-optimality and global near-optimality in cases when very different competing paths are not nearly equal in cost or neither of the paths is through a too-narrow region. Narrow regions are hard to sample, and hence it is hard to compute paths going through them. The RRTs are known to have a *Voronoi bias* or, in other words, the expansion tree grows more in wide-open areas than in narrow congested areas. Hence, a longer path going through open areas is usually the output as compared to a short path which goes through narrow areas. RRT is probably the best trade-off between optimality and computational expense, keeping completeness as a mandatory requirement.

The use of RRT for the problem is presented in two parts. The basic use of RRT for the problem is presented in Sections 7.4 and 7.5. In Sections 7.6 and 7.7, the approach is extended to the use of RRT-Connect along with some other modifications which make the approach even better. Both approaches use a naive implementation of priority-based coordination.

Segments of the chapter including text and figures have been reprinted from Kala and Warwick (2011a).

Segments of the chapter including text and figures have also been reprinted from Kala and Warwick (2011b).

7.2   A Brief Overview of Literature

An interesting work (Kuwata et al., 2009) uses RRT for the specific problem of navigation of an autonomous vehicle. For general details regarding the vehicle, refer to Leonard et al. (2008). The authors used a biased sampling technique, whereas the RRT was generated using the vehicle-specific control signals. The algorithm hence lacks global optimality as well as treats vehicles and obstacles alike which can lead to loss of completeness. Anderson et al. (2012) solved the problem of trajectory generation for a single vehicle using constrained Delaunay triangles which were fitted in the entire map from which the resultant trajectory was computed. The approach, however, is applicable only for structured environments.

RRTs are also widely used in mobile robotics. The general problems associated with RRTs are the same as the optimization-based methods. There is a requirement of an offline phase. Further, most multirobot planning techniques are noncooperative. Raveh et al. (2011) executed multiple instances of RRT in parallel to get optimal subpaths, which were integrated using a dynamic programming-based algorithm to get the overall optimal path. A similar technique of using multiple runs is used here. Due to the architecture of the road, it is simple to get optima by a few runs without requiring intelligent integration of the individual runs. A generalized version of the RRT and Probabilistic Road Map (PRM) algorithm was presented by Chakravorty and Kumar (2011). A connection between any two samples or points is done on the basis of Monte Carlo simulations applied in a vehicle-specific control phase. In this chapter, the conventional space would be used to allow integration of traffic heuristics, which is a prime motive of the chapter. Carpin and Pagello (2009) studied the problem of decentralized multirobot motion planning using priority-based coordination based on the metrics of computational time and optimality of the solution. The prioritized approach is noncooperative.

Various heuristic methods have been integrated with this class of approaches. The goal of this work was the integration of traffic-inspired heuristics, which was obviously not the goal of these works. Jaillet et al. (2008) used transition tests to disallow addition of nodes in the tree which increase the cost considerably. This saves significant computation time at the same time adding to optimality. Urmson and Simmons (2003) heuristically select nodes for expansion of the road map along better areas making the resultant plans near optimal. Kalisiak and van de Panne (2006) proposed the use of a flood-fill algorithm for expanding the tree while being biased towards the goal. Strandberg (2004) used local trees to traverse the difficult regions, which were connected to the rest of the road map. Similarly, Ferguson and Stentz (2006) proposed an anytime version of RRT. To solve the problem of low probability of generation of samples in a narrow corridor, Zhang and Manocha (2008) proposed retraction of samples inside the obstacle to the nearby obstacle-free space.

7.3 A Primer on Rapidly Exploring Random Trees (RRT)

Rapidly Exploring Random Trees (RRT) is a widely used method for the problem of motion planning. The algorithm initiates and extends a *search tree* covering the configuration space in pursuit of a solution from the source to the goal. The intent is that a complete search in the entire configuration space is not possible in short execution times, hence the configuration space is sampled and the samples are stored in a tree-like manner. The tree is rooted at the source and grows randomly till the goal is found. To limit the search it may be better to grow the tree more towards the goal.

7.3.1 Rapidly Exploring Random Trees

This subsection presents the basic RRT algorithm (LaValle and Kuffner, 1999; Kuffner and LaValle, 2000) used for motion planning. Initially, a tree is generated *rooted at the source*. The algorithm proceeds in iterations with every iteration responsible for growing the tree by *extending* one of its nodes and creating a new leaf node. To extend the tree, first a *random sample* is generated. The sample may be generated by a uniform sampling technique, or specialized sampling techniques which generate samples in a narrow corridor or near the obstacles. The generated sample must be collision free, otherwise it may be moved to a neighbouring collision-free region or discarded altogether. The collision-free sample (say q) represents an area which the search tree must try to seek in the current iteration. Hence, an attempt is made to grow the tree towards q. To grow or extend the tree, the sample existing in the tree (p)

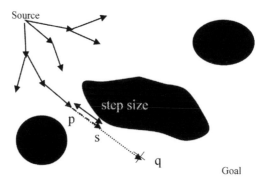

Figure 7.1 RRT algorithm.

which is at the *least distance* from q is chosen. A new sample (s) is generated at a distance of *step size* from p towards q. Therefore, it can be said that the nearest sample in the tree is extended towards q by a small step. The notations and extensions are shown in Fig. 7.1.

The newly generated node is added on the tree if it is collision free and the path from p to s is collision free. Every sample adds computation time in the processing of the algorithm; hence, it is good to limit the number of samples. Therefore, the new sample is accepted and added in the search tree only if it is not nearly at the same place as a preexistent node in the search tree. The new node is inserted with the line joining p to s as an edge. When the goal node is inserted into the tree, the search stops. Upon termination, because a tree is maintained rooted at the source and goal as the leaf node, the path can simply be found by traversing the parent of each node from the goal to the source.

The search tree in this manner grows completely randomly. Normally, the search tree would go ahead with filling up nearly the entire configuration space as no information about the goal is provided. The more the configuration space is searched, the more are the number of nodes in the search tree and the greater is the computation time. The configuration space can be very large restricting getting results in a short computation time. Hence, it is preferred to *bias* the generation of the search tree more towards the goal. This can be done by simply taking the goal as the sample for *bias* percentage of iterations, rather than randomly generating the sample. By doing so, the goal tries to pull the tree towards itself, or the tree tries to push itself in the direction of the goal.

The pseudocode of the approach is presented in Algorithm 7.1.

Algorithm 7.1: RRT(source, goal)

```
root ← source
repeat for a maximum of maxiter
    generate random sample q
    p ← node nearest to q in tree
    s ← node by extension of p in direction of q
```

```
      if no point close to s lies in tree and s is collision
      free and line from p to s is collision free
        add s to tree with parent p
        if s is close to goal, return path from root to s
      end if
  end repeat
  return null
```

7.3.2   RRT-Connect

A common problem with the basic RRT is that each iteration adds a small edge for the extension of the tree. This can be time-consuming. Hence, a variant called *RRT-Connect* is commonly used. The only difference with RRT is that instead of taking a single step towards the randomly generated sample (q), this approach takes multiple such steps. Steps are taken until an obstacle is encountered and further extension is hence not possible, or by multiple extensions q is reached. So by generation of 1 sample, multiple nodes are added to the tree. Normally, the continuous extension stops because of an obstacle in between p and q. The last sample would be near the obstacle and such samples are helpful in devising an obstacle avoidance strategy. However, many times, especially for simple and sparely occupied configuration spaces, connection between p and q is made and the continuous extension stops. The approach is summarized in Fig. 7.2. The pseudocode is given by Algorithm 7.2.

Algorithm 7.2: RRT-Connect (source, goal)

```
  root ← source
  repeat for a maximum of maxiter
    generate random sample q
    p ← node nearest to q in tree
    while true
      s ← node by extension of p in direction of q
      if no point close to s lies in tree and s is collision
      free and line from p to s is collision free
        add s to tree with parent p
        if s is close to goal, return path from root to s
        if s is close to q, break
      else break
      end if
    end while
  end repeat
  return null
```

This approach is more exploitative as it readily moves multiple steps in the direction of q. The intermediate nodes generated in making a connection to q are not very helpful, whereas having more nodes readily increases the computation time. The intermediate points may not be stored to save computation time and memory.

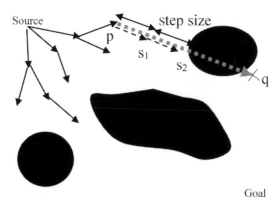

Figure 7.2 RRT-Connect algorithm.

7.3.3 Multitree Approaches

It is also possible to initialize and extend two trees instead of just one. Such an algorithm is called a *bidirectional RRT*. One tree goes from the source while searching for the goal. The second tree starts from the goal and searches for the source. This assumes that the actions are reversible or knowing the plan from *a* to *b* means the inverse path from *b* to *a* can be computed. The root of the first tree is the source whereas the root of the second tree is the goal. The expansion of both the trees is done one after the other; each may be expanded by generation of a random sample or biasing the extension by taking the corresponding goal as the sample. The algorithm terminates when both the trees meet, or in other words, when extension of one tree generates a new sample which already exists in the second tree. The concept is shown in Fig. 7.3.

The algorithm is better than a single-tree approach in which the search could only stop on attaining a goal. Here, each of the trees has a variety of options to exit, that is, when any one of the nodes of the other tree is met. Normally, merging takes place at

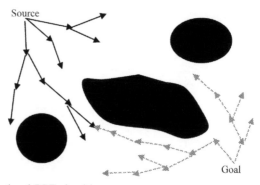

Figure 7.3 Bi-directional RRT algorithm.

the leaf nodes which denote the active frontier of the search tree or the boundary between explored and unexplored areas of the configuration space. The nonleaf nodes normally denote areas of configuration space which have been appreciably explored. However, instead of looking up for a single node as the goal or terminating condition, a number of nodes need to be searched. All nodes of the second tree are valid goal points for the first tree and vice versa. This may take time. The algorithm is summarized by Algorithm 7.3.

Algorithm 7.3: Bi-directional-RRT (source, goal)

```
root₁ ← source
root₂ ← goal
repeat for a maximum of maxiter
    generate random sample q
    p ← node nearest to q in tree 1
    s ← node by extension of p in direction of q
    if no point close to s lies in tree and s is collision
    free and line from p to s is collision free
      add s to tree with parent p
      if s is close to a node t in tree 2,
        return path connecting source to s in tree 1 and t to
        goal in tree 2
      end if
    end if
    swap tree 1 and tree 2
end repeat
return null
```

More generally, instead of two, a number of trees may be searched, each rooted at any characteristic point. Places like narrow corridors or places where sharp turns are required due to obstacles are good candidates for search trees. The trees may merge while they are extended. When a path from the source to goal is possible, the search stops.

RRTs are not optimal as they stop on reaching the goal. The path to goal is affected by *Voronoi bias*, or the path through open areas is more likely to be found in contrast to the optimal path. RRT* (Karaman and Frazzoli, 2011) algorithm does not stop on reaching the goal; however, it continues execution. If a better path is found, the node corresponding to the better path is added, whereas the node corresponding to the older path is deleted to maintain the tree-structure. RRTs also face the problem of optimality because characteristic positions of nodes in the tree can limit the generation of future samples considering that the sample generation is constrained by the selection of the nearest sample and extension by a specific step size. To aim at optimality, it is possible to run multiple RRTs in parallel until all of them give a path. The individual RRTs can then be intelligently merged to give a near-optimal path (Raveh et al., 2011). Further, it is also possible to maintain a graph structure throughout rather than a tree structure aiming at optimality. The resultant algorithm is called Rapidly Exploring Random Graphs (Kala, 2013), wherein the graph grows and extends even after a path to the goal is found. The extra edges making cycles help in optimality of the solution.

7.3.4 Role of Parameters

The most important parameter of the algorithm is the *step size* or the distance by which a node is extended to generate a new sample. High values of this parameter result in the tree growing by large steps and hence the algorithm cannot model fine turns, normally requiring the robot to go in-between obstacles creating a narrow corridor. Even though RRTs are not optimal, the solution quality is generally better with small step sizes. This is because small step sizes can facilitate finer turns, whereas the larger step sizes may make the robot travel far away from the obstacle to avoid it by taking large steps. The smaller step sizes, however, result in too many nodes being produced and hence have a very long computational time.

Considering these limitations, *heuristics* are often used to adaptively set the step-size factor. Near the obstacles, the factor may be kept small whereas away from the obstacles the factor may be kept large. Further, the factor can be initially kept large to produce some initial tree, whereas in the later iterations the step-size factor can be reduced.

The other parameter is the *bias* or how much effort must be made to bias the generation of a tree towards the goal. A high value is exploitative wherein the algorithm tries to travel straight to the goal, without much exploration. A low value, on the other hand, is explorative wherein the algorithm spends far too much time in exploring the configuration space rather than marching towards the goal. Exploitative settings may result in short computational times while compromising on the optimality. Explorative settings may have too long a computational time while generating good results subject to the absence of *Voronoi bias*, considering that the RRTs love and extend more towards open spaces. A problem with high bias value is also that the nearest node may not be extendable to the goal because of an obstacle, in which case the extension repeatedly fails. In this case, it is said that the bias element has *struck at a minimum* and the component will not respond until an alternate node is later generated closer to the goal.

7.4 Solution With RRT

The next approach used for solving the same problem is RRT. To better motivate the choice of the algorithm, the pros and cons of RRT are summarized in Box 7.1. The algorithm is ideally suited for such problems in which completeness is a must, whereas suboptimality to some extent would not hurt. This section presents a basic implementation of RRT for the problem which is further extended in Section 7.6. The chief focus in use of RRT is to make a *strategic expansion* such that the search completes with the least expansions as well as to integrate *path smoothing* using spline curves which makes the trajectory navigable for the vehicle.

Box 7.1 Pros and Cons of RRT

Pros
- Computational expense (better than GA)
- Probabilistically complete
- Can navigate through multiple obstacles with too many turns.

Concerns
- Suboptimal
- Does not work well in narrow corridors
- Voronoi Bias, prefers path through open spaces

The basic problem is to plan the paths of a number of vehicles in a traffic scenario. Each vehicle is assumed to be rectangular in shape with its own length and breadth and has its own maximum allowable speed. The generated plan needs to take into account that the operational speed of the vehicle must never exceed this limit. Each vehicle enters the planning scenario at some time. Only on its emergence is planning for a particular vehicle performed. The vehicles must not collide either with each other or with the static obstacles. Vehicles are nonholonomic in nature and hence only smooth travel paths may be traversable.

The approach uses the Cartesian coordinate system for curve generation and plan specification. The road coordinate system is used for the generation of samples. Interconversion between the two schemes is important. Knowing all the coordinates of both boundaries, it is fairly easy to convert from the road coordinate system to the Cartesian coordinate system. For the opposite conversion, a small search technique is used which attempts to find the best match.

The key features of the approach using RRT are summarized in Box 7.2, whereas the key concepts are summarized in Box 7.3.

7.4.1 RRT Expansion

In this problem, the source is the point from which a vehicle enters the segment being planned. The goal is any point in the segment end. It is important to make a trajectory

---

Box 7.2 Takeaways of Solution Using RRT

- Inspired by the general motion of vehicles in traffic, a planning strategy is proposed which is *biased towards a vehicle's current lateral position*. This enables better tree expansion and connectivity checks.
- RRT generation is integrated with spline-based curve generation for *curve smoothing*.
- The approach is designed and tested for *many and complex obstacles* in the presence of multiple vehicles.

Box 7.3 Key Aspects of Solution Using RRT

- First expansion in the *heading direction* of the vehicle
- Samples generated using *road coordinate axis system*
- Sampling *biased towards current lateral position* of the vehicle
- Splines used for curve smoothening, only curves allowing minimum speed are regarded as feasible.
- Priority-based coordination
- Speed *iteratively reduced* till RRT finds a feasible path

---

that takes the vehicle at the end of the road segment. The lateral position at the end of the road segment is not important. The RRT planning algorithm attempts to connect the source to the goal by a suitable path. At each iteration, the algorithm results in either the addition of a new node to the tree or no addition if the added node was infeasible. A maximum of *maxiter* iterations is used. If the planning algorithm still does not succeed in finding a solution, it is assumed that no feasible path exists for the set criterion.

It is important that the vehicle initially moves in the current direction of orientation. In other words, the path generated needs to start at an initial angle of heading that is equal to the vehicle's current angle. Hence, the root of the tree (source) has just one child which is a point along the vehicle's length in its current heading direction. This is the *direction maintenance child* for the RRT approach.

RRT by default generates samples randomly, which leads to the tree to be expanded in any direction. This may take the entire algorithm a significant amount of time, and it would eventually explore the entire road, till it reaches the road end or the goal. It is, hence, preferred to orient the search process towards the goal and centric towards a narrower region, while still having some exploratory potential to search distant sections of the road. This is controlled by sampling. It is assumed that the vehicle needs to travel smoothly, hence sampling is *biased towards points which have the same lateral position* as the current position of the road, or more generally the same value of Y coordinate in the road coordinate system. A number of random samples are generated with *probabilistic selection*, the probability being proportional to deviation of the Y coordinate of the generated sample from the current Y coordinate of the vehicle's position. Once the probabilities are computed, a Roulette Wheel Selection scheme is used for generation of the sample. To ensure all samples are generated within the road, the road coordinate axis system is used. This also takes into account variable road widths.

7.4.2 Curve Generation

The path of the RRT, which is a collection of nodes, cannot be a straight line joining the nodes as the resultant path would not be smooth. Hence, a *curve smoothening* technique is built into the algorithm framework. Curve smoothening techniques take as input a path formed by such line segments and fit a smooth curve which approximately resembles the original path. Smoothness and fitting of the original path are contradictory objectives which are controlled by the complexity of the smoothening technique. Here, splines (de Boor, 1978) are used, which take a set of points which are used as control points and return a smooth curve. Every node addition is followed by the generation of curve. Firstly, this ensures that the path formed by the addition of a node is feasible such that a vehicle lying at any point in the path does not collide with any obstacle. Secondly, it means that, at every point, the curve is smooth enough to allow the vehicle to travel at the set speed and at no point does it have to reduce speed. Thirdly, vehicles have to check for possible collisions with other vehicles and, for this, exact curve information is required. The feasibility of the curve is additionally checked by placing the vehicle at every point with its orientation as the immediate angle of the curve. The vehicle must not collide with any obstacle or other vehicle for which planned trajectories of other vehicles are queried.

For the edge or curve to be called feasible, it must be *traceable by the current speed* of the vehicle set as v_i. If the curve is not smooth enough to allow this speed, the edge is not added. The speed of the vehicle at any point (Xidias and Azariadis, 2011) is given by Eq. [7.1].

$$v = \min\left(\sqrt{\frac{\rho}{k}}, v_i \right) \qquad [7.1]$$

Here, ρ is a constant the value for which depends on the friction between the vehicle and road and v_i is the maximum speed specified for the running of the vehicle. k is the curvature of the curve which may be approximately given by Eq. [7.2].

$$k = \| \tau(t + d) + \tau(t - d) - 2\tau(t) \| \qquad [7.2]$$

Here, $\tau(t)$ is a point in the generated curve τ at a distance t from start, d is a small constant and $\|.\|$ denotes the Euclidian norm. For the speed of the vehicle to be feasible, v must be equal to v_i. Note that the higher speeds would require a smoother curve.

The generated curve is a set of points that the vehicle follows. Because there are multiple vehicles, each of them queries the others to ensure there is no collision. To reduce the computation, the travel plan is represented as a hash map which maps the vehicle position against travel time using time as the hashing function. This means any vehicle's position at a given time can be computed within unit time.

7.4.3 Coordination

A *priority-based coordination* approach (Bennewitz et al., 2001, 2002) is used for vehicle coordination. This approach assumes that each vehicle has a priority attached to it. A higher-priority vehicle never considers possible collisions with lower-priority vehicles. The priority of the vehicle is taken from its *time of emergence* into a segment, with a vehicle earlier into the planning scenario having higher priority. This ensures that a vehicle already planned need not be replanned as a new vehicle enters the segment.

The speed of a vehicle plays a major role in multivehicle planning and if an entering vehicle has a high speed with no room to overtake other vehicles, a collision is unavoidable. Hence, if the RRT planner of the vehicle fails to find a feasible travel plan, the *speed of the vehicle is reduced* by some amount Δ. Hence, the vehicle's speed is constantly reduced till the algorithm is able to generate a travel plan. In the worst case, the speed of the vehicle is reduced to the speed of the slowest vehicles in the segment in which case it simply *follows* them. If no feasible path is generated, it is regarded as a blocked route, and the vehicle would require *rerouting* of its entire journey similar to the rerouting used in GA.

The general outline of the algorithm is given by Algorithms 7.4 and 7.5.

Algorithm 7.4: Plan (vehicles, map)

```
while not end of simulation
   for i=1 to number of vehicles
      if τᵢ is null
         vᵢ ← vmax
         while true
            τᵢ ← RRT(Sᵢ, segment)
            if τᵢ is not null
               represent τᵢ as a hash map of time
               break while
            else vᵢ ← vᵢ − Δ
            end if
         end while
      end if
   end for
   move vehicles as per generated plan
end while
```

Here, τ_i denotes the plan, S_i denotes the source or the current position, v_i denotes the speed and v_{max} denotes the maximum attainable speed of vehicle i.

Algorithm 7.5: RRT (source, segment)

```
root ← source
rootChild ← point at distance of vehicle's length from root
at current angle of orientation
parent(rootChild) ← root
repeat for a maximum of maxiter
```

```
Samples ← Generate random samples in road coordinate
system such that probability of selection of sample q =
|Y(q)-Y(source)|
Select a sample q by roulette wheel selection
p ← node nearest to s in tree
s ← node by extension of p in direction of q
if no point close to s lies in tree ^ and vehicle placed
at s with direction p to s is obstacle free
     generate curve τᵢ to s
     check for collision with higher priority vehicles,
     static obstacles, and minimum travel speed vᵢ
     if τᵢ is feasible
        parent(s) ← p
        tree ← tree ∪ s
        if s is close to segment end, return path from
        root to s
     end if
   end if
end repeat
return null
```

7.5   Results

The discussed algorithm was developed and tested by means of simulations over an engine developed in Matrix Laboratory (MATLAB). The planning scenario is initiated by specifying the speeds, dimensions, entry times, entering orientations of all the vehicles. Different modules are made for RRT planning, curve generation, collision checking etc. The results are discussed separately for single-vehicle and multivehicle scenarios.

7.5.1   Single-vehicle Simulations

The algorithm was tested on a variety of maps. In all the maps the vehicle was generated on the left side of the road segment and was supposed to travel to the other end of the road segment. Three simulations are discussed in detail. The first scenario consists of a curved road. Two simple obstacles are placed one after the other. The map and the path traced by the vehicle are shown in Fig. 7.4A. It can be seen that the vehicle was able to traverse to its goal in a fairly simple path. The path may not be so good at the very end, but as stated the vehicle enters the next segment before finishing its journey. Hence, a replanned path is effectively followed rather than merely the path planned solely in this segment.

The second scenario consists of a straight road with a complex grid of obstacles. The map and the path traced are shown in Fig. 7.4B. Although multiple paths were possible, the continuous iterative expansion resulted in a path reaching the

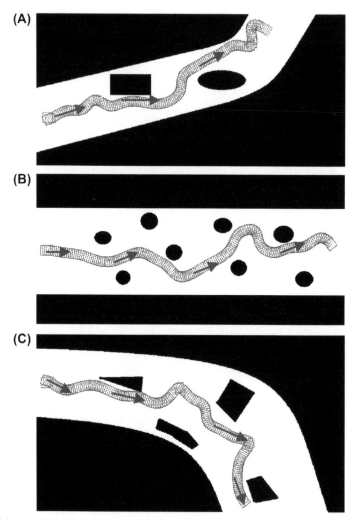

Figure 7.4 Path traced by vehicle in multiple scenarios using RRT. (A) Scenario 1, (B) Scenario 2, (C) Scenario 3.
© 2011 IEEE, Reprinted from Kala and Warwick (2011a).

segment end. It cannot be ascertained that the traversed path is optimal; however, in such driving, rapid decisions are more essential than spending high computation in ensuring the smallest path length.

The last scenario of study is again a curved road with variable-sized obstacles. The characteristic placement of obstacles is such that avoiding an obstacle leaves the vehicle in a harder position to avoid the obstacles that lie ahead, still maintaining smoothness and high speed. The map and the path traced by the vehicle are shown in Fig. 7.4C. The vehicle succeeds in finding a path to reach the segment end. It is visible that in the middle the vehicle had to traverse large distances, this was due to

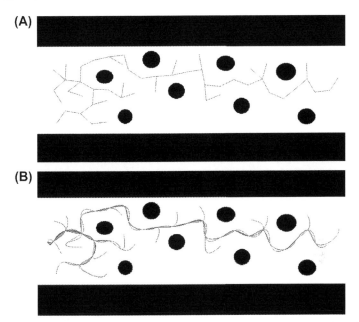

Figure 7.5 Tree generated in planning by RRT. (A) Tree, (B) Smoothened Trajectories.
© 2011 IEEE, Reprinted from Kala and Warwick (2011a).

a large step size which resulted in lower computational effort. Again, it may be seen that excessive computation is not spent in trying to optimize the trajectory, rather it is to generate a feasible trajectory.

The basic planning algorithm is the RRT. For the second scenario, the basic tree with various nodes connected by straight lines is shown in Fig. 7.5A. The tree produced by curve smoothing by splines is shown in Fig. 7.5B. The basic methodology of the algorithm can be seen in the figures. As per the design, the intent is to rapidly reach the goal from the source, rather than to explore the complete area. Hence, there are areas where the search did not proceed at all. A large step size results in the goal being found after a small number of iterations. This results in a lower computational time for the algorithm. Further, nodes very close to other nodes were not allowed, thus giving the entire tree a simpler structure which further helps in shortening the computational time of the algorithm.

7.5.2 Multi-vehicle Simulations

Single-vehicle scenarios enable a good understanding of the manner in which the algorithm works as well as in testing the algorithm for both simple and complex scenarios. The experimentation is further extended to scenarios involving multiple vehicles. In these scenarios, a vehicle enters the map, plans and traverses as per its plan. In the middle, another vehicle is generated which also needs to plan to avoid collision either with the earlier vehicle or with static obstacles. Three scenarios are discussed in detail.

The first scenario consists of a curved road where two vehicles navigate. The first vehicle travels by almost the same path as shown in Fig. 7.5A. A second vehicle is generated which is capable of travelling at higher speed. However, the second vehicle has no space to overtake the first vehicle and is forced to follow the first vehicle. The speed of the second vehicle drops to the speed of the first vehicle and the paths traced are similar. The scenario when the second vehicle enters is shown in Fig. 7.6A and the scenario at a random time in the vehicle chase is shown in Fig. 7.6B. This constitutes vehicle-following behaviour which is commonly seen in everyday driving.

The next scenario consists of a straight road. Again, the second vehicle is generated after the first. In this scenario, there was plenty of scope for the second vehicle to avoid the first vehicle and overtake it such that both vehicles traverse to the segment end. The scenario at the time of emergence of the second vehicle is shown in Fig. 7.7A. The scenario at a random time in the vehicle motion is given in Fig. 7.7B. This constitutes overtaking behaviour of the vehicles which is again common, especially when vehicles differ greatly in speeds.

In the last scenario, the vehicles emerge simultaneously from either side of the road. Here, initially the first vehicle plans and then the second vehicle plans. The second vehicle in its planning needs to account for the coming vehicle and must avoid it. The scenario at the time the two vehicles have just avoided each other is shown in Fig. 7.8A. The rest of the journey is given in Fig. 7.8B. The second vehicle had to align

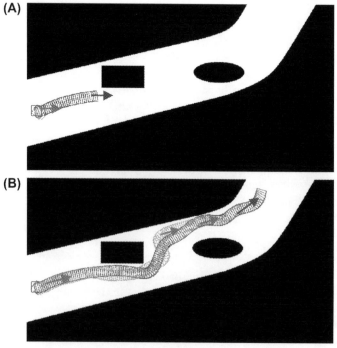

Figure 7.6 Vehicle-following behaviour exhibited by vehicles using RRT. (A) Start, (B) End. © 2011 IEEE, Reprinted from Kala and Warwick (2011a).

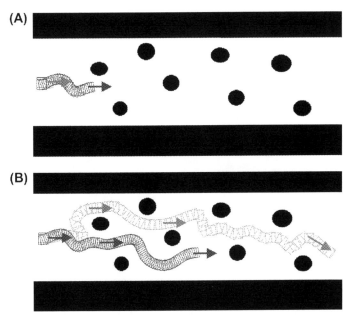

Figure 7.7 Overtaking behaviour exhibited by vehicles. (A) Start, (B) End.
© 2011 IEEE, Reprinted from Kala and Warwick (2011a).

itself in a manner to avoid the first vehicle, the plan for which was already decided. The presence of obstacles made the task more difficult. The RRT planner still succeeded in generating a feasible trajectory which the second vehicle could follow. This constitutes the avoidance behaviour of vehicles.

7.5.3 Analysis

Further, the effect of step size used as a parameter of the RRT in the planning of the vehicles is studied. The quality of solutions is judged by the path length and the time needed to generate the solution. Path length is of little importance, because any path constructed would have a path length almost equivalent to the road length. However, the ability to generate a solution, if one exists, is of high importance. Hence, the purpose of the analysis is to judge the parameter to allow rapid generation of feasible results in complex scenarios. A large step size makes the algorithm increment by large steps towards the goal. As a result, the total number of nodes in the RRT is small. Larger steps mean nodes are fairly wide apart and, hence, the total number of nodes in the tree is smaller. However, it may take time to generate all these nodes, which requires more iterations of the algorithm. Hence, a large step size does not mean a shorter computation time.

The effect of change in step size is shown in Fig. 7.9A for the total number of nodes in the RRT and Fig. 7.9B for the total number of iterations used by the algorithm. The number of iterations for a small step size is high as the number of nodes generated

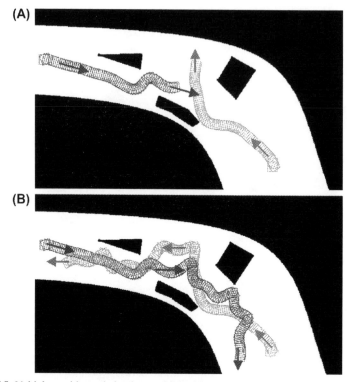

Figure 7.8 Vehicle-avoidance behaviour exhibited by vehicles. (A) Middle, (B) End.
© 2011 IEEE, Reprinted from Kala and Warwick (2011a).

in the tree is high. The number of iterations for a large step size is due to the inability to generate feasible nodes. By further increasing the step size, the algorithm is unable to generate a feasible path as the turns cannot be modelled.

7.6 Solution With RRT-Connect

RRT algorithm under the stated design is capable of generating paths reasonably early. However, the algorithm can be made even more computationally inexpensive by making some design changes which would be dealt with in this section. One of the design changes is the use of RRT-Connect algorithm over RRT. The computational gain so obtained can be used to solve the greatest problem of RRT which is its optimality.

This section introduces builds over RRT algorithm and proposes four changes which result in either computational efficiency or optimality. These are:

1. The procedure to use spline curves from source to the extended node and hence checking the feasibility and nonholonomicity of the (so far prospective) path has been replaced by an *approximate algorithm* that checks for these validity constraints only at the extended node

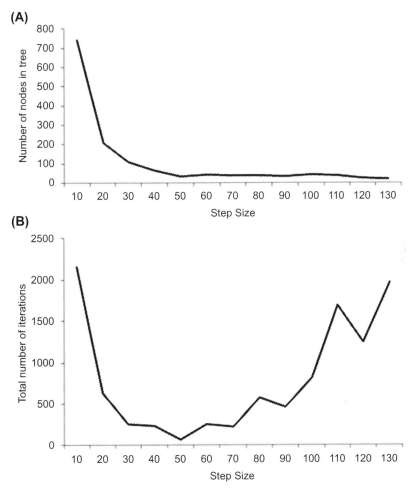

Figure 7.9 Analysis of the step-size parameter of RRT. (A) Number of nodes in tree versus step size, (B) Total number of iterations versus step size.
© 2011 IEEE, Reprinted from Kala and Warwick (2011a).

without the generation of splines. Considering the large number of failed and successful node extensions, this saves a significant amount of time, though the solution generated is approximate.

2. An additional *local optimization* algorithm has been added after the approximate path is generated, which checks the validity at every point and hence returns valid exact paths. This replaces the need to run an entire time-costly algorithm for a number of times for optimality considering that optimality is not intrinsic in RRT functioning. The resultant algorithm is hence fast and better in terms of path length.

3. RRT is replaced by *RRT-Connect* which is more suited for trivial *vehicle-following behaviour* for which it takes less, and indeed practically no, computation time. Considering that a significant amount of any journey involves vehicle following, this is a big boost to overall computation time.

Box 7.4 Takeaways of Solution Using RRT-Connect

- The planning algorithm can be used with very low *computational requirements* for very simple behaviours, whereas higher computation may enable near-optimal performance.
- A *decision-making module* is designed for choosing between *vehicle-following* and *overtaking* behaviours. The module relies on a fast planning lookup.
- The algorithm uses the notion of first building an *approximate path* and then *optimizing* it which induces an iterative nature to the algorithm, unlike the standard RRT approaches which invest computation to build a precise path.
- The algorithm uses *multiple RRT* instances to be assured of being near-global optima, which is largely possible due to the fast approximate path construction.

---

4. The planning algorithm needs to alter the travel speeds of the vehicles in case it is not possible to generate a feasible plan with all vehicles travelling at their preferred speeds. The earlier method used an iterative decrease in speed at each iteration, which resulted in a computationally heavy algorithm. In this section a more *heuristically intelligent* approach is taken in which the speed is guessed depending upon the possibility to avoid, overtake or follow another vehicle.

Each of these points is highlighted in the subsequent subsections. The key takeaways of the algorithm are given in Box 7.4 which supplement the key takeaways of the basic RRT approach given in Box 7.2. The key concepts are also summarized in Box 7.5.

Box 7.5 Key Aspects of Solution Using RRT-Connect

- RRT-Connect
  - For every node generated, connectivity till road segment end is checked.
  - No curve smoothing used in RRT generation to save computations, smoothness approximately checked
  - Vehicle-following behaviour takes unit computation, only one node expanded which has a direct connectivity to the goal
- Optimization
  - RRT-Connect called multiple times for *global optimality* — best solution worked further
  - Local optimization used on the best solution induce *local optimality*
  - Spline curves are used in local optimization
- Priority-based Coordination
- Speed Setting
  - For vehicles in the same direction: If you cannot overtake, follow — Speed equal to the speed of the heading vehicle
  - For vehicles in the opposite direction: Decrease speed iteratively till a feasible plan is reached

7.6.1   RRT-Connect

The basic path is generated using RRT-Connect algorithm. The hypothesis is that a vehicle prefers to keep itself intact in the *same lateral position* or the same position in the Y axis of the road coordinate axis system. This means that if the road is obstacle free and there are no other vehicles, the planning algorithm would generate a trajectory parallel to the road (assuming a smooth X axis), such that the distance of the vehicle from the two boundaries (as a ratio) is always constant. While driving in the presence of lanes this corresponds to keeping to one's own lane.

The RRT-Connect algorithm starts with the initial state or the root of the tree as the source or current position of the vehicle. The vehicle is already facing in a given direction and hence the first step it must take is in the same direction. This fixes the first extension of the root which is also the only child of the root. In subsequent iterations random samples are generated using the road coordinate axis system. The tree formulated so far is searched and the closest node to the generated sample (p) is found. A new node (s) is generated by extending p towards the random sample by a magnitude of *step size* which is a parameter of the algorithm. s is added to the tree with p as the parent if (1) s is not already in the tree. Nodes very close in the solution space are taken as being the same node. (2) s ensures that the vehicle can be safely placed at it. This means that a vehicle placed with its centre at s aligned in the direction of line p to s should be obstacle free and should not collide with any other vehicles. (3) s ensures kinematic constraints when travelling from a parent of p (say r) to s via p. The curved structure formed by these three points gives an indication of the maximum speed at which the vehicle may travel in the region considering its kinematic constraints (Xidias and Azariadis, 2011). The maximum speed for a vehicle over a curvature is already explained in Eq. [7.1]. Here, the curvature k is the curvature that the set of points (that is r, p and s) which may approximately be given by Eq. [7.3].

$$k = \|L_1(d) + L_2(d) - 2p\| \qquad\qquad [7.3]$$

Here, $L_1(d)$ represents a point at a distance of d from p along the line L_1 connecting p to r. Similarly, $L_2(d)$ represents a point at a distance of d from p along the line L_2 connecting p to s. The value of d is chosen so that it is large enough to approximate the prospective curve.

For the node to be regarded as feasible it is important that the vehicle need not reduce its speed. Hence, v given by Eq. [7.1] should be equal to the preferred driving speed v_i.

After adding every node to the tree, it is checked if it would be possible to reach the end of the road segment being planned by *travelling parallel to the road* by keeping the same relative position laterally. In the case when this travel is collision free and obeys the kinematic constraints, the algorithm is terminated. This results in a great saving of time while travelling on an obstacle-free road segment or, while travelling in *vehicle-following mode*, the algorithm needs to do minimal computations and the path may be returned with the first addition of a node in the RRT. Note that this condition has been used to refer to the Connect variant of RRT, rather than the variant discussed in Section 7.3.2. Customizing the definition specifically to the problem domain helps in achieving a better performance of the algorithm.

The algorithm stops when the goal state has been found. In this algorithm, the goal is taken to be any point at the road segment end. The algorithm proceeds for a total number of *maxiter* iterations which includes iterations in which the generated node could not be added due to infeasibility.

It may be computationally very expensive to search the entire solution space (which is the entire road segment) in trying to reach the goal from the source. Consider the *ideal path* that the vehicle would have followed in the absence of obstacles and other vehicles and kinematic constraints as per the hypothesis. This is going straight down the road without changing lateral positions or steering unnecessarily. The path would be a line parallel to the X axis in the road coordinate axis system with the same position laterally along the Y axis of the road coordinate axis system. The algorithm is *biased* to search in the region around this ideal path. In most simple scenarios involving a single obstacle or vehicle, it should be possible to generate feasible paths by minor deviations from this path. Hence, at every iteration, some points are generated and weighted by their deviation from the ideal path or simply their deviation in the Y axis to the current position of the vehicle in the road coordinate axis system. One of these points is selected using a Roulette Wheel Selection. This ensures bias towards the ideal path, at the same time the search process has exploratory potential for scenarios in which a valid trajectory may only be possible after large deviations. The RRT-Connect algorithm is given as Algorithms 7.6 and 7.7.

Algorithm 7.6: RRT-Connect (source, time, v_i)

```
root ← source
time(root) ← time
rootChild ← point at distance of vehicle's length from root
at current angle of orientation
parent(rootChild) ← root
time(rootChild) ← time + length/vᵢ
if (τᵢ=checkConnect(tree, rootChild))≠ null
    return τᵢ
end if
repeat for a maximum of maxiter
    samples ← generate random samples in road coordinate
    system such that probability of selection of sample q =
    |Y(q)-Y(source)|
    select a sample q by roulette wheel selection
    p ← node nearest to q in tree
    s ← node by extension of p in direction of q
    t₂ ← time(p) + stepsize/vᵢ
```

```
      calculate v from equation (7.1) and (7.3)
      if no point close to s lies in tree ∧ vehicle placed
      at s is obstacle and collision free at time t₂ ^ v = vᵢ
         parent(s) ← p, time(s) ← t₂
         tree ← tree ∪ s
         if (τᵢ=checkConnect(tree, s)) ≠ null
            return τᵢ
         end if
      end if
   end repeat
   return null
```

Algorithm 7.7: CheckConnect (tree, node)

```
p₁ ← node
p₂ ← point at a distance of stepsize along X axis from p₁
t₂ ← time(p₁) + ||p₂-p₁||/vᵢ
calculate v from equation (7.1) and (7.3)
while p₂ is not out of road segment
   if vehicle placed at p₂ is obstacle and collision free at
   time t₂ ^ v = vᵢ
      parent(p₂) ← p₁, time(p₂) ← t₂
      tree ← tree ∪ p₂
   else return null
   end if
   p₁ ← p₂
   p₂ ← point at a distance of stepsize along X axis from p₁
   t₂ ← time(p₁) + || p₂-p₁||/vᵢ
   calculate v from equation (7.1) and (7.3)
end while
return path from node to p₂
```

7.6.2 Local Optimization

The RRT by design focuses upon rapidly finding a valid path, but it gives no concern to the optimality of the generated path. Hence, optimality needs to be added externally in the algorithm. Both global and local optimizations are of interest here. Here, *global optimization* is a term referred to finding the correct strategy for a vehicle to avoid obstacles and other vehicles or finding the optimal homotopic group. It suggests for each obstacle or vehicle encountered whether to avoid it from its left side or right side. *Local optimization* is a term referred to computing exact points and magnitudes of turns or finding the optimal trajectory in the homotopic group.

The RRT algorithm outputs a path which is further optimized by a *local optimization* algorithm. The goal of the optimization algorithm is to minimize the path length while still trying to keep the vehicle separated from other vehicles and obstacles by some distance. The algorithm searches through a number of iterations and stores the best path generated so far. At every iteration, the algorithm attempts to deviate from the points represented at that time as the best path (excluding source, source's child and goal) by a factor and evaluates the resultant path cost. If the resultant path has a lower overall total path cost, the algorithm replaces the best path with that newly calculated. The deviation factor is chosen from a *Gaussian distribution*, the maximum deviation being taken as an algorithm parameter.

The path generated by the RRT or the one being optimized by local optimization is a set of points which gives an approximate idea of the trajectory. The trajectory is generated using these points by spline curves (de Boor, 1978). The path points are used as control points to which the output is a smooth and exact path that the vehicle follows. The cost is computed for this smoothed trajectory. The cost of any trajectory is given by Eq. [7.4].

$$\text{Cost}(\tau) = ||\tau|| + c_1 n_1(\tau) + c_2 n_2(\tau) \qquad\qquad [7.4]$$

Here, $||\tau||$ is the path length and $n_1(\tau)$ is the number of points in the trajectory which lie at a distance of less than *mindis* from any position of the vehicle. For this factor to be computed the vehicle (extended by a factor of *mindis* from all its sides) is traversed on the trajectory τ and checked for collision. $n_2(\tau)$ is the number of points in the trajectory in which the vehicle actually collides with some obstacle on the road, or a point in which the kinematic validity constraint as stated in Eqs [7.1]–[7.3] does not hold. This factor is again measured by simulated traversal. c_1 and c_2 are constants. A trajectory with infeasible points cannot be returned, whereas points with less than the desired clearance distance should be avoided as much as possible. Hence, $1 < c_1 \ll c_2$.

The first term in Eq. [7.4] is introduced to minimize the path length. The second term is to maximize the distance from obstacles (including from other vehicles). The last term is to penalize invalid solutions. The original path formed by RRT is approximate, it is important therefore to use a set of points such that the final trajectory is feasible. As the optimization is local to the range of paths nearby, it may or may not be possible to have a feasible path, and it may or may not be possible to find a path that has sufficient separation from obstacles. The local optimization algorithm cannot and does not search for the path in the global search space as this is computationally costly and as a result extremely time-consuming.

Here, the problem of *global optimization* is difficult to solve considering the computational costs of the algorithm. However, considering the characteristic nature of the vehicle-planning problem it is evident that there are limited options due to the narrow width of the road. The length of the unoptimized path returned by the RRT is indicative of the cost for travelling in that area. The RRT algorithm is called a few times and the shortest path length for valid paths is accepted and subsequently further optimized.

The local optimization algorithm is given by Algorithm 7.8.

Algorithm 7.8: LocalOptimization(τ)

```
τ_best ← τ
Cost_best ← Cost(Spline(τ))
repeat for limit iterations
    τ' ← Deviate(τ_best)
    cost' ← Cost(Spline(τ'))
    if cost' < cost_best, cost_best ← cost', τ_best ← τ'
end repeat
if cost_best < c₂ return Spline(τ_best) else return null
```

7.6.3   Coordination

Different vehicles need to ensure that they do not come in each other's way and/or collide with each other on their way. This problem is solved by devising a coordination technique between the vehicles. Considering the real time nature of the algorithm, it is not possible to have a centralized or tightly bound technique for coordination. Therefore, a priority-based coordination technique is chosen as detailed in Section 7.4.

When computing the path of a vehicle amidst vehicles with higher priorities, there is a choice to either alter the vehicle's path keeping the speed constant, alter its speed keeping the path constant or to simultaneously compute both path and speed to avoid collision. The RRT search algorithm keeps the *vehicle's speed constant and computes the path*. If the RRT fails to find a feasible path which is marked by the algorithm failing to reach the goal within the set number of iterations, it may be assumed that no feasible path is possible. Hence, the algorithm then needs to alter the vehicle's speed. It is not possible to experiment for a large number of speeds to select the highest speed which generates a feasible path due to large computational costs. For similar reasons the speed cannot be reduced by small magnitudes iteratively as was done in Section 7.4. Hence, the traffic scenario is assessed and a good *guess* is made to alter the vehicle's speed.

In light of the behaviour of the algorithm to a variety of maps, it is seen that when a higher-speed vehicle enters the planning scenario, it sees an array of vehicles in front. These vehicles may be travelling on the same side of the road as the vehicle or on the opposite side. If the vehicle is travelling inbound on the road, all inbound vehicles are said to be on the same side and all outbound vehicles are said to be on the opposite side. For the array of vehicles travelling on the *same side*, the vehicle may attempt to go by its preferred speed which, if it is higher than the traversal speeds of vehicles in front, leads to *overtaking behaviour*. Alternatively, it may choose to *follow* the array of vehicles.

Consider that the vehicle being planned is capable of high speed and that at this speed the vehicle is unable to find a feasible path. In this case, overtaking behaviour is not possible. Hence, the speed of the vehicle being planned is set to the *speed of the higher-priority vehicle which is travelling at a lower speed*. This causes the vehicle being planned to follow the higher-priority vehicle. The alteration goes on iteratively till the speed of the vehicle being planned is reduced to equate to the speed of the slowest vehicle in the array.

However, if the vehicle is likely to have a collision with a vehicle travelling in the *opposite direction*, the task is to alter the speed of the vehicle in such a way that a collision is avoided. For this, there is no choice but to *iteratively decrease its speed in small steps*. In the worst case, the speed of the vehicle is reduced to 0 and it stands still. In such a scenario, it may prefer to wait for other vehicles to leave a clear way. In this way continuous replanning by small speed changes in pursuit of a feasible path is a valid strategy. If the road is *blocked*, the vehicle may, as a result, wait indefinitely. Hence, after some time it may additionally prefer to choose another route, considering the present route blocked. The route planner algorithm is called again to compute the new route.

Because the higher-priority vehicles (if any) travelling in the opposite direction did not themselves account for the presence of a waiting vehicle either at the start or just before the road segment, they may have to validate their paths. It is possible for *deadlock* to occur in such situations, in which case vehicles must be reprioritized and replanned to realize a feasible travel plan.

The general planning algorithm is given by Algorithm 7.9.

Algorithm 7.9: Plan (road segment, time)

```
while true
    τbest ← null
    repeat for small number of iterations
        τ ← RRT-Connect(Si, segment)
        if (τbest is null ∨ cost(τ)<cost(τbest)) ∧ τ is feasible
            τbest ← τ
        end if
    end repeat
    if τbest ≠ null, τbest ← LocalOptimize(τbest)
    if τbest ≠ null
        represent τbest as a hash map of time and return τbest
    else if vi>0, vi ← vi-Δ
    else stop and wait
    end if
end while
```

7.7 Results

The algorithm was extensively tested by MATLAB simulation. The algorithm took as input an image file in bitmap (BMP) format which was a black-and-white pictorial of the road map. The scenario specifying the time of emergence of vehicles, their speeds, positions and orientations was fed separately. Various aspects of the algorithm

RRT-Connect, coordination conversions, spline generation, cost computation etc. were developed as separate modules. The outputs of the algorithm were displayed as an animation which indicated the manner in which the various vehicles entered and moved. To fully test the algorithm, a number of experiments were conducted ranging from simple to complex scenarios. Each scenario was selected to display some particular characteristic of the algorithm.

7.7.1 Single-vehicle Scenarios

First, the experiments are discussed which involve a single vehicle moving on a straight or curved road, with or without obstacles. Various scenarios are shown in Fig. 7.10. Fig. 7.10A is the simplest case wherein the vehicle emerges into the centre of the road and sees no obstacles on the road. The only thing it needs to do is to drive along a small curve. The RRT-Connect needed to expand a single node for the scenario and the completion was done by the connect part of the algorithm. The solution was generated in negligible time. The generated solution was feasible and local optimization was in this case unnecessary, though carried out. This shows that for most general driving, algorithmic effort would be minimal, which also displays the present state of planning with speed lanes. Fig. 7.10B represents another scenario extensively discussed in the literature. A single obstacle is added which the vehicle needs to avoid optimally. The RRT needed to expand a few nodes until a reasonable node was found after which the connect algorithm could make the vehicle traverse. It may be seen that vehicle keeps to the right side of the road and does not come to the (normal) left side.

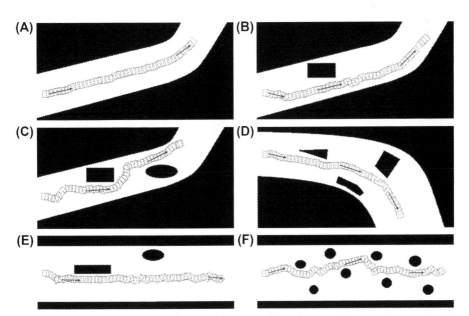

Figure 7.10 Simulation results using RRT-Connect with a single vehicle. (A) Scenario 1, (B) Scenario 2, (C) Scenario 3, (D) Scenario 4, (E) Scenario 5, (F) Scenario 6.

A couple of iterations of local optimization were needed to rectify the path for feasibility. After a few iterations, the exact point of obstacle avoidance could be determined, so that separation from the obstacle was greater than the threshold value.

Fig. 7.10C shows a reasonably complex formulation in which two obstacles are closely packed and the planner needs to make a fine trajectory to avoid both of them. RRT expansions were large considering the need to try out a large number of possibilities, before one turned out to be feasible. It can be seen that the algorithm could successfully draw out a feasible and optimal trajectory. A similar situation occurred in Fig. 7.10D. Here, obstacles are not tightly packed enabling local optimization to fine-tune the path. Fig. 7.10E displays a very complex obstacle framework. The algorithm attempts to find a path which is as central as possible, and which also leads to the goal. Considering path smoothness, it is difficult to find points for making turns. Similarly, local optimization here has a huge task to do in optimizing the trajectory for minimum separations from the variety of obstacles. Fig. 7.10F shows the scenario in which two paths were possible. The algorithm chose the simpler and shorter one. It cannot be ascertained whether every time the same path is chosen, but the algorithm is certainly more prone to choose the globally optimal path, as the path length is approximated by RRT runs and the best are taken.

7.7.2   Two-vehicle Overtaking and Vehicle-following Scenarios

In the next category of experimentation the same maps were used as before, but two vehicles were employed in the scenario rather than a single vehicle. The first vehicle was modelled as a slow-moving vehicle which entered the scenario unaware of the second vehicle. It formulated its trajectory and continued its motion. After some time the second vehicle entered the scenario and this was capable of moving at higher speed. The second vehicle hence had a choice to either attempt to overtake the first vehicle or to follow it. It first plans to overtake it, and if a feasible path is not found, it reduces its speed and simply constructs a trajectory by which it follows the first vehicle. Some of the experiments are shown in Fig. 7.11 indicating the scenario a little time after the emergence of the second vehicle and the final trajectories.

Fig. 7.11A shows a road with no obstacle in which the first vehicle was driving in the middle of the road. The second vehicle naturally has a lot of room to simply overtake it which it does as shown in Fig. 7.11B. The obstacle landscape for the second vehicle which accounts for the first vehicle is simply a single elongated obstacle, which is easy to overcome. However, overtaking in the presence of an obstacle is difficult and can only be done if sufficient space is available. The scenario shown in Fig. 7.11C and D clearly shows that the second vehicle had no space to overtake the first vehicle considering its obstacle landscape is filled up by both a static obstacle and a moving vehicle. Hence, as a result, it had to follow the first vehicle.

Fig. 7.11E and F show another scenario in which large spaces were available and hence overtaking was possible. The first vehicle took the straighter route. The second vehicle now had a choice to overtake the first vehicle either on the left or the right. If it does so on the left, it encounters an additional obstacle which it must also avoid. Hence, the second vehicle decides to overtake on the right. Both these possible trajectories have roughly the same cost, hence a number of experiments resulted in both

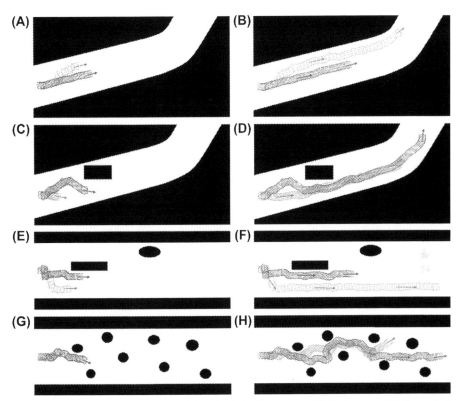

Figure 7.11 Overtaking and vehicle-following behaviours using RRT-Connect. (A) Scenario 1 middle, (B) Scenario 1 end, (C) Scenario 2 middle, (D) Scenario 2 end, (E) Scenario 3 middle, (F) Scenario 3 end, (G) Scenario 4 middle, (H) Scenario 4 end.

these paths being selected by the second vehicle on different runs. In the scenarios shown in Fig. 7.11G and H, a large amount of space was available, although it was still not large enough for a complete vehicle to fit in and overtake. Both vehicles preferred to remain roughly in the centre of road as per their initial generation. Once the second vehicle decided to follow the first vehicle, it reduced its speed and, from then on, the presence of the first vehicle barely affected the motion of the second.

7.7.3 Vehicle-avoidance Scenarios

In all these experiments, because the vehicles were assured of no vehicle approaching on the other side of the road, they could make use of the entire road width. In real situations, however, the same road would be used for both inbound and outbound traffic which necessitates proper coordination between the two sets of vehicles. This ability of the algorithm was also therefore tested via experiments on the same set of maps. In these experiments, two vehicles were generated, one closely after the other, from two opposite ends of the segment and facing each other. The algorithm had to devise trajectories for both these vehicles such that they avoided each other. This task is twofold. First, the algorithm needs to decide what the highest possible speeds of the vehicles are so that any likely point of collision is at a place where sufficient width is available for the vehicles to avoid each other. Then the planning algorithm needs to compute the trajectory. If at some speed the algorithm determines there is likely to be a collision and no feasible path, it reduces the travel speed. This shifts the place where vehicles meet to one at which it may be possible to avoid a collision. If a collision is still planned to occur or no feasible path may be computed, the process repeats. Experiments in a few scenarios are shown in Fig. 7.12. The figures show how the vehicles avoid each other, and henceforth how they complete their journeys.

Fig. 7.12A shows the first vehicle travelling straight and hence the second has to make a small turn to avoid a collision. The planning of the second vehicle is also simple because once the collision-avoidance point is found, the RRT-Connect algorithm proceeds to complete the trajectory. The vehicle which enters first gets the straighter path (in the road coordinate axis system). By the choice of speeds which were kept equal, the scenario shown in Fig. 7.12C and D was simple. The vehicles meet at a point where the second vehicle has a lot of space to modify its trajectory to avoid collision. Planning for the second vehicle is, however, slightly difficult as it first has to avoid the

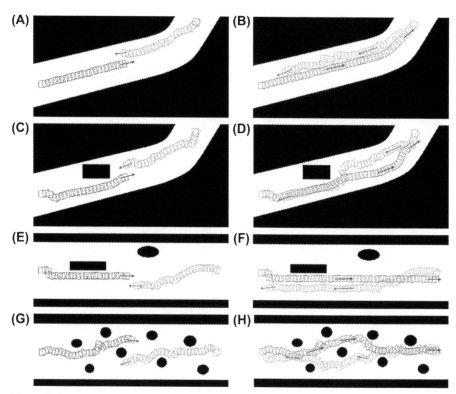

Figure 7.12 Vehicle-avoidance behaviours using RRT-Connect. (A) Scenario 1 middle, (B) Scenario 1 end, (C) Scenario 2 middle, (D) Scenario 2 end, (E) Scenario 3 middle, (F) Scenario 3 end, (G) Scenario 4 middle, (H) Scenario 4 end.

vehicle and then the static obstacle. Similar comments hold for the scenario shown in Fig. 7.12E and F. The scenario shown in Fig. 7.12G and H is more complicated, however, as the second vehicle has to not only escape from the static obstacle framework, which is itself complicated, but also avoid the first vehicle. This makes the obstacle landscape very complex. It may be seen how the vehicles can coordinate to travel on opposite sides of the obstacles to avoid collision.

7.7.4 RRT Analysis

It is important for RRT-Connect to generate a trajectory within a short execution time. Local optimizations are an optional feature which may be timed to assess their performance. This means that it must be possible to reach the goal by means of minimal expansions of the tree even in complex environments. The RRT generated for a static scenario is shown in Fig. 7.13A. It can be seen that effectively the algorithm only generates RRT until a point from which a straight path (in the road coordinate axis system) will reach the goal. Hence, complexity is proportional to the number of obstacles. Because in usual driving the number of obstacles is low, the algorithm is likely to generate paths with short overall execution times. Further, even with a high number of obstacles, the algorithm tries to restrict itself to good expansions, rather than

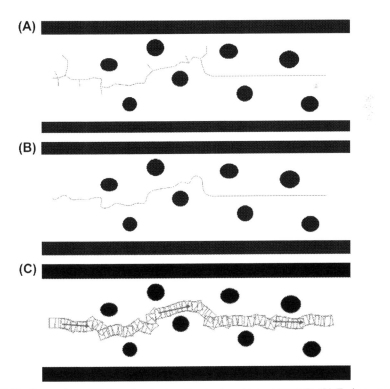

Figure 7.13 Generation of RRT and trajectory for single vehicle. (A) RRT, (B) Trajectory, (C) Trajectory traced by the vehicle.

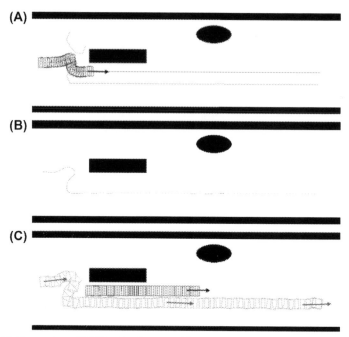

Figure 7.14 Generation of RRT and trajectory with multiple vehicles. (A) RRT, (B) Trajectory, (C) Trajectory traced by the vehicles.

expanding the entire road available. The path found for the map is shown in Fig. 7.13B, which, after local optimizations, results in the trajectory as shown in Fig. 7.13C.

The scenario is more interesting in the presence of other vehicles, the planned trajectory for which is known and which the algorithm attempts to cause the vehicle to avoid. The same set of figures with a single vehicle is shown in Fig. 7.14. The figure shows a case in which space was available and the vehicle could overtake. Fig. 7.14A shows a region in between the red trajectory of the slow vehicle already planned and the green RRT generated which appears as an obstacle area. This region is where the first vehicle's motion, as per its space—time, results in a collision with the second vehicle, which hence had to take a longer route to avoid a collision. It is worth noting that even in the presence of other vehicles the number of nodes expanded is fairly low which means that computation time is minimal. The curve generated for two vehicles is shown in Fig. 7.14B. The final trajectory after local optimization is shown in Fig. 7.14C.

7.7.5 Local Optimization Analysis

The other important aspect is to analyse the local optimization performed by the algorithm. Initially, the RRT generates a plan which ensures that the vehicle would not collide with any obstacle when placed at the nodal points. However, in a continuous domain in which the path is smoothed by splines, there may be collisions. This is therefore the first task performed by the local optimization algorithm. Then the local optimization tries to increase separation distances so that the vehicle's distance from all obstacles is greater than the specified threshold. Following this, the algorithm attempts to minimize the path length as the last criterion, which is considered of least importance. This is shown in Fig. 7.15A and B for optimization on a random scenario with only static obstacles, and Fig. 7.16A and B for optimization with an additional vehicle. In both these cases, it may be seen that the path returned by the RRT-Connect algorithm was actually infeasible, but could be rectified in a single iteration. The distance was slowly increased and later the path was optimized in both these approaches. The deviation factor accounting for the magnitude of deviation was kept a little high as compared to other approaches. This was done to ensure that the algorithm quickly made a feasible path, with all other objectives being secondary. This ensured that relatively few iterations of the local optimization routine were sufficient for trajectory generation.

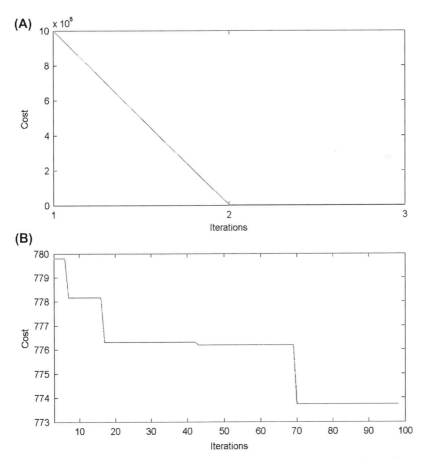

Figure 7.15 Performance of local optimization algorithm for single vehicle. (A) Smaller iterations, (B) Larger iterations.

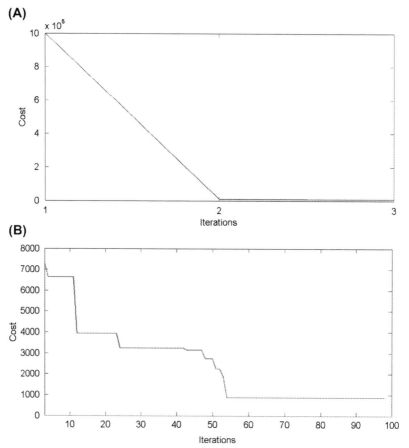

Figure 7.16 Performance of local optimization algorithm in the presence of multiple vehicles. (A) Smaller iterations, (B) Larger iterations.

7.8   Summary

The chapter touched upon the general class of sampling-based approaches to solve the problem of navigation of vehicles within the road segment under the assumption of communication. Given the reasonably large size of maps and the associated problems with a reduction in resolution, sampling becomes a natural choice to control dimensionality while still pressing on near-optimal searches to solve the problem. The chapter used the RRT algorithm to solve the problem, which was also extended to RRT-Connect.

In RRT, planning was done in road segments. Prioritization was used as a means of coordination between vehicles in which vehicles that entered the planning scenario earlier had a higher priority. An attempt was made to generate more samples so that vehicles occupy the same lateral positions. Experimental results over a number of scenarios show that vehicles were able to navigate in fairly complex environments. Behaviours of vehicle following, overtaking and vehicle avoidance were also displayed.

The approach was extended to RRT-Connect in which the intent was to generate a collaborative collision-free travel plan involving all vehicles on the road. The developed solution makes use of priority-based coordination between vehicles. Each vehicle plans using RRT-Connect and local optimization. In case a vehicle is unable to generate a feasible trajectory considering the motions of higher-priority vehicles, it may need to reduce its speed and replan. To generate a feasible plan within a short execution time, the algorithm attempts to make the vehicle follow the vehicle in front if one exists, failing which its speed is reduced by a fixed magnitude. The RRT-Connect algorithm ensures that feasible trajectories are generated by minimal expansions, the result of this being that algorithm execution times are low.

The algorithm was experimented on with a variety of scenarios ranging from simple to difficult. The purpose was to test whether the algorithm is able to generate a feasible trajectory even for scenarios in which it may be difficult to manoeurvre the vehicle. The results showed that a vehicle could be easily navigated with plans returned reasonably early. The plans could be locally optimized using few iterations of a local planner for shorter and safer paths. Every feasible overtaking of the faster vehicle was carried out by the slower vehicle. Further, vehicles could mutually avoid each other.

References

Anderson, S.J., Karumanchi, S.B., Iagnemma, K., 2012. Constraint-based planning and control for safe, semi-autonomous operation of vehicles. In: Proceedings of the 2012 IEEE Intelligent Vehicles Symposium, Madrid, Spain, pp. 383−388.

Bennewitz, M., Burgard, W., Thrun, S., 2001. Optimizing schedules for prioritized path planning of multi-robot systems. In: Proceedings of the 2001 IEEE International Conference on Robotics and Automation, Seoul, Korea, pp. 271−276.

Bennewitz, M., Burgard, W., Thrun, S., 2002. Finding and optimizing solvable priority schemes for decoupled path planning techniques for teams of mobile robots. Robotics and Autonomous Systems 41 (2−3), 89−99.

de Boor, C., 1978. A Practical Guide to Splines. Springer Verlag, Berlin-Heidelberg.

Carpin, S., Pagello, E., 2009. An experimental study of distributed robot coordination. Robotics and Autonomous Systems 57, 129−133.

Chakravorty, S., Kumar, S., 2011. Generalized sampling-based motion planners. IEEE Transactions on Systems, Man and Cybernetics Part B: Cybernetics 41 (3), 855−866.

Ferguson, D., Stentz, A., 2006. Anytime RRTs. In: Proceedings of the IEEE International Conference on Intelligent Robots and Systems, Beijing, China, pp. 5369−5375.

Jaillet, L., Cortes, J., Simeon, T., 2008. Transition-based RRT for path planning in continuous cost spaces. In: Proceedings of the IEEE/RSJ International Conference on Intelligent Robots and Systems, Nice, France, pp. 2145−2150.

Kala, R., 2013. Rapidly-exploring random graphs: motion planning of multiple mobile robots. Advanced Robotics 27 (14), 1113−1122.

Kala, R., Warwick, K., 2011a. Planning of multiple autonomous vehicles using RRT. In: Proceedings of the 10th IEEE International Conference on Cybernetic Intelligent Systems, Docklands, London, pp. 20–25.

Kala, R., Warwick, K., 2011b. Multi-vehicle planning using RRT-connect. Paladyn Journal of Behavioural Robotics 2 (3), 134–144.

Kalisiak, M., van de Panne, M., 2006. RRT-blossom: RRT with a local flood-fill behaviour. In: Proceedings of the 2006 IEEE International Conference on Robotics and Automation, Orlando, FL, pp. 1237–1242.

Karaman, S., Frazzoli, E., 2011. Sampling-based algorithms for optimal motion planning. The International Journal of Robotics Research 30 (7), 846–894.

Kuffner, J.J., LaValle, S.M., 2000. RRT-connect: an efficient approach to single-query path planning. In: Proceedings of the IEEE International Conference on Robotics and Automation, vol. 2, pp. 995–1001.

Kuwata, Y., Karaman, S., Teo, J., Frazzoli, E., How, J.P., Fiore, G., 2009. Real-time motion planning with applications to autonomous urban driving. IEEE Transactions on Control System and Technology 17 (5), 1105–1118.

LaValle, S.M., Kuffner, J.J., 1999. Randomized kinodynamic planning. In: Proceedings of the IEEE International Conference on Robotics and Automation, pp. 473–479.

Leonard, J., How, J., Teller, S., Berger, M., Campbell, S., Fiore, G., Fletcher, L., Frazzoli, E., Huang, A., Karaman, S., Koch, O., Kuwata, Y., Moore, D., Olson, E., Peters, S., Teo, J., Truax, R., Walter, M., Barrett, D., Epstein, A., Maheloni, K., Moyer, K., Jones, T., Buckley, R., Antone, M., Galejs, R., Krishnamurthy, S., Williams, J., 2008. A perception driven autonomous urban vehicle. Journal of Field Robotics 25 (10), 727–774.

Lian, F.L., Murray, R., 2003. Cooperative task planning of multi-robot systems with temporal constraints. In: Proceedings of the 2003 IEEE International Conference on Robotics and Automation, pp. 2504–2509.

Raveh, B., Enosh, A., Halperin, D., 2011. A little more, a lot better: improving path quality by a path-merging algorithm. IEEE Transactions on Robotics 27 (2), 365–371.

Strandberg, M., 2004. Augmenting RRT-planners with local trees. New Orleans, LA. In: Proceedings of the IEEE International Conference on Robotics and Automation, vol. 4, pp. 3258–3262.

Urmson, C., Simmons, R., 2003. Approaches for heuristically biasing RRT growth. Las Vegas, Nevada. In: Proceedings of the 2003 IEEE/RSJ International Conference on Intelligent Robots and Systems, vol. 2, pp. 1178–1183.

Xidias, E.K., Azariadis, P.N., 2011. Mission design for a group of autonomous guided vehicles. Robotics and Autonomous Systems 59 (1), 34–43.

Zhang, L., Manocha, D., 2008. An efficient retraction-based RRT planner. In: Proceedings of the IEEE International Conference on Robotics and Automation, Pasadena, CA, pp. 3743–3750.

Graph Search-Based Hierarchical Planning

8.1 Introduction

The problem of navigation of autonomous vehicles within a road segment without any lanes is a rather open-ended problem. Ideally, no assumption can be made regarding the placement of obstacles and motion of other vehicles which gives rise to infinite possibilities which an algorithm needs to address. With such an open-ended problem it is impossible to simultaneously cater to both optimality and completeness, whereby making the execution time very short. Some practical assumptions or heuristics need to be injected which may give good returns with small costs.

Characteristic placement of obstacles and other vehicles might make an optimal trajectory of any vehicle difficult to produce. *Sampling-based approaches* work well only in cases in which the optimal trajectory does not pass through regions of narrow passages, whereas the computational time may become very large if the optimal trajectory takes large number of turns. In both cases, the completeness of the approaches is questionable. *Reactive techniques* are clearly not complete. In cases with narrow passages, sampling-based techniques may either die out (converge) without producing a result or may produce a much worse path corresponding to a local optimum.

Graph-based techniques (Cormen et al., 2009; Langsam et al., 2009) can compute all possible transitions of the vehicle and are hence both complete and optimal. The approach, however, works in a discrete space rather than a continuous space and the completeness and optimality would hold only for a *high-resolution* discretization of space along with a rich action set. The limitation of the approaches is that they are highly computationally expensive. The computational cost is proportional to the number of states which has to be kept large for the generation of an optimal trajectory, especially when the optimal trajectory goes through congested regions. The basics of graph search are covered in Section 8.3. The pros and cons in the use of graph-search algorithms are summarized in Box 8.1.

This chapter is devoted to the use of *heuristics* to intelligently pick up points in the road segment to produce a limited graph which can be used to quickly return the optimal path or trajectory. The approach is hence near complete and near optimal when the loss of completeness and optimality is dependent upon the quality of heuristics floated. The chapter advocates how the road architecture can be used to make a small set of possible discrete actions which lead to minimal loss of completeness, whereas local optimality may be somewhat questionable.

The other major factor associated is *coordination*. A complete base planner when working under a incomplete coordination strategy gives a incomplete system.

Box 8.1 Pros and Cons of Graph Search

Pros
- Completeness subjected to resolution of selection of states and actions
- Optimality subjected to resolution of selection of states and actions

Issues
- Computational Complexity

Key Idea
- Hierarchies

Optimality is hard to achieve in decentralized planners whereas a centralized set of planners may be computationally very expensive. A similar challenge, hence, lies in making a coordination strategy which is complete and near optimal for a large variety of scenarios. Because the problem definition is rather open-ended, any number of vehicles may interact in any number of ways showcasing behaviours previously seen or unseen in a road network. A *strong coordination strategy* is one that allows for all possible combinations of ways in which the vehicles can avoid each other. The coordination shown in this approach is not based on the simple notions of vehicle avoidance, vehicle following and/or vehicle overtaking; but deliberates to *intelligently* place the vehicles so that they best avoid each other.

This chapter solves the problem in a hierarchical manner in which first the attempt is made to decide the route of the vehicle, which is the case with all approaches used and was centrally taken in chapter 'Introduction to Planning'. The next hierarchy deals with the optimal manner of avoiding the obstacles. This strategy is used to decide the manner in which a vehicle would avoid the other vehicles which are planned in a prioritized manner. The resultant strategy is used for computing the actual trajectory which the vehicle may take. Such a *hierarchical decomposition* of the problem creates a simpler problem to solve at a higher level, and once solved extensively simplifies the problem at the lower end. Of course, the higher-level plans are approximate or vague and hence optimality cannot be guaranteed, but it would be presented later how the heuristics make it very likely for the approach to be near optimal. However, the decomposition makes the algorithm very fast while still being complete.

Segments of the chapter have been reprinted from Kala and Warwick (2013) with kind permission from Springer Science + Business Media: Journal of Intelligent and Robotic Systems, Multi-Level Planning for Semiautonomous Vehicles in Traffic Scenarios Based on Separation Maximization, Vol. 72, No. 3, 2013, pp. 559–590, R. Kala, K. Warwick, Text and All Figures, © Springer Science + Business Media Dordrecht 2013.

8.2   A Brief Overview of Literature

This section discusses the approaches using graph search either on the entire state space, on a sampled state space produced out of the entire state space (or on a roadmap) or using hierarchical decomposition of the problem for making the resultant approach computationally efficient. The general problem of graph search is a high computational cost, whereas the roadmap requires an offline construction phase.

Multilayered planning is widely used in robotics wherein the map is represented using different cell shapes like triangular, hexagonal (Hou and Zheng, 1991), Voronoi-based cells (Takahashi and Schilling, 1989), rectangular cells with quad-tree map representation (Hwang et al., 2003; Urdiales et al., 1998), framed quad-tree (Chen et al., 1997; Yahja et al., 1998) etc. Each technique enables representations of the map in different resolutions. One of the popular methods is to represent the map in different levels, each level having a different resolution (Kambhampati and Davis, 1986). Segments of maps may be at different resolutions depending upon the need. The search is applied with different segments of maps at different resolutions. The search may delve into finer resolutions if needed.

In another related work, Zhang et al. (2007) used a hybrid approach of approximate cell decomposition and probabilistic roadmap. The two approaches were used one after the other. The process was repeated till a feasible solution was computed. The work of Lu et al. (2011) decomposed the cell and searched iteratively for a solution, till a reasonable solution was computed. Lifelong Planning A* algorithm was used for the search. Because intelligent vehicles are not associated with open spaces, a large number of decompositions may be needed before obtaining an estimate regarding the feasibility of overtaking, requiring a large computation time.

It is common to restrict the problem division to two levels, creating a coarser and a finer level. In a similar work, Chand and Carnegie (2012) used A* algorithm for coarser and finer planning. The finer-level A* algorithm was used for cost computations of the coarser roadmap. Kala et al. (2010) also solved the problem in two hierarchies using Genetic Algorithm at both levels. The finer-level Genetic Algorithm was designed to have less computational time to react to dynamic obstacles. Cowlagi and Tsiotras (2012) allowed for multiple costs between states depending upon the kinematics of the vehicle. Model Predictive Control was used for local trajectory planning. For the problem considered in this work, it would be impossible to compute the possibility of overtaking using a coarser or a lower-resolution map. Hence, different ways are constructed to deal with the problem of resolution.

8.3   A Primer on Graph Search

Many everyday problems can be converted into graphs in which the *state* of the system is represented by the *vertices* and all the possible *actions* are presented by the *edges* (Konar, 1999; Russell and Norvig, 2009). The actions take the system from one state to the other and thus connect two vertices in the problem graph. As an example, a game of chess has the vertices as any state of the game including the positions of all the chess pieces, whereas the edges are the moves that the player makes. Similarly, while planning for a project, one may have different states of the projects as different vertices, each denoting the modules completed or yet to be completed. The actions of completing a module or making some changes to the module are the edges, which make the project go from one state to the other. In the problem of autonomous vehicle navigation, the road-network graph of the city is used for searching a route from the source to the goal. Here, all intersection points are the vertices, whereas the roads are the intersections.

A graph $G <V,E>$ is a collection of vertices (V) and edges (E). Each edge $<u,v>$ connects any two vertices u and v. A graph is called a *directed graph* when the edges have directions denoting some phenomenon of real life associated with the edge. In transportation, if the roads are one-ways wherein the road can be used to travel inbound, but not outbound, the roads constitute directed edges of the road-network graph. An undirected graph is the case in which an edge from u to v means that the same edge can be used to go from u to v and from v to u with the same cost. *Undirected edges* between vertices u and v may be modelled by two directed edges, one going from u to v and the other going from v to u. For these graphs, if an edge $<u,v>$ exists, the edge $<v,u>$ must also exist.

The edges of the graphs may have *weights* associated with them denoting the weight of travelling by that edge as per real life. Let $w(u,v)$ denote the weight of the edge $<u,v>$. For edges denoting actions, the weight is normally the cost incurred in taking that action. In transportation, the roads act as edges whereas the road length acts as the weight. If all actions have equal or unitary weights, the weights of the edges may be ignored and hence the graph is regarded as unweighted.

The graphs are stored as a collection of vertices and a collection of edges, each edge joining two vertices. The data structures used to store the vertices are trivial. The edges may be stored as an *adjacency matrix*, Adj: $V^2 \rightarrow \{0,1\}$, which is a matrix of size $|V| \times |V|$, in which $|V|$ is the number of vertices of the graph. The entry a_{uv} in the graph of row u and column v may be 0, denoting no edge from vertex u to vertex v or 1 denoting an edge from vertex u to vertex v. In the case of an undirected graph, if the edge $<u,v>$ exists, then the edge $<v,u>$ must also exist. This means the adjacency matrix is the same as its transposition, Adj = Adj^T, meaning only half of the matrix may actually be stored. A separate *weight matrix* is needed to store the weights for weighted graphs. The adjacency and weight matrices may be combined into a single matrix which stores the edge weight, if one exists, and infinity otherwise. If the weight can be derived from the knowledge of the vertices, the weight may not be stored.

An alternative way of storing the edges is by an *adjacency list*, wherein a list of vertices is stored corresponding to every vertex v. The list stores all the vertices to which the vertex u is connected to by an edge. The weight information may be appended to the vertex making it a list containing pairs of vertex and edge weight. The adjacency matrix takes a lot of space, especially for *sparse graphs* wherein the adjacency matrix is a sparse matrix with most entries as 0s. On the contrary, the adjacency list method is more memory efficient in the case of sparse graphs. The overhead of the data structure is larger and the method is hence not useful for dense graphs. A common operation in graphs is to get the *list of edges* from a vertex u, which is computationally very efficient with an adjacency list as compared to an adjacency matrix. However, adding edges and checking existence of a particular edge is easier for the adjacency matrix. Overall, for most real-life problems which have sparse connectivity, the adjacency list method is preferable.

These graphs may not always be stored or represented as a set of vertices and edges, because the number of vertices and edges may be infinite or extremely large. The number of states are usually very high and thus initialization of all the vertices is also computationally infeasible. Sometimes the graphs are defined by *rules* governing the representation of vertices and edges, with vertices denoting *legal states* and edges denoting *legal actions* as per the problem definitions. Consider the game of chess in which every valid position of every valid chess piece defines a state and valid actions are defined by the chess rules. States are added and processed dynamically as they are generated. Every state or vertex is initialized when it is first seen in the search process and maintained thereafter as per the search process. Chiefly, only states near the path from the source to the goal of the problem are generated and worked upon, whereas the actual domain of the problem may be infinite.

8.3.1 Graph Search

Graph search is the process of finding a *path*, preferably the least-cost path, from a given vertex or the *source* to any other vertex which satisfies a *goal-test* criterion, through a path which adheres to any *path constraints*. Let the source vertex be S (or v_0). Let GoalTest(v) be the function which checks whether a vertex v can be used as a goal. For the single-goal problem, one unique vertex satisfies the goal test. For a multigoal problem, a number of vertices will satisfy the goal test. The path is a set of vertices starting from the source and ending at the vertex v_n which satisfies the goal test. The path can be given by $\tau = v_0 \rightarrow v_1 \rightarrow v_2 \rightarrow v_3 \rightarrow \ldots v_n$, in which GoalTest($v_n$) is positive. Here, every pair of consecutive vertices must be a valid edge, or $<v_i, v_{i+1}> \in E$. The total cost of the path is given by Eq. [8.1].

$$\text{cost}(\tau) = \sum_{i=0}^{n-1} w(v_i, v_{i+1}) \qquad [8.1]$$

The optimal path $\tau*$ is the path which minimizes the path cost given by Eq. [8.2], which the graph-search algorithms try to find.

$$\tau* = \arg \min \text{cost}(\tau) \qquad [8.2]$$

8.3.2 States and Actions

To solve any problem using graph search, two steps must be followed. First, the problem needs to be *modelled as a graph* and then any standard graph-search algorithm can be used to solve the graph. The innovation in the solution of the problem using graph search lies in efficient modelling of the graph as well as making the graph search more efficient.

States represent the vertices of the graph. Anything which gets changed by actions or in the course of moving from source to goal must be included as a variable in the state representation. Anything which does not affect the system or remains constant in the course of action from source to goal should not be included in the state representation. A *good state representation* must be able to represent all the necessary information in a compact form. *Abstraction* may be used to filter out unnecessary information by making suitable assumptions, while not losing much on solution optimality. The complete problem may be solved in multiple hierarchies in which each hierarchy stores information related to itself. All the states which the system can represent as per the state modelling are known as the *state space*. Many states representable may be infeasible or unreachable. In motion planning of vehicles, the position and orientation of the vehicle is enough to represent the state. The states which make the vehicle collide with an obstacle and the states outside the road boundary are infeasible.

The edges of the graphs represent the actions. Every action takes the system from one state to the other. The action set defined for any problem must be rich enough to represent the path from the source to goal without losing on the solution optimality. However, only those actions which help reach the goal from any valid state must be considered. The actions which have no contribution towards reaching a solution should not be modelled to reduce computational time. Only actions resulting in a legal state as per problem definition may be applied. Similarly, problems may impose constraints to which the actions must adhere. Illegal actions violating constraints are not considered as a part of the graph. Every action may have a cost associated with it, which becomes the weight of the edge. These costs are also known as the *step costs*. The *path cost* is usually the summation of all step costs incurred in the path and denotes the cost of the entire solution path.

The number and type of actions need to be judiciously chosen to act as edges. Problem-specific heuristic techniques can be used in proper selection of both states and actions, or vertices and edges. Different graph-search algorithms differ in the order in which they process the nodes in pursuit to find a solution from the source to the goal. Graph-search algorithms may be modified to heuristically select nodes for processing in an order that gives a short computational time while not adversely affecting the optimality for most typical and common scenarios of operation. *Pruning* of nodes or not processing some states and actions can result in a big computational boost.

Graph search can only be applied for *discrete* problems or when the number of states is discrete and the number of actions is discrete. Continuous states and actions need to be decomposed into discrete states and actions by fitting in grids. Each grid piece denotes a small continuous spectrum of states or actions, and the mean value can be used to represent the entire spectrum. The larger is the number of grids, the finer is the *resolution*. Finer resolutions give better results, but at the same time may take significantly long. These search techniques are therefore called *resolution optimal* and *resolution complete*, because the optimality and completeness can only be guaranteed for the current resolution of operation. Changing the resolution will affect the optimality and completeness. Searching for a trajectory in any road segment can also be done by fitting in grids in the road segment which become the vertices, whereas navigation from one grid to the other constitutes the edges.

8.3.3   Uniform Cost Search

The first algorithm discussed to solve the problem of finding a solution from the source to the goal is *Uniform Cost Search* (UCS) (Felner, 2011; Russell and Norvig, 2009). The algorithm gives optimal solutions when the edges have *nonnegative weights*, which is a practical assumption for most real-world problems. The algorithm comes under the class of *uninformed* or unintelligent searches, as no problem-specific information is used in the search process. The algorithm is similar to the commonly used *Dijkstra's algorithm*.

The algorithm maintains a priority queue of nodes called a *fringe*, in which the computed cost of the node from the source acts as a node's priority. The fringe stores all the nodes which have been discovered but have not yet been processed. At every iteration the node with the least priority is taken out from the priority queue and processed. The fringe is implemented using a Heap or any balanced tree, giving logarithmic time inserts and deletes. It can optionally maintain a *closed* collection of processed nodes. All the nodes, once they are processed, are added to this collection. The collection is normally implemented using a Hash map, giving constant time inserts and query.

For every node (n), a number of attributes are stored, which include the cost from the source ($g(n)$), and the parent of the node ($\pi(n)$) or the penultimate node in the path from the source to the node n. The variable $g(n)$ is also called the *historic cost* of the node n. Initially, the fringe stores just one element, which is the source S with $g(S) = 0$ and $\pi(S) =$ NIL. At the first iteration, the source is removed from the fringe and inserted into the closed set of nodes, whereas all children of the source are inserted into the fringe. The search goes on in this manner.

Every time a node is extracted from the fringe, all its children are inserted back into the fringe only if they are not already existing in the fringe or in the closed collection of processed nodes. Let the node extracted from the fringe be n with cost $g(n)$, which is connected to the node n' by an edge $<n,n'>$. A path of optimal length $g(n)$ is known till the node n, whereas a straight-line connection between n and n' exists as a result of this edge with cost $w(n,n')$. Hence, out of all possible paths from source to n', one of them has a cost of $g(n) + w(n,n')$, whereas better ones may still exist. In case the node n' already exists in the fringe and it is rediscovered, the metrics relating to the better cost is retained. In case n' already exists in the closed set of processed nodes, it can be ascertained that the rediscovered node has a worse cost and the rediscovered node can be discarded. When a node is processed, it is removed from the fringe and added in the set of closed or processed nodes.

Every node taken out from the fringe and processed has attained its optimal cost. This is because the nodes are extracted in the order of their cost from the source.

The search operates in *contours of increasing costs* from the source. At any iteration, if the node with cost $g(n)$ is extracted, it means that all nodes with a cost smaller than or equal to $g(n)$ have already been extracted and no node can now take a cost less than $g(n)$. On the other hand, no node with a cost greater than $g(n)$ has been extracted from the fringe. Hence, when a node is taken out of the fringe which tests positive on the goal test, the search can stop and the optimal path cost is obtained. On the contrary, the queue becomes empty when all the nodes reachable from the source have been processed. If the goal is not found, it means that the goal is nonreachable from the source, and no search path exists that links the source to the goal.

If and once a goal is found, the path can be printed by traversing from the goal to the source using the parent information stored in all nodes. The source is the node the parent of which is NIL. A *search tree* is formed by the vertex with an edge between the vertex and its parent. The search tree is rooted at the source. The search tree stores the optimal path from the source to any node. The general algorithm is given by Algorithm 8.1. Algorithm 8.2 prints the path from the source to the goal. By iterating from the goal to the source, the reversed path is obtained which starts from the goal. Hence, the path is reversed before returning.

Algorithm 8.1: Uniform Cost Search (G<V,E>, S, GoalTest)

```
g(S)←0, π(S) ←NIL
fringe ← empty priority queue
closed ← empty set of nodes
add S to fringe with priority g(S)
while fringe is not empty
        n← extract node with least priority
        closed← closed Un
        if GoalTest(n), break
        for all n': <n,n'>∈E
                if n'∉fringe ∧ n'∉closed
                        g(n') ← g(n)+w(n,n')
                        π(n') ← n
                        add n' to fringe with priority g(n')
                else if n'∈fringe ∧ g(n)+w(n,n')<g(n')
                        g(n') ← g(n)+w(n,n')
                        π(n') ← n
                        update priority of n' to g(n')
                end if
        end for
end while
if fringe is not empty, printPath(n)
else print "No Path Exists"
end if
```

Algorithm 8.2: PrintPath(n)

```
τ ← φ
while n ≠ NIL
        τ ← τ∪n
        n ← π(n)
end while
return reverse(τ)
```

8.3.4 A* Algorithm

The *A* algorithm* (Hart et al., 1968; Russell and Norvig, 2009) comes under the class of *informed* or *intelligent* search algorithms. The basic principle is same as the UCS. However, here, additional information is available specific to the problem domain. This information is called a *heuristic* and is denoted by a function $h: V \rightarrow R^+$. The *heuristic* cost $h(n)$ of a node n, is an estimated distance of the node n to the goal. This estimate may be made by a general knowledge of the problem. Because of this definition, the $h(\text{Goal}) = 0$ or the estimated distance of goal from the goal is 0. The heuristic value increases as one travels further away from the goal. The estimates strictly need to be positive as no edge has a negative weight. The A* algorithm attempts to improve the UCS by making use of this additional estimate. The basic working philosophy is the same as the UCS. A fringe is maintained to store the nodes which are seen but are yet to be processed. A closed collection of processed nodes may be maintained. The difference with the UCS is that the fringe is prioritized based on the *expected cost from source to the goal via the node*.

Let $g(n)$ be the cost from the source to the node n as computed in the UCS. This cost is obtained by expansions of the nodes of the graphs and adding their step costs or the edge weights. A cost of $g(n)$ assures that at least one path of this cost exists from the source to the node n. The heuristic cost $h(n)$ is an estimated cost of reaching the goal from the node n. This is just an estimate the correctness for which is not guaranteed. The expected cost $f(n)$ from source to goal via node n may thus be given by Eq. [8.3]. The fringe is prioritized based on the cost $f(n)$.

$$f(n) = g(n) + h(n) \tag{8.3}$$

Initially, the fringe contains only the source (S) with a historic cost of 0 and a heuristic cost of $h(S)$, making a total cost of $f(S) = 0 + h(S)$. At any instance the node with the smallest f cost is taken from the fringe and is processed by generating all its children which are nodes connected to n by an edge. Let n' be one such node connected by an edge $<n,n'>$ with weight $w(n,n')$. The cost from source $g(n')$ is computed using the same principles as in UCS and the cost is given by $g(n) + w(n,n')$. The heuristic cost $h(n')$ is problem specific and the heuristic function to calculate this cost is supplied in the problem specification. The estimated cost $f(n')$ is the sum of the two costs. If the node n' does not exist in the fringe and the closed set of nodes, it is added. However, if the node does exist, it is only added if the new cost estimates are better than the previous estimates, in which case the metrics corresponding to the new estimates are used. If the algorithm is optimal (Section 8.3.5), it is not possible to rediscover a node with a better cost which already exists in the closed set of nodes.

In A* algorithm, the search happens by expanding nodes in the order of their expected costs. The nodes promising a very small cost of the entire path are explored first, before the nodes which indicate a large path cost. The search happens by expanding in *contours of expected cost of the path*. Hence, on expanding a node n with cost $f(n)$, all nodes which promise a path with a cost less than $f(n)$ have already been explored.

A* algorithm is optimal as a graph search making use of both a fringe and a closed set of nodes when it guarantees the properties of *admissibility* and *consistency*. *Admissibility* means that the heuristic function should never overestimate the cost to goal of a node n. It must always underestimate the cost. Let $h^*(n)$ be the optimal cost of the node n to the goal, which may be computed by using any graph search with n as the source. For a heuristic function to be admissible, it must satisfy the inequality given by Eq. [8.4].

$$h(n) \leq h*(n) \, \forall n \tag{8.4}$$

Consistency of a heuristic function means that it must satisfy the inequality given by Eq. [8.5]

$$h(n) \leq h(n') + w(n, n') \tag{8.5}$$

The inequality given by Eq. [8.5] is also called the triangle inequality, because it intuitively states that the sum of two sides of a triangle must be larger than the third side. Intuitively, if $h(n')$ is the estimate of going from node n' to the goal, and one can travel from n to n' using the edge $<n,n'>$ of cost $w(n,n')$, there must be a path from n to goal via n' of cost $h(n') + w(n,n')$, whereas a path with a better cost may still exist. Consistency is a stronger criterion than admissibility as admissibility can be derived from the property of consistency by replacing all pairs of edges $<n,n'>$, starting from the goal the heuristic value for which is always 0.

The pseudocode of the A* algorithm is given by Algorithm 8.3.

Algorithm 8.3: A* Search (G<V,E>, S, GoalTest)

```
g(S)←0, f(S)=g(S)+h(S), π(S) ←NIL
fringe ← empty priority queue
closed ← empty set of nodes
add S to fringe with priority f(S)
while fringe is not empty
        n← extract node with least priority
        closed← closed Un
```

```
        if GoalTest(n), break
        for all n': <n,n'>∈E
                if n'∉ fringe ∧ n'∉closed
                        g(n') ← g(n)+w(n,n')
                        f(n') ← g(n')+h(n')
                        π(n') ← n
                        add n' to fringe with priority f(n')
                else if n'∈fringe ∧ g(n)+w(n,n')+h(n')<f(n')
                        g(n') ← g(n)+w(n,n')
                        f(n') ←g(n')+h(n')
                        π(n') ← n
                        update priority of n' to f(n')
                end if
        end for
  end while
  if fringe is not empty, printPath(n)
  else print "No Path Exists"
  end if
```

8.3.5 Heuristics in Search

It has already been discussed that the design of states and actions is of extreme importance in solving any problem using graph search. For informed search, a *good design of heuristic function* is equally important. If optimality is desired, it must be guaranteed that the heuristic function is admissible and consistent. A good heuristic function must generate estimates as close as possible to $h^*(n)$, or it must generate as large values as possible, strictly below $h^*(n)$. The computational time for the algorithm is proportional to the difference between the heuristic estimate and the optimal heuristic value (h^*-h). For admissibility, $h \leq h^*$, which gives an upper bound to the metric. However, computation of these heuristic values should not be computationally expensive. Therefore, using UCS to actually compute $h^*(n)$ for all nodes and using it as a heuristic function can result in getting a solution in minimum computation time with search following the optimal path rather than processing unnecessary nodes. However, the computation of $h^*(n)$ by such a means will take extremely long.

A common heuristic widely used in robotics is the *Euclidian distance* or the straight-line distance between the node and the goal. This distance does not consider any obstacles and is hence bound to be smaller than the actual optimal distance which will encounter obstacles and take an elongated route around the obstacles. The general mechanism of neglecting problem constraints and estimating heuristic distances by travelling from the node to goal without considering problem constraints is called a *relaxed problem* method. For most problems some problem constraint can be overlooked and by doing so the (usually invalid) solution can be computed in constant or small computation times, which acts as the heuristic function. The approach always gives heuristic estimates which are admissible and consistent.

Sometimes it may be acceptable to generate solutions which are suboptimal to a small degree of ε, if the same gives an acceptable computational speed-up. Hence, the cost function is given by Eq. [8.6].

$$f(n) = g(n) + (1 + \varepsilon)h(n) \qquad\qquad [8.6]$$

The more is the heuristic contribution to the cost function, the faster will the search perform, whereas the more suboptimal will be the path. In the worst case, the cost function is taken as $f(n) = h(n)$, which is called the *heuristic search*. The search normally proceeds very fast, while giving no guarantees on optimality.

8.3.6   D* Algorithm

A* algorithm assumes a static environment. The environment must be static and no changes may happen till the environment is perceived, a plan is made, the plan is executed and the plan safely completes. Any change invalidates the plan and the previously computed plan may no longer be optimal or even feasible. A basic mechanism to deal with the environmental changes is to constantly *monitor* the environment and on sensing any change in environment, the entire plan is recomputed. The recomputation is only possible for algorithms which have very short computation times. *Replanning algorithms* try to reuse the past computation to speed up the generation of results. D* is one such algorithm which stands for *Dynamic A* algorithm* (Stentz, 1994).

The D* algorithm is only valid when the changes are very few and slow, as the magnitude of computation is exponentially proportional to the magnitude of changes made. The speed of rectification of the plan per changes in the environment must obviously be faster than the speed by which the environment changes. The D* algorithm tries to *reuse* the computation done before the change of environment to ease the generation of a valid and optimal plan after the change in the environment. The summarized computation for this algorithm is present in the form of the data stored in the fringe and the closed nodes.

Whenever any change in the environment occurs, the costs associated with these nodes get changed. Say a state previously collision prone now becomes collision free. In this case, the grid may now have a finite cost, which can be estimated from the costs of the neighbours. The costs of the state are updated and the state is reinserted into the fringe even if it had been processed and stored in the closed set of nodes. In an A* algorithm, the closed collection of nodes contains the nodes with optimal costs. However, the costs were regarded as optimal only when the environment does not change. In D* algorithm, a node in the closed set of nodes can have invalid cost and therefore it needs to be reinserted into the fringe if a changed cost is computed. Every node reinserted into the fringe may trigger insertion of more nodes in the fringe or alteration of costs of nodes already in the fringe. The changed costs correspond to the changes due to the change in the environment. If the cost of the node increases due to addition of obstacles, it is important to reinsert the affected nodes and further all the nodes they are previously connected to, as the prior cost is no longer valid and the prior path may no longer be feasible. In case the cost reduces due to removal of an obstacle, the costs should be recomputed to cater to optimality, whereas the prior path may still be feasible though suboptimal.

Only nodes the optimal cost for which changes due to the changed environment are reworked. Hence, if the changes in the environment are very small, the optimal costs of only some nodes are affected which can be recalculated in short computation times. However, if the changes in the environment are large, a significant recomputation is needed, which would require the environment to remain static and not show more changes till the new plan is computed.

In robotics, wherein a robot moves from the source to the goal using a path initially computed by an A*-like algorithm, a common requirement is that the environment may change while the robot is in motion. The recomputation must happen as the robot moves. Therefore, it is preferable to compute the inverse path starting from the goal and going towards the source. This makes the costs invariant to the source, but dependent on the goal which is static. The motion of the robot does not change the costs. The recomputation should still be faster than the changes in the environment. The algorithm replanning may be made focussed towards the source while computing from the source by the use of heuristics. The resultant algorithm is called *focussed D** (Stentz, 1995). Further, the optimality may be compromised to computational time by increasing the heuristic contribution, to react to large changes in the environment. The resultant algorithm is called *D* lite*.

8.3.7 Problems With Graph Search

The graph search looks like an excellent tool to solve any real-life problem. However, a number of assumptions must be met. The environment must be *fully observable* to start with. The actions need to be *deterministic*. The environment must be *static*. D* algorithm can incorporate some dynamicity in the environment, but the changes must be small.

The greatest problem with graph-search methods is that they only work for *discrete spaces*. Hence, the map needs to be made discrete by some *resolution*. The optimality and completeness are guaranteed only for high resolutions. Higher resolutions take a significant amount of time and hence cannot be used for solving near-real time problems like the motion planning of autonomous vehicles. Lower resolutions may be tried, which will work for most scenarios of mobile robotics having wide-open spaces on maps. However, in autonomous vehicles, the lateral width of the road is usually small and the vehicles are closely packed in small lateral widths. Precise lateral space must be known to make correct assessments of overtakes. This necessitates a high *lateral resolution*. Otherwise, close overtaking, which is predominantly found in traffic scenarios, cannot be modelled. Such close overtakings are a key behaviour which must be modelled. Hence, a graph search cannot be implemented in a naïve manner. The trick is to use abstractions and intelligent selection of successor states and actions, which is done as a solution to the problem in this chapter.

8.4   Multilayer Planning

The task of planning a vehicle involves the computation of a smooth trajectory for the vehicle to reach a predetermined goal position. The presence of multiple vehicles on the road at the same time stresses the design of effective coordination techniques such that no collision occurs between any two vehicles. Conventionally, a similar problem has been studied as multirobot motion planning. Research in this domain usually does consider nonholonomic constraints, static and dynamic obstacles, as well as coordination between vehicles. However, planning autonomous vehicles differs in terms of the existence of *roads* which gives rise to specific problems, the presence of road *boundaries*, different times of emergence of different vehicles, vehicle size, diverse speed capabilities amongst vehicles and the twin problems of higher speeds and difficult or error-prone steering mechanisms.

Multirobot motion planning techniques may be centralized or decentralized. Centralized planning approaches result in a complex configuration space which is time-consuming to search and knowledge must be assumed about the robot sources and goals a priori. This eliminates their direct use in the planning of autonomous vehicles. Decentralized approaches meanwhile stress the building of effective coordination strategies, without which the resultant plan may be erroneous or may make some vehicle travel by unreasonably long paths. Even with decent cooperation strategies, individual algorithms invariably have their own limitations which include the discrete space and time-consuming nature of graph-search approaches, the time-consuming nature of the evolutionary approaches, loss of completeness and related problems of the potential, fuzzy and neural approaches etc.

In this section, the problem and the solution requirements specific to planning for autonomous vehicles are studied. In particular, the approach used involves a *layered architecture*. This mechanism differs, in several ways, from the available hybrid architectures like ACTRESS (Asama et al., 1989, 1991) and those used in the Defense Advanced Research Projects Agency (DARPA) Grand Challenge (Crane et al., 2007; Montemerlo et al., 2008). ACTRESS (and related architectures) provides a generalized sensing, planning, manipulation, coordination, communication and control architecture for mobile robotics. The complete hierarchy of the system includes static path planning, the use of a local algorithm, mobile rules, task prioritization, deadlock avoidance by low-level and high-level deadlock solvers and human operator expertise. A multilevel planning architecture on these lines was presented by Vendrell et al. (2001) concentrating upon planning, sensing and replanning for multiple robots. The hierarchy consisted of mission, task, action, motion, trajectory and robot-order types of planning.

Architectures exhibited in the DARPA Grand Challenge usually do not account for coordination between vehicles (because the attempt is to win the challenge, the strategy is rather not to let other vehicles gain benefit) and assume the presence of predefined-speed lanes. Speed lanes reduce the planning problem significantly to one of decision making regarding the change of speed lane. Whilst this makes the problem simpler in most respects, it may not be prevalent or mandatorily followed on all roads. In a situation in which vehicles have varying widths, ranging from big cars to two-wheelers, predefined-speed lanes will almost surely not lead to an optimal utilization of road space.

The basic motivation behind the approach is the naturally observable behaviour that humans tend to drive in a manner to *maximize the separation* between obstacles, other vehicles or road endings. While driving on a broad road, humans tend to perform a comfortable overtake of a vehicle, whilst on a narrower road overtaking takes place, but with greater caution and narrower gaps. This is indeed another difference between the presented approach and conventional mobile-robot motion planning, in which the task is usually to generate a smooth trajectory of the smallest length. This makes the separation between vehicle and obstacle minimal, which is completely contradicted by maximizing the distance. Here again, the presence of roads is responsible for the difference.

Obstacles may have multiple meanings. On English roads, parked vehicles are primarily obstacles, and these can be multiple in number. In an autonomous scenario, nonautonomous vehicles also act as obstacles. In the case of a broken-down vehicle or in an accident scenario, the affected vehicles then become obstacles. In some situations, very slow-moving vehicles may themselves be better modelled as obstacles. Besides these situations, the conventional meaning of obstacle holds.

The heuristic of separation maximization may certainly seem to be sub-optimal in certain cases. However, traffic efficiency is mainly characterized by optimal route selection. The *path length* is approximately the *length of the road*, which is almost the same for all lateral positions on roads. Lateral distances mainly dictate the *safety* associated with the drive. From a statistical point of view, certainly it would be best if vehicles just kiss each other, so-to-speak, and go on. The drop in efficiency is regarded as insignificant. It is assumed that the road scenario does not have very wide roads, on the order of many lanes together with small traffic density. This is a valid assumption considering such roads are themselves on many occasions broken up into multiple smaller-width roads.

A glimpse of this heuristic holding in traffic is visible in present traffic. Attention here is drawn to a traffic scenario in which speed lanes are not implemented − the Indian traffic system. Vehicles nicely place themselves in between other vehicles to comfortably fit in, or, in other words, exactly between adjoining vehicles. As per assumptions the traffic density is high or roads are not very wide.

In mobile robotics, the map is *largely bounded* or *completely unbounded* and hence the objective is the computation of smallest distance. In autonomous vehicles, however, any path may have a path length roughly equivalent to the length of the road. Hence, safe driving (considering differences between mobile robots and the bulky vehicles) is more important. The attempt to maximize separation plays a major role in the entire planning process. First, it ensures no collision and further sufficient room for corrections for the robotic control or speed-and-steering mechanism. At the same time, the approach acts as a predefined heuristic to make planning fast, near-real time and still complete. Equivalent graph-search approaches (Kala et al., 2009), evolutionary approaches (Xiao et al., 1997; Kala et al., 2011) or other similar approaches would take a long time to generate paths that (based on this heuristic) can be generated in a reasonably short time.

An assumption here is that different vehicles on the road have different importances which reflect the need for them to overtake. It is imperative for the higher-importance vehicles to *overtake* low-importance ones, usually because they are capable of higher speeds. Hence, overtaking must happen whenever feasible. This, in concept, is similar to the application of social potential in mobile robotics (Reif and Wang, 1999). Gayle et al. (2009) also used the concept of a generalized social potential field for the motion coordination of robots. Here, every moving agent or robot had a social behaviour, based on which was applied a potential that affected the motion of the other robots. However, the mere use of a *prioritized* approach may not solve the problem of overtaking as it may cause one vehicle to drive straight with no *cooperation* with another vehicle which may, as a result, have a very poor path or no feasible path at all. Further, decision making as to when and where to overtake is important. In any case, there may be a requirement for the *adjustment of speeds*. Here again, the approach differs from methods in which higher-priority vehicles plan and lower-priority vehicles adjust their speed (and path) in response to avoid collision (Todt et al., 2000; Kant and Zucker, 1986) or approaches in which different vehicles cooperatively decide a collision-free velocity (Wilkie et al., 2009), neglecting each other's velocity limits or importance.

The task considered here is to develop a system that plans and moves autonomous vehicles as per the given scenario. A total of four levels of hierarchies have been worked out, each hierarchy contributing in its own way to collision avoidance and distance maximization. The key takeaways of the work are summarized in Box 8.2. The basic concepts are also summarized in Box 8.3.

Every vehicle R_i has a maximum permissible speed (called *maximum speed*) as per its design and safety regulations. Every road has an associated speed restriction (called *restricted speed*) disallowing a vehicle to drive at speeds which may make travel risky. Hence, maximum speed with which the vehicle may travel on road is given by the least of the two values. However, on many occasions if a vehicle travels at the maximum or restricted speed this may possibly lead to a collision and hence a vehicle may be

---

Box 8.2 Key Takeaways of the Approach

- A general *planning hierarchy* in an assumed complex modelling scenario is presented, in which any algorithm may be used at any level of hierarchy.
- To use simple heuristics such as *separation maximization*, *vehicle following* and *overtaking*, to plan the trajectories of multiple vehicles in real time.
- An emphasis is placed on the *width* of feasible roads as an important factor in the decision-making process.
- The developed *coordination strategy* is largely cooperative, at the same time ensuring near-completeness of the resultant approach and being near-optimal for most practical scenarios.

Box 8.3 Key Concepts of the Approach

- Route selection hierarchy
  - Select the shortest route to destination
- Pathway selection hierarchy
  - Obstacle avoidance strategy
  - Select widest and shortest-length pathways
- Pathway distribution hierarchy
  - Arrange vehicles projected to lie in a pathway segment
  - Prioritization to decide vehicle relative order
  - Separation maximization to decide vehicle position
- Trajectory generation hierarchy
  - Spline curves
  - Feasibility check
  - Local optimization
- Coordination
  - Layer by layer
  - Each level shares its result with same level of the other vehicle
  - A vehicle can ask any other to replan at any level depending upon priorities

---

planned to reduce its permissible speed to a lower limit. The planned speed limit is called the *bounded speed*. A typical example of bounded speed is having a slow vehicle in front with no room to overtake. At any time, the vehicle must travel at a speed lower than the bounded speed. Let the bounded speed of vehicle R_i be denoted by v_i^b.

In the toughest scenario, a road would be embedded with multiple obstacles. Each of these obstacles will have its own shape and size. A *convex hull* algorithm is applied to get rid of all concave regions in the obstacle, as it is unlikely that a vehicle wants to place itself inside an obstacle segment. The *free road* excludes all regions occupied by obstacles or their concave regions. Free road (along with *free crossing region*) is the modified map used for collision checking by the algorithm.

A few terms are defined, which will be used throughout the work and will help in better understanding of the algorithm.

Definition 1 Pathway: The presence of multiple obstacles in a road segment necessitates the generation of an optimal strategy in which every obstacle must be avoided. This can be fundamentally broken down into the decision to steer the vehicle to the left or right of each of these obstacles. A *Pathway* is a closed region of free road such that no obstacle (or its concave region) lies inside it.

Definition 2 Pathway segment: A pathway may be broken down into a number of (mostly fixed length) segments along the length of the road, in which each segment is known as *pathway segment*. These segments are nonoverlapping for a pathway. The point at which the pathway segment ends is important for algorithmic working and is taken to be the midpoint of the extremities of the two boundaries bounding the pathway segment. The end location known as *pathway segment end centre* represents the entire pathway segment for algorithmic computations. In the rest of the discussions, the terms pathway segment and pathway segment end centre would be used interchangeably.

Definition 3 Distributed Pathway: At any instance of time, a number of vehicles may lie or may be projected to lie over a pathway segment. The complete segment space (along the width) needs to be *distributed amongst the individual vehicles.* As per design a vehicle may not directly lie in front of another vehicle in a pathway segment, due to its small length.

The basic problem is to plan and move a number of robotic vehicles in a given map. Let the problem have a total of N robotic vehicles. The map consists of a number of *roads* that crisscross each other in *crossings.* It is assumed that the road boundaries and crossing positions are known a priori. Each vehicle R_i has its own source point S_i and goal point G_i and emerges into the planning scenario at time T_i. The sources and goals of vehicles are known a priori, but not before their emergence into the planning scenario. The task is to simultaneously plan and move these vehicles, such that no collision occurs. On top of this, the kinematic constraints of individual vehicles are considered. At any instance of time t, let the vehicle R_i be at position $L_i(t)$ ($x_i(t)$, $y_i(t)$, $\theta_i(t)$) and moving at a speed of $v_i(t)$ in the direction $\theta_i(t)$. Hence, $L_i(T_i) = S_i$. It is assumed that the vehicles are rectangular or may be bounded inside a rectangle of length l_i and width w_i.

As per the definitions, the algorithm needs to be optimal at various levels. If seen by a macroscopic view, the algorithm operates multiple vehicles in an optimal route, and from a microscopic view, the algorithm generates feasible and optimal trajectories for motion of each of the vehicles. At the highest level of planning, the algorithm needs to judiciously select the roads to make an optimal path. At the next level, it needs the selection of an optimal strategy to avoid the static obstacles considering the complex obstacle network and the presence of other vehicles in the scenario. At the next level, it also needs an optimal coordination strategy for avoidance of collision between any two vehicles. At the finest level, it needs to check the feasibility of trajectory.

The basic algorithm developed for the purpose has a four-tier planning architecture. Each of the levels works at a different level of *abstraction* of the complete map, with the topmost layer being the most abstract layer. Unlike the general mechanism of decomposition of map used in mobile robotics, the decomposition here is on logical grounds. The topmost layer works over the complete map and uses Dijkstra's algorithm to select the path or the roads and crossing regions that the vehicle, the route for which is being planned, needs to take to reach the goal from the source by the shortest distance. This algorithm is common to all approaches and was centrally taken in chapter 'Introduction to Planning'. Planning thereafter is done for the road segment that the vehicle enters.

The next level of planning is used to decide the *pathway* planned for the vehicle. Here again, Dijkstra's algorithm is used; however in this level, the attempt is to additionally select *wide-pathway* segments for travelling. The next layer of planning involves *pathway distribution* that directs the vehicles to the (approximately) correct location at every pathway segment. Prioritization and distance maximization is used for the task. The distributed pathway so generated is smoothed in the last level of

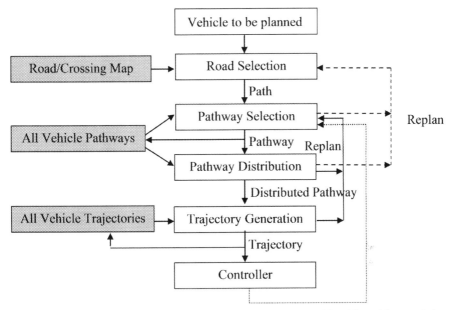

Figure 8.1 General algorithm framework. The algorithm consists of four hierarchies consisting of selection of roads to travel from source to goal (hierarchy 1), having a proper strategy to avoid all obstacles also called a pathway (hierarchy 2). The hierarchy computes the position of the vehicle within a pathway so that separations between vehicles are maximized, also called distributed pathway (hierarchy 3) and computing a smooth and feasible trajectory for vehicle motion (hierarchy 4). The four hierarchies of all vehicles interact with each other for coordination.

planning in that the *trajectory generator* uses BSpline curves to generate motion curves to avoid collision and at the same time allow maximum permissible speeds. The complete algorithm has been summarized in Fig. 8.1.

Coordination is an important feature into the domain of multirobots. This task is performed at different levels in the approach. Each vehicle publishes its pathway and the trajectory that is currently being followed for coordination with other vehicles. Hence, a *layer-by-layer coordination* is performed. *Priority-based approaches* are used at individual levels, in which distance traversed within the segment (measured in terms of the number of pathway segments) is indicative of the vehicles' priority. In simple terms, vehicles ahead have higher priority. This is unlike traditional priority-based approaches, higher-priority vehicles in this approach do create as much space as possible for lower-priority vehicles, thereby making possible trajectories likely for all vehicles.

Planning is not strictly a layer-by-layer mechanism in the developed algorithm. For any vehicle the route for which is being planned, a planning hierarchy can decide to *restart* the planning operation (due to a change in map or bounded speed of itself or another vehicle) or instead to *replan* the route of some other vehicle. Hence, this is a *feedback system* that operates between layers.

Communication between vehicles is not restricted to information of their published pathways and trajectories, but rather extends to the ability to rectify speeds and force a change of paths and replans. Solving a problem in which the vehicle emergence and travel plan is not known in advance (any vehicle can emerge, modify its plan or necessitate replan due to accumulated errors at any time), such a mechanism becomes inevitable. Each of the hierarchies, from top to bottom, is explained in the following subsections.

8.5   Hierarchy 1: Path Computation

The first task is to plan the path of the vehicle, in other words, to select the set of roads that the vehicle R_i will use to reach its goal. Planning is done when a vehicle emerges into the planning scenario at time T_i and location S_i, having its goal as G_i. Because the roads and crossings are known a priori, a road-network graph is built. Source S_i and Goal G_i are added as additional vertices, connected by edges to the crossings they lead to in the graph. Dijkstra's algorithm, a single-source shortest-path algorithm, is used for the planning purpose.

Planning steps henceforth work over roads (or road segments). It is essential to get rid of crossing regions that may lie between any two roads and to replace them with architecture similar to roads that can then become part of the plan. This is done by introducing virtual boundaries across the points where the roads end. In case of the presence of a roundabout, the region marked by virtual boundaries should exclude the roundabout region. The complete path may now alternatively be assumed to be a single road characterized by the two boundaries formed by joining the boundaries of the individual roads and the virtual boundaries of the corresponding crossing regions. The road may be further broken down into overlapping segments over which the subsequent levels operate. It must be noted here that for optimal performance, the algorithm presented needs to be assisted by some intelligent algorithm for planning in scenarios of crossing, diversions, mergers, parking etc. The focus here is on single-road scenarios only.

The result of the use of Dijkstra's algorithm over a sample map is given in Fig. 8.2. Dijkstra's algorithm is called only once during the motion of the vehicle. However, it may be additionally called on when a blockage is detected or if traffic is extremely slow on some road, which should be avoided.

8.6   Hierarchy 2: Pathway Selection

The next level of planning deals with the selection of the optimal *pathway*, given that the vehicle is travelling on a road segment. This hierarchy is an *obstacle-avoidance strategy* that defines the manner in which all the obstacles in the road segment are to be avoided. The output is a pathway which is part of a free road segment within which the vehicle is supposed to be moving. The pathway is bounded by obstacles or road boundaries on two sides and hence no collision with static obstacles is possible

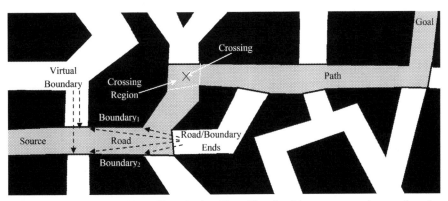

Figure 8.2 Path generated by Dijkstra's algorithm. The algorithm computes the set of roads which lead to the goal by the shortest path. Other roads in the crossing region are isolated by virtual boundaries. The resultant road is further broken into road segments.

whenever a vehicle is completely within the pathway. The first task is to find out all the pathway segments, and then to select the correct sequence of traversal of these pathway segments that makes the overall route. First, planning with a single vehicle is discussed, and later the approach is generalized for more vehicles. The overall approach is also summarized in Box 8.4.

8.6.1 Pathway Segments Computation

The *pathway segments* are first assumed to be of equal length, measured along the direction of the road. A sweeping line is made to traverse from one extremity of the road segment and to the other extremity. Traversal is made in step sizes of Δ, the length of the pathway segment. A point Y is said to be accessible by vehicle R_i if vehicle R_i placed at Y in direction of the road has every part of vehicle at free road.

A collection of all pathways may be extracted out for the given road segment as given by Algorithm 8.4. The sweeping line starts from segment start to segment end (Line 2), the extremities of which, at any time, lie at two boundaries (Line 3). All transitions from feasible to infeasible states are recorded, which mark either pathway segment start (Lines 5–6) or pathway segment end (Lines 7–8). Pathways that are too narrow are eliminated (Line 9), as the vehicle may not be able to fit in, else the pathway segment and its end centre are added (Line 10). Here, w_i denotes the width of vehicle R_i.

Algorithm 8.4: getPathwaySegments

```
Line 1  PaAll ← NULL
Line 2  for l ← start_of_road_segment to end_of_road_segment in
                steps of Δ
Line 3          sweeping_line ← line from Boundary₁ to Boundary₂ at l
Line 4          for Y ← all points along sweeping_line
```

Box 8.4 Pathway Selection Procedure

- Traverse a sweeping line across the road length in small steps
- Find areas (**Pathway segments**) without obstacles in this line
  - **Pathway segment end centre:** Centre of the sweeping line in the obstacle-free region
  - **Pathway segment:** Area bounded by the consecutive line sweeps in the same obstacle-free region
- Connect the obstacle-free areas to produce a graph
- Search this graph for widest and smallest path (**Pathway**) to the end of the road
  For multiple vehicles:
  - For every edge/pathway segment
  - Extrapolate the motion of the other vehicles by their pathways
  - List vehicles using the same pathway segment at the same time
  - Classify the vehicles into higher priority and lower priority
  - For every higher-priority vehicle, subtract its width from the segment width
- Replan lower-priority vehicles at the pathway level
- Replan lower-priority vehicles at the distributed pathway level

Coordination

- **Prioritization:** R_i is said to have a higher priority over R_r if
  - R_i and R_r are driving in the *same direction* and R_i lies ahead of R_r, or
  - R_i and R_r are driving in *opposite directions* point of collision lies on the *left side* of the complete road (*because R_r is on the wrong side*)
- **Speed Adjustment**
  - If unable to generate a feasible pathway: Find the higher-priority vehicle ahead blocking the road segment and *follow it* (reduce speed)
  - Else select a new *route* −blockage avoidance

Line 5	if Y makes transition from inaccessible to accessible
Line 6	$X_1 \leftarrow Y$
Line 7	else if Y makes transition from accessible to inaccessible
Line 8	$X_2 \leftarrow Y$
Line 9	if $\| X_1 - X_2 \|$ is not too narrow
Line 10	PaAll $\leftarrow$ PaAll∪path segment (X_1, X_2) with end centre $= (X_1+X_2)/2$ and width $= \| X_1 - X_2 \| + w_i$
Line 11	end if
Line 12	end if
Line 13	end for
Line 14	end for

For algorithmic purposes, *pathway segment end centre* represents the pathway; for theoretical reasons, pathway may be visualized knowing the sweeping line and the segment end centres. For the same reasons the width of the pathway segment is assumed to be the width at the pathway-segment end.

8.6.2   Graph Conversion and Search

Once the pathway segments are formulated that exist in any road segment, these pathways must be used to find the optimal pathway for the vehicle. Dijkstra's algorithm is again employed to use the individual pathway segments for constructing the pathway. A *graph* is constructed consisting of each pathway segment as vertices. Two pathway segments are said to be connected to each other if a vehicle can traverse from their end centres, by any straight or curved path, without going through any inaccessible region or any other pathway segment. Note that traversal of a vehicle as a whole is considered, and hence if any part of the vehicle is in an inaccessible region, the connection is nonexistent. The weight of the edge is taken as the Euclidian distance between the pathway segment end centres.

Unlike Dijkstra's algorithm used for road selection, at this level the emphasis is not only on finding the shortest distance, but also on the width of the pathway segments. As per the distance maximization hypothesis, a wider pathway segment which may lead to longer distances may be preferable to a narrower pathway segment but with less distance. However, this attempt to maximize width has an upper threshold w_{max}. Further, the width factor considers quality of a potential pathway as the *lowest width* the vehicle would encounter in its possible journey by that pathway. Let the pathway segment sequence (making a complete pathway Pa) be Pa(m), in which Pa(m) physically denotes the pathway segment end centre. The length of the pathway (to be minimized by the algorithm) and the width of the pathway (to be maximized by the algorithm) is given by Eq. [8.7].

$$\text{length}(\text{Pa}) = \sum_m \|\text{Pa}(m+1) - \text{Pa}(m)\|$$
$$\text{width}(\text{Pa}) = \min_m(\min(\text{width}(m), w_{max}))$$

[8.7]

The cost of any vertex (pathway segment) v_2, while expanding v_1 using Dijkstra's algorithm, may be given by Eq. [8.8].

$$\text{ds}(v_2) = \text{ds}(v_1) + \|v_2 - v_1\|$$
$$\text{mwidth}(v_2) = \min(\text{width}(v_2), \text{mwidth}(v_1), w_{max})$$
$$\text{cost}(v_2) = \text{ds}(v_2) - \alpha \, \text{mwidth}(v_2)$$

[8.8]

Here, ds(x) measures the distance of vertex x from source. ds(source) = 0. mwidth(x) measures the minimal width in travelling from source to x. mwidth(source) = min(width(source), w_{max}). The cost(x) is the total cost of vertex x. Variable α may be interpreted as a penalty constant for lower widths or an objective weight. In the current implementation, the main intent is to find wide roads (even if they make overall route long) and hence $\alpha \gg 1$. If the resultant pathway has a width less than the width of the vehicle, it is assumed that no pathway is possible and path-selection algorithm needs to be computed again with the particular road blocked for traversal by this vehicle.

For this algorithm, the source is the pathway segment the centre of which lies closest to the current position of the robotic vehicle $L_i(t)$. The goal is any one of the pathway segments (the one which eventually obtains lowest cost) produced at the last scan of the sweeping line, when it is at the end of the road segment being planned. The result of applying Dijkstra's algorithm is a list of pathway segments Pa(m) that the vehicle must traverse to travel on the road.

The pathway segments generated in a sample road segment, along with the optimal pathway planned (as a line joining pathway segment end centres) is shown in Fig. 8.3. The pathway (as a region) is approximately computed and shown in same figure. A single pathway segment is shown in yellow.

The approximate motion of the vehicle (at the pathway hierarchy) may be visualized to be in a straight line joining the pathway segment ends Pa(m), which indicates the approximate planned position $L_i^{\mathrm{pa}}(t)$ of the vehicle R_i as given by Eq. [8.9]

$$L_i^{\mathrm{pa}}(t) = \mathrm{Pa}(m_1) + \frac{\mathrm{Pa}(m_2) - \mathrm{Pa}(m_1)}{\|\mathrm{Pa}(m_2) - \mathrm{Pa}(m_1)\|} v_i^b \left(t - \frac{\mathrm{ds}(m_1)}{v_i^b} \right) : \frac{\mathrm{ds}(m_1)}{v_i^b} \leq t \leq \frac{\mathrm{ds}(m_2)}{v_i^b}$$

$$[8.9]$$

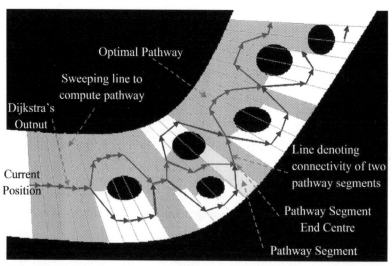

Figure 8.3 Pathway selection. A sweeping line is made to pass the entire road segment to mine out feasible pathway segments. Connectivity between segments is determined. Dijkstra's algorithm mines out the pathway which is widest and shortest to the end of the road segment.

8.6.3 Presence of Multiple Vehicles

The next task is to generalize the algorithm for the presence of multiple vehicles. The set criterions of short overall path and wider pathway segments are independent from the presence of other vehicles. However, it is possible that *too many vehicles get scheduled in the same pathway segment region*. Every pathway has a fixed bandwidth of traffic which it can handle, and hence planning algorithm cannot schedule more vehicles to this number at any time. Discussions henceforth cater to this constraint.

In this case, the specification of pathway segments is different for the different vehicles as they are all planned independently by their planning hierarchies. Because the coordination is level by level, every vehicle needs to publish its own resultant planned pathway.

In planning using Dijkstra's algorithm, the approximate time interval that the vehicle R_i stays at the pathway segment Pa(m_2), which lies after pathway segment Pa(m_1), is given by Eq. [8.10].

$$[t(m_1), t(m_2)] = \left[\frac{\mathrm{ds}(m_1)}{v_i^b}, \frac{\mathrm{ds}(m_2)}{v_i^b}\right] \tag{8.10}$$

At every pathway segment (physically its end centre), while planning in a multivehicle scenario, the first step is to check *which other vehicles are using pathway segment* Pa(m_2) in the same time interval as the vehicle R_i. Vehicles close by a distance of η are also included to account for approximation and measurement errors. Each of the other vehicles is projected to move from its current position in a direct path along the consecutive pathway segments that together form its pathway (which is not the same as the actual path traversed by the vehicle as per its smooth trajectory).

Once the vehicles using a particular pathway segment Pa(m_2) at the same time as the vehicle R_i are known, along with their priorities relative to R_i, the next important question is to decide whether the vehicle R_i would be able to use the particular pathway segment itself, or if too many vehicles are already scheduled and hence there is no available width in which vehicle R_i could fit. This decision making is done on the basis of priorities.

Consider a vehicle R_r which uses the same pathway segment. R_i is said to have a *higher priority* compared to R_r if (1) R_i and R_r are driving in the same direction (eg, both going from one end of a road to the other end) and R_i lies ahead of R_r, or (2) R_i and R_r are driving in opposite directions on a road and a point of collision lies on the left side of the complete road. The second point is inspired by a 'to-left' driving rule, in which a possible collision on the right side of the road usually implies that the vehicle being driven on the wrong side is the defaulter. A vehicle need not consider vehicles of lower priority in its course of planning, but it does need to consider higher-priority vehicles.

The general rule is that if R_r has a higher priority and it needs to go through a pathway segment, vehicle R_i attempts to check the *available remaining width* in the pathway segment. If R_r has a lower priority, R_r needs to be *replanned* to account for the motion of R_i. In replanning, vehicle R_r would check if there is enough width to accommodate it along with vehicle R_i, whilst R_i travels through the pathway segment, or alternatively it may attempt to find a broader (and thus better) pathway.

Giving higher priority to vehicles *ahead* ensures that, upon the arrival of new vehicles at the rear, there is no change in the plans of any of the vehicles ahead. This saves a lot of computation by minimizing the total number of replans. In a scenario in which the traffic density is large, it is barely possible for a vehicle to overtake another vehicle. In such a scenario, in effect, only the movement of the vehicle arriving needs to be planned, and this vehicle would necessarily have to follow other vehicles. Hence, the computational effort becomes very small, even though the total number of vehicles is high and the computational complexity is (theoretically) proportional to the number of vehicles.

The other reason for vehicles ahead being given a high priority is vehicle dynamics. Consider as a converse example that the movement of a vehicle at the rear is planned first, followed by the movement for a vehicle at the front. By doing so, the vehicle at the back computes a good path (neglecting the vehicle in front of it), essentially seeing an empty road ahead, usually therefore driving as fast as possible. If the movement of the vehicle in front is then planned later it may be that the front vehicle is unable to compute a reasonable path and it may slow down or even come to a standstill. However, the vehicle at the rear did not originally consider a waiting vehicle in front of it. Hence, the originally planned path for the rear vehicle would have become collision prone.

Consider the reverse scenario. The vehicle in front first comes with a good travel plan, and later the vehicle at back is planned. A possible travel plan for the vehicle at the back may be to follow the vehicle in front. In such circumstances the vehicle to the rear is far more likely to be able to come up with an overall feasible travel plan. From a practical perspective, consider the case of a slow-moving vehicle travelling ahead of a fast-moving vehicle. If the fast-moving vehicle cannot overtake the vehicle ahead, it will follow it. It does not accelerate and collide into it.

Consider that the pathway segment $\text{Pa}(m_2)$ sees a scheduled vehicles $\text{high}(\text{Pa}(m_2))$ which have *higher priorities* and the set $\text{low}(\text{Pa}(m_2))$ which have *lower priorities*. The two sets are given by Eq. [8.11]

$$\text{high}(\text{Pa}(m_2)) = \bigcup_r \text{Separ}\big(L_i^{\text{pa}}(t), L_r^{\text{pa}}(t)\big) \leq \eta \, \forall t(m_1) \leq t \leq t(m_2), \text{priority}(R_r)$$
$$\geq \text{priority}(R_i), r \neq i$$

$$\text{low}(\text{Pa}(m_2)) = \bigcup_r \text{Separ}\big(L_i^{\text{pa}}(t), L_r^{\text{pa}}(t)\big) \leq \eta \, \forall t(m_1) \leq t \leq t(m_2), \text{priority}(R_r)$$
$$< \text{priority}(R_i), r \neq i$$

$$[8.11]$$

Here, $\text{Separ}(x,y)$ computes the separation between two vehicles placed at x and y with a value of 0 in the case that the vehicles overlap each other/collide. $L_i^{\text{pa}}(t)$ is the projected motion of vehicle R_i at time t at the pathway level, at which vehicles occupy the complete pathway segment width and move as per pathway segment directions. This ensures collision is recorded if vehicles lie in same pathway segment.

The cost function of Dijkstra's algorithm is modified to account for high-priority vehicles, which it must consider in planning. It is assumed that every high-priority vehicle must be given w_{max} of space, without which the path of the vehicle may become suboptimal. The modified cost function for any vertex v_2 is hence given by Eq. [8.12].

$$ds(v_2) = ds(v_1) + \|v_2 - v_1\|$$

$$\text{mwidth}(v_2) = \min([\text{width}(v_2) - \text{size}(\text{high}(v_2)) \cdot w_{max}], \text{mwidth}(v_1), w_{max})$$

$$\text{cost}(v_2) = ds(v_2) - \alpha \, \text{mwidth}(v_2)$$

$$[8.12]$$

The final output of this algorithm is the pathway segments Pa(m) giving the pathway Pa. For every pathway segment Pa(m), all the collision-prone vehicles with lower priorities are further set for replanning at the *pathway level*, and all the collision-prone vehicles with higher priorities for replanning at the *distributed pathway level*. This ensures all vehicles register presence of R_i in their travel plans.

8.6.4 *Heuristics in the Presence of Multiple Vehicles*

In the case when a vehicle computes a feasible pathway for its motion, the work of this level is done and the computed pathway is returned. However, in case when the pathway computation fails to return a feasible pathway, *heuristics* are used for further decision making. Because the generation of a pathway was infeasible, the vehicle's *speed needs to be altered* (most likely decreased, but could be increased) to a value which makes the pathway generation feasible.

If a vehicle R_i fails to find a pathway the width of which is more than the required width, and it observes that it has encountered a collision with a vehicle R_r travelling in the same direction with higher priority, it may choose to reduce its speed to equal the speed of the vehicle with which it collided (or $v_i^b = v_r^b$). Reduced speed would result in the vehicle not colliding (at the pathway level) and thus the width being reserved for the higher-priority vehicle would be eliminated, making generation of a pathway feasible. This is a display of the *vehicle-following* behaviour. If, however, the collision does not occur with a higher-priority vehicle, then the vehicle R_i must recall the road selection algorithm with the current road marked as blocked. This is display of being *unstuck in blocked route* behaviour. In any case, reducing speed v_i^b does not mean that the vehicle's speed has been permanently limited. Speeds are restored to their highest possible value (minimum of maximum and restricted speed) at the next planning of pathway level (which may again be reduced if shortage of width or related problems occur).

Replanning at a level means that all published planning at that and the lower levels is invalidated and needs to be recomputed. All the registered replans of pathway levels by any vehicle must be performed before a vehicle's motion may be planned at the next level, as pathway-level planning of a vehicle impacts the pathway-distribution planning of other vehicles, irrespective of priorities.

The general algorithm employed here is given by Algorithm 8.5. The first task is computing all possible pathway segments (Line 1) and converting them into a graph (Line 2). The graph is searched using Dijkstra's algorithm (Line 3) which gives the pathway. If no pathway is possible for the vehicle as per the dynamics of other vehicles (Line 4), the possibility to compute a plan with reduced speed is checked (Line 6), the condition for which is that the vehicle must be colliding with another vehicle on the same side (Line 5). If it is not possible to come up with a feasible pathway, replanning is done at the level 1 hierarchy (Line 8) in which the vehicle attempts to travel by a completely new road. In the case when a feasible plan is found, the list of vehicles affected and which need to be replanned at pathway level (Line 11) and distributed-pathway level (Line 12) are found.

Algorithm 8.5: getPathway

```
Line 1  PaAll ← getPathwaySegments
Line 2  make graph using all pathway segments PaAll(m) in pathway PaAll
Line 3  Pa ← pathway search using Dijkstra's algorithm
Line 4  if Pa = null
Line 5          if unavailability of pathway segment due to collision
                with a higher priority vehicle Rr in same side
Line 6                  v_i^b = v_r^b, replan R_i at pathway level
Line 7          Else
Line 8                  re-call path computation (hierarchy 1)
Line 9          End
Line 10 Else
Line 11         mark replans of all vehicles high(Pa(m)) at pathway
                level for all Pa(m)∈Pa
Line 12         mark replans of all vehicles low(Pa(m)) at distributed
                pathway level for all Pa(m) ∈Pa
Line 13         return pathway Pa
Line 14 End
```

8.7   Hierarchy 3: Pathway Distribution

Any pathway at any time could be occupied by multiple vehicles travelling in either direction. The purpose of this hierarchy is to *distribute width amongst the vehicles*. In other words, whereas the pathway planning ensures no collision between a vehicle and static obstacles, the pathway-distribution planning level attempts to *maximize the separation* and to ensure that there is no collision between vehicles.

Box 8.5 Pathway Distribution Procedure

- For every pathway segment in pathway
- Extrapolate and list vehicles using the same pathway segment at the same time
- Classify the vehicles into higher priority and lower priority
- Keep relative placing: In a width of obstacle-free segment
  - Keep high-priority vehicles on the left
  - Keep the vehicle being planned in the centre
  - Keep low-priority vehicles on the right
- Divide segment width equally amongst vehicles and hence compute position
  - Attempt to tune infeasible paths for feasibility
  - If still infeasible, replan lower-priority vehicle at pathway selection level
  - If still infeasible, reduce speed and follow

Coordination

- **Prioritization:**
  - Design of priority scheme such that higher-priority vehicles are relatively on left and lower ones on the right
  - R_i has a higher priority if:
    - It lies *ahead* of R_r with R_i and R_r going in the *same direction*, or
    - R_r and R_i are travelling in different directions
    - Scheme is consistent with implementation of behaviours of *overtaking* on the right, *being overtaken* on the right and drive left

Prepreparation and Postpreparation

- *Prepreparation:* Rather than going very near to a vehicle and then aligning to avoid it, take relative position well in advance
- *Postpreparation:* Rather than quickly returning to the centre after having avoided a vehicle, stay at the same relative position for some time
- Both strategies are followed in case no other vehicle is present

---

This is done by strategically placing vehicles at every pathway segment used by them. The approach is summarized in Box 8.5.

8.7.1 Order of (Re-) Planning

The planning of a pathway for a vehicle R_i leads to other vehicles being forced to replan at the distributed-pathway level. Hence, the task of this layer is to plan the vehicle R_i along with all the other vehicles requiring replanning. The order of planning is important. The vehicles are planned strictly as per their *priorities*, in which the priority for sorting is taken as the distance travelled in the particular road segment. Vehicles that have traversed a greater distance are ahead and therefore have higher priorities. In the case of two vehicles travelling in the opposite direction, the one which has just entered the segment is assigned the lower priority.

8.7.2   Heuristic Placement of Vehicles

The main activity carried out by this layer of planning is the relative placement of vehicles inside each pathway segment. In other words, the vehicle the trajectory for
which is being planned needs to be positioned inside the pathway segment by considering the other vehicles which are present in the pathway segment at the same time. For
every pathway segment $Pa(m)$ in the pathway being planned Pa for vehicle R_i, first the
vehicles are found with which the vehicle R_i would collide or would be separated from
by less than η. This is done in exactly the same manner as is performed in the
pathway-selection step. Then this information is used for the placement of the vehicles.

Consider any one vehicle R_r, which is found to collide in the pathway segment
$Pa(m)$ within the time interval R_i, is at the segment. R_i has a *higher priority* if (1) it
lies ahead of R_r with R_i and R_r going in the same direction, or (2) R_r and R_i are travelling in different directions. However, R_i has a lower priority if it lies behind R_r with
R_i and R_r travelling in the same direction. In a similar fashion to Eq. [8.11], the sets
high($Pa(m)$) and low($Pa(m)$) are computed containing vehicles with high and low priorities. It can be observed that vehicles with higher priorities high($Pa(m)$) must lie to
the *right* of R_i and those with lower priorities low($Pa(m)$) must lie to the *left* of R_i in the
distribution of the pathway segment $Pa(m)$, along its width. This conforms to the traffic
behaviour of '*overtaking on the right*', '*being overtaken on the right*' and '*drive left*'.
Note that the second criterion of prioritization is changed from the previous step, as
the vehicles are not allowed to change pathways at this level, and the fact that R_r
has the same action (of being placed on the right) that needs to be performed for
higher-priority vehicles.

Let the points at the two boundaries at which the pathway segment ends, be P_1 and
P_2, respectively. The planned approximate position of the vehicle $L_i^{\text{pda}}(m)$ when R_i is
expected to cross pathway segment $Pa(m)$ is given by Eq. [8.13].

$$L_i^{\text{pda}}(m) = P_1 \cdot (1 - p(Pa(m))) + P_2 \cdot p(Pa(m)) \qquad [8.13]$$

in which $p(Pa(m))$ is the relative position of R_i in the pathway segment $Pa(m)$ given by
Eq. [8.14].

$$p(Pa(m)) = \frac{2\text{size}(\text{high}(Pa(m))) + 1}{2(\text{size}(\text{low}(Pa(m))) + \text{size}(\text{high}(Pa(m))) + 1)} \qquad [8.14]$$

The approximate motion of the vehicle at this level is in a straight line between the
points $L_i^{\text{pda}}(m)$ taken in a continuous order.

8.7.3   Prepreparation and Postpreparation

In many cases, size(high($Pa(m)$)) and size(low($Pa(m)$)) may both be zero, making the
value of $p(Pa(m)) = 0.5$, which would mean that R_i lies in the centre of the road. In
fact, this is just fine when there are no other vehicles around; one may drive in the

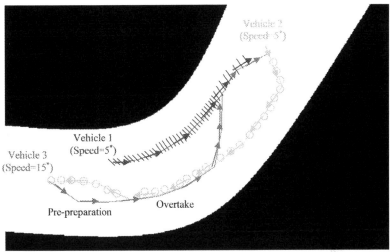

*Speed measured as unit distance per unit time

Figure 8.4 Pathway distribution. Given the various vehicles that occupy a pathway segment at any given time, the desired position of vehicle within pathway can be determined. This may result in an overtaking behaviour subjected to possibility.

centre of the road as per the separation maximization hypothesis. However, if there are any other vehicles recorded ahead or behind, one may attempt to prepare for them. In other words, it means that if another vehicle lies in front, one may attempt to align the vehicle in a correct fashion prior to entering the pathway segment in which the distribution was actually computed (called *prepreparation*). Further, a vehicle may attempt to wait for some time in a wrong distribution after a vehicle has overtaken or passed by, and it has crossed the pathway segment in which the overtaking took place, before returning to a central position (called *postpreparation*). Both of these strategies contribute towards the overall separation maximization hypothesis. Hence, the value of all $p(\text{Pa}(m))$ are changed to account for this factor. At every possible point, the vehicle tries to perform prepreparation or postpreparation, whichever is closer in terms of the number of segments.

Fig. 8.4 shows the distributed-pathway segment for a sample map.

8.7.4   Feasibility Checks of Hierarchy

Recall that the pathway connection definition demanded that a vehicle in its entirety must be able to reach one pathway from another. Similarly, it is necessary for the distributed pathway to be such that any part of the vehicle does not lie in an inaccessible region whilst traversing. However, although the pathway distribution strategy ensures no collision as well as distance maximization from dynamic obstacles, it does not ensure avoidance of static inaccessible regions on the road. Hence, the distributed-pathway points may require to be tuned, to ensure a safe journey for the vehicle during driving. For any two consecutive points $p(\text{Pa}(m_1))$ and $p(\text{Pa}(m_2))$, the minimal distance of either of these points in the direction of a sweeping line is computed, such that the vehicle does not pass through an inaccessible region. The corresponding point is then moved by an equivalent amount.

At any pathway segment there are a total of high(Pa(m)) + low(Pa(m)) + 1 vehicles that possibly lie width to width in the pathway segment. In case the pathway segment width is such that it cannot accommodate this many vehicles considering their widths, then some correction must be applied. Again, the same concept of *priorities* comes into play. If while planning for a vehicle R_i, a pathway segment is found to contain an excessive number of vehicles, including any vehicle R_r with a lower priority than R_i, then R_r must replan at its *pathway-selection level*. This means that the original planned trajectory and pathway of R_r would be invalidated and replanned. As a consequence, the replanning of R_r may invalidate the trajectory of other vehicles. If, however, a pathway segment contains vehicles, all of which have higher priorities, then R_i must itself replan with its bounded speed v_i^b reduced to the speed of the vehicle just ahead of it. This reduction in speed ensures that excessive vehicles are not found at the pathway segment again.

The general method of this layer is given by Algorithm 8.6. For every segment, first the values for high(Pa(m)) and low(Pa(m)) are computed as per definition (Line 2) which gives the position of the vehicle in the pathway segment (Line 3). The position may be tuned to account for feasibility (Line 4). If a feasible plan is found it may be returned (Line 5). In case a plan is infeasible, an attempt may be made to make the plan feasible. If infeasibility is due to a potential collision with a lower-priority vehicle, the decision as to how to make the plan feasible is to be taken by the other vehicle which is replanned (Lines 7−8). In case the collision is with a higher-priority vehicle, the speed is altered to make the vehicle being planned follow the other vehicle (Line 10).

Algorithm 8.6: getDistributedPathway

```
Line 1  for every pathway segment Pa(m) in pathway Pa
Line 2          calculate high(Pa(m)) and low(Pa(m))
Line 3          calculate position by equation (8.13), accounting for
                   pre-preparation and post preparation
Line 4          feasibility checks and distributed pathway tuning
Line 5          if resultant distributed pathway is feasible return
                   Lᵢᵈᵖᵃ
Line 6          else
Line 7                  if collision with lower priority vehicle Rᵣ
Line 8                          replan Rᵣ at pathway level
Line 9                  else
Line 10                         vᵢᵇ = vᵣᵇ, replan Rᵢ at pathway level
Line 11                 end if
Line 12         end if
Line 13 end for
```

8.8 Hierarchy 4: Trajectory Generation

The generated distributed pathway is a curve consisting of line segments that may be used for navigation of the vehicle R_i. Ideally, if the vehicle travels along this path with bounded speed v_i^b, no collision would occur and all the individual vehicles would always be separated by a decent distance. This would happen due to the already computed safety measures for avoiding an excessive number of vehicles in a pathway segment and the attempt to maximize separation. However, vehicles are unable to take sharp turns and hence there is a need to *smooth* the curve. The smoothing or trajectory generation operation is done here by means of *BSpline curves* (Bartels et al., 1987; de Boor, 1978). Initially, all the points in the distributed pathway $L_i^{pda}(m)$ serve as the control points for the generation of the spline curve, the resulting curve thus becomes the prospective trajectory of the vehicle $\tau_i(t)$.

8.8.1 Collision Checking and Replanning

It is important to coordinate the different vehicles at this level as well, which is done by checking possible collisions between vehicles. Because the coordination is level by level, this hierarchy considers the published trajectories of the other vehicles. If any collision is found between the trajectory $\tau_r(t)$ published by a vehicle R_r, conflict resolution is performed using priorities. In this hierarchy, prioritization is carried out only for vehicles travelling in the same direction, with vehicles ahead having higher priorities. For the case of a collision with a lower-priority vehicle, the lower-priority vehicle must replan. For the case of a collision with a higher-priority vehicle, the speed v_i^b of the vehicle R_i is reduced to the speed v_r^b of the higher-priority vehicle R_r and R_i then replans. However, if a collision occurs with a vehicle moving in the opposite direction, then the speed is reduced by a fixed amount and vehicle R_i is replanned. In replanning if a collision is again encountered, the reduction-in-speed procedure is repeated until no collision is apparent.

8.8.2 Curve Smoothing

When a generated smooth trajectory is collision free, a further attempt can be made to make the trajectory even smoother. The maximum allowable speed of the vehicle is given by its curvature, which needs to be lower than the bounded speed v_i^b. This is given by Eq. [8.15].

$$v_i(t) = \begin{cases} \min\left(\sqrt{\dfrac{\rho}{Cu(t)}}, v_i^b\right) & Cu(t) \neq 0 \\[2em] v_i^b & Cu(t) = 0 \end{cases} \qquad [8.15]$$

in which ρ is the frictional constant the value for which depends upon the friction on the road and the value of acceleration due to gravity. $Cu(t)$ is the curvature. It is not possible to work over a continuous curve in the model and hence the curve is described in a piecewise discrete manner. To calculate the curvature in this discrete model at a position $\tau_i(t)$, two adjacent points are taken at a small distance of δ at either side of $\tau_i(t)$. The curvature $Cu(t)$ is given by Eq. [8.16]

$$Cu(t) = \|\tau_i(t + \delta) + \tau_i(t - \delta) - 2\tau_i(t)\| \qquad [8.16]$$

The curve $\tau_i(t)$ must be smoothed as much as possible to allow the vehicle to possess a speed v_i^b throughout its journey. It can be stressed that the objective of this hierarchy is not to minimize distance, but rather to maximize the separation. *Curve smoothing* is done by relocation of the control points that ultimately change the spline or the trajectory. Every point in the curve is checked for the maximal allowable speed. Points, for which this speed is less than v_i^b, need to be modified. An iterative-smoothing algorithm is implemented that first finds a point $\tau_i(t)$ at which the speed is less than the desired speed v_i^b. The algorithm then finds the closest control point which needs to be modified. To best smooth the curve, this point is placed at the midpoint of the control point just in front and to the back.

This value is admitted only if the resultant trajectory produced is valid ie, the complete trajectory lies on the free road and the vehicle does not collide with other vehicles. The optimization algorithm will stop if all points reach the specified speed threshold v_i^b. To avoid the optimization being carried out indefinitely if the road segment disallows maximum speed, the optimization can be time limited or the iteration limited — in practice, this does not present a problem. A trajectory generated for sample scenarios is given in Fig. 8.5.

The general approach for trajectory generation is given by Algorithm 8.7. The algorithm first checks whether the trajectory computed so far (Line 1) is collision free (Line 2). In the case of a collision (Lines 3−8), the algorithm makes a decision based on the priority of the vehicle with which the collision has occurred. In the case of a collision with a lower-priority vehicle (Line 3), the other vehicle must replan. In the case of a collision with a higher-priority vehicle (Lines 4−8), the speeds are altered to avoid a collision. The next step, in the case of a collision-free plan, is local optimization (Lines 10−22). At every iteration, the algorithm proceeds by relocating a control point c, which is closest to the point on the curve in which the speed of the vehicle is computed to be lower than desired (Lines 13−14), to the midpoint of adjoining control points (Line 19). If this relocation results in an infeasible curve, the change cannot be made (Line 21). Lines 15−17 avoid the same point c being selected at every iteration, which would continually produce an infeasible curve.

(A)

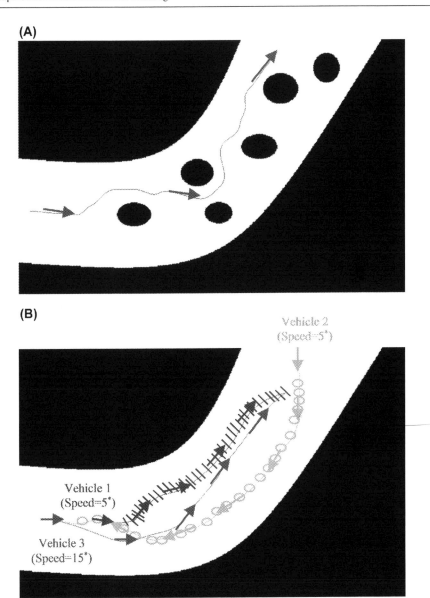

(B)

*Speed measured as unit distance per unit time

Figure 8.5 Trajectory generation. The indicative motion of the vehicle as stated by distributed pathway is smoothed to generate a feasible trajectory over which the vehicle moves. (A) Single vehicle scenario, (B) Multiple vehicles scenario.

Algorithm 8.7: getTrajectory

Line 1   generate curve τ using distributed pathway L_i^{dpa}(m) as control points

Line 2   if curve τ has collision

Line 3   if collision occurs with a lower priority vehicle R_r, replan R_r at pathway level

Line 4   else if collision occurs with higher priority vehicle R_r moving in same side

Line 5 $v_i^b = v_r^b$, replan R_i at pathway level

Line 6   else if collision occurs with higher priority vehicle R_r moving in opposite side

Line 7   reduce v_r^b by small amount, replan R_i at pathway level

Line 8   end if

Line 9   end if

Line 10 while stopping criterion

Line 11   generate curve τ using distributed pathway L_i^{dpa}(m) as control points

Line 12   p ← point in curve τ such that the speed of vehicle is less than v_i^b

Line 13   if no such point p, break

Line 14   c ← closest control point to p

Line 15   if c was used in previous iteration which did not produce a valid curve

Line 16   chose a different p and c, till a new c is selected

Line 17   end if

Line 18   if no such point c, break

Line 19   temporarily move c to mid-point of next and previous control points

Line 20   generate curve τ using modified distributed pathway as control points

Line 21   if resultant curve τ is feasible with no collision with static or dynamic obstacles, commit the change

Line 22 end while

8.8.3   Vehicular Movement

Once the trajectory of the vehicle is known, it may be simply used for traversal of the vehicle. A simple control algorithm may be used for the same. The control algorithm sends control signals to the vehicle to trace the planned trajectory $\tau_i(t)$. Along with the position, the controller also updates the vehicular speed. The speed update is performed assuming the vehicle is capable of infinite acceleration (as in an instantaneous model, Peng and Akella, 2005). Clearly, this is not a valid assumption in the case of vehicles, however, it is argued that the planner mostly generates smooth trajectories, for which the speed change requirement would be minimal and hence within finite acceleration limits. As the vehicle is planned to remain separated from other vehicles and obstacles, there is a larger scope for correcting errors in speed control of the vehicle without any collision.

8.9 Algorithm

Based on the discussion, the general procedure is written as given in Algorithm 8.8. It is assumed here that the route-selection algorithm has already been called and the road has been divided into a number of segments. This algorithm is called upon a vehicle R_i reaching any road segment entry. Starting from the vehicle R_i, the algorithm has two main tasks, to plan all vehicles at the pathway level (Lines 4−9) and at the distributed-pathway and trajectory level (Lines 10−18). Planning any vehicle at the pathway level generates as output (Line 5) a pathway-level travel plan along with a list of all vehicles which are disturbed by this generated plan and need to be replanned. The replanning may be at pathway level (Line 6) or at distributed-pathway level (Line 7). Planning at distributed pathway happens in strict order of priorities (Line 10). For every vehicle, first the distributed pathway (Line 12) is generated and then the corresponding trajectory (Line 17). Planning at this level may invalidate many vehicles at the pathway (Line 15) or distributed-pathway (Line 14) levels. As the distributed-pathway planning only makes sense if all vehicles agree to a common pathway plan for all vehicles, replanning is done if any vehicle changes its pathway (Line 16).

Algorithm 8.8: RoadSegmentPlan

```
Line 1    pathwayQ ← Rᵢ
Line 2    distributedPathwayQ← NULL
Line 3    Main Loop: while true
Line 4        while pathwayQ ≠ NULL
Line 5            getPathway(pathwayQ->top)
Line 6            pathwayQ ← pathwayQ ∪ all vehicles Rᵣ marked for
                  re-planning at pathway level
Line 7            distributedPathwayQ ← distributedPathwayQ ∪
                  pathway->top ∪ all vehicles Rᵣ marked for
                  re-planning at distributed pathway level
Line 8            pathwayQ ← pathwayQ − { pathway->top }
Line 9        end while
Line 10       sort distributedPathwayQ
Line 11       while distributedPathwayQ ≠ NULL
Line 12           getDistributedPathway(distributedPathwayQ ->top)
```

```
Line 13              distributedPathwayQ ← distributedPathwayQ - {
                     distributedPathwayQ ->top }
Line 14              distributedPathwayQ ← distributedPathwayQ ∪ all
                     vehicles Rr marked for re-planning at distributed
                     pathway level
Line 15              pathwayQ ← pathwayQ ∪ all vehicles Rr marked for
                     re-planning at pathway level
Line 16              if pathwayQ ≠ NULL continue main loop
Line 17              τi ← getTrajectory(Ri)
Line 18       end
Line 19       break
Line 20  end
```

8.10   Results

As per the problem formulation, the main task associated with the algorithm is to generate feasible paths for every vehicle in every scenario that is presented. This is the main intention behind the testing of the algorithm. Hence, the algorithm must be able to navigate multiple vehicles from both sides amongst complex obstacle frameworks. Further, optimality is of concern. Hence, overtaking is of main interest, and whenever possible a faster vehicle should be able to overtake a slower vehicle that lies ahead of it. The algorithm further stresses that the paths constructed for the vehicles be fairly wide and short. If all these conditions are met, the algorithm is stated to perform well for the scenario presented.

The experimentation methodology employed was layered in nature, as per the basic architecture of the algorithm. A different set of scenarios were designed to focus on different levels of the algorithm, whereas, in each case, the overall complete motion was guided by all the levels. First, experimentation with only single or double vehicles is presented — this is partly to indicate some of the features of the algorithm. Then more complex behaviour of overtaking and vehicle following are presented using multiple vehicles. Then the analysis of the algorithm parameters is presented.

8.10.1   Single- or Two-Vehicle Scenarios

This set consists of scenarios with single or two vehicles and their ability to avoid static obstacles and each other. The first experiment tests the highest level of planning or the ability of vehicle to find the optimal path from its source to its destination. A roadmap is given to the algorithm for computation of the path, which is then used for vehicle navigation. As per the algorithm guidelines, whereas the path computation using Dijkstra's algorithm takes place only once, the planning of pathway, distributed pathway and trajectory is done at every road segment. The resultant trajectory of the vehicle is shown in Fig. 8.6. Here, the road was broken down into nine overlapping segments. The vehicle displayed motion with a constant fixed speed of 10 unit distances per unit time.

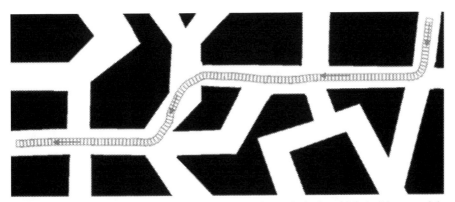

Figure 8.6 Path traced by the vehicle for a sample roadmap. A single vehicle is able to reach its destination starting from its source by following a smooth and feasible trajectory using the shortest path.

The next experiment was focussing upon the ability of the vehicle to steer through static obstacles. For better visibility a section of road is taken over which the vehicle was supposed to move. To make planning difficult, the maps used for the purpose had curves which forced the vehicle to turn, while manoeuvring through the obstacles. The pathway-generation algorithm is the key to this scenario, which works by emphasis being placed on maximization of widths. It is evident from this that placement of multiple obstacles on narrow roads with large vehicles, still making some trajectories feasible, is extremely difficult and in some cases simply not possible. This is because insertion of two or more obstacles close by can eliminate any possible feasible trajectory, as the obstacles are practically seen as a road blockage. Hence, here in the first instance a wide road is taken with a small vehicle that can easily steer around the obstacles presented. For the same reasons, the speed was limited to five unit distances per unit time.

The obstacle set consisted of both regularly and irregularly shaped obstacles, each of which was placed strategically to either make a complex network of pathway segments with smaller obstacles or a simple network with more elongated obstacles. The resultant trajectories traced by the vehicle are given in Fig. 8.7. In all these cases, the vehicle is able to compute the largest-width path of the shortest length. In these cases, the path was traced using the maximum speed of the vehicle. Sometimes it can be observed that the vehicle gets relatively close to obstacles, especially during the trajectory directly before entering an obstacle region and directly after leaving an obstacle region. This happens as the ideal path of the vehicle, without any trajectory optimization, may be reasonably sharp, and the trajectory optimizer results in a trajectory which is smooth enough and does not lie on the obstacle. This behaviour also ensures the vehicle does not stray too far to avoid a small obstacle.

The next set of experiments focussed upon the ability of a vehicle to steer though obstacles, at the same time avoiding a vehicle coming towards it. This required simultaneous working of pathway-determination and pathway-distribution modules. At the same time, this tested the ability of vehicles to coordinate such that each vehicle is driven on the left side of the road. The resultant trajectories (with possible

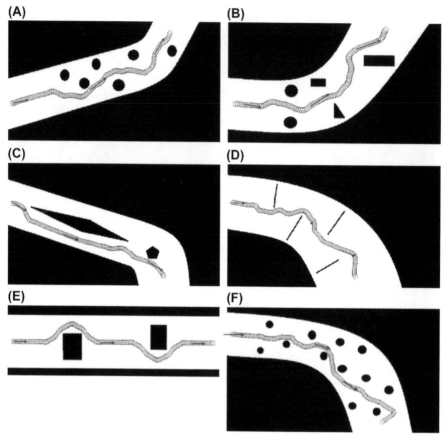

Figure 8.7 Path traced by a single vehicle. (A) Scenario 1, (B) Scenario 2, (C) Scenario 3, (D) Scenario 4, (E) Scenario 5, (F) Scenario 6.

points of collision) are shown in Fig. 8.8. It may be observed that in all these cases the vehicles maintained a large separation from each other at the possible point of collision, at the same time avoiding the many static obstacles.

The map for Fig. 8.8A consisted of multiple obstacles, in which the collision between vehicles was about to take place at a narrow point on the road. In planning, the first vehicle reduced its speed by two units to delay collision, making the resultant passing at a place with much greater width as shown in Fig. 8.8A. In all other cases, the vehicles travelled at their maximum set speed. Similarly, Fig. 8.8B shows that the vehicle followed a different route (as compared to the route in Fig. 8.7B). This was because a single pathway could comfortably accommodate a single vehicle only and a left driving of the vehicle was preferable. In the same figure, the second vehicle did travel on the right side of the smaller obstacle. This is because the algorithm perceives the width of a pathway segment end as the smallest width. As the widths are scanned at intervals of Δ, it cannot be ascertained that the minimal width of a pathway would be recorded in cases in which it lies in a very small region.

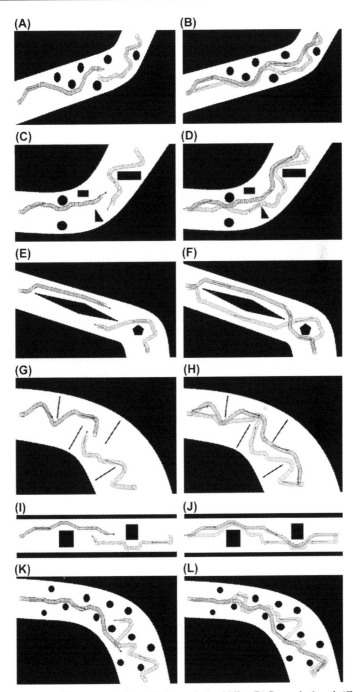

Figure 8.8 Path traced by two vehicles. (A) Scenario 1 middle, (B) Scenario 1 end, (C) Scenario 2 middle, (D) Scenario 2 end, (E) Scenario 3 middle, (F) Scenario 3 end, (G) Scenario 4 middle, (H) Scenario 4 end, (I) Scenario 5 middle, (J) Scenario 5 end, (K) Scenario 6 middle (L) Scenario 6 end.

8.10.2 Multivehicle Scenarios

The complexity was further extended to three vehicles with two vehicles being generated on one side of the road and the other vehicle on the other side. One of the vehicles was generated after 10 units of time to ensure a decent initial separation between the vehicles involved. The trajectories of the vehicles at points of closest distance are plotted in Fig. 8.9A. It may be seen that this case is an extension of the previous case in which the vehicles need to avoid collision first from one passing vehicle and then from the other. The two vehicles generated

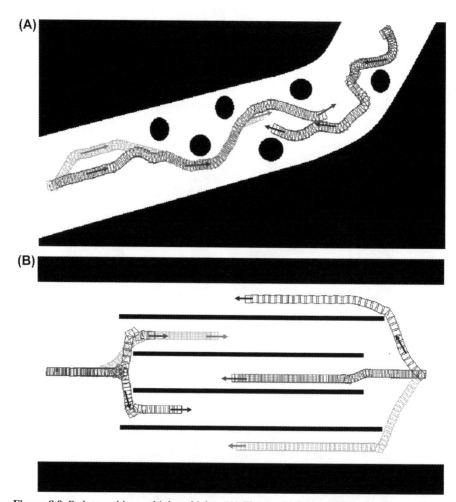

Figure 8.9 Path traced by multiple vehicles. (A) Three vehicles coordinate with each other to meet at a reasonably sparse region and further to maximize separation between each other. (B) Each vehicle chooses to go by a new lane to avoid overtaking the earlier vehicles within a small lane or to follow a slower vehicle in some lane. The last vehicle does not get a free lane and has to follow an earlier vehicle.

on the same side of the road followed almost the same trajectory. As they had the same speed, collision between them was not possible.

Focussing upon the pathway generation algorithm while dealing with multiple vehicles, one of the important characteristics associated is the ability of the algorithm to distribute pathways amongst vehicles in the case when a single pathway may not have sufficient width to simultaneously accommodate multiple vehicles. Hence, a map is generated with five different pathways. Vehicles were generated on either side of the road, each vehicle emerging once the previously emerged vehicle is sufficiently ahead. Speeds were decided such that the vehicles needed to pass each other within any pathway (if they decided to use the same pathway). After tuning of the emerging times and speeds in this manner, the behaviours of the vehicles were noted. It was found that different vehicles occupied different pathways. This is shown in Fig. 8.9B. Another vehicle was added to the scenario, which had no free pathway left. As a result, it decided to use the central pathway again. As the pathways were set not to allow two vehicles, it had to lower its speed and chase the vehicle in front.

The next experiment was designed to showcase the ability of vehicles to overtake when sufficient width was not available on the road to accommodate three vehicles simultaneously. The design involved a straight road and two vehicles heading (on opposite sides of the road) towards each other. It can be seen in Fig. 8.10 that

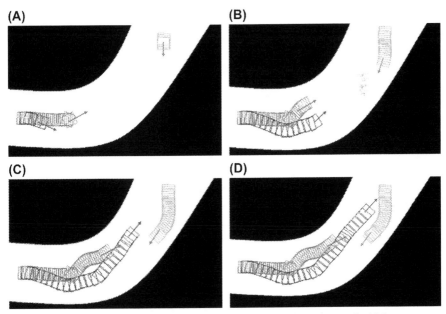

Figure 8.10 Overtaking behaviour. (A) The rear vehicle at left is generated which computes a feasible overtake and proceeds for the same. (B) The two vehicles earlier in scenario adjust each other to allow overtake to happen, as the overtaking vehicle proceeds with overtaking trajectory. (C) The overtaking vehicle proceeds to surpass the vehicle being overtaken. (D) The overtaking vehicle successfully avoids the vehicle at the other side.

when the third vehicle enters the scenario, it needs to move onto the other side of the road, pass the first vehicle and return to its original side, while the two vehicles are coming towards each other. This is the classic mechanism of overtaking that involves good judgement from the vehicles concerned.

The vehicle size was, in fact, increased to make the task more difficult. The allowed speeds of the first two vehicles were 5 unit distances per unit time, whereas the speed of the vehicle overtaking was 15 unit distances per unit time. The simulation shows that when the third vehicle was generated, it employed a mechanism to avoid the other two vehicles maintaining the same speed. It is clearly visible in Fig. 8.10 that all the vehicles attempted to align themselves to maintain the largest possible separation.

Whilst overtaking was feasible for the scenario generated in Fig. 8.10, it may not always be the case. The scenario for Fig. 8.10 was computed to make the overtaking procedure feasible. Changing the time or speed may easily make overtaking infeasible. In such a case, the vehicle travelling behind may decide to simply follow the vehicle in front. Its speed is reduced to the speed of vehicle in front. As soon as the other vehicle passes from the opposite direction, the vehicles attempt to return to central position. This event is premarked as a replanning event by the earlier computation of the planning algorithm. Hence, the vehicles replan and an overtaking procedure can then take place. For this example, the speed of the vehicle is changed to 8 unit distances per unit time. The resultant trajectories are shown in Fig. 8.11.

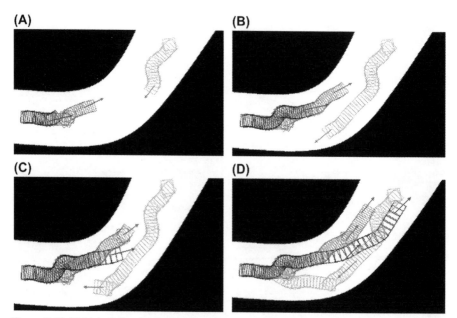

Figure 8.11 Vehicle-following behaviour. (A) The rear vehicle at left is generated which cannot overtake and hence decides to follow the vehicle in front. (B) The vehicle at the other side passes which was a barrier to a successful overtake. Overtake is now feasible and initiated. (C) The vehicle continues its overtaking trajectory. (D) Overtake is completed.

8.10.3 Algorithmic Parameter Analysis

One of the major factors associated with the algorithm is the frictional parameter ρ that effectively decides the instantaneous speed of the vehicle as per Eqs [8.15] and [8.16]. A very high value of this factor means that the vehicle is capable of making sharp turns at high speed. This may mean that the distributed pathway can be easily converted to a suitable trajectory with only minor work over the edges that the splines are capable of. This happens with mobile robots which display a dramatic change in directions while maintaining a constant predefined speed. A very low value of this factor might, however, mean that a small turn can be made at a reasonably low speed, which would in turn influence the trajectory-optimization algorithm in generating paths which are as straight as possible.

A curved path with few obstacles was used for experimentation and the distance traversed by the vehicle, vehicle speeds during the journey (minimum, maximum and average), time of journey and time of optimization were plotted for various values of parameter ρ. The corresponding graphs are shown in Fig. 8.12. The graph with

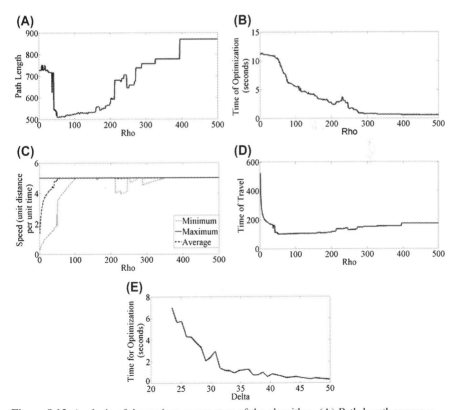

Figure 8.12 Analysis of the various parameters of the algorithm. (A) Path length versus ρ (B) time required for optimization versus ρ. (C) Speed of traversal of vehicle versus ρ. (D) Time of travel of vehicle versus ρ (E) time of optimization versus Δ.

distance traversed is given in Fig. 8.12A. This graph can be segmented into three regions, suboptimized (1—50), optimized (50—400) and constant (400 and beyond). The suboptimized region consists of very small values of ρ, for which the optimizer cannot draw a trajectory, keeping the instantaneous speed close to bounded speed, at the same time the complete trajectory is feasible. In this set of points, the optimizer is stopped long before it can safely terminate either reporting all points having reached bounded speed, or reporting that no further optimization is possible (as further movement of any point results in infeasibility).

The optimizer starts from points near to the current position of the vehicle working to the points farther off. This is because the points farther off may, in practice, be optimized at the next road segment, if they belong to the same. In the suboptimized region, the earlier part of the trajectory is optimized and the latter is not. For the same reasons, the distance keeps reducing (so more of trajectory is optimized) as ρ is increased. Fig. 8.13A shows the trajectory generated at a value $\rho = 25$. A further increase in ρ (in the optimized region) enables every point of the not-so-smooth trajectory also to be within the bounded speed, which is acceptable as optimized by the algorithm because it leads to separation maximization. Hence, the distance of travel increases

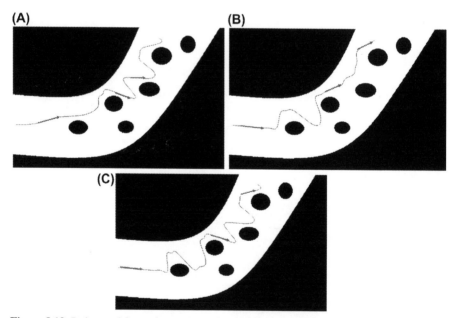

Figure 8.13 Path traced for various values of ρ. (A) Small values of ρ in which the curve needs to be extremely smooth for the vehicle to travel. Hence, the optimization algorithm fails after a few iterations and only the starting few regions of the curve are smoothed. Trajectory is infeasible. (B) Intermediate values of ρ in which initial curve generated based on separation maximization needs to be moderately smoothed to enable vehicle to travel it. (C) High values of ρ in which case it is possible to drive at high speeds even with sharp turns. Hence, main attempt becomes separation maximization, knowing any curve would be traceable.

with an increase of ρ, as a vehicle is capable of maintaining high speeds even during sharp turns. Fig. 8.13B shows the trajectory generated at a value $\rho = 250$. Increasing ρ further than this leads to no change in the distance of travel. This is into the constant region of the graph, in which the produced spline from the distributed pathway is considered smooth enough for the vehicle to travel at maximum speed. The trajectory of any vehicle in this region is shown in Fig. 8.13C.

Based on the discussion concerning Fig. 8.12A, it is reasonably straightforward to analyse the time taken for optimization of the algorithm for increasing values of ρ shown in Fig. 8.12B. For the suboptimized region $(1-50)$, the algorithm stops because of the number of iterations criterion. Hence, the computation time is uniform. The time taken then decreases sharply in the optimized region $(50-400)$. This decrease is because more curved paths are acceptable, which reduces the number of iterations required for optimization with increasing ρ. In the constant region (400 and over), the time of optimization is constant as it involves only the single computation of the spline produced by the distributed pathway. Slight increases that can be witnessed around some regions (eg, $200-250$) are characteristic of the position of obstacles — making a part of the curve extra-smooth automatically avoids some obstacles.

The speed graph (shown in Fig. 8.12C) shows similar trends. The maximum speed is a constant magnitude equivalent to the bounded speed. Consequently, some regions of the trajectory are straight with infinite curvature. The average speed shows a sharp increase in the suboptimal region. This is because as the value of ρ is increased, a vehicle is capable of possessing high speeds through small curves, up to the bounded speed. Further, increases in ρ result in more parts of the trajectory being optimized, which further increase the average speed. The average speed henceforth is almost constant, equivalent to the value of the bounded speed. This means that most regions of the trajectory are smooth enough to exhibit maximum speed.

The graph in Fig. 8.12A showing path length did display a large variation in this region, whereas the average-speed graph does not. This is because a sharp turn may result in a low speed over a small region but a high speed in all neighbouring regions. The curve of the average speed shows a similar trend to that of the average speed, except for some lower magnitudes in certain regions, which again are characteristics due to obstacle placement. It is possible in pursuit of optimization for a point to move in a manner such that any further motion results in infeasibility with the speed being somewhat low at the point. The time of travel is a simple ratio between the distance of travel and the average speed, the graph for which is given in Fig. 8.12D. The sharp decrease in the suboptimal region is due to an increase in speed in the same region. The curve thereafter shows the same trends as the distance-of-travel graph.

The next parameter used in the algorithm was Δ, the distance traversed by the sweeping line while recording the pathway segments. The ideal value of this parameter depends on obstacle size. A large value of this parameter would result in large pathway segments, the ends for which may not represent the complete pathway segment. Hence, it would not be possible to compute pathway connectivity in a short time span. Too-small values of this parameter, however, result in an excessive number of pathways, thereby increasing the computational time of the algorithm. On top of this, they cause the vehicle to make sharp transitions (turns) between pathways for navigation.

The time needed for optimization is analysed for increasing values of Δ. It is evident that Δ needs to be large enough to allow a vehicle to traverse from one pathway to another pathway without any end lying on some further pathway. In addition, values of Δ that are too large would result in the algorithm being unable to compute connectivity being pathways by any simple method. Loss of connectivity would mean no feasible path being reported in pathway selection. Hence, the graph only shows values within these two extremes. The corresponding plot is shown in Fig. 8.12E. Note that the complexity of every level (except road selection) directly depends on the number of pathway segments. Because the value of Δ directly affects the placement of pathway segment ends, which drastically influences the ease of optimization, there are some irregular trends visible.

8.10.4 Algorithmic Scalability Analysis

The last aspect of the simulation exercise was to judge the scalability of the algorithm. The scalability of Dijkstra's algorithm is clearly far more than the present road-network graph available for any country. The scalability of other hierarchies of the algorithm depends on the number of obstacles and the number of vehicles. The increase in time of computation due to both of these are studied separately. It is evident that, because every vehicle plans on its entry to every road segment, only vehicles and obstacles within the segment are considered in any practical scenario. Hence, the number of both obstacles and vehicles are limited to the maximum that may be accommodated within a segment.

To test scalability of the number of obstacles, a simple curved-road scenario is taken and road blockages are placed alternatively on either side of the road, forcing the vehicle to steer away from them. It is worth noting that every obstacle necessitates the planner to realize a path which avoids the obstacles. Although the extra computation at the pathway generation is minimal, there is increased computation required for trajectory generation. However, as only minor tuning of the pathway segment points needs to be done, the increase is minor. This is unlike evolutionary approaches in which an increased number of obstacles in the path of the vehicle produces an extremely large increase in planning complexity. The graph is shown in Fig. 8.14A.

Scalability issues due to the number of vehicles were tested as well. Every vehicle which plans its path thereby affects other vehicles which need to replan as a consequence. Replanning every vehicle may be time-consuming and hence the planning time increases as the number of vehicles in the scenario increases. For experimentation, a curved road is taken with vehicles being generated from both ends at regular intervals. Every vehicle being generated had to plan its trajectory and possibly replan the trajectory of other vehicles as well, in case of a potential collision. The corresponding graph is shown in Fig. 8.14B. It can be seen that there is a small increase in the computational time with increasing number of vehicles.

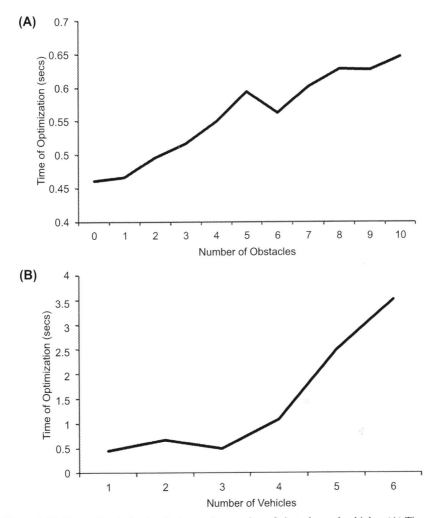

Figure 8.14 Time of optimization for increasing number of obstacles and vehicles. (A) Time of optimization versus number of obstacles. (B) Time of optimization versus number of vehicles.

8.11   Summary

The multitude of scenarios, which a vehicle may be exposed to, make the problem of motion planning for autonomous vehicles a difficult task. These difficulties are increased considerably with the presence of multiple vehicles and other obstacles. The real-time nature of the problem eliminates the possibility of using numerous algorithms which, although they can provide encouraging results, are costly in terms of computation time. Humans are prone to make incorrect decisions, many times possibly because of emotional pressures. Machines, after an optimal setting of risk factors associated with decision making, are likely to be more accurate in such decision making.

This chapter was devoted to the use of graph-search methods for solving the problem. A general graph search is computationally intensive and hence cannot be used. The approach made the problem computationally efficient and hence solvable by the introduction of hierarchies. A total of four hierarchies were proposed for route selection, obstacle avoidance, vehicle placement and trajectory generation. The vehicles were placed such that the separation of the vehicles from the obstacle or the other vehicles is maximized.

Coordination between the different vehicles was based on a layer-to-layer manner, and any vehicle could ask any other vehicle to replan. A shortcoming of the presented approach is the requirement for a relatively large separation between vehicles. Having vehicles already close by while the planning starts may cause the vehicles to make large deviated movements which aim towards separation maximization. However, the algorithm means that for much of the time each vehicle would be able to travel at speeds close to their respective bounded speed — although this is not possible if a vehicle passes through a region where speed is, for some reason, to be slowed by a large amount.

References

Asama, H., Matsumoto, A., Ishida, Y., 1989. Design of an autonomous and distributed robot system: ACTRESS. In: Proceedings of the IEEE/RSJ International Workshop on Intelligent Robots and Systems, Tsukuba, Japan, pp. 283–290.

Asama, H., Ozaki, K., Itakura, H., Matsumoto, A., Ishida, Y., Endo, I., 1991. Collision avoidance among multiple mobile robots based on rules and communication. In: Proceedings of the IEEE/RSJ International Workshop on Intelligent Robots and Systems, Osaka, Japan, pp. 1215–1220.

Bartels, R.H., Beatty, J.C., Barsky, B.A., 1987. An Introduction to Splines for Use in Computer Graphics and Geometric Modelling. Morgan Kaufmann, San Francisco, CA.

de Boor, C., 1978. A Practical Guide to Splines. Springer Verlag, Heidelberg.

Chand, P., Carnegie, D.A., 2012. A two-tiered global path planning strategy for limited memory mobile robots. Robotics and Autonomous Systems 60 (3), 309–321.

Chen, D.Z., Szcerba, R.J., Uhran, J.J., 1997. A framed-quadtree approach for determining euclidean shortest paths in a 2-D environment. IEEE Transactions on Robotics and Automation 13 (5), 668–681.

Cormen, T.H., Leiserson, C.E., Rivest, R.L., Stein, C., 2009. Introduction to Algorithms, third ed. MIT Press, Cambridge, MA.

Cowlagi, R.V., Tsiotras, P., 2012. Hierarchical motion planning with dynamical feasibility guarantees for mobile robotic vehicles. IEEE Transactions on Robotics 28 (2), 379–395.

Crane, C., Armstrong, D., Arroyo, A., Baker, A., Dankel, D., Garcia, G., Johnson, N., Lee, J., Ridgeway, S., Schwartz, E., Thorn, E., Velat, S., Yoon, J., Washburn, J., 2007. Team Gator Nation's autonomous vehicle development for the 2007 DARPA Urban challenge. Journal of Aerospace Computing, Information and Communication 4 (12), 1059–1085.

Felner, A., 2011. Position paper: Dijkstra's algorithm versus uniform cost search or a case against Dijkstra's algorithm. In: Proceedings of the Fourth International Symposium on Combinatorial Search. AAAI, pp. 47–51.

Gayle, R., Moss, W., Lin, M.C., Manocha, D., 2009. Multi-robot coordination using generalized social potential fields. In: Proceedings of the 2009 IEEE International Conference on Robotics and Automation, Kobe, Japan, pp. 106—113.

Hart, P.E., Nilsson, N.J., Raphael, B., 1968. A formal basis for the heuristic determination of minimum cost paths. IEEE Transactions on Systems Science and Cybernetics 4 (2), 100—107.

Hou, E.S.H., Zheng, D., 1991. Hierarchical path planning with hexagonal decomposition. In: Proceedings of the IEEE International Conference on Systems, Man, and Cybernetics, pp. 1005—1010.

Hwang, J.Y., Kim, J.S., Lim, S.S., Park, K.H., 2003. A fast path planning by path graph optimization. IEEE Transactions on Systems, Man and Cybernetics Part A: Systems and Humans 33 (1), 121—128.

Kala, R., Shukla, A., Tiwari, R., 2009. Robotic path planning using multi neuron heuristic search. In: Proceedings of the ACM 2nd International Conference on Interaction Sciences: Information Technology, Culture and Human, Seoul, Korea, pp. 1318—1323.

Kala, R., Shukla, A., Tiwari, R., 2010. Dynamic environment robot path planning using hierarchical evolutionary algorithms. Cybernetics and Systems 41 (6), 435—454.

Kala, R., Shukla, A., Tiwari, R., 2011. Robotic path planning using evolutionary momentum based exploration. Journal of Experimental and Theoretical Artificial Intelligence 23 (4), 469—495.

Kala, R., Warwick, K., 2013. Multi-level planning for semiautonomous vehicles in traffic scenarios based on separation maximization. Journal of Intelligent and Robotic Systems 72 (3—4), 559—590.

Kambhampati, S., Davis, L.S., 1986. Multiresolution path planning for mobile robots. IEEE Journal of Robotics and Automation 2 (3), 135—145.

Kant, K., Zucker, S.W., 1986. Toward efficient trajectory planning: the path-velocity decomposition. International Journal of Robotics Research 5 (3), 72—89.

Konar, A., 1999. Artificial Intelligence and Soft Computing: Behavioral and Cognitive Modeling of the Human Brain. CRC Press, Boca Raton, FL.

Langsam, Y., Augenstein, M.J., Tenenbaum, A.M., 2009. Data Structures Using C and C++, second ed. PHI Publishers.

Lu, Y., Huo, X., Arslan, O., Tsiotras, P., 2011. Incremental multi-scale search algorithm for dynamic path planning with low worst-case complexity. IEEE Transactions on Systems, Man and Cybernetics Part B: Cybernetics 41 (6), 1556—1570.

Montemerlo, M., Becker, J., Bhat, S., Dahlkamp, H., Dolgov, D., Ettinger, S., Haehnel, D., Hilden, T., Hoffmann, G., Huhnke, B., Johnston, D., Klumpp, S., Langer, D., Levandowski, A., Levinson, J., Marcil, J., Orenstein, D., Paefgen, J., Penny, I., Petrovskaya, A., Pflueger, M., Stanek, G., Stavens, D., Vogt, A., Thrun, S., 2008. Junior: the Stanford entry in the Urban challenge. Journal of Field Robotics 25 (9), 569—597.

Peng, J., Akella, S., 2005. Coordinating Multiple Robots with Kinodynamic Constraints Along Specified Paths. International Journal of Robotic Research 24 (4), 295—310.

Reif, J.H., Wang, H., 1999. Social potential fields: a distributed behavioral control for autonomous robots. Robotic and Autonomous Systems 27 (3), 171—194.

Russell, R., Norvig, P., 2009. Artificial Intelligence: A Modern Approach, third ed. Pearson, Harlow, Essex, England.

Stentz, A., 1994. Optimal and efficient path planning for partially-known environments. In: Proceedings of the International Conference on Robotics and Automation, pp. 3310—3317.

Stentz, A., 1995. The Focussed D* Algorithm for Real-Time Replanning. In: Proceedings of the International Joint Conference on Artificial Intelligence, Vol. 95, pp. 1652−1659, Chicago.

Takahashi, O., Schilling, R.J., 1989. Motion planning in a plane using generalized Voronoi diagrams. IEEE Transactions on Robotics and Automation 5 (2), 143−150.

Todt, E., Raush, G., Sukez, R., 2000. Analysis and classification of multiple robot coordination methods. In: Proceedings of the 2000 IEEE International Conference on Robotics and Automation, San Francisco, CA, pp. 3158−3163.

Urdiales, C., Bantlera, A., Arrebola, F., Sandoval, F., 1998. Multi-level path planning algorithm for autonomous robots. IEEE Electronics Letters 34 (2), 223−224.

Vendrell, E., Mellado, M., Crespo, A., 2001. Robot planning and re-planning using decomposition, abstraction, deduction, and prediction. Engineering Applications of Artificial Intelligence 14 (4), 505−518.

Wilkie, D., van den Berg, J., Manocha, D., 2009. Generalized velocity obstacles. In: Proceedings of the 2009 IEEE/RSJ International Conference on Intelligent Robots and Systems, St. Louis, USA, pp. 5573−5578.

Xiao, J., Michalewicz, Z., Zhang, L., Trojanowski, K., 1997. Adaptive evolutionary planner/navigator for mobile robots. IEEE Transactions on Evolutionary Computation 1 (1), 18−28.

Yahja, A., Stentz, A., Singh, S., Brumitt, B.L., 1998. Framed-quadtree path planning for mobile robots operating in sparse environments. In: Proceedings of the IEEE International Conference on Robotics and Automation, pp. 650−655.

Zhang, L., Kim, Y.J., Manocha, D., 2007. A hybrid approach for complete motion planning. In: Proceedings of the IEEE/RSJ International Conference on Intelligent Robots and Systems, pp. 7−14.

Using Heuristics in Graph Search-Based Planning

9

9.1 Introduction

Graph-based techniques (Cormen et al., 2009; Langsam et al., 2009) are widely used to solve a variety of problems from different domains. Using a graph search-based methodology to solve the problem of navigation of multiple autonomous vehicles is thus worthy of investigation. The real-time nature of the problem restricts the use of computationally expensive techniques. Most autonomous vehicles come with heavy computational units onboard to execute vision, sensor fusion, localization, mapping, control and planning algorithms in parallel. Thus, powerful hardware capabilities can be used to some extent to speed up computation. However, even in the worst situations, the algorithm must be of low computational complexity and should facilitate near-real time decision making. This stresses devising ways to make the traditional graph-search algorithm computationally efficient.

A standard way to do so is to reduce the resolution of the map (Tiwari et al., 2013), thus making a lower-resolution graph. The algorithms thus become resolution optimal and resolution complete, which are acceptable provided one is ready to compromise on optimality and completeness to some extent. However, traffic scenarios are largely marked by close overtaking. The overtaking procedures can only be assessed when the resolution is very high, as the lateral separation is very small during overtaking. Lower-resolution maps give vague information about the separations and thus cannot be used for accurate feasibility assessment. This motivates exploiting the road architecture and to devise heuristics for limiting the resolution by carefully selecting the vehicle states.

The chapter delves into the notion of lane as being composed of a large number of lanes, all with *variable widths* and *distributed along the road segment*. This notion is used to iteratively plan the vehicles in a prioritized manner. Each iteration tries to insert a vehicle amidst a pool of planned vehicles. The vehicle being planned attempts to place itself by moving around other vehicles as long as possible. Hence, *state representation* of a vehicle not only includes its state, but also includes the trajectories of the other vehicles. The *action set* of a vehicle's attempts to best place itself at the next time step in-between the obstacles. In case of the presence of other vehicles, an attempt is made to aptly move the other vehicles while the vehicle being planned attempts to place itself amidst a vehicle pool. Heuristics of vehicle following, vehicle overtaking and waiting for a vehicle are also used for limiting state expansions.

This algorithm solves the problem as a *general graph-search problem*. The problem is not decomposed into hierarchical layers as was the case with the approach discussed in chapter 'Graph Search-based Hierarchical Planning'. However, this approach chiefly relies upon *state reduction*. The approach deals with devising *heuristics* and methods to intelligently pick states and actions which lead to a near-optimal trajectory, giving a high computation boost with the least loss of optimality or completeness. The approach does all this while staying within a single hierarchy (or the finest hierarchy), unlike the previous approach in which different layers of hierarchy were used. The chief purpose behind this approach is that all decision making, be it regarding avoiding an obstacle or overtaking a vehicle, can be best done when precise information is available, which was not the case with the previous approach in which all such decision making was done at higher hierarchies with a vague idea of self or the other vehicles.

Centralized approaches are nonimplementable in real time, due to which planning needs to be performed in a decentralized manner. Pure decentralized approaches cannot model strong cooperation between the vehicles which is only possible with the centralized approaches. Hence, the use of a decentralized approach must ensure that the resultant system can showcase all types of desirable behaviours between vehicles, eg, in prioritized approaches, lower-priority vehicles always lose out, which can lead to overtaking becoming infeasible due to a lack of cooperation.

The general aspects of solution using graphs are described in Box 9.1. The key takeaways are noted in Box 9.2. For the purpose of this approach, the notion of *generalized lane* is floated. A (generalized) lane is defined as the portion of road (without obstacles) occupied exclusively by a vehicle for a specific duration of time. The vehicle may be found at any location inside a portion of road with a guarantee that it will not collide with any other vehicle. For a road with predefined-speed lanes or lane-oriented traffic, this would specialize to the general concept of speed lane. Some characteristics of lanes, which must be considered to make an optimal travel plan, are:

1. *Distributed* – The number of speed lanes and their widths may change in different segments of the road. Both number and widths of speed lanes at any particular segment (called *pathway*) depends upon the demand of the width of the road, presence of obstacles, demand of the pathway at any instance of time and widths of vehicles demanding the pathway.

Box 9.1 Pros and Cons of Graph Search

Pros
- Completeness subject to resolution of selection of states and actions
- Optimality subject to resolution of selection of states and actions

Issues
- Computational complexity

Key Idea
- State reduction: Carefully select the states to expand

Box 9.2 Key Takeaways of the Approach

- The generalized notion of *lanes* is defined.
- The generalized notion is used for *planning* and *coordination* of multiple vehicles.
- A *pseudocentralized* coordination technique is designed which uses the concepts of *decentralized coordination* for iteratively planning different vehicles but empowers a vehicle to move around the other vehicles. The coordination is hence better in terms of optimality and completeness than most approaches (discussed so far), although being somewhat computationally expensive in the worst cases.
- The concept of one vehicle *waiting* for another vehicle coming from the other direction is introduced, when there may be space enough for only one vehicle to pass.
- *Heuristics* are used for *pruning* the expansions of states, which result in a significant computational efficiency while leading to a slight loss of optimality.

---

2. *Dynamic* — The speed lanes change along with time for every pathway as vehicles pass by. For every vehicle, the speed lane changes along with time as the vehicle discovers more vehicles it may possibly overtake, completes overtake, is overtaken, decides to travel aside some vehicle and similar situations.
3. *Single vehicle* — No two vehicles may occupy a speed lane side by side.
4. *Variable width* — every lane has a different width that depends upon the width of the vehicle that uses it and the total available width excluding obstacles.

Although the definition of speed lane holds for all the approaches discussed so far, this is being specifically brought here as the approach to be presented specifically uses these properties for planning, whereas the earlier approaches dealt the entire road as a free space for planning. The problem is first solved with the assumption of a single vehicle in the entire road in Section 9.3. The approach is then generalized for more than one vehicle occupying the road in Section 9.4.

Segments of the chapter have been reprinted from Kala and Warwick (2014) with kind permission from Springer Science + Business Media: Applied Intelligence, Dynamic distributed lanes: motion planning for multiple autonomous vehicles, Vol. 41, No. 1, 2014, pp. 260–281, R. Kala, K. Warwick, Text and All Figures, © Springer Science + Business Media New York 2014.

9.2 A Brief Overview of Literature

Some of the very relevant approaches using Graph Search and roadmaps have already been discussed in chapter 'Graph Search-based Hierarchical Planning'. The section extends the discussions of chapter 'Graph Search-based Hierarchical Planning', concentrating on the approaches using roadmaps for motion planning for mobile robots. Yao and Gupta (2011) presented an algorithm to make the roadmap in an entirely distributed manner with different sensors at different locations making a local roadmap, which can be integrated to make the overall global roadmap. In a road scenario, the vehicles need to largely follow the road and hence the sensors need not be deployed for guidance. Further, the use of sensors on road for producing the roadmap is not a practical approach. Clark (2005) solved the problem of robot-motion planning using Probabilistic Roadmaps. The robots communicated each other's position to make the complete roadmap. Such a communication framework to form a rich roadmap within the considered road segment distributed between the robots is a presumption of some of the algorithms presented in this book. Kala et al. (2011) solved the problem by using a multiresolution approach. At every iteration, a path was computed and the resolution was increased around the areas of the path, which initiated a search at the next iteration. The solution was anytime in nature and the path improved with time.

The roadmaps used for navigation can be made to react to the changes in the environment, in which case the roadmap is always updated as per the current obstacles. As the obstacles move, the vertices and edges are also made to adjust. The edges and vertices may move, get formed or break as a result of addition and deletion of obstacles. Gayle et al. (2007, 2009) presented an algorithm for reactive deformation of the roadmap and used it for navigation of a large number of virtual agents. The offline phase of roadmap construction is the major disadvantage. Further, having a single road used for traffic from both the sides faces the problem of accidentally having two vehicles facing each other head to head, which is a problem with such an approach. Kala (2013) presented an approach to iteratively build a roadmap by first making a roadmap around the most important parts of the map, and then extending the same roadmap to cover most of the search space.

Some approaches (Bhattacharya et al., 2012; Demyen and Buro, 2006; Schmitzberger et al., 2002) aim to discover all possible homotopic groups in the map. The homotopy-based roadmap construction ensures representation of all possible paths between all pairs of sources and goals, with a very small number of vertices and edges. This significantly reduces the search time. Similarly, the roadmap may first be constructed, which may be very large to perform motion planning in small computation times. Graph-spanning algorithms (Dobson and Bekris, 2014; Dobson et al., 2013; Littlefield et al., 2013) aim to delete the vertices and edges which do not affect the cost of the path by a large factor, thus reducing the computation time during searching of a path in the roadmap, with a very small loss of optimality. Similarly, Visibility Roadmaps (Siméon et al., 2000) restrict the number of vertices by iteratively growing the roadmap and only allowing the vertices which are either not in the vicinity of any vertex or result in a connection between two disjointed subgraphs. This produces a roadmap with a very small number of vertices and edges.

9.3 Dynamic Distributed Lanes for a Single Vehicle

The first problem considered is planning of a single vehicle. A road segment (part of the entire road), characterized by its two boundaries, is given; which may consist of any number/type of static obstacles. Let R_i be the vehicle to be planned, initially located at position (x_i^s, y_i^s) with orientation θ_i^s; and having current speed v_i^s ($\leq v_{max}^i$, the maximum permissible speed). Further, consider that the vehicle is a rectangular grid of known size $len_i \times wid_i$. Nonrectangular vehicles are expanded to the nearest rectangular shape or handled as per the bounding box approach. For algorithmic purposes, the X axis is taken as the longitudinal axis, along the direction of the road; and the Y axis is taken as the lateral axis.

Let the *free-configuration space* of the vehicle be given by ζ_{static}^{free}. The initial configuration of the vehicle S is known. The algorithm computes the trajectory $\tau_i(t)$ for the duration $0 \leq t \leq T_i$. Here, T_i is the time for the vehicle to reach the planned end point. The objective of the algorithm is to make the vehicle travel as far as possible in the road segment (or *maximize* $\tau_i(T_i)[X]$). For trajectories that reach equally as far, the one that takes the shortest time is selected (or *minimize* T_i). If the vehicle travels farther, there is less chance of it being struck in an obstacle framework with no way out other than reversing, if there is no single collision-free trajectory which can overcome all obstacles as per the limited view of the current road segment being planned.

The planning algorithm is a *uniform cost search* using which all possible expansions are made. A state or node is represented by the vector $L_i<x_i, y_i, \theta_i, v_i, t, \tau_i, lane_i>$ in which (x_i, y_i) denotes position, θ_i denotes orientation, v_i denotes speed (constant), t is the time of arrival at the state, τ_i denotes trajectory from the source that leads to the state and $lane_i$ denotes the *generalized lane* from the source to the state. To make discussions simple, in this text the term state is used to refer to both a state of the configuration space (consisting of $<x_i, y_i, \theta_i>$ only) as well as a node of the graph search. The cost of a state L_i is taken to be its time, or $L_i[t]$. Once all the nodes are expanded, the best trajectory is selected. The general algorithm is given in Algorithm 9.1. Unlike Algorithm 8.1 of chapter 'Graph Search-based Hierarchical Planning', rediscovery or discovery with a better cost is not checked as rediscovery of a position will mostly happen at a different time which is taken as an integral part of the state. The general aspects of the approach are summarized in Box 9.3.

Algorithm 9.1: Uniform Cost Search for a Single Vehicle

Plan(R_i)
```
    Source ← Sᵢ
    source[τᵢ] ← NULL, source[laneᵢ] ← Sᵢ, source[t] ← 0
    Add source to fringe with priority source[t]
    closed ← NULL
    while fringe is not empty
      n ← extract node with least priority
      closed ← closed U n
      N' ← Expand(n)
      For n' ∈ N, Add n' to fringe with priority n'[t].
    end while
    qp ← best plan in closed
    return qp[τᵢ], qp[laneᵢ]
```

9.3.1 State Reduction

For a rich set of possible actions, the graph search is computationally expensive and cannot be used. Hence, heuristic means are used to carefully select the actions

---

Box 9.3 Key Concepts of the Approach for a Single Vehicle

- Use a sweeping line (Z) for obstacle analysis in large steps
- Select regions of line (**Pathways**) in obstacle-free regions
- Select **best** point inside every pathway (L_3)
  - **Objectives:**
  - Vehicles maximize lateral separations
  - Vehicles prefer to travel at the same lateral positions along the road
- Select similar point (L_2) at an earlier sweeping line (Z_2)
- Feasibility analysis
  - Vehicle expected to align itself at L_2 for obstacle seen at L_3
  - If vehicle can go from L to L_3 via L_2, it is safe at L_2
  - Curve from L to L_2 may be traversed if optimal
  - Curve from L_2 to L_3 is only for connectivity check/obstacle analysis
  - L_3 acts as a forerunner for L_2 informing it about any obstacle well in advance before the vehicle actually lands.
  - There may be an obstacle just before L_3 not giving enough time for the vehicle to steer, hence landing at L_3 may not be safe.
- Make these children and connect to parent (L) by a smooth trajectory
  - Connect L to L_2 by a clothoid curve
  - Connect L_2 to L_3 by curve along the road length
  - Check feasibility
  - If infeasible, perform local optimization
  - If still infeasible, reduce speed/sweeping-line step size and repeat
- Apply a search on the graph so produced
- The points at the sweeping line, along with the guaranteed width of the road without obstacles or other vehicles, constitutes the vehicle's lane.

---

or states to generate in expansion. Two hypotheses are used regarding the motion strategy of the vehicle. (1) Every vehicle attempts to move in the road such that its relative position (on the lateral axis) remains the same. This is naturally seen while one drives on roads, when no steering is performed while travelling on a straight road. Required turns are made in a manner to keep oneself at approximately the same relative position compared to others. (2) The vehicle always attempts to move such that its separation from obstacles or road boundaries is as large as possible. However, the attempt to maximize the separation is limited to a value of sm. This is again visible in everyday driving in which in the presence of any obstacle one attempts to align the vehicle, maximizing the separation from the obstacle.

Expansion of a state $L_i < x_i, y_i, \theta_i, v_i, t, \tau_i, \text{lane}_i >$ of the graph search is done in a manner that the vehicle moves forwards by a magnitude of $\Delta(v_i)$ in the X axis. The value in the Y axis is chosen by the above-mentioned hypotheses. The choice of new position for the vehicle needs to be done in a manner to maximize the separation of the vehicle in both the lateral (X) and longitudinal (Y) axes. *Lateral separation* is maintained by the hypothesis whereas *longitudinal separation* is maintained by $\Delta(v_i)$. If an obstacle happens to lie just ahead of the vehicle, it may not be able to further continue its journey without collision. The aim is to keep a minimal distance of $\Delta(v_i)$ from the obstacle in front. $\Delta(v_i)$ (Eq. [9.1]) must be large enough so that the vehicle can steer itself a significant amount laterally, while travelling longitudinally on the road. Higher speeds would naturally need larger longitudinal distance (along the X axis) to steer across the Y axis.

$$\Delta(v_i) = c \cdot v_i \qquad\qquad [9.1]$$

Here, c is a constant called the *longitudinal separation constant*.

The first step in expansion is to analyse the obstacles in the region. The analysis of the obstacles is done at a distance of $2\Delta(v_i)$ from the current position (to guarantee a longitudinal separation of $\Delta(v_i)$ from the expanded state). The *obstacle analysis line Z* is considered as the Y axis at a longitudinal distance of $2\Delta(v_i)$ from x_i (Fig. 9.1). Vehicles mostly align themselves along the road, and hence while crossing the line Z, the ideal vehicle's position should be perpendicular to Z. Valid states P are given by Eq. [9.2].

$$P = \bigcup_l l\langle x_i, y_i, \theta_i, v_i, t\rangle : (x_i, y_i) \in Z, \theta_i \perp Z, l \in \zeta_{static}^{free} \qquad [9.2]$$

Minimal states are selected from the set P as the *expanded states*. The line Z may cross obstacles, which means all possible states P would be disjointed into smaller sets $P_a (P_a \subset P, P_a \cap P_b = \varphi \; \forall \; P_a, P_b, P_a \neq P_b)$ (Fig. 9.1). The states formed by expansion are reduced to one per disjoint set. Fig. 9.1 also shows the importance of selecting states from each set, as two of the three sets are later discovered to be suboptimal (or blocked).

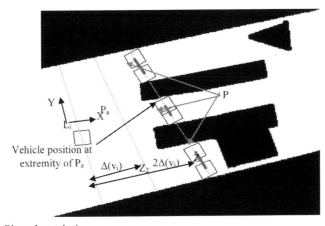

Figure 9.1 Obstacle analysis.

9.3.2   State Selection

The selection of a state to be expanded (l) in the set P_a is based on the set hypothesis. Let $d(l_3)$ be the distance of l_3 lying on Z measured from the road boundary. Let s_1 and s_2 be the points in Z which have the least and the highest value of $d(.)$ such that the line s_1 to s_2 is collision free and includes P_a (Fig. 9.2). Going by the set hypothesis, disregarding the obstacles, the vehicle may attempt to keep itself in the same lateral position on the road which gives its preferred position pr such that pr[Y] = y (or $L_i[Y]$) through which Eq. [9.3] is obtained.

$$d(\text{pr}) = y \cdot \text{rl} \tag{9.3}$$

Here, rl is the road width at Z. pr may though be such that a vehicle placed at pr is not feasible and within P_a or may not keep a distance of sm from obstacles or road boundaries, which if introduced gives the preferred position l as in Eq. [9.4].

$$d(l) = \begin{cases} d(\text{pr}) & d(s_1) + \text{sm} + \text{wid}_i/2 \le d(\text{pr}) \le d(s_2) - \text{sm} - \text{wid}_i/2 & \text{(a)} \\ d(s_1) + \text{sm} + \text{wid}_i/2 & d(\text{pr}) < d(s_1) + \text{sm} + \text{wid}_i/2 \wedge d(s_2) - d(s_1) \ge 2\text{sm} + \text{wid}_i & \text{(b)} \\ d(s_2) - \text{sm} - \text{wid}_i/2 & d(\text{pr}) > d(s_2) - \text{sm} - \text{wid}_i/2 \wedge d(s_2) - d(s_1) \ge 2\text{sm} + \text{wid}_i & \text{(c)} \\ (d(s_1) + d(s_2))/2 & \text{otherwise} & \text{(d)} \end{cases}$$

$$\tag{9.4}$$

Eq. [9.4a]−[9.4c] are applied when it is possible for the vehicle to have a margin of sm on both sides. Eq. [9.4a] states the condition when the vehicle as per $d(\text{pr})$ already maintains a wide margin at both sides. Eq. [9.4b] and [9.4c] apply for conditions when the vehicle has a wide margin at one side but not on the other side and needs to be moved so that a margin is produced on both sides. Eq. [9.4d] is an attempt to maximize the separation as much as possible by placing the vehicle in the middle of the segment. Various cases that arise in selection of l are shown in Fig. 9.3. For details, refer to Eq. [9.4a]−[9.4d].

However, the state l lies on Z which may be quite close longitudinally to an obstacle, even though lateral separation has been maximized. The purpose was to

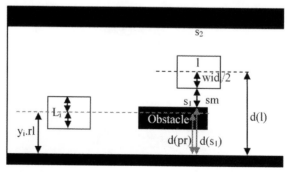

Figure 9.2 Computing state for expansion.

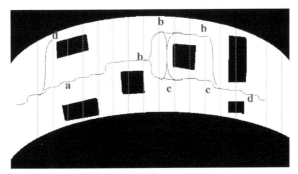

Figure 9.3 Various cases in selection of state for expansion.

find a valid state at a longitudinal distance of $\Delta(v_i)$, say l_2. Let Z_2 (*vehicle placement line*) be the Y axis at a distance of $\Delta(v_i)$. l_2 (Eq. [9.5]) is selected as a state such that the vehicle can steer itself and orient itself parallel to the road attaining the correct relative (lateral) position of $l[Y]$ when it crosses the line Z_2. It may then later travel parallel to the road to state l (Fig. 9.4).

$$l_2\langle x_i, y_i, \theta_i, v_i, t\rangle : (x_i, y_i)\in Z_2, y_i = l[Y], \theta_i \perp Z_2 \qquad [9.5]$$

The *obstacle analysis line* (Z) keeps the algorithm informed about the obstacles it may face in the future, and the vehicle must receive its correct orientation well in advance. The motion takes place only till the *vehicle placement line* (Z_2), after which (in the next expansion) Z is moved farther to keep the vehicle informed about still farther obstacles. In this way, Z acts as a forerunner for the vehicle, whereas Z_2 acts as the line up to which the vehicle can confidently travel.

9.3.3 Curve Generation

The point L_i is connected to l_2 using *clothoid curves* (Nutbourne et al., 1972; McCrae and Singh, 2009) denoted by $\text{connect}_2(L_i, l_2)$. The purpose of clothoids is to generate a

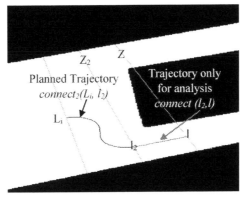

Figure 9.4 Object analysis and state expansion.

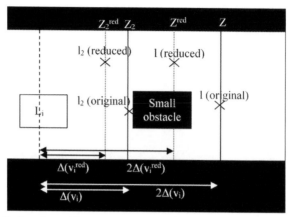

Figure 9.5 Small-obstacle problem.

curve that makes the vehicle correct its angle of orientation, *change its lateral position in the road* and generate a curve for the vehicle to traverse on a curved road. l_2 is connected to l by a curve parallel to the road denoted by connect(l_2, l) (Fig. 9.4). The feasibility of the two curves is checked such that all intermediate points belong to $\zeta_{dynamic}^{free}$. A small local search around l is performed to assess if either of the curves is infeasible.

Infeasibility may be due to a *small-obstacle problem* shown in Fig. 9.5. The line Z completely misses citing the small obstacle, and hence the computations are erroneous. In such a case, feasibility may only be returned by Z discovering the small obstacle for which $\Delta(v_i)$ is avoided by a reduction of speed. Reduction of speed by a discrete amount at a time instant is assumed possible.

If the curves are feasible, l_2 becomes the expanded state with parent L_i and is added in the processing queue of the uniform cost search. The region in which l can be moved in Z and correspondingly l_2 can be moved in Z_2 without causing a collision (under the threshold of sm at both sides from l) constitutes a vehicle's lane at l_2. The *lane definition* for tentative motion from L_i to l_2 can be visualized by connecting the lane definition at L_i to the lane definition at l_2 along the longitudinal axis. The points L_i and l_2 characterize the lane, and these are stored for algorithmic purposes. The expansion is given as a pseudocode in Algorithm 9.2. The graph generated for a synthetic problem is shown in Fig. 9.6.

Algorithm 9.2: Expansion for a Single Vehicle

Expand(L_i)

```
A ← NULL

for vᵢ ← Vⁱₘₐₓ to 1 in small steps
  Calculate P using equation (9.2)
```

```
for all Pₐ ∈ P

  calculate l, l₂ for Pₐ using equations (9.4) and (9.5)

  [l, l₂] ← apply small deviations to l till connect₂(Lᵢ, l₂) ε ζ_dynamic^free ∧

  connect(l₂, l) ∈ ζ_dynamic^free, break Pₐ if still infeasible

  l₂[τᵢ] ← Lᵢ[τᵢ] ∪ connect₂(Lᵢ, l₂))

  l₂[laneᵢ] ← Lᵢ[lane] ∪ l₂

  l₂[t] ← Lᵢ[t] + ‖connect₂(Lᵢ, l₂))‖/vᵢ

  A ← A ∪ l₂
  end Pₐ
  if A ≠ NULL, return A
end vᵢ
return NULL
```

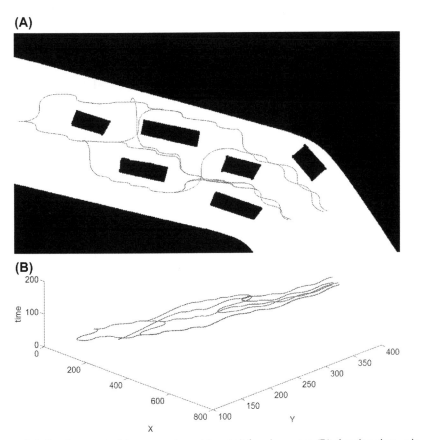

Figure 9.6 Graph generated for a sample problem (A) for given map (B) showing time axis.

9.3.4 Results

The algorithm was developed in MATLAB with the map being given as a bitmap (BMP) image file. The aim of testing was to uncover the potential of the algorithm to compute optimal paths in a simple to complex grid of obstacles. The first scenario (Fig. 9.7A) was created to test the ability of the vehicle to enter narrow regions. Multiple such placements at small distances stress a clever steering strategy which is more difficult if it is to be done at high speeds. Another scenario was made to test the ability of the vehicle to pass through a complex network of obstacles (Fig. 9.7B). It may be seen that the vehicle could do well in the scenario making only a few steering attempts.

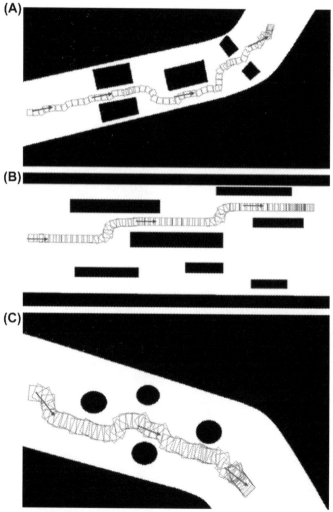

Figure 9.7 Experimental results with single vehicle. (A) Scenario 1, (B) Scenario 2, (C) Scenario 3.

In the third scenario (Fig. 9.7C), some simple obstacles were placed at random places. It may be seen that the vehicle could detect all the obstacles and pass by them, either by a reduction of speed or by local optimization.

9.4   Dynamic Distributed Lanes for Multiple Vehicles

The problem consists of N vehicles, each vehicle R_i with its own time of emergence e_i in the planning scenario. The source S_i which is the state of the vehicle at e_i is known, but only after vehicle emergence. Each vehicle is planned upon its emergence. When R_i arrives, the trajectories τ_j of vehicles R_j are already known for $e_j < e_i$. The aim is to generate a plan τ_i for R_i. The plan τ_i is called *admissible* in the planning scenario τ if no collision occurs between any two vehicles or static obstacles. If R_i fails to generate an admissible plan, the plans of all vehicles are nullified, and replanning is done. The order of planning may be altered and any combination that results in a valid plan may be selected.

Uniform cost search is used to plan the vehicle. The first task is *obstacle analysis* using the *obstacle analysis line* (Z), which gives a set of states P in disjoint sets of P_a, disjoint by static obstacles. This is used to select the expanded state. Here, however, the other vehicles need to be considered as well. The general algorithm is the same as Algorithm 9.1.

R_i may not be able to construct a feasible plan without moving the other vehicles by modifying their plans. R_i is *allowed to alter the lanes* of the other vehicles, which correspondingly changes their trajectories. The lane of any vehicle R_j is in the form of a set of points $lane_j[u]$, whereas the trajectory τ_j is a result of connecting these points by a smooth curve. The points $lane_j[u]$ act as controls available with R_i, which may alter any of these to produce a different lane for R_j and hence change the trajectory.

Every graph node (state) maintains trajectories τ detailing all alterations made to the trajectories of the vehicles in the path from the source to the state in the graph expansion; and *lane* detailing all alterations made to the lanes of vehicles. Hence, every child takes the plan (trajectories and lanes) specified by the parent, makes modification to it as per its desires and stores the resultant plan which becomes the plan specification given to all its children which may modify it further.

Consider any graph node or state $L_i < x_i, y_i, v_i, \theta_i, t, \tau,$ lane$>$ which needs to be expanded using the search technique. Three expansion strategies are formulated and any one or more of the strategies may be used for the expansion. These are *free-state expansion*, *vehicle following* and *wait for vehicle*. Each of these is discussed in the following subsections.

9.4.1   Free-State Expansion

The *free-state expansion* strategy attempts to place every vehicle laterally with R_i moving with the highest possible speed. In the absence of other vehicles, this strategy is similar to the expansion strategy discussed in Section 9.3. The expansion might lead the vehicle to overtake any slow-moving vehicle it encounters. Hence, the strategy may also be termed as an *overtaking strategy*.

The problem is the expansion of state L_i for the valid states P_a returned by static obstacle analysis. Here, the task is not only to select a single position in P_a for the vehicle R_i, that was done in planning for a single vehicle. Rather, P_a (specifically, complete *obstacle analysis line* segment Z bounded by road boundary or obstacle from both ends) is *divided and distributed* between the vehicles which plan to move parallel to R_i in its motion till Z, and must hence occupy distinct lanes. Let there be n such vehicles. The task is the selection of states $l^1, l^2, \ldots l^n$ in which l^j denotes the state of R_j on crossing Z. The task is hence twofold, first to *find the vehicles* that need to occupy distinct lanes, and second to *identify the lane* for each of these vehicles (l^j is known as the lane for R_j at the particular state).

Imagine a thick slice of road with a vehicle travelling through the slice amidst other vehicles moving under their own plans. If there seems to be a likely collision between any two vehicles, one would like to alter the travel plans of the two vehicles such that they lie in different lanes. Carrying out and completing the alteration of the plan of the new vehicle can span the considered road slice and hence the slice may need to be widened. If the changed plan results in a different collision, one would like to include the new colliding vehicle and assign all three vehicles different lanes. *Alterations may result in further widening of the slice.* The altered plan may further cause more collisions, in which case the participating vehicles are added. The addition is done till all vehicles can harmoniously cross the road slice. Once the vehicles are identified, the (obstacle-free) road slice can be divided between the vehicles, such that each occupies an independent lane across the width of the road slice. This example illustrates the two tasks, to identify the vehicles requiring independent lanes and to divide the road slice into different lanes between these vehicles.

9.4.1.1 Computing the Number of Lanes Required

The first task is *calculating the vehicles that need separate lanes* while R_i crosses Z. As per the hypotheses, the lateral separation of R_i from any other vehicle or obstacle should be maximized, subjected to a maximum of sm, whereas *longitudinal separation* may preferably be $\Delta(v_i)$. As per the free-state expansion strategy, all vehicles that as per their plan lie at a longitudinal separation of less than $\Delta(v_i)$ need an independent lane. This, however, excludes the *following vehicle behaviour*, when the vehicle ahead moves with a greater or equal speed. Giving an independent lane assures that no collision happens as long as the vehicles stick to their allocated lanes, irrespective of what longitudinal separation they maintain. Suppose a vehicle R_j needing a separate lane is selected, which would be required to alter its plan. Modification of the plan of R_j might make the longitudinal separation between R_j and any other vehicle R_k less than the required $\Delta(v_j)$, threatening a collision between R_j and R_k. In this manner, vehicles are added continuously until a point is reached at which modification of lanes does not affect the other vehicles.

Let set H denote the vehicles (including R_i) requiring independent lanes. Initially, the set H contains only R_i. Suppose at time t, R_j as per its planned trajectory is at position $\tau_j(t)$. Let $S(S_a, S_b)$ denote the segment of road under consideration for computing lanes, in which S_a is the start position in the X axis and S_b is the end position in the X axis. Within the region S, vehicles must occupy independent lanes, unless a vehicle follows another. Because the prospective motion of R_i from its current position x_i to its final position at a longitudinal separation of $2\Delta(v_i)$ is being studied, the initial value of the segment S is $(x_i - \text{len}_i/2, \ x_i + 2\Delta(v_i) + \text{len}_i/2)$, accounting for the entire coverage of R_i in this motion (Fig. 9.8).

The sets H and S are iteratively grown until no further vehicle qualifies to be added to H. At any time, every R_k is checked such that motion of any R_j in H within S may result in a collision between R_j and R_k. This separation check is only done at points denoting lanes of R_j. If the separation is less than $\Delta(v_j)$, there is a possible collision. In such a case, R_k is added to H and accordingly S is modified to account for the segment S_2 that is the region affected by R_k being given an independent lane. The modified S is given by $S \cup S_2$. The procedure is repeated until no vehicle in the planning scenario qualifies to be added in H. The two subproblems that arise are: calculate

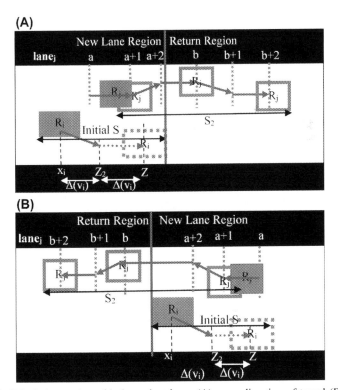

Figure 9.8 Calculating region of independent lanes (A) same direction of travel (B) opposite direction of travel.

S_2 if some R_j is given an independent lane, and calculate S if some R_j in H threatens some R_k, requiring it to have an independent lane, each of which is dealt with one after the other. The general algorithm is given in Algorithm 9.3. The process is diagrammatically presented in Fig. 9.9.

Algorithm 9.3: Getting Number of Vehicles Requiring Independent Lanes

getLaneVehicles(L$_i$)

```
A ← Rj: ej < ei,  H ← Ri

S(Sa,Sb) ← (xi-leni/2,  xi + 2Δ(vi)+leni/2)
do

    for Rj ∈ A — H

      if τj(t) ⊗ Rj ∩ (Sa,Sb+d) ≠ φ

        S2 ← (lanej[a+1] — lenj/2, lanej[b + 2] + lenj/2)

        S(Sa,Sb) ← S U S2
        H ← H U Rj,  A ← A - Rj
        break Rj
      end if
    end Rj
while no change in H
return H
```

Getting into a new lane means R_j needs to steer to get to the new lane and later *steer back to the original lane* (Fig. 9.8); the positions where both these changes happen need to be calculated. Both turns require a distance of $\Delta(v_j)$ along the X axis. New lanes are brought into effect immediately, or R_j is immediately asked to go to the new lane. Suppose R_j at time t has left the point lane$_j[a]$ and moves towards the point lane$_j[a + 1]$

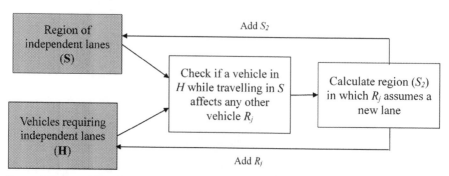

Figure 9.9 Computing the vehicles requiring independent lanes.

(or its motion has been confirmed) as per its planned trajectory. In such a case, the motion of the vehicle from $lane_j[a]$ to $lane_j[a+1]$ cannot be altered. Modification of $lane_j[a+1]$ would mean a different trajectory connecting $lane_j[a]$ to $lane_j[a+1]$. The part of trajectory that vehicle has already travelled from $lane_j[a]$ to $lane_j[a+1]$ may not be the same as the one desired with an altered $lane_j[a+1]$. However, lanes $lane_j[a+2]$ and onwards can be certainly modified (Eq. [9.6]).

$$a: lane_j[a][T] \leq t \wedge lane_j[a+1][T] > t \tag{9.6}$$

Here, $lane_j[a][T]$ denotes the time R_j arrives at the lane, whereas $lane_j[a]$ denotes the position of the lane.

At the time of expansion of L_i, the algorithm assumes that R_i suddenly disappears after it achieves the expanded state at the *vehicle placement line* Z_2, whereas no collision would be recorded in moving from L_i to Z. The subsequent graph expansions cater for the latter motion. Hence, after R_i attains the expanded state, other vehicles must return to their original lanes. R_i ends its journey at a longitudinal distance $\Delta(v_i)$ from x_i, after which vehicles must maintain an additional longitudinal safety distance of $\Delta(v_i)$ before going to their original lanes (Fig. 9.8). Vehicles may be driving inbound or outbound. Considering both cases, R_j can return to its original lane after it is completely out of the region of considered motion of R_i (considering lengths of both vehicles contribute to their longitudinal occupancy) at $lane_j[b]$ (Eq. [9.7]). For $lane_j[b+1]$ onwards, R_j has normal lanes.

$$b: lane_j[b] \otimes R_j[Y] \cap (x_i - len_i/2, x_i + 2\Delta(v_i) + len_i/2) \neq \varphi, lane_j[b][T]$$
$$> lane_j[a][T] \tag{9.7}$$

Here, $lane_j[b] \otimes R_j[Y]$ denotes the longitudinal coverage of R_j placed at $lane_j[b]$.

No vehicle must lie at a separation of $\Delta(v_j)$ from $lane_j[b+1]$, which is the location of $lane_j[b+2]$. Hence, S_2 is given by $(lane_j[a+1] - len_j/2, lane_j[b+2] + len_j/2)$. The notations are illustrated in Fig. 9.8. An assumption here is that after the vehicle R_i reaches the line Z, all vehicles may more or less simultaneously travel back to their original lanes. This may lead to collisions when one vehicle readily moves back to its original lane, whereas another vehicle takes a little time to start moving back. These are handled by optimization during trajectory generation.

The last task is to check if the changed dynamics of any R_j cause a potential threat with any general R_k, indicating the set H to be grown to include R_k. Because the changes in lanes of R_j are not yet computed (Section 9.4.1.2), the time when R_j crosses $lane_j[b]$ is unknown. Due to this, a threat to any R_k cannot be determined. This issue may be handled by heuristics. Knowing the positions of the vehicle at time t, the separation of the vehicles is decided. The decision is divided into two cases. These are whether R_j and R_k are travelling in the same directions (both inbound or outbound) or different directions (one inbound and the other outbound). As a general rule, R_k needs to be included in H if any part of R_k, while at $\tau_k(t)$, lies within $(S_a, S_b + d)$.

In the case when vehicles are travelling in the same direction, d is taken to be 0. The worst case is when R_j reaches its normal lane at lane$_j[b + 1]$ and R_k has not moved much making the separation between R_j and R_k as small as possible. The inclusion of extra distance in computing S_2 for R_j ensures this distance is wide enough. Any motion of R_k would make the separation more than the minimum amount of $\Delta(v_j)$. The notations are shown in Fig. 9.10A. In the case when vehicles are travelling in different directions, d is taken to be $2\Delta(v_k)$. Here, because R_j races towards R_k as R_k travels, the distances get shorter very quickly. The smallest distance would be recorded when R_j returns to its normal lane. Because the time required for the same, and also the distance travelled by R_k in the process is not known, an approximate value is calculated assuming longitudinal travel for both vehicles (minimum of $2\Delta(v_k)$), as the maximum distance that can be covered by R_k, whereas R_j turns to a modified lane and again back to its normal lane. The notations are shown in Fig. 9.10B.

9.4.1.2 Lane Distribution

The task is *division of the entire available width of the free road* into lanes between the vehicles demanding the same. The same notations is used as in planning for a single vehicle. Let s_1 and s_2 be the points in Z which have the least and highest value of $d(.)$ (distance from boundary) such that the line s_1 to s_2 is obstacle free and includes P_a. Let rl be the road width at Z. The preferred position of any R_j in H is given by Eq. [9.8].

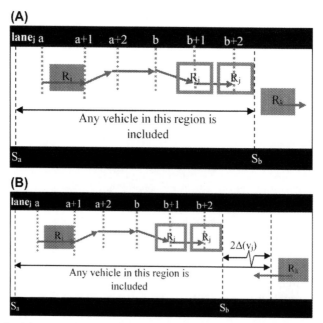

Figure 9.10 Decision regarding inclusion of R_k in H (A) same direction of travel (B) opposite direction of travel.

$$d(\text{pr}_j) = \begin{cases} \text{lane}_j[a][Y] \cdot \text{rl} & j \neq i \\ y \cdot \text{rl} & j = i \end{cases} \tag{9.8}$$

The different pr_j may be infeasible or without maximum separations. The vehicles in H are sorted as per their $d(\text{pr}_j)$ values. pr_i may not be within the segment, which is corrected by giving it the highest or lowest possible value. Working with any R_j to create maximum separation is similar to the approach used in the single-vehicle case, with the addition that R_j might move other vehicles to create extra space for itself. Let $\text{left}(\text{pr}_j)$ (Eq. [9.9]) be the point to which R_j placed at pr_j may be moved leftwards without colliding with any other vehicle, road boundary or obstacle. The value of $\text{left}(\text{pr}_j)$ is the distance of that point from the boundary. Similarly, for right, $\text{right}(\text{pr}_j)$ is given by Eq. [9.10]. Notations are shown in Fig. 9.11.

$$\text{left}(\text{pr}_j) = \begin{cases} d(\text{pr}_{j+1}) - \text{wid}(\text{pr}_{j+1})/2 & j < \text{size}(H) \\ s_2 & j = \text{size}(H) \end{cases} \tag{9.9}$$

$$\text{right}(\text{pr}_j) = \begin{cases} d(\text{pr}_{j-1}) + \text{wid}(\text{pr}_{j-1})/2 & j > 1 \\ s_1 & j = 1 \end{cases} \tag{9.10}$$

The first case that arises is the possibility of placing pr_j such that no movement of other vehicles is required, and it would maintain a distance of sm from both its ends. Eq. [9.11] gives the precondition and Eq. [9.12] gives the placement.

$$C = \text{left}(\text{pr}_j) - \text{right}(\text{pr}_j) - \text{wid}_j \geq 2\text{sm} \tag{9.11}$$

$$d(\ell) = \begin{cases} d(\text{pr}_j) & \text{right}(\text{pr}_j) + \text{sm} + \text{wid}_j/2 \leq d(\text{pr}_j) \leq \text{left}(\text{pr}_j) - \text{sm} - \text{wid}_j/2 \\ \text{right}(\text{pr}_j) + \text{sm} + \text{wid}_j/2 & d(\text{pr}_j) < \text{left}(\text{pr}_j) - \text{sm} - \text{wid}_j/2 \wedge C \\ \text{left}(\text{pr}_j) - \text{sm} - \text{wid}_j/2 & d(\text{pr}_j) > \text{right}(\text{pr}_j) + \text{sm} + \text{wid}_j/2 \wedge C \end{cases} \tag{9.12}$$

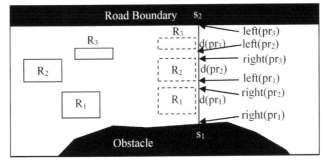

Figure 9.11 Distribution of lanes.

Eq. [9.12] in structure represents Eq. [9.4]. However, in case condition C is not met, the task is to move the vehicles to have more space for accommodating R_j. A vehicle is only moved if the vehicle already has a distance excess of sm from both its ends. This is the cooperation shown by the vehicles wherein they allow R_j to have sufficient separation by effectively lessening their own excessive separation. The total separation needed by R_j is s (Eq. [9.13]). The spare separation that R_k has and is ready to offer is given by Eq. [9.14].

$$s = \max\left(2sm - \left(\text{left}(\text{pr}_j) - \text{right}(\text{pr}_j) - \text{wid}(\text{pr}_j)\right), 0\right) \qquad [9.13]$$

$$s_k = \max(\text{left}(\text{pr}_k) - \text{right}(\text{pr}_k) - \text{wid}(\text{pr}_k) - 2sm, 0) \qquad [9.14]$$

Vehicles R_k one by one on either side of R_i are moved, till the required separation s is created. Vehicles on the left of R_j are moved towards the left and those on the right of R_j are moved towards the right to get the necessary separation; that is, movements are centred along R_j. Notations are shown in Fig. 9.11.

In case, for any vehicle, the required separation is unavailable, the total separation available $\left((s_2 - s_1) - \sum_j \text{wid}(\text{pr}_j)\right)$ is evenly distributed between the vehicles. If all vehicles get sufficient space without overlapping each other, the planned distribution may be used for the planned movement of the vehicles. The complete algorithm is given by Algorithm 9.4.

Algorithm 9.4: Division of the Road Into Lanes

getLanePositions(H, L$_i$)

```
calculate pr by equation (9.8), sort(H) by d(pr_j)

for R_j ∈ H

    if equation (9.11) then equation (9.12), pr_j ←1^j, continue R_j
    calculate s from equation (9.13)

    mov[R_k] ← 0 ∀ R_k ∈ H

    for R_k ∈ H in increasing deviation from R_j in H
        calculate s_k from equation (9.14)
        mov[R_k] ← max(s, s_k), s ← s − s_k

        if s ≤ 0, break R_j
    end R_k
    if s > 0

        space_per_segment ←((s_2 - s_1) - ∑_o wid(pr_o))/(size(H) + 1)
        for R_k ∈ H, 1^k ← space_per_segment + [∑_o(space_per_segment +
        wid(pr_o)) ∀ R_o < R_k] + wid(pr_k)/2

        break R_j
```

```
else
   for Rₖ < Rⱼ in H, pr'ₖ ← prₖ -     Σ      mov[proo]
                                    Rₖ≤Roo<Rⱼ

   for Rₖ > Rⱼ in H, pr'ₖ ← prₖ +     Σ      mov[proo]
                                    Rⱼ<Roo≤Rₖ

   pr'ₖ ← left(pr'ₖ) — sm - widₖ/2

   pr ← pr', 1←pr
 end if
end Rⱼ
return 1
```

9.4.1.3 *Vehicle Trajectory Generation*

In planning the *trajectory* for the expanded state, the task is twofold. First, R_i needs to be moved from the state L_i to the state l_2^i which is obtained from state l^i as per the terminology followed in planning for a single vehicle. Then all vehicles in H-$\{R_i\}$ must be modified to occupy their modified lanes and to further ensure that no collision is recorded between any two vehicles.

First, motion of R_i is studied. A state l_2^i is selected at Z_2, with orientation perpendicular to Z_2 and the same lateral position. The clothoid curve is drawn from L_i to l_2^i and a curve parallel to the road is drawn from l_2^i to l^i. Both curves are checked such that all intermediate points in both curves belong to $\zeta_{\text{dynamic}}^{\text{free}}(\tau^2)$. Here, $\zeta_{\text{dynamic}}^{\text{free}}(\tau^2)$ denotes the free configuration space considering the dynamics of vehicles specified in plan τ_2 (Eq. [9.15]). At time t, the space is given by Eq. [9.16].

$$\tau^2 = \cup \tau_j \forall j, e_j < e_i, R_j \notin H \tag{9.15}$$

$$\zeta_{\text{dynamic}}^{\text{free}}(\tau_2, t) = \zeta_{\text{static}}^{\text{free}} - \underset{j \in \tau^2}{\cup} \tau_j^2(t) \otimes R_j \tag{9.16}$$

Local optimizations are done which include varying state l^i along Z to make the path feasible. On computing a feasible plan, the plan vector τ^2 is updated to account for the computed trajectory of R_i.

The next task is to *plan each of the vehicles* excluding R_i in H. For any vehicle R_j, the lane information $lane_j$ is first updated, and then a trajectory is attempted by using the updated lane. The change is made by altering the lateral axis values of lanes. All lanes from $lane_j[a + 2]$ to $lane_j[b]$ are given a lateral value of l^j. This lane is then used for the motion of the vehicle. Vehicles may be taken in any order in H. However, once the trajectory of a vehicle is computed, it is added in the plan τ_2. The complete algorithm is given by Algorithm 9.5. The expansion strategy is given in Algorithm 9.6. The *number of lanes required* are counted and the *available space is distributed* between the vehicles. This is followed by the *generation of the necessary trajectories*. The approach is summarized in Box 9.4.

The approach is illustrated in Fig. 9.12. Fig. 9.12A shows the discovery of another vehicle the travel plan for which is shown in the same figure. Fig. 9.12B shows the generation of the lanes. The planned trajectories are shown in Fig. 9.12C. It may be noted that both the lane and trajectory of R_i would not be as shown in Fig. 9.12B and C. This is because presently it was assumed that R_i stops after the planned path at l_2^i.

Box 9.4 Free-State Expansion Strategy Procedure

- Use a sweeping line for obstacle analysis
- For every pathway
- Compute vehicles requiring an independent lane
  - Find vehicles (R_b) affected by the navigation of vehicle under planning (R_a) to Z and hence require a separate lane
  - Get the affected region (S) in which every vehicle has an independent lane
  - Find vehicles (R_c) affected by altered navigation of R_a or R_b and hence also require a separate lane
  - Modify the affected region (S)
  - Find vehicles (R_d) affected by altered navigation of R_a, R_b or R_c and, hence, also require separate lane
  - Modify the affected region (S)
  - And so on till no vehicle is further affected
  - Vehicle being followed or going ahead by a higher speed not considered
- **Distribute pathway** width amongst the vehicles
  - Get preferred position for the vehicle being planned
  - Check if placement possible without moving other vehicles
  - If not, check if placement possible with maintenance of maximum separation threshold
  - If not, compute total free space and distribute evenly between the vehicles

---

When the state of R_i is further expanded, R_j would give space for R_i to pass by, considering the extended motion as a result of subsequent expansions. In this manner, R_i would keep moving with R_j to the side, as planning proceeds denoting motion of R_i along the road. This stops when R_i is sufficiently ahead of R_j.

Algorithm 9.5: Trajectory Generation From the Current State to the Expanded State

GenerateTrajectory(l, L$_i$)
   Calculate τ^2 using Eq. [9.15]

Algorithm 9.5.1: GenerateTrajectorySelf()

$[1^i, 1^i_2] \leftarrow$ apply small deviations to 1^i till connect$_2$(L$_i$, 1^i_2) $\in \zeta^{free}_{dynamic}(\tau_2) \wedge$ connect$_2$(1^i_2, 1^i)

$\in \zeta^{free}_{dynamic}(\tau_2)$. return NULL if still infeasible

$\tau^2_j \leftarrow$ L$_i$[τ_i]$\cup$ connect$_2$(L$_i$, 1^i_2d)

1_2[lane$_i$] $\leftarrow$ L$_i$[lane$_i$] $\cup$ 1^i_2

1_2[t] $\leftarrow$ L$_i$[t] + ||connect$_2$(L$_i$,1^i_2))||/v$_i$

for R$_j \in$ H
   lanej[1][Y] $\leftarrow$ 1j $\forall$ a + 2 $\leq$ 1 $\leq$ b

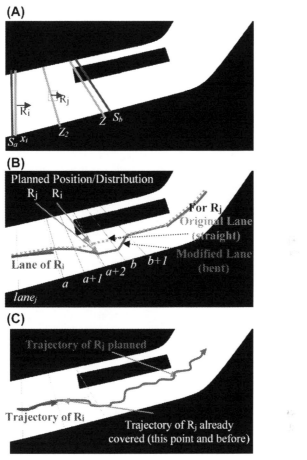

Figure 9.12 (A) Computing lane vehicles, (B) distribution of lanes amongst vehicles (label colours correspond to line colours), (C) generated trajectories for vehicles.

Algorithm 9.5.2: GenerateTrajectoryElse()

```
τ²ⱼ ← τⱼ(t) ∀ t ≤ laneⱼ[a + 1][T]
if τⱼ(t) ∉ ζfree_dynamic(τ², t) ∀ Lᵢ(t) ≤ t ≤ laneⱼ[a + 1][T], return NULL

for l ← a+2 to b
    laneⱼ[l] ←apply small deviations to laneⱼ[l] till
    connect₂(laneⱼ[l-1],

    laneⱼ[l]) ∈ ζfree_dynamic(τ₂), return NULL if still infeasible

    τ²ⱼ ← Lᵢ[τⱼ] ∪ connect₂(laneⱼ[l-1], laneⱼ[l])

    laneⱼ[l][T] ← laneⱼ[l-1][T] + ‖ connect₂(laneⱼ[l-1], laneⱼ[l]) ‖/vⱼ

    end l
end Rⱼ
l₂[τ] ← τ₂
return l₂
```

Algorithm 9.6: Free-State Expansion Strategy

FreeStateExpansion (L_i)
```
H ← getLaneVehicles(Lᵢ)
l ← getLanePositions(H, Li)
return GenerateTrajectory(l, Lᵢ)
```

9.4.2 Vehicle Following

The free-state expansion attempts to make a vehicle move at its highest possible speed, which may not always result in an optimal plan. Consider that a vehicle R_i capable of high speeds finds a vehicle R_j moving slowly in front. R_i would naturally attempt to overtake R_j. However, overtaking may not be possible due to various reasons like narrowing of road width, discovery of another vehicle, obstacle discovery etc. In all these cases, R_i has to slow down by significant amounts to avoid a collision. Slowing the vehicle iteratively to compute a possible trajectory is itself a time-consuming activity.

The ideal travel plan in the above-presented scenario would be planning R_i to *follow* R_j until the conditions become favourable for overtaking and to accommodate both vehicles in parallel for the required time. Keeping to the general algorithm methodology, a state L_i has to be expanded into state l_2^i using the *vehicle-following strategy*. The implementation of this behaviour of the vehicle is simple. It involves two parts. *Selecting the vehicle to follow* and *following the selected vehicle*.

A vehicle aims to follow the vehicle going in the same direction that is *closest to it*. Separation is measured along the Y axis. It is, however, important that the selected vehicle R_j must be passing through P_a. Further, the search for a potential R_j is always carried within a longitudinal distance $2\Delta(v_i)$. The other thing important in designing the vehicle-following strategy is that vehicle R_i would ultimately be attempting to overtake R_j. For overtaking to be possible, R_i must have a sufficient longitudinal distance from R_j, which, as per the modelling, is $\Delta(v_i^{max})$. Vehicle following is done by reducing the speed of the vehicle R_i to a value of v_i^{red} given by Eq. [9.17].

$$v_i^{red} = \begin{cases} v_j & |\tau_j(t)[X] - L_i[X]| > \Delta(v_i^{max}) \\ v_j - \delta & \text{otherwise} \end{cases} \qquad [9.17]$$

Here, δ is a small factor maintaining the difference in speeds. The value of speed v_j can be found by looking at R_j's travel plan at the current time t.

The next task is making the vehicle *move forwards*. Exactly the same algorithm is used as in the free-state expansion. The only difference is that the obstacle analysis line Z is constructed at a longitudinal distance of $2\Delta(v_i)$. However, the vehicle placement line Z_2 is constructed at a longitudinal distance of $\Delta(v_i^{red})$. Because the motion of R_i is with a speed of v_i^{red}, it is evident that all the steering needed would be possible. The task of obstacle analysis at Z is to ensure that R_i avoids all static obstacles and other moving vehicles other than R_j. Because the width of R_i is different from R_j, the manner of avoiding vehicles and obstacles would be different for the two vehicles. Although the vehicles are considered for the lane requirements, R_j having greater or equal speed to R_i would not qualify for an independent lane. The possibility of a vehicle following another vehicle was accounted for in the lane computation.

Consider the same situation as in Fig. 9.8. Here, R_i also needs to take the option of following R_j. As at the time of planning the distance between R_i and R_j is less than $\Delta\left(v_i^{\max}\right)$, the speed of R_j is taken to be slightly slower than the speed of R_i. In planning R_i does not consider R_j and travels as shown in Fig. 9.13. Fig. 9.13 also shows the travel plan of R_j. No collisions happen because R_i is travelling at a lower speed than R_j. The general algorithm is given in Algorithm 9.7. The process is also summarized in Box 9.5.

Algorithm 9.7: Vehicle-Following Expansion Strategy

VehicleFollowExpansion(L_i)
```
Select closest Rⱼ in Y axis, passing through Pₐ in the same direction
Calculate v_i^red from equation (9.17)
Return FreeStateExpansion(Lᵢ) with vᵢ = v_i^red and Z at xᵢ+Δ(vᵢ), Z₂
at xᵢ+Δ(v_i^red)
```

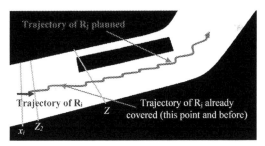

Figure 9.13 Trajectories of vehicles for vehicle-following behaviour.

Box 9.5 Vehicle-Following Expansion Strategy Procedure

- **Select the vehicle to follow**
  - Laterally closest vehicle is selected
  - Speed is fixed to allow following with some distance
- **Follow the selected vehicle**
  - Z is taken as per normal speed
  - Z_2 is taken as per reduced speed
  - Rest expansion is the same as free-state expansion

9.4.3   Wait for Vehicle

The travel plan of multiple vehicles may not always involve all vehicles moving slowly or quickly in the road segment. It may in fact require a vehicle to *wait* for some time. As an example, consider the situation of a *narrow bridge*. Because vehicles have variable width, the number of vehicles that may pass the bridge at same time without collisions is variable. Consider the case when a large vehicle R_j is already on the bridge, making it impossible for another vehicle R_i of decent width to fit if coming from the other side of the road. In such a case, R_i needs to stop at a location before the bridge and wait for R_j to leave the bridge. The strategy is inspired by a general driving scenario in English traffic, wherein parked vehicles (obstacles) on small roads reduce the number of lanes from two to one, requiring inbound and outbound vehicle drivers to coordinate between themselves deciding who goes first.

This stresses the need to make a vehicle wait at some point for some duration. It is computationally impossible to compute the time a vehicle must wait at every location by trying all possibilities in the search task. A simple heuristic is used that a vehicle must always *wait for a vehicle to pass by*. Using this heuristic the task of waiting may be broken down into steps: *deciding the vehicle to wait for*, *adjusting the position for the other vehicle to smoothly pass by*, *modifying other vehicles' trajectories* to account for the vehicle waiting, and *computing the state after wait* which becomes the expanded state as per this expansion strategy.

The general expansion algorithm expects the expansion strategy to expand the state L_i towards P_a. In this approach, no restriction is placed on the selection of the vehicle and all vehicles R_j that are tentative to pass through P_a at a later stage in their travel plan are worked for. Consider that R_i is waiting for R_j to pass through. Consider the set H containing all vehicles R_k (including R_j) that pass through P_a from the other direction, earlier than R_j. The time when a vehicle R_k passes Z can be known by iterating through its travel plan.

9.4.3.1   Calculating Preferred Positions

It is known that a stream of vehicles H would cross P_a, after which R_i may continue its journey. The next major issue is adjusting the position of R_i. Here, it needs to be decided whether R_i would *wait left or right* of the stream of incoming vehicles. Similar situations in everyday life have vehicles waiting towards the left (countries having the left-side driving rule). However, in certain cases of road and obstacles, there may be a need for vehicles to wait on the other side of the road, making the entire travel plan optimal. Assuming R_i would be able to maintain a distance of sm, for every R_k in H, the ideal positions of R_i along the Y axis are computed.

It is assumed that the obstacle analysis line (here also called the *inside bridge line*) Z lies in the region which was narrow and hence could not accommodate R_i. It is further assumed that the vehicle placement line (here also called the *vehicle waiting line*) Z_2 lies in the region which is wide enough for R_i to wait while the other vehicles in H pass by. Let l^k be the position of R_k when it crosses the line Z as per its plan. It is assumed in the modified plan of any R_k in H, R_k would like to stay in the same lateral position till it crosses the waiting R_i. This means that any interesting behaviour of R_k would be suspended till it crosses R_i and it would need to travel parallel to the road while overcoming the waiting R_i. Hence, while R_k is crossing the line Z_2 at a point l_2^k it is to be found at the lateral position $l^k[Y]$. The preferred position of R_i considering only R_k is given by Eqs [9.18] and [9.19]. The values denote the distance measured from the road boundary along Z_2. Here, rl is the road width at Z_2.

$$\text{right}_k = l^k \cdot \text{rl} - \text{wid}_k/2 - \text{sm} - \text{wid}_i/2 \tag{9.18}$$

$$\text{left}_k = l^k \cdot \text{rl} + \text{wid}_k/2 + \text{sm} + \text{wid}_i/2 \tag{9.19}$$

Whether R_i attempts to wait on the left or right side of the stream, it needs to leave enough space for each vehicle in the stream to pass through, and hence the preferred points on both sides may be given by $\min(\text{right}_k)$ and $\max(\text{left}_k) \ \forall \ R_k \ \varepsilon \ H$, out of which R_i chooses the one which requires the least steering (Eq. [9.20]).

$$d(l_2^i) = \begin{cases} \min(\text{right}_k) & \text{abs}\left(\dfrac{\min(\text{right}_k)}{\text{rl}} - y\right) \leq \text{abs}\left(\dfrac{\max(\text{left}_k)}{\text{rl}} - y\right) \\[2em] \max(\text{left}_k) & \text{abs}\left(\dfrac{\min(\text{right}_k)}{\text{rl}} - y\right) > \text{abs}\left(\dfrac{\max(\text{left}_k)}{\text{rl}} - y\right) \end{cases} \tag{9.20}$$

In case $d\left(l_2^i\right)$ lies outside the road, it is made to lie inside with a separation of sm. Local optimization may be performed to induce feasibility, which gives the state to be expanded l_2. The state l_2, however, does not guarantee a distance of less than $\Delta(v_i)$ from the obstacle ahead. It means further expansion of l_2 may require reducing the vehicle's speed.

9.4.3.2 Generating Trajectories

The next task is *modifying the trajectory* of each vehicle in H to account for the waiting R_i. Consider the segment of road along the X axis $S = (x_i + \Delta(v_i) - \text{len}_i/2 - \Delta(v_k),$ $x_i + \Delta(v_i) + \text{len}_i/2 + 2\Delta(v_k))$. This is the region, at a distance of $\Delta(v_k)$ on either side of l_2^i, which is affected by R_i waiting at l_2^i including the length of the waiting vehicle. In other words, R_i is considered as an obstacle waiting at l_2^i and the attempt is to avoid R_k at a distance of $\Delta(v_k)$. Extra space is kept to correct the position of R_k before surpassing R_i at l_2^i and further space after it has crossed waiting R_i to avoid collisions due to the immediate correction of position. Any lane of R_k that lies within this region is modified to account for the waiting R_i. As per the calculations of l_2^i, it is expected that R_k will have the same lateral position that it had when it crossed the line Z. Let R_k enter the region S at lane$_j[a]$ and leave at lane$_j[b]$. The manner of changing the lane and then the trajectory is similar to free-state expansion.

The *order of planning* of vehicles in *H* specifically takes planning of the vehicle being waited for R_j ahead of the other vehicles. This is because this planning gives us the time (tw) when R_j crosses its lane$_j[b]$ and hence R_i can continue its journey from this time onwards. Other vehicles in *H* consider R_i to be waiting only till the time tw. Hence, any of their lanes are only modified if the vehicle crosses the position before this time.

The next task is generating a trajectory for R_i from L_i to l_2^i. The intention behind first generating trajectories of the vehicles in *H* and later R_i is that, as trajectories are generated, they are added to the travel plan $\zeta_{dynamic}^{free}(\tau_2)$ which the vehicles being planned later must avoid. Local optimizations are performed on all vehicles in *H* and R_i. By this scheme it is made clear that the path of vehicles in *H* must be as unchanged as possible, with R_i compromising as much as possible in terms of separation availability. Local optimizations in trajectory generation of R_i ensures that any possibility of feasible curve generation of R_i returns a trajectory, in case there is insufficient space. Computing the clothoid from L_i and l_2^i, and adding it into the expanded state of L_i is similar to free-state expansion. The major difference is that the time a vehicle reaches l_2^i is taken to be tw and not as computed by the general algorithm.

Fig. 9.14A shows the stage when R_i discovers the sudden change in the width of the road, which initiates this expansion technique as per the heuristic (Section 9.4.4). The region *S* for R_k is shown in the same figure. Fig. 9.14B meanwhile shows the modification of the lane of R_k. Because a large amount of space was available and the vehicle was travelling in a straight line, no change was recorded. The trajectories of R_i and R_k are shown in Fig. 9.14C. The final waiting point for R_i becomes the initial state for further expansions. Note that in this figure the direction of motion of the vehicle is in the opposite direction as compared to the earlier figures. The general structure of the algorithm is similar to the free-state expansion and is given as Algorithm 9.8. The process is also summarized in Box 9.6.

Algorithm 9.8: Wait for Vehicle Expansion Strategy

WaitForVehicleExpansion(L_i)
```
l₂s ← NULL
for Rⱼ: Rⱼ crosses Pₐ in opposite direction to Rᵢ
  calculate H
  calculate τ² using equation (9.15)
  calculate l₂ⁱ using equation (9.20)
  calculate laneⱼ[a] and laneⱼ[b]
  laneⱼ[l][Y] ← lʲ[Y] ∀ l ∈ (a,b)
  success ← GenerateTrajectoryElse()
  if not success, continue Rⱼ
  tw ← laneⱼ[b][T]
  for Rₖ ∈ H − {Rᵢ, Rⱼ}
```

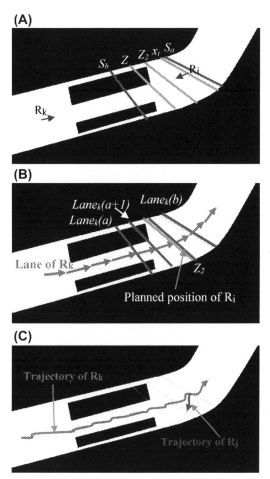

Figure 9.14 (A) Detection of sudden narrowing of width of road, (B) modification of lane of R_k, (C) the generated trajectory for vehicles.

```
calculate lane_k[a] and lane_k[b]
lane_k[l][Y] ← l^k[Y] ∀ l ∈ (a, b)
success ← GenerateTrajectoryElse()
if not success, continue R_j
end R_k
success ← GenerateTrajectorySelf() from L_i to l_2^i only
l_2[T] ← tw
if not success, continue R_j
l_2s ← l_2s ∪ l_2
end R_j
return l_2s
```

Box 9.6 Wait for Vehicle Expansion Strategy Procedure

- Consider a *narrow bridge* with a stream of vehicles coming from one side
- Plan to wait for some of the vehicles in the stream to clear and then go forwards

Procedure

- Decide the vehicle to wait for
  - Attempt waiting for all vehicles in the crossing stream of vehicles
  - Number of state expansions = number of vehicles in the stream
- Compute the position/trajectory to wait
  - Assume obstacle analysis line (Z) lies inside the narrow bridge
  - Assume previous line (Z_2) lies outside the narrow bridge in a wide region
  - Maximum separation threshold is left
  - Two options to wait: *Left* of the leftmost vehicle in the stream, or *Right* from the rightmost vehicle in the stream
  - Whichever has minimum deviation from current position is selected
- Modify other vehicles' trajectories
  - Modify trajectory for the vehicle being waited for
  - Calculate time when it crosses the vehicle being planned
  - Compute trajectory of the vehicle being planned
  - Modify trajectories of all other vehicles till this time
  - Vehicles travel straight in this region, all overtaking and like behaviours get cancelled
  - Get a snapshot of state after wait and use as the expanded state
- Compute state after wait

9.4.4 Expansion Strategy

An easy method of implementation would be to use all three expansion strategies in all the states of the planning. However, this results in a significantly large computational time. To reduce the complexity, *heuristics* were used to *prune* the expansions. Along with a natural driving analogy as well as general modelling of the planning scenario, this helps us to mine out good scenarios. A heuristic rule is used common for the *free-state* and *vehicle-following expansion*, based on which either or both techniques may be used; and one for the wait for vehicle technique.

The decision whether to use the free-state expansion strategy or a path following strategy is done on the basis of the vehicles in the *critical region* (say H) that have been encountered in the expansion of state L_i. The vehicles that were encountered in a similar region when the state L_i was the expanded state are known, say $L_i[H]$. If the two sets $L_i[H]$ and $L_i[H]$ match, it means that planning is encountering the same vehicles again. However, in case the two sets do not match, it means that *a new vehicle has been encountered* or R_i has *completely passed* some R_j. Now consider a road segment. By natural driving it is known that once a vehicle R_i decides to overtake another vehicle R_j, it continues to do the same till overtaking is successful. Similarly, if R_i decides to follow R_j, it continues to do so, till some vehicle R_k leaves, which was not giving space for R_i to overtake R_j. Three observations are made here. First, the *decision whether to overtake* a vehicle or to follow it is made on its first emergence. Second, once the decision has been made, it is *unaltered* as the vehicles go forwards. Third, a change of decision is always marked by a *vehicle entering or leaving*. Vehicles entering and leaving in the algorithm are marked by a difference in the two sets. The strategy, out of the two strategies, that resulted in the expansion of state L_i is known, say strategy(L_i). A small assumption here is that every R_k is recorded at the critical region set. Alarmingly high differences in speed may make R_k unnoticed anytime in the critical region. This assumption can be improved by increasing the size of the critical region. The selected strategy of expansion is given in Algorithm 9.9. The selection is also summarized by Table 9.1.

Algorithm 9.9: Selection of Expansion Strategy

Expand(L_i)

```
A ← NULL
for vᵢ ← Vᵢₘₐₓ to 1 in small steps
  calculate P using equation (9.2)
  for all Pₐ ∈ P
    calculate H, width(Pₐ)
    lookup H(Lᵢ), width(Lᵢ), strategy(Lᵢ)
    a, b, c ← NULL
    if H(Lᵢ) ≠ H
      a ← FreeStateExpansion (Lᵢ)
      strategy(a) ← FreeStateExpansion
      b ← VehicleFollowExpansion(Lᵢ),
      strategy(b) ← VehicleFollowExpansion
    else
      if strategy(Lᵢ) = FreeStateExpansion
        a ← FreeStateExpansion (Lᵢ)
        strategy(a) ← FreeStateExpansion
```

Table 9.1 Selection of Expansion Strategy

S. No.	Strategy	Precondition
1.	Free state expansion	• Parent state was expanded by free-state expansion and no new vehicle is encountered/left • A new vehicle is encountered/left
2.	Vehicle following	• Parent state was expanded by vehicle following and no new vehicle is encountered/left, • A new vehicle is encountered/left
3.	Wait for vehicle	• A sudden change in road width used for obstacle assessment

```
      if strategy(Lᵢ) = VehicleFollowExpansion
        b ← VehicleFollowExpansion(Lᵢ)
        strategy(b) ← VehicleFollowExpansion
    end if
    if width(Lᵢ)- width(Pₐ) > width_threshold
      c ← WaitForVehicleExpansion(L)
      strategy(c) ← WaitForVehicleExpansion
    A ← A U a U b U c
   end Pₐ
   if A ≠ NULL, return A
  end vᵢ
  return NULL
```

Similarly, a heuristic rule is placed for deciding whether the expansion by *wait for vehicle* technique needs to take place or not. Consider the obstacle analysis line Z. Let the length of obstacle-free line in P_a of state L_i be width(P_a). Let width(L_i) be the length of the obstacle free line when L_i was produced. The technique of wait for vehicle expansion is only used if there is a significant (and sudden) decrease in the width. Hence, a drop in values between width(L_i) and width(P_a) triggers this expansion technique. This is given in Algorithm 9.9.

9.5 Results

The complete system discussed was experimented with for a variety of scenarios. First, the aim is understanding of the vehicle dynamics in the presence of multiple vehicles. The focus is on the three strategies presented, and hence the map taken does not have any obstacle. Then the aim is understanding of the vehicle dynamics within an obstacle network. For this part, it is important to ensure that the vehicles do not collide or lie close to obstacles while displaying any behaviour.

9.5.1 Simple Road Experiments

A curved road is taken with no obstacles. A number of vehicles enter the scenario from both ends at specified timings. The lengths and widths of all the vehicles were different. The first experiment involved studying the free-state expansion strategy which leads to overtaking. To make the scenario complicated, an additional vehicle appears from the opposite end. The additional vehicle puts a strict restriction that the overtaking completes on time, else a collision is likely. The instance of time when overtaking completes is shown in Fig. 9.15A. The trajectories denote the manner in which the vehicle emerges and goes forwards with overtaking. It is clearly visible that the vehicle had to steer enough to avoid any collision and then move forwards with its natural speed till it was sufficiently ahead of the vehicle.

The other behaviour was a vehicle-following scenario which is also studied by similar experiments. Here again, three vehicles are used. The oncoming vehicle allowed no possibility for the faster vehicle to overtake the slower vehicle.

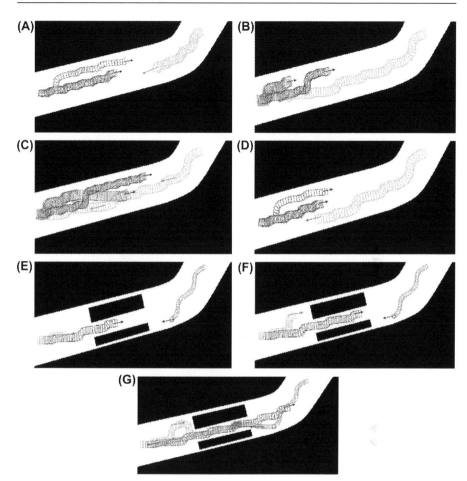

Figure 9.15 Experimental results with multiple vehicles without obstacles. (A) Overtaking scenario (B) Vehicle following scenario middle (C) Vehicle following scenario end (D) Parallel placement of vehicles scenario (E) Narrow bridge scenario start (F) Narrow bridge scenario middle (G) Narrow bridge scenario end.

However, after the vehicle crossed, overtaking was possible. Fig. 9.15B shows the situation in which the oncoming vehicle crosses the two vehicles. Note that both vehicles drifted leftwards to ensure that no collision was recorded as well as a maximum separation being available. Soon after this vehicle crossed, the faster vehicle decided to overtake the slower vehicle. The intent to overtake is shown in Fig. 9.15C.

The different vehicles may need different lanes. Though the intent is to align them in a way such that after steering and attaining their orientations they distribute the available road width amongst themselves, it is important to see whether the distribution takes place neatly. The next scenario presented three vehicles, two originating from one side of the road while the third from the other. The speeds were arranged and emergence timed so that the three met at some point in between. The result is shown in Fig. 9.15D. It can be seen that the separation between any two vehicles and road ends is kept large enough.

The last experiment was for the wait-for-vehicle strategy which requires buildup of a narrow bridge in the middle of the road. Here, three vehicles are chosen, two coming from one side and the other coming from the other side. The scenario and the results are shown in Fig. 9.15E–G. The first vehicle emerges and moves almost straight to cross the bridge. The second vehicle precomputes that it needs to wait to avoid collision; it moves forwards to take its waiting position as shown in Fig. 9.15E. Even before the first vehicle crosses, the third vehicle emerges into the planning scenario. It computes that it cannot cross the bridge before the first vehicle or just after the first vehicle, avoiding a collision with the second vehicle. Therefore, it also decides to wait for the second vehicle to cross the bridge by taking its position as shown in Fig. 9.15F. After the first vehicle has crossed, the second vehicle starts and similarly after the second vehicle has crossed, the third vehicle starts. This is shown in Fig. 9.15G.

9.5.2 Experiments of Multivehicles in an Obstacle Network

The first scenario in this category was designed to look at the manner in which multiple vehicles select their paths. A broad straight road was divided into three lanes by physically placing elongated obstacles. The vehicles were generated from either side of the road. The scenario and the solution are given in Fig. 9.16A. The first vehicle naturally selected the middle road ahead. As the second vehicle generated was wide, it could not travel straight in the middle road and hence selected the bottom road. Now the other two vehicles generated preferred to stay in the middle road and occupy different lanes within this middle road, even though they could just fit in. At the time of their planning the top road was empty which was hence occupied by the next vehicle. Every vehicle was assigned a speed such that all the vehicles met or collided in the middle if travelling throughout by their speeds. This ensured no vehicle-following behaviour was encountered.

In the experiments of Section 9.5.1, vehicle behaviours in the absence of obstacles were discussed. However, it may be essential for overtaking behaviour to complete within some obstacle subnetwork. If a collision is encountered, the plan may alternatively choose vehicle-following behaviour which is collision free. The scenario was set up with a barrier dividing a curved road. The vehicles were generated from the same side of the road. Scenario and results are shown in Fig. 9.16B and C. The first vehicle took the narrower segment and the second the broader one. The third vehicle had a higher speed and hence decided to overtake the slower vehicle in the broader segment. The intent to overtake is shown in Fig. 9.16B. Soon another vehicle was added in the scenario. Even though it had a high speed, it could not construct any overtaking strategy that completed within the segment with all the preplanned behaviours of the other vehicles intact. Therefore, it decided to follow the front vehicle, while the earlier vehicle had almost overtaken. The situation is given in Fig. 9.16C.

The last scenario is to see if multiple behaviours can be simultaneously displayed. A curved road was taken with two vehicles of varying widths originated at different times

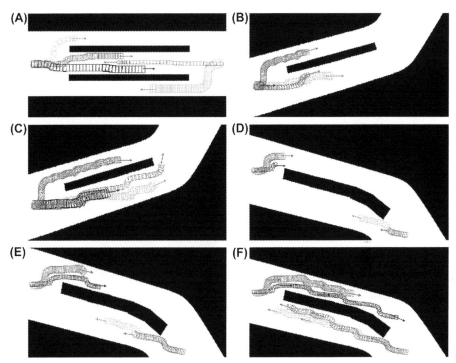

Figure 9.16 Experimental results with multiple vehicles in presence of obstacles. (A) Scenario 1, (B) Scenario 2 middle, (C) Scenario 2 end, (D) Scenario 3 start, (E) Scenario 3 middle, (F) Scenario 3 end.

from either side of the road. The scenario is displayed in Fig. 9.16D—F. On the first part of the road the vehicle generated later was of high speed and small size and was generated when the two vehicles were reasonably close by. The intent to overtake is shown in Fig. 9.16D. The vehicle is completely overtaken as shown in Fig. 9.16E. However, on the other side of the road, the two vehicles were separated by a large margin and the difference in speeds was kept low. As a result, the second vehicle followed the first vehicle at high speed for a while as shown in Fig. 9.16E. Later, it attempted to overtake it as displayed in Fig. 9.16F.

9.5.3 Parameter Analysis

Giving effective results as early as possible is one of the prime requirements of any planning algorithm. It is, hence, important to analyse the complexity of the algorithm, based on which the road-segment length considered for planning may be decided. Too large a length may make the algorithm unable to give results within time, and too small a length would make planning ineffective, wherein different vehicle behaviours cannot be displayed and the vehicle can possibly be trapped in an obstacle network.

For n static obstacles on the road, going left or right of the obstacle may result in two different plans, indicating the total number of plans possible in the worst case to be 2^n which the graph search explores. Consider a number of states produced by the graph search. A state may be called connecting state if it has one child or no children. These states do not multiply the search space or do not lead to new paths. Every obstacle approximately adds one nonconnecting state corresponding to the state when it was first discovered in all possible paths. Hence, for n obstacles, total paths in the worst case are 2^n each approximately of length l, which makes the complexity $O(2^n l)$. The effect of factor $\Delta\left(v_{max}^i\right)$ and connecting states on complexity is together communicated by the factor l. If some of these obstacles are small obstacles, there would be additional cost due to the reduction in speed of a vehicle which is done in steps.

The algorithm may also have multiple vehicles. Let a be the number of vehicles travelling in the same direction (inbound or outbound) as the vehicle being planned and b the number of vehicles travelling in the opposite direction. For a vehicle going in the same direction, it may either follow it or overtake it. Both possibilities need to be tried. The total possibilities to travel are 2^a in the worst case. As per heuristics, both possibilities were checked on encountering any vehicle. The possibilities are tried both when the vehicle arrives as well as when it departs. This may sometimes lead to a multiplicity of cases. The maximum number of possibilities in the worst case hence are $2^a(a + 2b)$. In addition, a vehicle may need to wait for another vehicle. Say the scenario has only one narrow bridge. At most the vehicle has to wait for all b vehicles coming, which means an additional b possibilities. However, these possibilities are only tried at one particular state in the road when the width changes dramatically. Hence, the addition of complexity is nonexponential. The complexity is hence $O(2^a(a + 2b))$, which shows it is exponential for the number of vehicles in the same direction and obstacles. This number is usually very small considering the small road segment taken, or the vehicles would be simply following each other, which is not considered.

The resultant complexity becomes $O(2^{n + a}(a + 2b)l)$, which means it is exponential in both n and a. Both factors of obstacles and vehicles are experimented by a more realistic experimental scenario. First, a straight roadmap is taken and experimented on the effect of the addition of obstacles. The intent was to expose the worst case that the algorithm may face, and hence all the obstacles were added one after the other. The road was significantly elongated so that lots of obstacles could be fit in. The effect of the increase in the number of connecting states s and nonconnecting states n is shown in Fig. 9.17. Because the vehicle was intentionally kept too small to allow avoiding all obstacles to be possible, many times after aligning to overcome an obstacle, in the next expansion stage it could further go to the other side of the obstacle, making possibilities per obstacles 3 rather than 2.

Similar experiments were performed with the number of vehicles. However, here, the worst case was found when the vehicle being planned, even if it decided to follow another vehicle, had possibilities of overtaking other vehicles while it was following. This is possible only when all a vehicles in front are arranged in increasing order of speeds with large gaps between them. The scenario was difficult to create even after

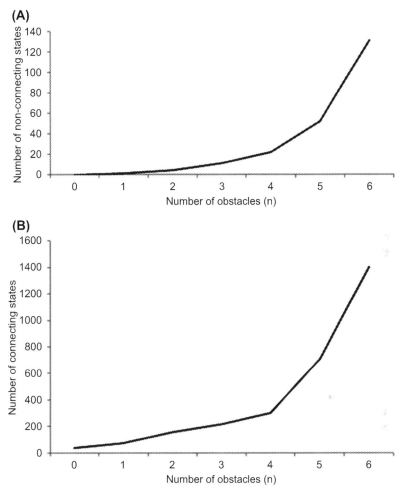

Figure 9.17 Increase in number of (A) nonconnecting states and (B) connecting states with increase in number of obstacles.

prolonged increase in the road width. Clearly, though theoretically possible, the situation is not practical. The practical worst case is to have a series of vehicles going at equal speeds separated by large gaps. For every vehicle, the vehicle being planned may decide to overtake or to follow. If it overtakes, the choices repeat with the other vehicle in front. If it follows, no further choice is available. Hence, the total number of cases possible are $a + 1$, in which a is the number of vehicles in front. Clearly, the practical worst case is linear. The number of connecting and nonconnecting states for an increasing number of vehicles is shown in Fig. 9.18. The connecting states are not linear, as the section in which the vehicle being overtaken is recorded, and the road length remaining makes the difference.

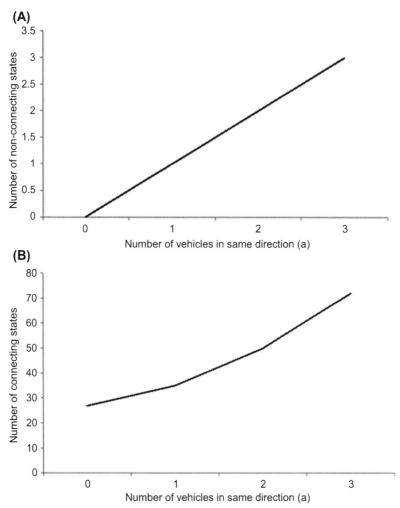

Figure 9.18 Increase in number of (A) nonconnecting states and (B) connecting states for increasing number of vehicles in the same direction.

The other factor is the longitudinal separation constant. This factor was kept as per the present notion that a vehicle must be able to steer significant distance in the Y axis. This puts a lower threshold on the value of the factor. Increasing this factor decreases steering and increases smoothness. However, on increasing the factor, a small obstacle problem becomes harder to detect and rectify, close overtaking cannot be performed and vehicles start aligning early on seeing an obstacle or some other vehicle. All these factors affect optimality. Too small a value, however, leads to too-sharp turns and large computational time.

From the software point of view, one of the greatest tasks ahead is to change the data structure in which paths and lanes are represented internally. The data structures must cater to fast query response. Heuristics can be added in the expansion for computational speedup. Postprocessing techniques may be added to optimize the trajectories as a whole and remove any oscillations visible in overtaking, lane changes and while travelling on a curved road. Displaying vehicle behaviour at traffic merging, diversions and crossing scenarios is further to be done. The simulations can then be extended to an entire transportation system. The simulations ultimately need to be validated on real vehicles and real roads to prove their effectiveness.

9.6 Summary

The approach intends to solve the problem using heuristics to intelligently select the best states for expansion. The central idea was to model the lanes as being dynamic and distributed along the length of the road. The planning performed was based on a prioritized Uniform Cost Search, affecting replanning of vehicles of higher priority which ensured cooperation. The algorithm could try all possibilities and come up with an optimal plan. The expansion in case of obstacles was done to maximize the separation.

The same rule was used in the case of other vehicles in which the vehicles were laterally placed at maximum separations. A vehicle altered the plan of the other vehicles to accommodate itself. A vehicle could decide to overtake another vehicle, follow it or wait for a vehicle coming from the other side to cross. Experiments over complex grids of obstacles and other vehicles showcased the effectiveness of the algorithm. Further, vehicle behaviours of overtaking, vehicle following and waiting for another vehicle ensured optimal performance for a variety of scenarios.

A major question that arises out of the research is the validity of the present traffic rules. In an autonomous scenario, should the vehicles be allowed to collaboratively construct optimal plans which do not adhere to static traffic rules, which was the case in this approach? This leads to possibilities of more driving behaviours. The algorithm used for planning should, however, ensure completeness and optimality in a collaborative manner.

References

Bhattacharya, S., Likhachev, M., Kumar, V., 2012. Topological constraints in search-based robot path planning. Autonomous Robots 33 (3), 273–290.

Clark, C.M., 2005. Probabilistic road map sampling strategies for multi-robot motion planning. Robotics and Autonomous Systems 53, 244–264.

Cormen, T.H., Leiserson, C.E., Rivest, R.L., Stein, C., 2009. Introduction to Algorithms, third ed. MIT Press, Cambridge, MA.

Demyen, D., Buro, M., 2006. Efficient triangulation-based pathfinding. In: Proceedings of the 21st National Conference on Artificial Intelligence. AAAI, pp. 942–947.

Dobson, A., Bekris, K.E., 2014. Sparse roadmap spanners for asymptotically near-optimal motion planning. International Journal of Robotics Research 33, 18–47.

Dobson, A., Krontiris, A., Bekris, K.E., 2013. Sparse roadmap spanners. In: Algorithmic Foundations of Robotics X, Springer Tracts in Advanced Robotics, vol. 86. Springer, pp. 279—296.

Gayle, R., Moss, W., Lin, M.C., Manocha, D., 2009. Multi-robot coordination using generalized social potential fields. In: Proceedings of the 2009 IEEE International Conference on Robotics and Automation, Kobe, Japan, pp. 106—113.

Gayle, R., Sud, A., Lin, M.C., Manocha, D., 2007. Reactive deformation roadmaps: motion planning of multiple robots in dynamic environments. In: Proceedings of the 2007 IEEE/RSJ International Conference on Intelligent Robots and Systems, San Diego, CA, pp. 3777—3783.

Kala, R., 2013. Rapidly-exploring random graphs: motion planning of multiple mobile robots. Advanced Robotics 27 (14), 1113—1122.

Kala, R., Shukla, A., Tiwari, R., 2011. Robotic path planning in static environment using hierarchical multi-neuron heuristic search and probability based fitness. Neurocomputing 74 (14—15), 2314—2335.

Kala, R., Warwick, K., 2014. Dynamic distributed lanes: motion planning for multiple autonomous vehicles. Applied Intelligence 41 (1), 260—281.

Langsam, Y., Augenstein, M.J., Tenenbaum, A.M., 2009. Data Structures Using C and C++, second ed. PHI Publishers.

Littlefield, Z., Li, Y., Bekris, K.E., 2013. Efficient sampling-based motion planning with asymptotic near-optimality guarantees for systems with dynamics. In: Proceedings of the 2013 IEEE/RSJ International Conference on Intelligent Robots and Systems, pp. 1779—1785.

McCrae, J., Singh, K., 2009. Sketching piecewise clothoid curves. Computer Graphics 33 (4), 452—461.

Nutbourne, A.W., McLellan, P.M., Kensit, R.M.L., 1972. Curvature profiles for plane curves. Computer Aided Design 4 (4), 176—184.

Schmitzberger, E., Bouchet, J.L., Dufaut, M., Wolf, D., Husson, R., 2002. Capture of homotopy classes with probabilistic road map. In: Proceedings of the 2002. IEEE/RSJ International Conference on Intelligent Robots and Systems, vol. 3, pp. 2317—2322.

Siméon, T., Laumond, J.P., Nissoux, C., 2000. Visibility-based probabilistic roadmaps for motion planning. Advanced Robotics 14 (6), 477—493.

Tiwari, R., Shukla, A., Kala, R., 2013. Intelligent Planning for Mobile Robotics: Algorithmic Approaches. IGI Global Publishers, Hershey, PA.

Yao, Z., Gupta, K., 2011. Distributed roadmaps for robot navigation in sensor networks. IEEE Transactions on Robotics 27 (5), 997—1004.

Fuzzy-Based Planning

10

10.1 Introduction

The approaches discussed in chapters 'Optimization-based Planning', 'Sampling-based Planning', 'Graph Search-based Hierarchical Planning' and 'Using Heuristics in Graph Search-based Planning' assumed *communication* between the vehicles. Although it is possible to make a standard protocol which all the vehicles on the road may follow, the assumption disregards the presence of any human-driven vehicle on the road. Communication has strong advantages of deliberating much into the future and creating a strongly bound coordination leading to the possibility of more efficient travel. However, it comes with the associated disadvantage of higher computational cost which further requires sophisticated computing devices installed in all the vehicles.

A better modelling framework is the one which does not assume communication between the vehicles. Not only does the resultant system allow for human drivers to occupy the road and autonomous vehicles to plan using their own algorithms, it also uses visual information for coordination which can be faster than the communication-based coordination having networking delays.

Not assuming lanes makes the problem of motion planning of autonomous vehicles very general. With such a generalized definition of traffic operating without lanes, the problem of trajectory planning of vehicles may be broadly seen as a problem of motion planning of a mobile robot. A variety of sophisticated planning algorithms have already been proposed for the same. These may be broadly classified into deliberative, reactive and hybrid algorithms. The *deliberative algorithms* (including algorithms of chapters: Optimization-based Planning and Sampling-based Planning) are time-consuming and, hence, work only for static environments. Meanwhile, *reactive algorithms* work for obstacles with known or unknown dynamics and can handle any sudden appearance of obstacles. *Hybrid algorithms* (Kala, 2014; Kala et al., 2010) assume some parts of the problem as static for offline planning, taking the other parts as dynamic for online planning. These algorithms may therefore be made to function at two different levels of planning.

Reactive planning techniques (Tiwari et al., 2013) assess the current situation based upon which the immediate action is produced. These methods do not deliberate much into the future and figure out or evaluate competing plans. The decisions are made using fast instincts and prior experience. Essentially, the distances to nearby obstacles are given as an input to an inference system, which produces as output the immediate action to be taken. The decisions are thus based only on the immediate scenario and immediate situations with a few variables representing the entire scenario. The approaches are, hence, faster and can swiftly react to sudden changes in the environment. However, neither optimality nor completeness can be guaranteed.

Before delving into reactive planning, it would be worthwhile to note if there is a significant loss incurred in the use of reactive planning for the problem. Most roads have no or a just a simple obstacle grid, mostly restricted to a single obstacle. Hence, it is highly unlikely that the vehicle would get trapped in some manner by the obstacle. In case of the presence of multiple vehicles, the vehicle can always follow the vehicles just in front as long as they do not stop without any reason. This does not happen in traffic systems. It may, hence, be stated that *completeness* is practically not a problem if the built reactive planner can assess and overcome a few obstacles.

The other issue is of *optimality*. The reactive techniques can be made near-locally optimal which in this case means that it is possible to make near-optimal obstacle avoidance and overtaking trajectories in the case that not many strategies exist to choose from. This is practically what driving is all about. One sees a vehicle in front and attempts to overtake (if possible) without worrying too much about what lies ahead of the vehicle being overtaken. Mostly, vehicles are following each other, which further means that knowledge of what lies ahead is of little use. Theoretically, though one would have preferred to delay overtaking or taken a different side of overtaking if a painfully slow vehicle lay on the other side from which overtaking was attempted. Global optimality is, hence, absent in reactive techniques but does not pose a serious threat.

A high magnitude of deliberation is in any case not useful in traffic-like systems in the absence of communication. Every vehicle perceives information about the other vehicle which may be different from the actual information which is forecasted into the future for deliberation. If deliberation is done much into the future, the uncertainties become too large to practically make any judicious decisions.

This chapter discusses an important approach that makes use of reactive planning, namely *fuzzy logic* (Ross, 2010; Shukla et al., 2010). The general pros and cons of the approach are discussed in Box 10.1. Many people learn driving with someone instructing them the actions to take for different kinds of situations. This makes a *rule base* for the person to follow. While driving without the instructor, the person may simply assess the situation and take actions as indicated by the rule base. This motivates

---

Box 10.1 Pros and Cons of Fuzzy Logic

Pros
- Less computation time
- Can handle uncertainties
- Can work in dynamic environments and partially known environments

Cons
- Not complete
- Not optimal
- Can easily get stuck in scenarios

the use of a logic-based navigation system. The problem with *discrete logic* (Russell and Norvig, 2014) is that the outputs may change sharply as the logic set changes, which further stresses judicious division of the continuous spaces into sets and logic rule formulation. Fuzzy logic-based inference techniques can associate every input with numerous sets by varying degrees and integrate the results from all possible logic rules fired with varying degrees depending upon the degree of association with the logic set.

The fuzzy-based planner is given as inputs the distances from various obstacles (including road boundaries and the other vehicles) that lay around the vehicle. Some other *decision-making variables* are also passed as additional inputs to the fuzzy inference system. These inputs categorize the scenario of operation. A rule set is used to decide the immediate action of the vehicle. *Overtaking* is a decision that cannot be made purely on reactive basis and requires critical decision making. The system extrapolates the motion of the other vehicles to decide the feasibility of overtaking. The feasibility decision is passed as additional input to the fuzzy system which implements the overtaking or vehicle-following behaviour based on the feasibility.

Segments of the chapter have been reprinted from Kala and Warwick (2015), Electronics, vol. 4, no. 4, R. Kala, K. Warwick, Reactive Planning of Autonomous Vehicles for Traffic Scenarios, pp. 739—762, MDPI Publishers.

Section 10.2 presents a very brief overview of the relevant literature. Section 10.3 presents a primer on problem solving with fuzzy logic. The design and working of the fuzzy system used in the approach is given in Section 10.4. The section illustrates the modelling of the inputs and outputs and formulating relevant rules for the system. It also presents the decision-making module to decide the feasibility of overtaking. The human-written rules may be suboptimal whereas the human designer may not be willing to devote much time and energy in the production of an optimal fuzzy inference system for the vehicle's navigation. Section 10.5 presents an evolutionary framework to tune the fuzzy inference system produced by the human designer. The results are given in Section 10.6 and concluding remarks are given in Section 10.7.

10.2 A Brief Overview of Literature

The general problem with the reactive methods is their *lack of completeness* and *global optimality*. In the work of Alvarez-Sanchez et al. (2010), every robot step was taken to maximize the distance from the obstacles, assessed using a circular scan. In a similar work, Sezer and Gokasan (2012) aimed to move the robot into the largest gap, whereas the choice was based on orientation and nonholonomic constraints. Here, a similar idea is used for the vehicles, although the maximization effort may be put under a threshold.

Ye and Webb (2009) selected subgoals for the motion of the robot. The selection of subgoals was based on a variety of factors including widths, orientation and location of the goal. The resultant approach is still not complete or trap free. Jolly et al. (2010) solved the problem of planning of robot for robot soccer. First, rules were made for a discrete field, which were later learnt using a fuzzy neural network. This work follows a similar methodology for the traffic scenarios in which the general behaviours are modelled and then optimized (learnt) or modelled in a way to clearly optimize an objective of choice.

Velocity obstacles (Fiorini and Shiller, 1998) are another concept wherein the speeds and orientations of all robots are considered to find the set of feasible velocities that the robot can take, based on which the immediate velocity of the robot is decided. Recent works include Hybrid Velocity Obstacles (Snape et al., 2011) which eliminate a common problem of oscillations in the velocity assignment using the method. Here, opposite corrections are applied to the two prospectively colliding robots. In another related work, Rashid et al. (2012) assigned corrective velocities by studying the robot's orientation and speed. Velocity obstacle-based approaches, however, do not account for the bounded nature of the map. The obstacles also need to be structured.

Fuzzy-based systems are also widely used in a road scenario. These, however, face problems similar to the potential approaches. Lack of completeness is a major problem. Kala et al. (2010) solved the problem of robot-motion planning by using probabilistic A* at the coarser level and fuzzy inference system at the finer level. However, the approach relies on decomposition, which is not an option with the road scenarios. Sgorbissa and Zaccaria (2012) used Voronoi graphs for the coarser level and potential approach for the finer level. The robot was not allowed to take positions which would prohibit it from continuing with the plan indicated by the coarser level planning, and such scenarios were called roaming trails. However, here, because the environment is assumed unstructured, making a complete deliberative plan is not an option.

Various approaches have been tried to enable fuzzy approaches to model complicated problems. However, the decision regarding whether or not to overtake can still not be taken by these approaches, which is a more deliberative decision to make. Motlagh et al. (2012) used fuzzy cognitive maps for making rules for the fuzzy inference system. Robot slippage was also modelled and actions could be taken which avoid robot slippage. Selekwa et al. (2008) designed a system wherein multiple fuzzy inference systems were used to compute the output for any scenario, each system denoting a particular robot behaviour. Weighted addition of the individual outputs was taken for the robot's motion.

10.3   A Primer on Fuzzy Logic

Most of the actions that we take in real life are guided by some kinds of *rules*. Most of these rules are logical and can be easily explained by *logical reasoning*. Many times these rules are the result of one's own experience, having tried different decisions in different situations. We learn the rules of the game from people and act as per the same rules. Humans are able to do remarkable things the foundations for which lie on such logical reasoning which is learnt and adapted with time. The set of rules which governs the actions and performance of an agent constitutes its *knowledge base*. In every situation, the agent analyses its situation of operation and fires appropriate rules as a response, which changes the state of the agent based on which next action is taken.

The example of driving is no different. There are predefined traffic rules which one must follow at all times. Similarly, there are best practices of driving which every good driver knows about. Every person may have his/her own driving patterns which one learns along with time and adapts as per experience. When one learns driving from an instructor, the instructor teaches all these rules, which are practiced and perfected. The rules here are nothing but a simple specification of a suitable action corresponding to every situation. The actions may be valid driving actions, denoting the way of handling steering and speed. The situation may be characterized by a set of variables denoting the current scenario. The rules may be prioritized to make some action take precedence in case two rules are applicable for a situation.

Consider a simple rule which states that 'If distance from obstacle in front is near and vehicle at right lane is far, then steering is steep-right', which is a very natural rule of obstacle avoidance while driving in lane-prone traffic. Here, the 'if' part is called the *antecedent* and the 'then' part is called the *consequent*. The inputs are 'distance from obstacle in front' and 'vehicle at right lane', whereas the output is 'steering'. Here, the inputs may be in any one of the discrete sets, say the input *distance from obstacle in front* could be *near* (0−5 m), *far* (5−10 m) and *very far* (more than 10 m). Each input participates in the rule through one of these sets. The logic is, hence, known as discrete logic. Upon perceiving any input, one could find out the set to which the input belongs to and, hence, the rules which are applicable. The rules the antecedents for which are true is fired. In case of multiple such rules, the *highest priority* or the *most specific one* is fired. The action corresponding to the fired rule is implemented.

Most of the real-life inputs and outputs are continuous. For the application of discrete logic the inputs and outputs need to be divided into discrete classes and rules are made between all combinations of input and output classes. The problem with this approach is that for a small change of input, the class to which the input belongs may change, changing the rule which gets fired and therefore produce a very large change in the output. In the above example, if *distance from front obstacle* changes from 5.1 to 4.9 m, the membership function changes from *far* to *near*, and thus the fired rule changes. This makes it very difficult to make logic sets and logic rules for the system.

10.3.1 Fuzzy Sets

Fuzzy logic is a more general solution to the same problem. The general concept of *fuzzy set* is that the same item or input may belong to numerous sets with varying degrees of association. The degree of association is called the *membership degree* of the input to a particular set, which can be between 0 and 1. A membership value of 0 denotes complete absence of the item from the set whereas a membership of 1 denotes a full presence at the set. The resultant set is called a fuzzy set.

A discrete set is denoted by the set of items that the set contains. For example, $A_1 = \{5, 6, 7, 8\}$ is a set containing four elements, whereas $\text{Near}_1 = [0, 5]$ is a set of all the real numbers between 0 (included) and 5 (not included). Correspondingly, a fuzzy set is denoted by the set of items that the set contains and the associated membership degree of that item. The fuzzy set $A = \{5/0.2 + 6/0.7 + 7/0.8 + 8/0.4\}$ denotes the item 5 with a membership of 0.2, item 6 with a membership of 0.7 and so on. Note that here '+' is not the arithmetic sum, but it denotes the collection of items of the sets. Similarly, '/' is not the arithmetic division, but denotes the membership values corresponding to every element. The set Near = { $\int x/\mu(x)$ }, $x \in R^+$, denotes the set of positive real numbers, in which every item x has the membership of $\mu_{near}(x)$. Here, $\int$ is the generalization of the sum used to separate items in a discrete set. Here, $\mu_{near}(x)$ is called the membership function and denotes the membership of item x in the set *Near*. Making a suitable function $\mu_{near}(x)$ is a part of the problem of modelling. A typical example is a Gaussian membership function given by Eq. [10.1] and shown in Fig. 10.1.

$$\mu_{near}(x) = e^{-(x-\mu_1)^2/2\sigma_1^2} \qquad\qquad [10.1]$$

here, μ_1 and σ_1 are the parameters of the membership functions which must be learnt or set by trial and error.

The typical choices of membership functions are triangular, Gaussian, bell-shaped, sigmoidal etc. The complete modelling of an input may be given by Fig. 10.2. For

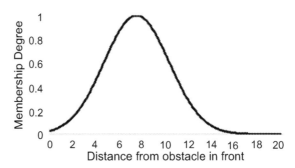

Figure 10.1 Gaussian membership function.

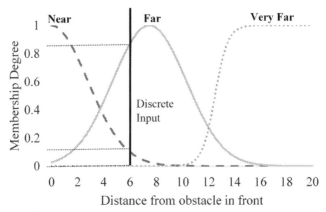

Figure 10.2 Modelling of a fuzzy input.

example, if the input is 6, the membership degrees can be given by $\mu_{near}(x) = 0.1$, $\mu_{far}(x) = 0.87$ and $\mu_{very\text{-}far}(x) = 0$. These three membership degrees characterize the input. It can be seen that the inputs have the largest association with the set *far*, whereas some association with the set *near* as well.

Discrete logic runs using discrete rules, wherein each input and output belongs to a discrete set. Once the notion of fuzzy sets is conceived, this fuzziness needs to be added into all the elements of reasoning of the otherwise discrete logic. In other words, all the unit procedures of discrete logic need to be modified to handle fuzziness, which is the aim of this section.

Every *fuzzy rule* consists of the antecedents and the consequents. The antecedents consist of different inputs and participating membership sets, separated by logical operators (and, or, not). Consider one such rule 'If distance from obstacle in front is near and vehicle at right lane is far then steering is steep-right'. The antecedent consists of the input *distance from obstacle in front* with participating set *near* and *vehicle at right lane* with participating set *far*. The membership of the given input to the class indicates the strength with which the rule must fire per that input. Because multiple inputs are connected by fuzzy operators, *fuzzy arithmetic* is used to compute the degree of membership of the inputs combined in that rule, which becomes the firing degree of the rule. The antecedents are related to the consequents by the implication operator, which is another fuzzy arithmetic operator. This transforms the inputs to the outputs.

Fuzzy logic consists of a set of fuzzy rules, each of which has inputs belonging to different sets by various membership degrees. This indicates the degree to which a particular rule fires, or the contribution of that rule in the overall output of the system. *Aggregation* is the process of taking fuzzy outputs of individual rules and aggregating them to make a single fuzzy output. The fuzzy output is a single fuzzy set with a range of outputs with different membership degrees. *Defuzzification* is the process of converting this fuzzy set into a discrete output, which becomes the output of the system.

Let I be the inputs in the system, O be the outputs in the system and R be the rules in the system. Let there be $|I|$ inputs, $|O|$ outputs and $|R|$ rules. Let the input I^i be divided into overlapping fuzzy sets $M_i^1, M_i^2, M_i^3, ..., M_i^{|M_i|}$, in which $|M_i|$ is the number of fuzzy sets into which the input is decomposed. Let the rule r be of the form 'If I^1 is M_1^{a1} and I^2 is M_2^{a2} and I^3 is M_3^{a3} ... $I^{|I|}$ is $M_{|I|}^{a|I|}$ Then o^1 is M_1^{b1}, o^2 is M_1^{b2}........'. Noting the membership degrees of the inputs and outputs, the rule can be easily reduced to 'If $\mu_1^{a1}(I^1)$ and $\mu_2^{a2}(I^2)$ and $\mu_3^{a3}(I^3)$ .... $\mu_{|I|}^{a|I|}(I^{|I|})$. Then $\mu_1^{b1}(o^1)$, $\mu_1^{b2}(o^2)$.........'.

10.3.2 Fuzzy Logical Operators

In any general rule the inputs and their participating sets are connected by some logical operators, typically *AND, OR, NOT* and *implication*. In discrete logic, wherein the membership of any input can only be 0 or 1, these logical operators were derived from their natural English meanings. With the introduction of fuzzy sets, these definitions need to be extended to work for fractional membership degrees for each of the inputs.

The first operator is *AND* or the *logical conjunction* ($\wedge$). In Boolean algebra, the binary operator produces an output of 1 (true) when both the inputs are 1 (true) and 0 (false) in all other cases. In fuzzy algebra, the operator needs to be applied to inputs with different membership degrees, say $\mu_A(x)$ and $\mu_B(x)$. The popular ways of modelling the *AND* operator are *min* and *product*. These are summarized in Eq. [10.2].

$$\mu_{A \wedge B}(x) = \begin{cases} \min(\mu_A(x), \mu_B(x)) & \text{for min method} \\ \mu_A(x) \times \mu_B(x) & \text{for product method} \end{cases} \qquad [10.2]$$

It would be worth verifying that for the Boolean case with inputs as 0 or 1, the fuzzy AND operator behaves exactly the same as the Boolean AND operator, which is a necessary condition to be met by all fuzzy arithmetic operators. The operator always produces outputs between 0 and 1. Further, both *min* and *product* modelling satisfies the properties of associativity, commutativity, distributivity, idempotency and existence of an identity (one).

The *min* method is easy to realize. Suppose a person is given two jobs, and the quality of work depends upon the person's expertise. If the jobs are from the same domain, a person who is able to do the more difficult job will be able to do the easier job. Hence, the chances of the person performing the tougher job only can be seen (which is the minimum of the chances of performing the tougher job and chances of performing the easier job). If the person can perform the tougher job, the easier job can definitely be done by the same assumptions. The *product* method is easy to derive from the theory of probability, although it must be stressed that fuzzy systems do not always behave similar to probabilistic systems. If p_1 is the probability of the first subcondition being true and p_2 is the probability of the second subcondition being true, then the probability that the condition is true is $p_1 \times p_2$.

A related operator is *OR*, that is a *logical disjunction* ($\vee$). In Boolean algebra the operator produces an output of 1 (true) when either or both of the inputs are 1 (true) and 0 (false) in all other cases. In fuzzy algebra, the operator can be applied by the *max* and *probabilistic or* method given in Eq. [10.3]. Here, $\mu_A(x)$ and $\mu_B(y)$ are the associated membership degrees.

$$\mu_{A \vee B}(x) = \begin{cases} \max(\mu_A(x), \mu_B(x)) & \text{for max method} \\ \mu_A(x) + \mu_B(x) - \mu_A(x) \times \mu_B(x) & \text{for probabilistic or method} \end{cases} \qquad [10.3]$$

It would be worth verifying that for the Boolean case with inputs as 0 or 1, the fuzzy *OR* operator behaves exactly same as the Boolean *OR* operator, which is a necessary condition. The operator always produces outputs between 0 and 1. Further, both *max* and *probabilistic OR* modelling satisfies the properties of associativity, commutativity, distributivity, idempotency and existence of an identity (zero).

The *max* method is easy to realize. Suppose a person is given two jobs, and the quality of work depends upon the person's expertise. The jobs are related requiring the same expertise. The person is required to do only one of the jobs. The person may opt to do the easier job, or the chances of the overall job happening is given by the maximum of the chances of the two jobs. Once the easier job is done, the tougher one need not be implemented. The probabilistic OR method can be easily understood from probability theory. Consider a condition with two subconditions separated by *OR*. If p_1 is the probability of the first subcondition being true and p_2 is the probability of the second subcondition being true, then the probability that the condition is true is $p_1 + p_2 - p_1 \cdot p_2$.

The next major operator is *NOT* ($\neg$). It is a unary operator which simply inverts the input. For the Boolean case the operator outputs 1 (true) for an input of 0 (false) and vice versa. For the fuzzy case the operator is given by Eq. [10.4].

$$\text{NOT } \mu_A(x) = 1 - \mu_A(x) \qquad [10.4]$$

The operator modelling can be easily understood from probability theory, the probability of a subcondition being false is 1 minus the probability of the subcondition being true.

The last major operator is implication($\Rightarrow$). Mathematically, $a \Rightarrow b$ is equivalent to $\neg a \lor b$ and the operator is studied similarly. In fuzzy logic, any rule consists of the *antecedents* and the *consequents*. The antecedents are input membership values separated by *AND*, *OR* and *NOT* operators which can be solved using the fuzzy arithmetic discussed. On solving the antecedents, a single value is obtained which also denotes the firing strength of the particular rule. It denotes, in general, how close the input is to the input class for which the rule is written. The implication operator maps the antecedents to the consequents. Multiple outputs in a fuzzy system are treated independently as multiple single-output systems.

The implication is a binary operator, which takes as input the membership degree of the antecedents and the output membership function. The operator gives as output a fuzzy set denoting the output of that particular rule. Let the inputs be $\mu_A(x)$ and μ_B. The operator is given by Eq. [10.5].

$$\mu_{A \Rightarrow B}(x) = \begin{cases} \min(\mu_A(x), \mu_B(x)) & \text{for min method} \\ \mu_A(x) \times \mu_B(x) & \text{for product method} \end{cases} \qquad [10.5]$$

The min operator is also called a *slicing operator*, because its slices the consequent membership function after the value of $\mu_A(x)$. The product operator is also called the *compression operator*, because it compresses the output membership function by a factor of $\mu_A(x)$.

10.3.3 Aggregation

Every rule is a function of logical operators which can solve the entire rule and produce a fuzzy output using the operators studied so far. However, the entire fuzzy system has a number of rules. Every rule produces a fuzzy output, which incorporates the output based on the firing strength of the rule for the particular input. Because there are multiple such rules, all the fuzzy outputs need to be integrated to produce the fuzzy output of the overall system. The problem is to integrate the fuzzy outputs of all rules. This can be done like addition, taking two outputs at a time. Let μ_A and μ_B be two fuzzy outputs corresponding to two rules. The resultant output of the fuzzy *aggregation* is given by Eq. [10.6].

$$\mu_{A \vee B}(x) = \begin{cases} \max(\mu_A(x), \mu_B(x)) & \text{for max method} \\ \mu_A(x) + \mu_B(x) - \mu_A(x) \times \mu_B(x) & \text{for probabilistic or method} \end{cases} \quad [10.6]$$

10.3.4 Defuzzification

The aggregated output of all fuzzy rules is a valid fuzzy output of the system. The fuzzy output denotes all possible outputs along with their membership in the output class. However, one may prefer to convert the fuzzy output into a real number that can be used by the applications. *Defuzzification* takes a fuzzy output in the form of different outputs along with their membership degrees and converts it into a real-numbered output. The process is converse of the process of *fuzzification* which takes real-numbered inputs and converts them into membership degrees.

Let the fuzzy output be $\{ \int x/\mu(x) \}$. Many methods are possible to convert this output into a real-valued output. The most common method is *centroid* which takes a weighted average or computes the centroid of the output membership function as the output given by Eq. [10.7].

$$x^* = \frac{\int \mu(x) \cdot x \cdot dx}{\int \mu(x) \cdot dx} \quad [10.7]$$

10.3.5 Fuzzy Inference Systems

Fuzzy Inference Systems take inputs and process them based on the prespecified rules to produce the outputs. Both the inputs and outputs are real valued, whereas the internal processing is based on fuzzy rules and fuzzy arithmetic. Let us study the processing of the fuzzy inference systems with a small example. To make things simple, let us consider a system with only two inputs and one output. Consider the inputs as *distance from obstacle in front* and *vehicle at right lane*. Consider the output as *steering*. The first thing to be done is to divide all inputs and outputs into membership functions. To make things very simple, let the input *distance from obstacle in front* have three membership functions which are *near*, *far* and *very far*. The first two membership functions are Gaussian, whereas the third is sigmoidal. Further, let the input *vehicle at right lane* have only two membership functions, which are *near* and *far*. Both the membership functions are taken as Gaussian. Let the output steering have five membership functions *steep left*, *left*, *no steering*, *right* and *steep right*. It must be stressed that the real-life systems may have a large number of membership functions depending upon the complexity. The membership functions are shown in Fig. 10.3.

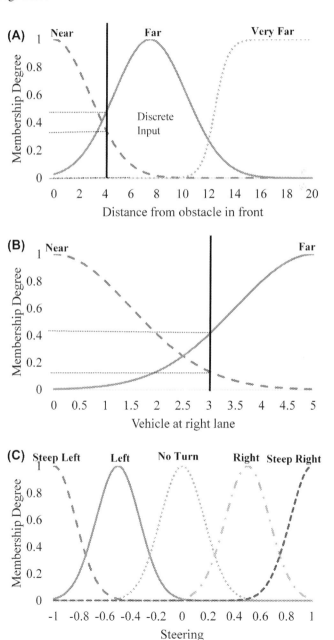

Figure 10.3 Modelling of fuzzy inputs and outputs (A) input 1 (B) input 2 (C) output 1.

Further, to make the example short, let there be only two rules.

Rule 1: If distance from obstacle in front is near and vehicle at right lane is far, then steering is steep right

Rule 2: If distance from obstacle in front is far and vehicle at right lane is far, then steering is right

Solving for this small problem, consider that the inference needs to be done for the inputs, *distance from obstacle in front* as 4 and *vehicle at right lane* as 3. The first step is fuzzification or assigning the membership degrees. From Fig. 10.3, it can be seen that the membership degrees are:

Input 1: Distance from obstacle in front
$\mu_{near}(4) = 0.36$, $\mu_{far}(4) = 0.46$ and $\mu_{very-far}(4) = 0$.

Input 2: Vehicle at right lane
$\mu_{near}(3) = 0.14$ and $\mu_{far}(3) = 0.41$

Then the rules need to be applied. The application of rules considers the computed membership values as inputs, which are mapped to the output membership function. The application of Rule 1 is given by Eq. [10.8] and Rule 2 is given by Eq. [10.9].

Rule 1:

$$\mu_{near}(4) \text{ AND } \mu_{far}(3) \Rightarrow \mu_{steep-right}(x)$$

$$= \min(\mu_{near}(4), \mu_{far}(3)) \Rightarrow \mu_{steep-right}(x)$$

$$= \min(0.36, 0.41) \Rightarrow \mu_{steep-right}(x) \qquad [10.8]$$

$$= 0.36 \Rightarrow \mu_{steep-right}(x)$$

$$= \min(0.36, \mu_{steep-right}(x))$$

Rule 2:

$$\mu_{far}(4) \text{ AND } \mu_{far}(3) \Rightarrow \mu_{right}(x)$$

$$= \min(\mu_{far}(4), \mu_{far}(3)) \Rightarrow \mu_{right}(x)$$

$$= \min(0.46, 0.41) \Rightarrow \mu_{right}(x) \qquad [10.9]$$

$$= 0.41 \Rightarrow \mu_{right}(x)$$

$$= \min(0.41, \mu_{right}(x))$$

The outputs of Rule 1 and Rule 2 are shown in Fig. 10.4. Everything above the rule firing strength has been omitted. The dotted and dashed lines show the complete output membership function for reference. The aggregated output by taking max as the aggregator is shown in Fig. 10.5. The defuzzification of the aggregated output produces the real output, which by the centroid method is computed as 0.578.

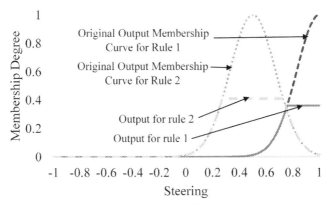

Figure 10.4 Output membership functions after implication.

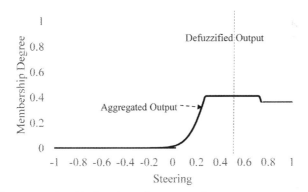

Figure 10.5 Output membership functions after aggregation.

10.3.6   Role of Parameters

As per the problem-solving methodology with fuzzy systems, first the inputs and outputs are identified, then each input and output is divided into a number of membership functions. Rules can be made for every combination of membership function between the inputs. The *number of membership functions* for every input and output is the most important parameter of the fuzzy inference system. A larger number of membership functions gives rise to the possibility of a greater number of rules. This makes the system very localized, wherein the decisions are made essentially based on the locally applicable rules. A large number of rules are impossible to train manually by a human expert or by any automated systems. If some historic data exist to *train* the system to maximize the performance over the training input, a large number of membership functions necessarily need a lot of such data. Further, the system operates on high-variance and low-bias settings, which are very likely to overfit and give a poor-generalization error. This has a larger effect of outliers and noise. On the other hand, too few membership functions per input may result in high bias and low variance, wherein the system may not be sufficiently complicated to model the problem. A trade-off must, hence, be maintained.

Every membership function may have parameters which can be learnt using learning algorithms or manually tuned by the designer. A large *number of rules* make the system operate in high-variance settings and vice versa. Hence, it is important to have a sufficient number of rules and membership functions to model the problem, but not overtrain to maintain a high generalization.

10.4 Fuzzy Logic for Planning

Fuzzy logic is a very intuitive choice for decentralized motion planning of autonomous vehicles. To simplify planning in the stated scenario, an analogy is made with the natural driving of professional drivers. These drivers always tend to use some instincts for driving. This includes their actions in making turns, avoiding general obstacles, overtaking, allowing other vehicles to pass, adjusting their driving as per the sharpness of turns in the road etc. Some *deliberative decision making* may though be required in deciding whether to overtake or not and how to avoid some obstacles, which is modelled separately in the algorithm. These instincts lie in terms of both steering and speed control, which are highly related to each other. Further, such drivers may not require extensively long training cycles to learn all these driving skills. Natural driving is simply a set of behaviours that generally follow *traffic rules*, using which a vehicle may be navigated along any path. This inspires the use of fuzzy logic for planning. The general structure of the rules may be the same, but mayhave small variations between vehicles. The same driver may well not be able to drive a big lorry and a small car with the same ease. It may further be argued that the drive may be region specific as well. The rules to drive on a hill will be slightly different from those on a flat road surface due to safety restrictions. Drivers usually *learn* the skills to drive, which may be particular to the vehicle and the region. In the same manner, the element of learning may be beneficial for autonomous vehicles as well.

The key takeaways of the algorithm are presented in Box 10.2. The approach, in a nutshell, is also summarized in Box 10.3. Most roads can comfortably accommodate two to three vehicles and are *relatively narrow*. Roads may be one-way or two-way, in which case half the road is usually for inbound traffic and the other half for outbound

---

Box 10.2 Takeaways of Solution Using Fuzzy Logic

- Design of a *Fuzzy Inference System* for the problem.
- Design of a *decision-making module* for deciding the feasibility of overtaking purely based on the vehicle distances and speeds.
- Design of an *evolutionary technique* for optimization of such a fuzzy system.
- Using the designed fuzzy system enabling vehicles to travel through a *crossing* by introducing a *virtual barricade*.

Box 10.3 Key Aspects of Solution Using Fuzzy Logic

- Codify the immediate scenario to a few *inputs*
  Continuous:
  - Angle deviation from road
  - Distance from left boundary/obstacle (similarly, right and front)
  - *Side:* distance of vehicle on wrong side
  Discrete:
  - *Steer:* Suggested turn to avoid obstacle/overtake the vehicle in front
  - *Requested turn:* Suggested turn to enable another vehicle to overtake
- Decide the actions to be taken and, hence, design the *outputs*
  - Steering
  - Speed
- Think of various scenarios and the associated inputs/outputs and generalize them as *rules*
  - Initially manually designed
  - Later tuned using *Evolutionary Algorithms*

---

traffic. The design is particularly aimed at such roads. Narrow roads pose a serious problem for the vehicles in that there is a smaller margin of error associated with the road geometry. A more serious problem associated with such roads is overtaking which is especially important if a high-speed vehicle is travelling behind a very slow vehicle ahead. A common sight in such roads is that a vehicle can partly, or for some time, drive on the wrong side of the road to overtake a slower vehicle in front. This is all modelled in the algorithm.

The basic problem is planning the motion of a number of autonomous vehicles. Let the vehicle V_i have a start position (say P_i) and a goal (say G_i). The complete map consists of a number of roads. Any point where three or more roads meet is considered a crossing. The complete route of the vehicle consists of a series of roads/crossings that it must follow in strict order. The route can therefore be characterized by Dijkstra's algorithm, on the basis of knowing the road network map. Any path may further have on it any regular/irregular static or moving obstacle. At any instance of time, t let the vehicle V_i be at position L_i and oriented at an angle θ_i. Let the linear speed of the vehicle be $S_i\left(\leq S_i^{max}\right)$ and angular speed be $\omega_i\left(\leq \omega_i^{max}\right)$, in which S_i^{max} and ω_i^{max} are the maximum linear and angular speeds, respectively. The linear speed is acceleration controlled with the maximum acceleration/deceleration of A_i^{max}. The orientation is controlled by the angular speed. The vehicle is assumed to be rectangular of dimensions $l_i \times w_i$ with four corners denoted by: front-left L_i^1, front-right L_i^2, rear-left L_i^3 and rear-right L_i^4. The various notations have been summarized in Fig. 10.6.

The basic problem is to plan the immediate movement of the vehicles involved. Every vehicle must avoid obstacles as well as any other vehicles it encounters on the way. In addition, the overall travel needs to be near optimal. There is a further set of preferences. It is expected that the vehicle maintains some minimal safe distances from both other vehicles and obstacles from the front and side. The maintenance

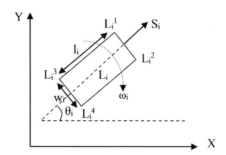

Figure 10.6 Basic notations of the vehicle.

of a safe distance from the front ensures that the vehicle would have a sufficient time to come to a halt or move to another lane in case the vehicle in front suddenly stops. Similarly, the side distances allow a sufficient safety margin for changes in road geometry, uncertain driving by another vehicle or uncertainties associated with the static obstacles. The safe distance required clearly increases with the vehicle's speed. The safe distance of front F_d and side F_s may be given by Eq. [10.10].

$$F_d = c_1 S_i^2$$ [10.10]

$$F_s = c_2 S_i^2$$

here, c_1, c_2 are constants. $c_1 > c_2$

It is further assumed that the same road is used for both inbound and outbound traffic with (in general) half the road for each side. It is assumed that the 'left side' driving rule is followed. However, the road may not have physical barriers separating the two sides and it is possible to partly occupy the wrong side of the road to perform an overtaking procedure.

10.4.1 Fuzzy Inference System

The first task associated with the implementation of the algorithm is to build a fuzzy inference system to be used as a motion planner. Here, a decentralized motion planning approach is taken. All the vehicles are planned independently from each other. Here, an overview of all the inputs and outputs of the fuzzy planner is given. The fuzzy rules can be built along the lines of the thought processes and instincts of professional drivers.

The fuzzy system is given the immediate scenario, for which it tries to plan the *immediate movement* of the vehicle. The fuzzy system developed is given a total of seven inputs. These include five continuous inputs and two discrete inputs, which give additional processing to the fuzzy system for decision making. Consider the vehicle V_i. Because the road on which the vehicle is to move is predetermined, the intention of the vehicle is to drive along the orientation of the road. The first input to the fuzzy system is the *angle between the road γ_i and the current orientation θ_i*. This is given

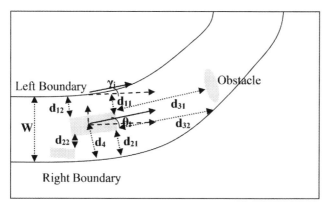

Figure 10.7 Various inputs to the fuzzy-based planner.

by Eq. [10.11]. The angle γ_i is computed as the slope of the boundary that lies nearer to the vehicle. The various notations are also given in Fig. 10.7.

$$I_i^1 = \gamma_i - \theta_i \qquad\qquad [10.11]$$

The next two inputs to the fuzzy system are the *distance of the vehicle from the left* boundary or the obstacle/other vehicle on the left and the *distance of the vehicle from the right* boundary or the obstacle/other vehicle on the right. The distance is measured in orientation to the vehicle. The distances are measured from both of the corners and the smaller distance is used as being descriptive of the closest point. The safe distance is subtracted. This encourages the vehicle to maintain the safe distance and gives a speed dependency to the obstacle distances. A vehicle may optionally lower its speed to reduce the effective distance giving it more time to turn to avoid the obstacles. Let d_{11} be the distance of the obstacle from the front-left corner of the vehicle (L_i^1). Let d_{12} be the distance from the rear-left corner of the vehicle (L_i^3). The input from the left (L_i^2) is then given by Eq. [10.12]. The various notations are shown in Fig. 10.8.

$$I_i^2 = \min\{d_{11}, d_{12}\} - F_s \qquad\qquad [10.12]$$

Similarly, let d_{21} be the distance of the obstacle from the front-right corner of the vehicle (L_i^2). Let d_{22} be the distance from the rear-right corner of the vehicle (L_i^4). The input from the right (L_i^3) is then given by Eq. [10.13].

$$I_i^3 = \min\{d_{21}, d_{22}\} - F_s \qquad\qquad [10.13]$$

The next input to the fuzzy system is the *distance from the road boundary or obstacle directly ahead*. A moving vehicle need only be considered if it has a negative relative velocity or there should be a potential for collision. Alternatively, the two vehicles should be moving on opposite sides of the road. The distance is measured

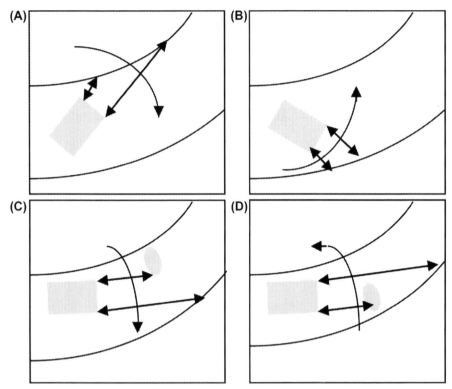

Figure 10.8 Decision making regarding steering left or right. (A) Right turn without obstacle (B) Left turn without obstacle (C) Right turn with obstacle (D) Left turn without obstacle.

from the front corners of the vehicle in orientation to the vehicle and the minimum of two is taken. Let d_{31} be the distance of the obstacle from the front-left corner of the vehicle $\left(L_i^1\right)$ and let d_{32} be the distance from the front-right corner of the vehicle $\left(L_i^2\right)$. The input from the front $\left(L_i^4\right)$ is given by Eq. [10.14]. Here, also the safe distance is subtracted.

$$I_i^4 = \min\{d_{31}, d_{32}\} - F_d \qquad\qquad [10.14]$$

The next input to the fuzzy system is a discrete input called *steer to avoid obstacle*. Assume that an obstacle (or moving vehicle) lies directly in front. Now, the vehicle needs to decide whether to overtake/avoid the obstacle on its left or right side. This input I_i^5 is given by Eq. [10.15].

$$I_i^5 = \begin{cases} -1 & \text{avoid obstacle by steering left} \\ 0 & \text{no steering required} \\ 1 & \text{avoid obstacle by steering right} \end{cases} \qquad [10.15]$$

Properly assigning this input is important because an inappropriate choice could lead to the vehicle being trapped in between obstacles or between an obstacle and a road boundary. Although the optimal choice of input can only be made by deliberative means, *heuristics* can be employed for decision making in real time and in partially known environments as considered in this work. In most general cases, an obstacle may be aligned to indicate the required steering correction. Road curvature, which for a disoriented vehicle would be reported as an obstacle in front, also indicates the steer necessary. The two distances measured from the forward obstacle are used to assess the obstacle orientation and, hence, the turn required. If the left distance is less, the obstacle may be aligned such that a right steer is preferable and vice versa. Numerous cases are shown in Fig. 10.8. Static obstacles are rarely seen in traffic as people tend to clear the road as soon as possible if something falls over in the road. Most such obstacles are small for which this heuristic holds. The heuristic may not hold for moving vehicles, for which an overtaking mechanism is used as is discussed in Section 10.4.2.

The next input to the system is *side* which tries to push the vehicle to the left side of the road. If the vehicle is already completely on the left side of the road, this input is given a value of zero. Otherwise, this input takes a value as the distance to which the (rightmost part of the) vehicle is shifted from the middle of the road (demarking the left and right sides) and, hence, by how much it needs to veer to get back to the middle of the road at least. Let W be the immediate width of the road and d_4 be the distance of the centre of the vehicle from the right boundary. The distance of the rightmost part of the vehicle from the left boundary can be approximated by subtracting half the vehicle's width (w_i). The value of the fuzzy input side I_i^6 may, hence, be given by Eq. [10.16].

$$I_i^6 = \begin{cases} W/2 - (d_4 - w_i/2) & \text{if} \quad d_4 - w_i/2 < W/2 \\ 0 & \text{otherwise} \end{cases} \qquad [10.16]$$

The next input to the system is another discrete input called *requested steer*. This is an input introduced to enable vehicles to coordinate. It may be beneficial, on many occasions, for a vehicle to the rear to have the vehicle in front steer to the left or right. This mostly happens in overtaking, when the steering of a vehicle in front may make an overtaking option easier or even may make an infeasible overtaking option feasible. A vehicle is thus allowed to request any other vehicle to steer on any side as an act of cooperation. The possible values of the input are given by Eq. [10.17].

$$I_i^7 = \begin{cases} -1 & \text{requested to steer left} \\ 0 & \text{no steer requested} \\ 1 & \text{requested to steer right} \end{cases} \qquad [10.17]$$

The planner performs two actions for the immediate motion of the vehicle which become the outputs of the fuzzy system. The first action is to generate a *steering mechanism* in which the fuzzy system tells the vehicle by what magnitude a steer needs to be performed. The steer may be left or right, and by any magnitude. The second action performed by the vehicle is its *acceleration and braking mechanism*. This mechanism tries to control the speed of the vehicle. It tells the vehicle how much it is supposed to either brake or speed up.

10.4.2 Overtaking

During driving, whether to overtake a vehicle in front or not is an important decision. In most cases once such a decision is made, it is adhered to until the situation changes by a large amount. A fuzzy inference system driven by its set of rules may not be able to model such a phenomenon to overtake or not, which is not purely reactive. Hence, a separate mechanism has been designed here for deciding on the *feasibility of overtaking* and this uses the fuzzy inference system to *carry out the overtaking procedure* if needed. The approach is summarized in Box 10.4.

Box 10.4 Key Aspects of Overtaking

- Check if the front vehicle (1) is slower and needs to be overtaken by the vehicle being planned (2), while any vehicle (3) is nearby as (2) overtakes (1)
- Check if overtaking is feasible
    Assumption:
    - Road not wide enough for multiple (>3) vehicles to lie side by side
    - The vehicles are projected to travel straight subsequently (else would require to adjust for overtaking)
    Conditions:
    - *Condition 1*: (1), (2) and (3) can simultaneously lie side by side along the road,
    - *or*
    - *Condition 2*: (2) can complete overtake (1) within the time (3) does not lie in the overtaking zone
- Initiate the overtake
    - (2) decides the side of overtake
    - (2) steers on the decided side
    - (1) steers on the opposite side
    - (2) can drive on the wrong side and does not consider the vehicle in front (both inputs disabled)
    - When (2) is ahead of (1), the inputs are again enabled
- If overtaking, check whether *cooperation* of (3) is required or not
    - Cooperation is required in case the vehicles need to align to fit within road width (condition 1)
    - (3) moves opposite to the locations of (1) and (2)
- If overtaking is infeasible
    - *Follow* the vehicle in front
    - If any oncoming vehicle/any other vehicle causing overtake infeasibility passes, overtaking feasibility is reassessed

---

Frequently, a slower vehicle may lie in front of a faster vehicle and may block out the way for the faster vehicle. In such a context, the faster vehicle may want to overtake the slower vehicle ahead. The faster vehicle should be able to overtake if there is enough space. In such a case, the faster vehicle would *detect the slower vehicle as an obstacle* and avoid it. A steer would be initiated, which would grow sufficiently till the slower vehicle was completely behind the faster vehicle. Postovertaking, a return to the original lane may take place if the vehicle used the opposite side of the road for overtaking. Unfortunately, if there was another slow vehicle in the possible overtaking lane, or an oncoming vehicle in the case of one-lane roads, then overtaking at that time would not be feasible. In such a case, it is expected for the vehicle to follow the slower vehicle in front for a while at least, correcting its speed and orientation.

The feasibility of overtaking can be decided on purely on the basis of the speeds and positions of the vehicles involved, which are assumed to be known for all the vehicles in the scenario. The positions are used to estimate the distance of the vehicles from the left and right boundaries. It is further assumed that the vehicles continue to travel straight with their same speeds if overtaking is attempted. Because the decision is to be made in real time, it may not be possible to simulate the system in any way to decide on the feasibility. Hence, an assumption is made that *the road is not wide enough to accommodate more than three vehicles simultaneously.*

Let the vehicle initiating the overtaking be V_i. Let the vehicle being overtaken be V_j. For the overtaking to be feasible, it is mandatory that V_i should be able to come with a trajectory such that while overtaking it does not collide with V_j. Further, it is mandatory that during the motion V_i should not collide with any other vehicle V_x in the scenario. The feasibility is given by Eq. [10.18].

$$\text{Feasible} = \text{Feasible}_2(V_i, V_j) \wedge_{x, x \neq V_i, V_j. V_x \text{ ahead of } V_i} \text{Feasible}_3(V_i, V_j, V_x)$$

$$[10.18]$$

$\text{Feasible}_2 (V_i, V_j)$ is a measure of the possibility of overtaking being performed between a faster V_i and a slower V_j ahead. For this it is essential that the road must be wide enough to accommodate both V_i and V_j during the overtaking period with some lateral safety margin φ on both sides along the width of the road. The point of overtaking may approximately be taken as the current location of V_j. Let the length of the two vehicles be l_i and l_j, widths be w_i and w_j and let the width of the road be W. The feasibility is given by Eq. [10.19].

$$\text{Feasible}_2(V_i, V_j) = w_i + w_j + 2\varphi \leq W \qquad\qquad [10.19]$$

$\text{Feasible}_3(V_i, V_j, V_x)$ measures the feasibility of overtaking by considering the three vehicles V_i, V_j and V_x ($\neq V_i, V_j$). For overtaking to be feasible, it is necessary that the road be wide enough that the three vehicles may all fit in easily along with safety margins on all sides. Alternatively, the longitudinal separation along the length of the road must be large enough that V_i can overtake V_j while V_x is not yet in the overtaking zone. Although the lateral separation condition simply specifies the road to be wide enough, the relative speeds and distances need to be measured for formulating the feasibility condition due to longitudinal separation. The total time taken for overtaking is given by Eq. [10.20].

$$t = \frac{\|L_i - L_j\|_L + l_i/2 + l_j/2}{S_i - S_j}$$ [10.20]

Here, $\|.\|_L$ measures the distances between the vehicles in the longitudinal direction, along the length of the road.

In case V_x is travelling in the same direction as V_i and V_j (all inbound or all outbound), the relative speed of V_x with respect to V_j is given by $S_x - S_j$. In case V_x is travelling in the opposite direction to V_i and V_j (overtaking is being attempted partly using the wrong side) the relative speed is given by $S_x + S_j$. The distance travelled by the vehicle V_x relative to the vehicle V_j may, hence, be given by Eq. [10.21].

$$D = \begin{cases} (S_j - S_x)t & \text{same side of travel} \\ (S_j + S_x)t & \text{opposite side of travel} \end{cases}$$ [10.21]

The vehicles are stated to be longitudinally well apart to avoid collision if the final separation between the vehicles V_j and V_x is large enough to sufficiently accommodate the complete length of the overtaking vehicle V_i. The net feasibility is then given by Eq. [10.22].

$$\text{Feasible}_3(V_i, V_j, V_x) = w_i + w_j + w_x + 6\varphi \leq W \vee \|L_j - L_x\|_L - D > l_i + 2\Delta$$ [10.22]

here, φ and Δ are the safety distances in the lateral and longitudinal directions. The various notations are illustrated in Fig. 10.9 for the case in which the width of the road is enough to enable overtaking for the three vehicles and Fig. 10.10 for the case in

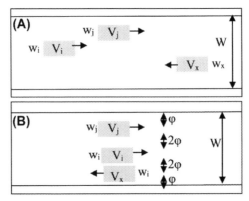

Figure 10.9 Notations for deciding feasibility of overtaking when three vehicles can be accommodated within the road width. (A) Initial positions, (B) final positions.

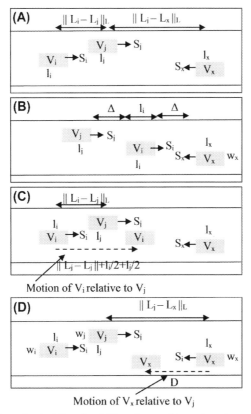

Figure 10.10 Notations for deciding feasibility of overtaking when three vehicles have enough longitudinal separation. (A) Initial positions, (B) final positions, (C) motion on V_i relative to V_j, (D) motion on V_x relative to V_j.

which the longitudinal separation of the three vehicles is enough to enable the overtaking to take place.

If overtaking turns out to be feasible, the next task is to *decide on the motion of each of the participating vehicles*. It is first decided whether V_i should preferably overtake V_j on its left side or right side. This is done by analysing the distance that V_i would be required to move for both the cases and the lesser one is used. The preferred direction of motion of V_j is set to the converse of that of V_i. For V_i the computed direction is given as the input to the *steer* variable of the fuzzy planner (I_i^5). For V_j, the computed direction is given as input to the *requested steer* variable (I_i^7). The task needs to be performed for all other vehicles with whom the feasibility was checked. First, it is checked if there is any requirement of the other vehicles to steer in a specific manner. In case there is no need to steer, V_x simply acts as an agent to check feasibility and the entire overtaking procedure is said to be noncooperative. However, in case V_x needs to move to give some extra space to the overtaking vehicle, overtaking is referred to as being cooperative. Overtaking is, hence, looked at as being cooperative or noncooperative.

Overtaking may be *cooperative* only when the longitudinal distance computed was not large enough. In case the road width is large enough, the vehicles would treat each other as obstacles and steer automatically. When the steer is cooperative, V_x steers in the direction opposite to the direction of V_i and V_j. The steer is uncooperative if there is enough space for both the vehicles to drive through without the movement of V_x. Consider the case that V_x is travelling in the opposite direction to V_i and V_j. The need for cooperation is given by Eqs [10.23] and [10.24].

$$\text{Coop}_1(V_i, V_j, V_x) = \left(e_x^l - w_x/2 < e_i^r + w_i/2 + \varphi \vee e_x^l - w_x/2 \right.$$
$$\left. < e_j^r + w_j/2 + \varphi \right) \wedge \left\| L_j - L_x \right\|_L - D < l_i + 2\Delta \qquad [10.23]$$

$$\text{Coop}_2(V_i, V_j, V_x) = \left(e_x^r - w_x/2 < e_i^r + w_i/2 + \varphi \vee e_x^r - w_x/2 \right.$$
$$\left. < e_i^l + w_j/2 + \varphi \right) \wedge \left\| L_j - L_x \right\|_L - D < l_i + 2\Delta \qquad [10.24]$$

Here e_x^l denotes the distance of the centre of V_x from the left boundary (relative to itself) and e_x^r denotes the distance of the centre of V_x from the right boundary (relative to itself). The closest distances to the vehicle from the boundaries are approximated by adding/subtracting half the vehicle's width. Overtaking is cooperative if either Eq. [10.23] or Eq. [10.24] computes to be true. In case Eq. [10.23] is true, the steer is cooperative and R_x is expected to steer towards its right. In case Eq. [10.24] is true, R_x is expected to steer left. The various notations are given in Fig. 10.11 for Eq. [10.23] and Fig. 10.12 for Eq. [10.24]. The notations of V_j (not shown in the figures) are similar to those of V_i.

On many occasions a faster vehicle may be forced to slow down and *follow a slower vehicle* if overtaking is not immediately possible. The above equations allow

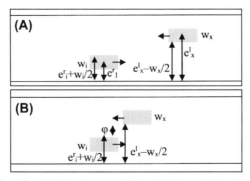

Figure 10.11 Notations for deciding the turn of V_x with V_i overtaking V_j. Case involving right turn. (A) Initial positions, (B) overtake positions.

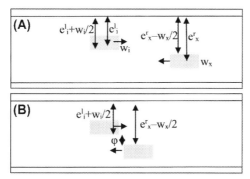

Figure 10.12 Notations for deciding the turn of V_x with V_i overtaking V_j. Case involving left turn. (A) Initial positions, (B) overtake positions.

overtaking to take place only if the vehicle initiating overtaking has a higher speed than the vehicle being overtaken. In such a scenario, overtaking still may not necessarily take place. For this an exception is placed that a vehicle capable of overtaking may overtake if no other vehicle lies ahead in the vicinity. A vehicle is capable of overtaking if its maximum speed is greater than the speed of the slower vehicle ahead. The condition Feasable$_2$(V_i, V_j) must, however, be true. The two vehicles would move in the opposite directions to enable overtaking. V_i in its course of overtaking would increase its speed. While overtaking is being initiated, the left-side-preferred rule is deactivated (by always giving an input of 0 to the corresponding fuzzy input I_i^6) as well as the front obstacle distance is considered infinitely large. This, however, is only for the initiation of overtaking. Once the vehicle has laterally crossed the vehicle being overtaken — in other words, when the two vehicles are in different lanes — both the features are again enabled.

The overtaking mechanism employed here is inspired by the general driving of today. If a vehicle sees a possibility of overtaking which requires the cooperation of other vehicles, it sends a signal in the form of a horn or the overtaking gesture is visually perceived by the other vehicles around. The other vehicles, hence, know about the overtaking attempt and can align themselves in a manner to allow overtaking to take place.

10.4.3 Going Through a Crossing

The last feature of the algorithm is the mechanism of passing over a road *crossing*. While approaching a crossing, the vehicle knows the next road to take. Virtual barricades are drawn over the road on which the vehicle is not supposed to move. This effectively converts the crossing region into a continuous (maybe curved) road segment bounded by real and virtual boundaries at both ends. Crossings with roundabouts are much simpler where such boundaries are prominent. This is shown by Fig. 10.13.

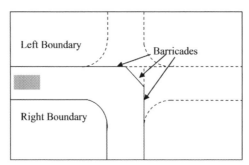

Figure 10.13 Use of virtual barricades for crossing scenario.

Coordination of vehicles in a crossing region is of importance because a simultaneous attempt to pass through the crossing by vehicles from different sides could lead to a deadlock. Traffic lights, if present, carry out this task of coordination, in which vehicles are supposed to simply obey the traffic lights. For crossings without lights, a first-come-first-served system of motion is maintained. Vehicles are allowed to cross through strictly in the order of their arrival at the crossing. This ensures a fair distribution of crossing amongst the vehicles. A vehicle, on being near a crossing, checks the possibility to cross through. If allowed to cross, normal motion continues, else the vehicle may have to slow down, stop and wait to be allowed to cross. When allowed, by its normal motion it crosses to connect to the desired road.

The other major restriction while at or near a crossing is that the overtaking module is completely disabled and vehicles need to follow each other, ie, no overtaking is allowed over a road crossing. Further, no vehicle may ask another vehicle to itself steer left or right. Allowing overtaking at or near a crossing scenario might result in travelling over the stop lines if overtaking is not completed in time or conversely stopping in an awkward pose blocking the way for the incoming vehicles.

10.5 Evolution of the Fuzzy Inference System

The fuzzy system with all the characteristics discussed in Section 10.4 was initially manually designed by heuristics and trial and error. It is natural that a complete *evolutionary system* may be capable of generation of a better fuzzy system, but there are heavy computational requirements and the overall problem of mapping the inputs to the outputs is very complex. Hence, first a prototype system is made using general human logic. The system was simulated on multiple paths with anomalies in the motion being noted and iteratively rectified by tuning or adding membership functions and rules. This was done until a fuzzy system was obtained that performs well over a diverse set of scenarios.

Evolution is then restricted around the prototype fuzzy system so produced. Evolution is used for the tuning of the rules and membership functions. Evolution is carried forwards in cycles. First, the evolutionary process tried to tune the rules for some generations. This produces a fuzzy system that has better rules. The next task is to optimize the membership function as per the requirements of the new set of rules generated. This is done by the evolutionary process for the next few generations. The evolutionary process further tries to again tune the rules as per the tuned membership functions. This complete cycle of tuning the membership functions and rules one after the other is carried out for some cycles. The *search space* of each of the two evolutionary processes is limited to enable some optimization in a massively complex search space. Each of the two evolutionary processes thus only returns the optimal solution in a small neighbourhood. At every cycle the location of the so far best fuzzy system changes by small amounts and, hence, is in the neighbourhood of the next evolutionary search step. Even though generation of the optimal fuzzy system is beyond computational limits, such an evolutionary technique can make the fuzzy prototype proceed towards a near-optimal solution. The complete evolutionary process is summarized in Fig. 10.14.

Tuning of the rules is the first half of the problem. A large number of rules are, however, possible with different combinations of membership functions in the antecedents and the consequents. The entire search space is, hence, very complicated. A very small part of this search space is extracted to make the search space simpler. Evolution is allowed only to increase or decrease the participating membership function of any antecedent or consequent in the best fuzzy system (so far) by a unit value. Hence, in the best fuzzy system for any rule, if an antecedent/consequent has a participating membership function 'high', the evolutionary process can change it to 'low' or 'very high', provided both are existent. The genotype consisted of a string of integers which may be either $-1, 0$ or $+1$. In all there are $R(I + O)$ integers. Here, R is the number of rules, I is the number of inputs and O is the number of outputs. These are in

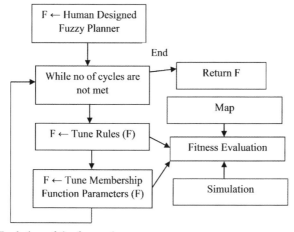

Figure 10.14 Evolution of the fuzzy planner.

distinctive sets of $I + O$ integers denoting a rule, each integer representing an antecedent or a consequent in order. A -1 means the corresponding input's/output's membership function is to be decreased by 1. Similarly, a $+1$ means that an input/output's membership function needs to be increased by 1. The presence of a zero means the membership function must remain unchanged. The negative membership function numbers denote use of a NOT operator.

Optimization of the membership functions is the second task. Here, as well the complete search domain is quite large. The various rules run over the symbolic meanings of the membership functions. Hence, it is preferable that in the entire optimization process the various membership functions retain their symbolic meanings. The rules may further be very sensitive to the locations and the shapes of the membership functions. Hence, again, to keep the fitness landscape limited, the search domain is restricted to a small percentage around the present value. The individual representation in this case is simple. All the membership functions are noted and their parameters appended one after the other.

Conventional genetic operators are used in both the evolution of the rules and the membership functions. The choice for the evolution is rank-based fitness scaling, stochastic universal selection, scattered crossover, Gaussian mutation and elite operator.

The last part left for the implementation of the evolutionary technique is formulation of the *fitness function* which is common for both the techniques. The evolution algorithm tries to generate the best possible trajectories for the vehicles. Evolution is performed on a set of *benchmark maps*. These maps cover a wide variety of scenarios that the vehicle may face in real life. The fitness is simply the mean performance of the vehicle in all these maps. It is aspired to make the vehicles travel the path in as little time as possible. The system is further penalized for any irregularities or mishap. Penalties are added for not maintaining safe distances from the front and side, for colliding and for driving on the wrong side of the road. The entire fitness function for any individual is given by Eq. [10.25].

$$\text{fit} = \cfrac{\sum\limits_{i=1}^{np} \left. \cfrac{\sum\limits_{j=1}^{nv_i}(t_{ij}+p_1 \cdot \text{collide}_{ij}+p_2 \cdot (\text{safe}1_{ij}+\text{safe}2_{ij})+p_3 \cdot \text{side}_{ij})}{np} \middle/ nv_i \right.}{np} \qquad [10.25]$$

here, np is the number of maps used for learning and nv_i is the number of vehicles used in the ith map. For the ith map and the jth vehicle in the map, t_{ij} is the time taken to travel; collide$_{ij}$ is the distance left for the vehicle to travel if the vehicle collides before completing the journey, 0 otherwise; safe1_{ij} and safe2_{ij} are the margins by which the front and side safety margins are disobeyed; and side$_{ij}$ is the average distance that the vehicle stays in the wrong side. p_1, p_2 and p_3 are the penalty constants. To compute these factors, the Eqs [10.12−10.14] and [10.16] are extended to compute the fuzzy inputs. safe1_{ij} and safe1_{ij} are given by Eq. [10.26] and side$_{ij}$ is given by Eq. [10.27].

$$\text{safe1}_{ij}(t) = \max(F_s - \min\{d_{11}, d_{12}\}, F_s - \min\{d_{21}, d_{22}\}, 0)$$

$$\text{safe1}_{ij} = \max_t(\text{safe1}_{ij}(t))$$

$$\text{safe2}_{ij}(t) = \max(F_d - \min\{d_{31}, d_{32}\}, 0)$$

$$\text{safe2}_{ij} = \max_t(\text{safe2}_{ij}(t))$$

[10.26]

$$\text{side}_{ij} = \frac{\int_{t=0}^{t_{ij}} I_6^i(t) \cdot dt}{t_{ij}}$$

[10.27]

10.6 Results

The problem in its entirety with the mentioned modelling scenarios, constraints and assumptions was tested using a simulation tool. The entire tool was developed in MATLAB. The simulation engine took a bitmap (BMP) image as an input. The various modules of the software developed included boundary determination, boundary tracking, distance computation, fusion of the different vehicles in the map, crossing detection and implementation of the move. The decision-making modules included both overtaking and obstacle avoidance mechanisms. A separate module was made for scenario specification including initial positions, speeds and orientations, speed thresholds and time of entry into the scenario. The algorithm parameters were specified in another similar module. The fuzzy planner was designed separately and included in the main function.

The prototype fuzzy system was manually designed. For the optimization of the prototype fuzzy system, five cycles were applied with a population size of 20 individuals and 20 generations as the stopping criterion. The crossover rate was 0.8 and an elite count of 2 was used. This produced the final fuzzy planner used for the simulations. The next task was to generate a number of scenarios and to simulate a group of vehicles. Each of the scenarios is discussed in the following subsections.

10.6.1 Single-Vehicle Scenarios

First, the system was simulated with a single vehicle. A number of roads of different curve directions, curve steepness, straightness etc. are generated. The vehicle needed to move in a manner to maintain an ideal speed and steering. The first path consisted of a general path with a slight leftwards curve. The vehicle was generated in the middle of the road with zero speed at an angle of 0 degree, which was not the direction at which the road was inclined. This meant it must adjust its speed and steering to first align itself with the direction of road, at the same time attempting to move to the left side of the road, accelerate from zero speed and then prepare to turn left. The trajectory tracked by the vehicle is shown in Fig. 10.15A. In the second scenario, the turn is made steeper and closer to the vehicle's origin. The trajectory traced is given in

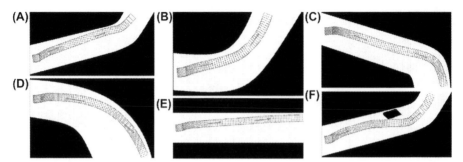

Figure 10.15 Simulation results with one vehicle. (A,B) Left turning road, (C,D) Right turning road, (E) Straight Road, (F) Road with obstacle.

Fig. 10.15B. In this case, it may be observed that the vehicle managed its acceleration to maintain a match between speed and side. Higher speeds would only be possible on the right side of the road, but this would defy the left-hand-side driving rule, at the same time making a possibility of collision with the right boundary.

In the next couple of scenarios the road was turned right in place of left. Now the vehicle had to be able to keep control over both speed and steering. Higher speeds might have been possible by sticking to the left boundary, where the paths would have been smaller keeping as far right as possible. The complete trajectory of the vehicles for the two scenarios is given in Fig. 10.15C and D. The next scenario was on a straight road. This situation was different as the vehicle was generated in the correct orientation, but on the wrong side of the road. The vehicle first concentrated on correcting its side, followed by accelerating and orienting along the road. The corresponding path is shown in Fig. 10.15E.

The next scenario in the approach was with the presence of an obstacle in the path of the vehicle. The vehicle was generated in the middle of the road, which meant the task was to correct the vehicular orientation, correct the side of travel, make a turn and avoid the obstacle. Initially, the obstacle was not in the visibility range of the vehicle which made the motion similar to that seen in the first scenario. This was because of the current orientation of the vehicle. The vehicle then saw the obstacle and was between two contradictory objectives of correcting the side and moving to avoid the obstacle. Because the obstacle was not in the immediate visibility of the vehicle and because the road was turning, the moment between obstacle discovery and obstacle avoidance was small. The vehicle was nevertheless still able to avoid the obstacle. Whilst doing so it needed to turn farther on the curved path and correct its side of motion. The trajectory is shown in Fig. 10.15F.

10.6.2 Double-Vehicle Scenarios

The next scenarios were with two vehicles travelling on opposite sides of the road, heading directly towards each other. The same maps were taken as were used in the earlier cases. The challenge for either of the moving vehicles was, hence, increased.

They needed to orient themselves as per the requirement of the road, adjust their speeds, move to the correct side and avoid each other on the way. In all the simulations the vehicles were generated with zero speed. In the first scenario, it was observed that the second vehicle, generated at the top of the map, had to take a steep turn, correct the side, orient itself and accelerate simultaneously. It chose to maintain a slow speed, thereby avoiding any potential collision with its left boundary. This, however, made the point of turn a potential collision point of the vehicles, thereby making it more difficult to avoid the collision. However, a collision in this case was primarily avoided due to the left-side rule, in which the vehicles were aligned towards their respective left side of the road. The trajectory is given in Fig. 10.16A.

In the next scenario, however, a mere alignment was not enough. The possible point of collision in this case happened just after the turn. While making the turn the first vehicle, generated in the left part of the map, had an intention to drive straight as it slowly corrected its orientation with respect to the road. However, on discovering the presence of the other vehicle, the two vehicles oriented against each other and moved on. The trajectories are given by Fig. 10.16B. A similar trend was observed in the next scenario in which the vehicle being generated at the bottom of the map tried to accelerate and slowly orient itself in the turn; however, in doing so it moved on the other side on discovering the other vehicle coming towards it. The trajectory for this vehicle is given in Fig. 10.16C and D. Planning in the case of straight roads was simple. The two vehicles moved to their own sides and passed each other easily. This trajectory is given in Fig. 10.16E.

10.6.3 Overtaking

The next set of simulations was used to test the overtaking mechanism of the vehicles. The overtaking of two vehicles and three vehicles were checked independently. In the first scenario, a faster vehicle was generated and made to move on a straight road. After some time a slower vehicle was then produced and was made to move on the same road, behind the faster vehicle. In this case, there was no possibility of overtaking taking place. The trajectories of the vehicles were exactly alike and the slower vehicle merely followed the faster vehicle. The trajectory of the vehicles is given in Fig. 10.17A. The situation, however, reversed when the speeds of the two vehicles were reversed. Now first, the slower vehicle was generated and allowed to move on the map, and later the faster vehicle was generated and subsequently also moved on the map. In this case, it may be easily seen that the faster vehicle soon saw the possibility of overtaking and initiated the same. The two vehicles steered in opposite (widthwise) directions to each other till there was sufficient distance for the faster vehicle to overtake the slower vehicle. The other task was that the faster vehicle needed initial acceleration as it entered the scenario with a speed of zero. It can be seen that the vehicles easily overtake each other. After the faster vehicle completely lies ahead of the slower vehicle, it returns to the original lane ahead of the slower vehicle. The trajectory of the two vehicles is shown in Fig. 10.17B.

The next experiments involve three vehicles. Here, the experimental setup was done to check the working of the overtaking feasibility equations. First, two vehicles entered the scenario at the start and travelled in opposing directions. The third vehicle was

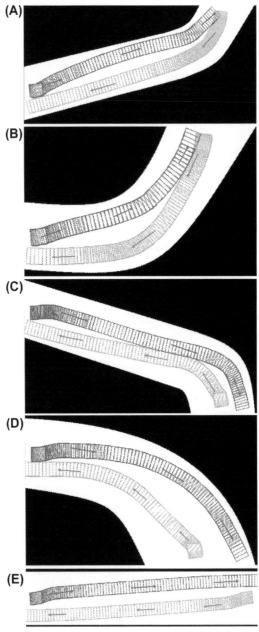

Figure 10.16 Simulation results with two vehicles. (A,B) Left turning road, (C,D) Right turning road, (E) Straight Road.

made to enter after some time. In the first scenario, a map was considered such that the road width was enough to accommodate the three vehicles with sufficient margins. In such a case overtaking should have been feasible. This was observed when the third vehicle was made to enter with a high velocity. Overtaking needed to be taken in a

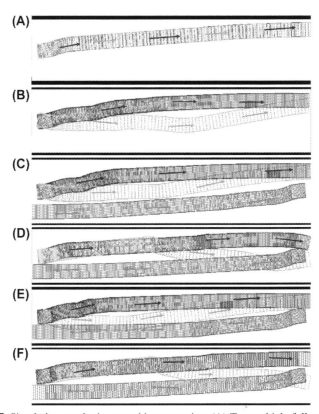

Figure 10.17 Simulation results in overtaking scenarios. (A) Two vehicle following scenario, (B) Two vehicle overtaking scenario, (C) Three vehicle overtaking scenario with large lateral separation, (D) Three vehicle following scenario with small lateral separation, (E) Three vehicle overtaking scenario with large longitudinal separation, (F) Three vehicle following scenario with small longitudinal separation.

cooperative mode and accordingly all three vehicles oriented themselves. Overtaking of the vehicle was successfully carried out by the overtaking vehicle, which later returned to its normal lane. The vehicles could have further adjusted their speeds and steering to make overtaking possible even in the case that the margin of overtaking was narrower. The trajectories of the vehicles are shown in Fig. 10.17C. The same experiment was repeated. But this time the width of the road was decreased. In this case, the vehicle decided not to overtake straightaway, but instead it initially followed the slower vehicle in front. This indeed wasted time for the faster vehicle which made the best attempt to adjust its speed to follow the slower vehicle. However, as soon as the oncoming vehicle passed by, the faster vehicle initiated an overtaking procedure which was safely executed. The trajectories are shown in Fig. 10.17D.

The next attempt was to test the overtaking decision made when the longitudinal separation between the vehicles was large enough. The same map with a narrow-width road was taken to ensure that overtaking was always returned as infeasible by the lateral distance factor. This left decision making entirely on the basis of the longitudinal distance between the vehicles. A number of speed combinations of the vehicles were tried. The cases revolving around the most critical overtakings with the smallest margins are presented. First, the case was examined in which overtaking was regarded as feasible. Here, the vehicle followed its own overtaking trajectory and overtaking was performed. The trajectories and the path traced are shown in Fig. 10.17E. The same experiment was repeated with the speeds of the two vehicles initially in the scenario slightly increased. In this case, overtaking was infeasible and the faster vehicle decided to follow the slower vehicle. Once the oncoming vehicle passed, overtaking was carried out. The trajectories are shown in Fig. 10.17F.

10.6.4 Crossing

The last simulation was applied to a crossing scenario. A crossing of four roads was considered. The attempt was to generate a complex crossing scenario with multiple vehicles all around. A total of eight vehicles were generated, two on each road. All these vehicles were initially travelling at high speed, which they had to reduce slowly after they entered the map. Motion of the vehicles was on a first-come-first-served basis. Hence, all vehicles had to stop except the first one which was allowed to pass through the crossing unhindered. First, four vehicles, one on each road, were made to enter the map. The other four vehicles entered after some time. These vehicles perceived the moving vehicle in the front as an obstacle and decreased their speeds. Each vehicle moved as per the schedule for crossing. The simulation showcased that all vehicles were able to pass each other in a reasonably short crossing time. The resultant path traced by the various vehicles is given in Fig. 10.18.

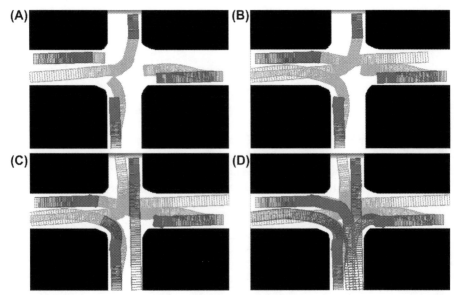

Figure 10.18 Simulation results in crossing scenario. (A) After one vehicle crossing, (B) After three vehicles crossing, (C) After five vehicles crossing, (D) After eight vehicles crossing.

10.6.5 Analysis

The next step involved with the algorithm was to analyse the working of the various mechanisms in the various scenarios. First, speed control was considered. For the study of this mechanism, the speed of the vehicle was captured at various stages in time and plotted graphically for a number of maps. The first class of maps involved simple maps with a single vehicle in the path. In all these maps, it was expected that the vehicle accelerates as fast as possible and then maintains its maximum speed as it traverses on its way. All units used were arbitrary and specific to the simulation tool. For all the experiments, the maximum speed was fixed at 10. The speed of the vehicle for various scenarios is shown in Fig. 10.19A. Based on the figure, it may be easily seen that the greatest acceleration was recorded by the scenarios in which the vehicle was supposed to turn left. This is because in none of the scenarios the vehicle could drive straight without turns as it was generated on the wrong side. The simultaneous actions of accelerating and correcting the side of travel put a threshold on the acceleration.

In the roads having a left turn, the leftwards-turning boundary gives an additional distance for the vehicle to accelerate, making the vehicle conscious of the left boundary by a lesser amount. Hence, the acceleration was higher. The acceleration was more on the map in which the curve was earlier on the map (scenario 2) for the same reasons. Similarly, it may be seen that the scenarios with a right turn made vehicles lower their speed to adjust to the curve and then again accelerate. This lowering of the speed enabled them to avoid potential accidents with their left end of the road as well as to maintain a safe distance on the left. Once they steered on their turning path to

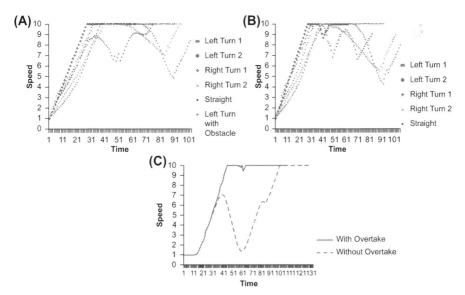

Figure 10.19 Relation between speed and time for (A) scenarios with one vehicle, (B) scenarios with two vehicles and (C) overtaking scenarios.

make enough safe distance, speeds were increased. Deceleration is more for the steeper turn which was available at the end of the path (scenario 3). This is because the vehicle needed to be slower to adjust to the sharper turn, as compared to a more uniform turn. The scenario involving an obstacle showed a decrease and increase in speed in the central part of the journey. This enabled the vehicle to overcome the obstacle at a lower speed, upon its sudden emergence.

The study was further extended to cases involving two vehicles. The corresponding graph is shown in Fig. 10.19B. The graph clearly shows that the presence of the other vehicle produced a change in the curves of all the scenarios, except for the one on a straight road; in which the two vehicles moved to the correct side early to completely avoid each other. For the vehicle turning left on an early uniform turn (scenario 2), there is a visible region in which the speed was reduced to some degree to avoid the other vehicle. The other scenario with a left turn (scenario 1) shows two regions in which the speed was dropped. The first drop was when the other vehicle was initially sighted (then at a far-off point) and accordingly the vehicle adjusted itself. The velocity was later increased. The second drop of speed was when the vehicle was actually being avoided on the turn. After the vehicle was completely avoided, the speed was further increased. For the right-turning path scenarios, it was seen that deceleration was done to an even greater extent as compared to the cases with a single vehicle. This further decrease of the speed was due to the presence of the other vehicle. Further, in both cases the vehicles were encountered more or less at the turns.

The last case of overtaking was between three vehicles. The manner, in how the speed changes, was examined while a vehicle was overtaking. Here as well, two scenarios were considered. The first scenario was when the vehicle was allowed to overtake and the second when the vehicle was not allowed to overtake. The speed of the overtaking vehicle in both these scenarios is presented in Fig. 10.19C. The figure shows that the vehicle more or less assumed the maximum speed, which means that overtaking happened at high speed. A small reduction in speed can be seen when the overtaking was just about to end, as the corners of the vehicle overtaking and the oncoming vehicle had come quite close;, however, as soon as they were avoided, the speed was again increased. However, in case overtaking was not allowed, the vehicle was supposed to reduce its speed and follow the vehicle ahead for some time, till the oncoming vehicle passed by and overtaking could be achieved. In this case, the decrease and reincrease in speed can be seen when overtaking was computed as infeasible. Later, the vehicle increased the speed during overtaking. The speed was in fact increased till it reached its maximum value.

The other part of the analysis was to measure the angle between the orientation of the vehicle and the orientation of the road. For normal driving this angle should be zero. This is the case when the vehicle aligns itself with the road and then drives forwards at its maximum possible speed. Measurement of this angle gives an idea of the performance of the steering mechanism of the system. The angle for various scenarios with a single vehicle is shown in Fig. 10.20A. To avoid the graph being too congested, as the various scenarios revolve around the same region around zero, only single cases of left and right turn have been plotted. These correspond to scenarios 1 and 3 as discussed. The figure shows that for most cases the angle revolves

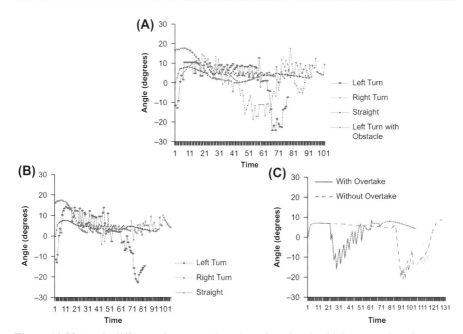

Figure 10.20 Angle difference between orientation of road and vehicle versus time for (A) scenarios with one vehicle, (B) scenarios with two vehicles and (C) overtaking scenarios.

around zero. There are some oscillations visible to the magnitude of about 5 degrees. These are produced and controlled by the steering mechanism, while trying to maintain an angle of perfect 0 degree. For the case of a left turn, a sharp increase in the angle is visible, which is eventually eliminated. This is produced by the sudden emergence of the turn. For the right turn, the angle quickly approaches zero and this is maintained within the small area around zero. For a straight road, the initial increase of the angle is due to the attempt to return to the correct side of the road. The last few oscillations on the straight road are when the vehicle tries to completely align itself with the road, producing a small error which the steering mechanism attempts to correct. For the scenario with an obstacle, the deviation increases even further at the moment the obstacle is being avoided. The angle completely reverses when the vehicle completes the obstacle avoidance. Then the steering mechanism tries to correctly orient the vehicle.

The same kinds of observations were visible when the experiments were repeated for the two-vehicle scenarios. The plot of angle for different times is given in Fig. 10.20B. Because the plots shown measure the difference between the orientation of the road and the orientation of the angle of the vehicle, the presence or absence of the other vehicle has no direct influence over the measurement of the angle. The similar nature of the plots indicates that the vehicles could perform some minor steering mechanisms to avoid the other vehicle. No major turn was taken that drastically changed the angle recordings from the previous experiment.

The trend, however, was not the same with overtaking. The scenario of overtaking with three vehicles was studied, when the overtaking vehicle attempted to overtake the vehicle in front of it. The plots for angle with time for the overtaking vehicle are given in Fig. 10.20C. It may be seen that the angle decreased by a large amount and then increased denoting the overtaking. The decrease of the angle below zero was when the vehicle initiated the overtaking and went on the other side of the vehicle being overtaken. Afterwards, the angle converged on zero, denoting the attempt to orient the vehicle with the road, while the vehicle being overtaken was still to the side. After overtaking, the vehicle attempted to come back to its original side, for which the angle became positive, which was controlled and brought towards zero. A similar trend may be observed for both the curves. It must be noted that although the title reads without overtaking, this means that overtaking did not happen while the oncoming vehicle was in front. In both cases, overtaking did take place, but at different times as can be seen from the figure.

10.7 Summary

The basic motivation is to plan autonomous vehicles for real-time traffic scenarios. The intention is to make a robust planning system for vehicles that can perform in cases of narrow roads with high diversity of traffic in terms of the travelling speeds with no pre-defined lanes. In such a scenario, driving becomes a difficult task. This chapter was dedicated to the use of reactive planning techniques, particularly fuzzy logic, for the planning of autonomous vehicles.

The fuzzy logic-based approach assumed the traffic on both sides could mingle and a vehicle could use a part of the lane normally used for the vehicles from the other side. Motivated by the basic instincts of professional drivers who drive in such scenarios, fuzzy logic was used for the design of the system. The fuzzy system took as input the status of the surroundings of the vehicle. The fuzzy planner controlled both the steering mechanism as well as the speed. The decision regarding overtaking was based on the projected motions of the other vehicles. Overtake was called feasible if the potentially colliding vehicles could fit themselves laterally on the road or overtake could be completed before any potential collision with an oncoming vehicle.

The fuzzy rules carried the task of understanding of the environment and producing the necessary control actions. The rules tried to increase the distance from a vehicle on all sides to obstacles and to orient the vehicle as per the road orientation. The fuzzy system was developed and used for planning for a number of roads that ranged from straight roads to roads with steep or smooth turns on either side. In all the cases, it was observed that the vehicle could drive itself as per the situations and follow the path.

References

Alvarez-Sanchez, J.R., de la Paz Lopez, F., Troncoso, J.M.C., de Santos Sierra, D., 2010. Reactive navigation in real environments using partial center of area method. Robotics and Autonomous Systems 58 (12), 1231−1237.

Fiorini, P., Shiller, Z., 1998. Motion planning in dynamic environments using velocity obstacles. The International Journal of Robotics Research 17 (7), 760−772.

Jolly, K.G., Kumar, R.S., Vijayakumar, R., 2010. Intelligent task planning and action selection of a mobile robot in a multi-agent system through a fuzzy neural network approach. Engineering Applications of Artificial Intelligence 23 (6), 923−933.

Kala, R., 2014. Navigating multiple mobile robots without direct communication. International Journal of Intelligent Systems 29 (8), 767−786.

Kala, R., Shukla, A., Tiwari, R., 2010. Fusion of probabilistic A* algorithm and fuzzy inference system for robotic path planning. Artificial Intelligence Review 33 (4), 275−306.

Kala, R., Warwick, K., 2015. Reactive Planning of Autonomous Vehicles for Traffic Scenarios, Electronics 4 (4), 739−762.

Motlagh, O., Tang, S.H., Ismail, N., Ramli, A.R., 2012. An expert fuzzy cognitive map for reactive navigation of mobile robots. Fuzzy Sets and Systems 201, 105−121.

Rashid, A.T., Ali, A.A., Frasca, M., Fortuna, L., 2012. Multi-robot collision-free navigation based on reciprocal orientation. Robotics and Autonomous Systems 60 (10), 1221−1230.

Ross, T.J., 2010. Fuzzy Logic: With Engineering Applications, third ed. Wiley, West Sussex, UK.

Russell, S., Norvig, P., 2014. Artificial Intelligence: A Modern Approach, third ed. Pearson, Essex, England.

Selekwa, M.F., Dunlap, D.D., Shi, D., Collins Jr., E.G., 2008. Robot navigation in very cluttered environments by preference-based fuzzy behaviors. Robotics and Autonomous Systems 56 (3), 231−246.

Sezer, V., Gokasan, M., 2012. A novel obstacle avoidance algorithm: follow the gap method. Robotics and Autonomous Systems 60 (9), 1123−1134.

Sgorbissa, A., Zaccaria, R., 2012. Planning and obstacle avoidance in mobile robotics. Robotics and Autonomous Systems 60 (4), 628−638.

Shukla, A., Tiwari, R., Kala, R., 2010. Towards Hybrid and Adaptive Computing: A Perspective. Studies in Computational Intelligence. Springer-Verlag Berlin, Heidelberg.

Snape, J., van den Berg, J., Guy, S.J., Manocha, D., 2011. The hybrid reciprocal velocity obstacle. IEEE Transactions on Robotics 27 (4), 696−706.

Tiwari, R., Shukla, A., Kala, R., 2013. Intelligent Planning for Mobile Robotics: Algorithmic Approaches. IGI Global Publishers, Hershey, PA.

Ye, C., Webb, P., 2009. A sub goal seeking approach for reactive navigation in complex unknown environments. Robotics and Autonomous Systems 57 (9), 877−888.

Potential-Based Planning

11

11.1 Introduction

Reactive planning techniques (Tiwari et al., 2013) form the simplest and the most natural way to solve motion-planning problems, especially in real-time scenarios like traffic systems. These techniques interpret the current scenario based on a set of very few inputs primarily encoding the distances from the obstacles and goals, and map them to the immediate output or the immediate action based on some simple formulation. The movement changes the environment, which is reinterpreted and correspondingly the next move is taken. The robot moves by taking such steps at each instant of time. The approach is very useful for real-time scenarios, scenarios wherein the environment may change very quickly and scenarios wherein the uncertainties may be very large.

It is very natural to use reactive algorithms for a real-time problem like the navigation of an autonomous vehicle. Essentially, one drives straight on the road while avoiding all obstacles around. Any reactive methodology is competent enough to display such primitive behaviours. An overtaking behaviour and cooperating for overtaking behaviour can be additionally modelled based on the specific choice of the reactive algorithm. Chapter 'Fuzzy-based Planning' proposed the use of *fuzzy logic* as the choice of the reactive system, which performed well over all experimental scenarios. The greatest problem with fuzzy logic is that a rule base needs to be constructed. Even though the rule base gives great flexibility to design the behaviours of the vehicle, for an overly complicated problem like navigation of an autonomous vehicle, the rule base can be very large. Developing and testing the fuzzy inference system by hand is not a very recommendable approach as humans may not be able to put in tremendous amount of time, energy and concentration to design the fuzzy system. Humans face a lot of difficulty when the magnitude of the system gets overly large. The humans may also not be able to consider all the cases, from practical to the rarely occurring ones. There may be a very limited to nonavailability of a data set to automatically train the fuzzy inference system. Manually built data sets, like the approach used in chapter 'Fuzzy-based Planning', question the validity of the approach only to the scenarios well represented in the training data set.

This motivates the use of more intuitive approaches in the design of the reactive systems, even if they come at the cost of some assumptions. An important assumption made in this chapter is that the traffic operates from one side only, all vehicles are either outbound or inbound. This helps to place vehicles anywhere on the road without worrying about the issues of completeness in the absence of static obstacles. In the worst case, the vehicle would end up following the vehicle in front, but would still be moving. Dual carriageways threaten a deadlock when a vehicle moves on the wrong

On-Road Intelligent Vehicles.
Copyright © 2016 Elsevier Inc. All rights reserved.

side for overtaking but faces a vehicle directly in front, the two vehicles disallowing each other to either move or change sides. A solution to this problem will be presented in chapter 'Logic-based Planning'.

This chapter first introduces the concept of *lateral potential* (LP, Sections 11.3–11.5). The *Artificial Potential Field* (APF, Khatib, 1985; Choset et al., 2005) method solves the problem by assuming all obstacles are a source of repulsive potential, with the potential inversely proportional to the distance of the robot from the obstacle. The goal attracts the robot by applying an attractive potential. The derivative of the potential gives the value of the virtual force which is applied to the robot, based on which the robot moves. The force tends to infinity as the robot approaches towards the obstacles, which applies a strong repulsive force away to the robot, meaning theoretically the robot cannot collide with the obstacles, although collisions may be possible due to discretization. The common problem with the method is getting struck at local minima causing an equipotential region, in which the attractive and repulsive forces balance each other out and the robot stands still.

APF cannot be directly applied to the problem as the method is known to cause *oscillations* in narrow passages which is the case for roads. Further, a vehicle following another vehicle is directly repelled backwards, even though it may have been possible to overtake. The solution applied in Section 11.4, hence, departs from the conventional APF modelling and models the problem as an LP problem. The solution measures potentials from a variety of indicators and the potentials are applied laterally. Coordination is modelled as a source of potential.

In chapter 'Fuzzy-based Planning' it was stated that, even though reactive approaches are neither optimal nor complete, there is no major loss in the use of these approaches in most realistic scenarios. Although loss of completeness and optimality is not practically significant in the use of reactive methods, it may be little matter of concern. The best trade-off is hence to use a *little deliberation* to induce near-completeness and near-optimality. In this chapter, this is done by the use of an Elastic Strip. An *elastic strip* consists of a strip which is acted upon by repulsive forces from the obstacles along with an internal spring force. The strip represents the vehicle's trajectory and due to the notion of forces can be quickly given a shape. The chapter explores the manner of quickly producing such a strip. The planning is hence between high-end deliberative methods (as discussed in chapters: Optimization-based Planning, Sampling-based Planning, Graph Search-based Hierarchical Planning and Using Heuristics in Graph Search-based Planning) and reactive planning methods.

The approach (Section 11.6–11.9) uses an elastic strip (Brock and Khatib, 2000) to solve the same problem. In the elastic strip, a trajectory acts as an elastic strip. The strip is repelled by all the obstacles around. The obstacles apply a repulsive potential which moves the strip. The strip has an internal force which attempts to make the strip contract to the shortest strip length. The strip or the trajectory in this manner adapts to the changing obstacles. The strip readjusts every time obstacles move, appear or disappear. The stable shape is when the forces balance each other out. This method is an extension of LPs and solves the problem of optimality and completeness. LPs are used to generate a bouquet of potential plans out of which the best is chosen and continuously optimized using the concept of an elastic strip.

Segments of the chapter including text and figures have been reprinted from Kala and Warwick (2012), IEEE, Proceedings of the 2012 IEEE Intelligent Vehicles Symposium, Alcalá de Henares, Spain, R. Kala, K. Warwick, Planning autonomous vehicles in the absence of speed lanes using lateral potentials, pp. 597–602. © 2012 IEEE.

Segments of the chapter including text and figures have also been reprinted from Kala and Warwick (2013), IEEE, IEEE Transactions on Intelligent Transportation Systems, R. Kala, K. Warwick, Planning autonomous vehicles in the absence of speed lanes using lateral potentials, Vol. 14, No. 4, pp. 1743–1752. © 2013 IEEE.

11.2 A Brief Overview of Literature

Potential methods, due to their ease of implementation, are widely used. These methods are cooperative. The problem of adoption of such approaches in traffic scenarios is that they are not applicable for a narrow road-like structure in which the robot may be found oscillating within the road, not allowing a possible overtake (in case of two robots symmetrically ahead and behind each other) or poorly allowing a possible overtake (in case the two robots are almost symmetrically ahead and behind each other). Further, unlike mobile robots, any turn or lateral movement (lane change) in a vehicle scenario threatens a collision with some other vehicle at the back. Baxter et al. (2006, 2009) used APF for planning of multiple robots. The perceptions were shared amongst robots to correct the sensor errors. Fahimi et al. (2009) used harmonic potential field with the concepts of fluid dynamics. In a similar work, Gayle et al. (2009) modelled the cooperation or interaction amongst the different robots as a social potential.

Ge and Cui (2002) included terms for the velocities of target and obstacles for the computation of the potential, which could be used by the robot to land at the target with a desired speed. Yin et al. (2009) also considered accelerations of the robot, target and obstacles for computing the potential values. Jaradat et al. (2012) used the fuzzy inference system to compute the attractive potential and adaptive neuro-fuzzy inference system to compute the repulsive potential. Taking additional measures is desirable for the road scenario, but the mentioned general limitations hold.

Parhi and Mohanta (2011) solved the problem of navigation of multiple robots using a hybrid of potential and fuzzy-based controller. Tu and Baltes (2006) used a fuzzy-based potential function. The gradient of the fuzzy potential fields was used for the robot's motion. Marchese (2006) demonstrated the use of Multicellular Automata. The automata modelling made the robot avoid the obstacles while being attracted to the goal.

11.3 A Primer on Artificial Potential Field

Because most of the robotics problem is real time, the need to have a very fast and simple motion planner is evident. The simplicity enables fast development and deployment on the robot, whereas the computationally inexpensive nature allows the algorithm to be implemented in robots with limited vision, memory and computational capabilities. APF is one of the simplest methods, and the method is capable of autonomously moving the robot in realistic obstacle frameworks.

11.3.1 Concept

Consider a negatively charged particle in an electrostatic field. The particle is attracted to positive charges and repels from negative charges. Consider that the goal is a strongly positively charged particle, whereas the obstacle boundaries are fitted with negatively charged particles. This causes electrostatic forces to act on the workspace which repels and attracts the negatively charged particle or the robot attempting to reach towards the goal. Because the repulsive forces between the robot and the obstacle tend to infinity as the robot approaches the obstacle boundaries, it is noteworthy that a collision with the obstacle is not possible (considering continuous dynamics, although discretization of time during simulation for some resolutions may encounter collisions due to the resolution loss). Further, the goal will always attract the robot, so unless the robot is very close to the obstacle and gets very strongly repelled by it, the robot will always attempt to go towards the goal. No trajectory is planned and the robot moves as per this dynamics alone. The motion is completely *reactive* in nature.

To compute the robot motion, it is important to compute the *potential* applied by the obstacles and goals at different points of the space. The potential corresponding to all the points in the workspace may not be computed in advance to construct the entire potential field. The potential at the current position of the robot is only of interest. The derivative of the potential gives the *force* which acts on the robot. Every goal and obstacle adds to the potential. Even though the potential and the forces are virtual, and not physically existent, the effect of this force is applied to the robot as a control signal and hence the robot reacts to the applied virtual potential. The motion of the robot causes the position and velocity to change, which changes the potential based on which the robot moves further, till it reaches the goal.

The general approach is very similar to fuzzy logic-based planning with the only difference that the *fuzzy arithmetic* and fuzzy rules are replaced by the *potential-based arithmetic*. There is no rule base in potential field and the only embedded rule as per modelling is that the robot is repelled from the obstacles and attracted to the goal. Absence of rules makes it easier to construct and optimize the APF, in contrast to fuzzy logic wherein framing membership functions and a rule base is very difficult. However, at the same time, a fuzzy logic-based robot can be better controlled by adding new rules, which the potential-based approach does not facilitate.

11.3.2 Attractive Potential

The *attractive potential* is applied by a single goal to attract the robot towards itself. There are a variety of ways in which the attractive potential may be modelled depending upon the application and environment. A general method is to keep the attractive potential directly proportional to the distance between the current position of the robot and the goal. This causes the potential to tend to zero as the robot approaches the goal and hence the robot slows down near the goal and the overshooting after reaching the goal is avoided. The degree of proportionality and the proportionality constant are also algorithm parameters which may be tuned for different purposes, like maintaining high clearance and short path lengths. The attractive potential may be given by Eq. [11.1]. This modelling is known as the *quadratic potential*.

$$U_{att}(x) = \frac{1}{2}k_{att}\|x - G\|^2 \tag{11.1}$$

Here, x is the current position of the robot and G is the goal. $\|.\|$ is the Euclidian distance function and k_{att} is the proportionality constant, whereas the degree is taken as 2.

The *force* is given by the vector which has a magnitude of the derivative of the potential function and direction as the line which maximizes the change in potential, which is the line directly connecting the robot to the goal. This may be given by Eq. [11.2]

$$F_{att}(x) = \nabla U_{att}(x) = k_{att}\|x - G\| \cdot u(x - G)$$

$$= k_{att}\|x - G\|\frac{(x - G)}{\|x - G\|} \tag{11.2}$$

$$= k_{att}(x - G)$$

Here, $u()$ is the unit vector.

An alternative way of potential modelling is the *conic potential* in which the degree of proportionality is given by 1 and denotes much smaller changes in the potential values as the robot approaches the goal. The potential and the force are given by Eqs [11.3] and [11.4].

$$U_{att}(x) = k_{att}\|x - G\| \tag{11.3}$$

$$F_{att}(x) = \nabla U_{att}(x) = k_{att} \cdot u(x - G)$$

$$= k_{att}\frac{(x - G)}{\|x - G\|} \tag{11.4}$$

Hybrid modelling uses quadratic potentials near the robot boundary to smoothly make the robot reach the goal while decreasing the potential uniformly as the robot approaches the goal; and conic potential far away from the robot so as not to cause a very high potential value when the robot is far away. The resultant potential using the hybrid scheme is given by Eq. [11.5].

$$U_{att}(x) = \begin{cases} \dfrac{1}{2}k_{att}\|x - G\|^2 & \text{if } \|x - G\| \le d' \\[2ex] d' \cdot k_{att}\|x - G\| - \dfrac{1}{2}k_{att}d'^2 & \text{if } \|x - G\| > d' \end{cases} \quad [11.5]$$

Here, d' is the distance after which the potential scheme changes from quadratic to conic. Here, the extra terms are added to ensure that the resultant potential function is continuous, differentiable and uniformly increasing as the distance from the goal increases.

Correspondingly, the force is given by Eq. [11.6].

$$F_{att}(x) = \begin{cases} k_{att}(x - G) & \text{if } \|x - G\| \le d' \\[2ex] d' \cdot k_{att}\dfrac{(x - G)}{\|x - G\|} & \text{if } \|x - G\| > d' \end{cases} \quad [11.6]$$

11.3.3 Repulsive Potential

The *repulsive potential* is applied by all the obstacles which try to repel the robot from coming close to them and potentially causing a collision. It is assumed that the boundaries of all obstacles are fitted with small obstacles, each of which contributes to a repulsive potential. Consider any one small obstacle positioned at o_i. The repulsive potential can also be modelled in a variety of ways. The potential is inversely proportional to the distance so that the potential tends to infinity if the robot comes near the obstacle and the robot is repelled in the same scale. The obstacles far away rarely cause a concern for the robot. Hence, the obstacles are considered only till a distance of d^*. Obstacles farther away have no affect and hence a potential of 0. This eases the computation. The repulsive potential is given by Eq. [11.7].

$$U_{repel}^i(x) = \begin{cases} \dfrac{1}{2}k_{rep}\left(\dfrac{1}{\|x - o_i\|} - \dfrac{1}{d^*}\right)^2 & \text{if } \|x - o_i\| > d^* \\[2ex] 0 & \text{if } \|x - o_i\| \le d^* \end{cases} \quad [11.7]$$

Here, x is the current position of the robot and o_i is the position of the obstacle. $\|.\|$ is the Euclidian distance function and k_{rep} is the proportionality constant, whereas the degree is taken as 2.

The force is given by the derivative of the potential with the direction opposite to the obstacle. This is given by Eq. [11.8]

$$\begin{aligned} F_{repel}^i(x) = \nabla U_{rep}(x) &= -k_{rep}\left(\dfrac{1}{\|x - o_i\|} - \dfrac{1}{d^*}\right)\dfrac{1}{\|x - o_i\|^2}u(x - o_i) \\[2ex] &= -k_{rep}\left(\dfrac{1}{\|x - o_i\|} - \dfrac{1}{d^*}\right)\dfrac{1}{\|x - o_i\|^2}\dfrac{(x - o_i)}{\|x - o_i\|} \quad [11.8] \\[2ex] &= -k_{rep}\left(\dfrac{1}{\|x - o_i\|} - \dfrac{1}{d^*}\right)\dfrac{(x - o_i)}{\|x - o_i\|^3} \end{aligned}$$

Here, $u()$ is the unit vector.

However, there are multiple such obstacles. There are two techniques widely used. The first technique is to consider the *nearest obstacle* and use its potential as the repulsive potential, given by Eq. [11.9]. The forces are similarly defined by Eq. [11.10].

$$U_{\text{repel}} = U_{\text{repel}}^{i}, i = \text{argmin}\|x - o_i\| \qquad [11.9]$$

$$F_{\text{repel}} = F_{\text{repel}}^{i}, i = \text{argmin}\|x - o_i\| \qquad [11.10]$$

The forces are taken as the derivative of the potential. This causes a sudden change in force vector when the robot moves from the area equidistant between obstacles. The derivative of the potential or the force vector at the equidistant point is undefined.

The other method is to take the resultant potential as the *sum of all individual potentials* and similarly for the force, given by Eqs [11.11] and [11.12]

$$U_{\text{rep}} = \sum_i U_{\text{rep}}^{i} \qquad [11.11]$$

$$F_{\text{rep}} = \sum_i F_{\text{rep}}^{i} \qquad [11.12]$$

11.3.4 Total Potential and Motion

The resultant potential is given by the sum of attractive and repulsive potential. Correspondingly, the force applied to the robot is given by the derivative of the resultant potential, which is simply the sum of the individual force vectors. This is given by Eqs [11.13] and [11.14].

$$U = U_{\text{att}} + U_{\text{repel}} \qquad [11.13]$$

$$F = \nabla U = \nabla U_{\text{att}} + \nabla U_{\text{repel}} = F_{\text{att}} + F_{\text{repel}} \qquad [11.14]$$

The robot moves to *minimize the potential*, or the direction of motion is given by the derivative of the potential vector. This desired change may be applied to the robot as a virtual force vector or an instantaneous speed vector, depending upon the robot and the modelling. The robot controller attempts to move the robot to make the robot move to lower potential areas. The robot stops when the robot reaches an *equipotential region* in which the forces cancel each other out. It can correspond to a maximum, minimum or a saddle point. The robot attempts to move while minimizing the potential, so a maximum is not possible unless the robot starts from the maximum. Due to the modelling, a saddle point also practically does not exist. Hence, stoppages of the robot correspond to minima.

It is not guaranteed that the robot gets to the goal. The robot may get trapped in a local minimum and get stuck there indefinitely. Consider Fig. 11.1 in which the trajectory of the robot has been shown. The point of rest is not the goal, however, the robot will not move as the forces cancel each other out, or the attraction balances the repulsion. Sometimes giving mild to severe *perturbations* may make the robot escape the local minimum. The perturbations are applied at random. Even though the robot

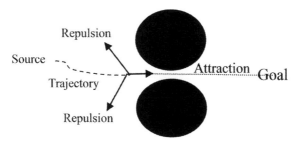

Figure 11.1 Equipotential region.

may escape the local minimum, it may again get trapped in another local minimum. It may not be possible to reach the goal at all. Local perturbations do not always succeed in getting the robot out of the minimum.

11.3.5 Working

The robot is initially located at the source, initially assumed at rest. Based on its position the robot computes the instantaneous gradient of the potential which gives the initial motion, usually towards the goal. The robot keeps getting attracted to the goal, until it comes in the vicinity of some obstacle. Now the forces of attraction towards the goal and the repulsion from the obstacle both play a role and the motion is given by the resultant vector. If the robot accidentally comes very close to the obstacle, it is strongly repelled and it may have to go back a little and take a route closer towards the goal. In this way, the robot mainly follows a smooth path from the source. If it does not get trapped in a local minimum, it approaches straight to the goal. A sample trajectory is shown in Fig. 11.2.

The gradient of potential at any time may be applied to the robot as an instantaneous acceleration or instantaneous speed depending upon the control of the robot. Based upon the dynamics of the robot, this change is applied, which moves the robot. If the robot is in motion, its current momentum will play a role, whereas the immediate motion will be based on the current speed and the desired speed. Even though the

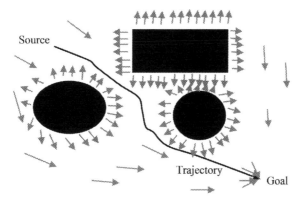

Figure 11.2 Working of artificial potential fields (APFs).

general theory of APF is for continuous spaces, wherein a continuous application of the potential gives the overall trajectory of the robot, the application of digital environments force a discretization. The computation takes place in small computation cycles, and the potential and therefore the control is computed at every cycle. Within the cycle, the robot moves within the same speed or control signal. If the robot accidently gets placed very close to the obstacles with a nonzero speed component towards the obstacle, it may in fact collide with the obstacle with a small speed. The dynamic constraints of the robot further prevent taking highly reactive measures for robot control as may be desired by the potential.

The computation of potential values relies upon distances. An important aspect is from whence do these distances come? The attractive potential is based on the distance of the robot from the goal. The *localization* module always gives the position of the robot, whereas the position of the goal is fixed, which enables computation of the attractive potential. For repulsive potential, the distances to the nearby obstacles are needed. Robots with *proximity sensors* get these distances via their proximity sensors. Light Detection and Ranging (lidar) gives a very high resolution, and all lidar scans can be used for the computation of the resultant potential. An array of Sound Detections and Ranging (sonars) or ultrasonics has a limited resolution, and still each of the distance scans can be a source of repulsive potential. Robots with overhead cameras may be able to detect obstacles using camera vision and estimate the distance from camera calibration or stereovision. Essentially, distance from obstacles in the front is important and a 360 degree view may not be desirous. Overhead cameras in the area of operation, providing overhead imagery also indicate the positions of the robot and obstacles, which may be processed rapidly to get distance estimates. To facilitate fast distance computations, the distances to obstacles at 8 or 16 directions may be taken. This is like fitting 8 or 16 virtual sonars and computing distances from them.

11.4   Lateral Potentials for Planning

Planning the motion of an autonomous vehicle may be broadly separated into longitudinal planning and lateral planning. *Longitudinal planning* deals with sticking to one's own speed lane. This involves speed control and steering control in case of curved roads. *Lateral planning* deals with deciding on lane changes and generating a feasible trajectory for a lane change. The planning primarily involves steering control.

APFs have been widely used in robotics for planning the motion of a mobile robot. Though the problem of robot motion planning closely resembles the problem of planning of an autonomous vehicle in the absence of speed lanes, the potential method cannot be directly applied to vehicles. The prime reason is the presence of roads within which vehicles need to be driven. In a road scenario with moving vehicles and obstacles, it would be evident to have too many *zero potential points*. Further, *cooperation* is weakly modelled in potential approaches, whereas in traffic scenarios it is important for a vehicle to cooperate and allow another vehicle to overtake it. The same holds true for elastic band approaches as well. The general pros and cons of APF are given in Box 11.1.

Box 11.1 Pros and Cons of Potential Fields

Pros
- Computational time
- Work with partially known environments

Cons
- Completeness
- Optimality
- Oscillations in narrow corridors (like roads)
- Getting stuck in equipotential regions
- A vehicle in front may be repelled by the vehicle at the back rather than allowing for cooperative overtaking
- Weakly modelled cooperation

It is assumed here that a map of a road segment is available in which the road is bounded by a road boundary on both sides. There can be a number of vehicles in the map at any time. The size, position and speed of nearby vehicles can be sensed by the vision system of the vehicle. There exist no speed lanes in the road and hence any vehicle can potentially drive anywhere in the road.

Let, at some time, the position of the vehicle being planned be $R(x', y', \theta')$. Here, the X' (or longitudinal) axis is taken as the heading direction of the road and $Y'(R)$ (or lateral) axis, at any point R, as the axis joining two boundaries normal to the X' axis. The angle θ' denoting orientation of the vehicle is measured as the angle from the $X'(R)$ axis at the point of measurement (R). Let the vehicle be of size $l \times w$. Let the corners of vehicle in cyclic order be C_1, C_2, C_3 and C_4. Let the vehicle's preferential speed of driving be vpref which is the speed by which the vehicle travels on a straight road in the absence of any other vehicle or obstacle. Let v ($\leq$vpref) be the current speed of the vehicle. Let the rotational speed of the vehicle be ω ($\leq\omega_{max}$). Here, v and ω are measured in the Cartesian coordinate system which is not the system used to represent vehicle position R. The maximum acceleration that the vehicle can have is accmax.

The objective of the algorithm is to move the vehicle at every instant of time such that the vehicle does not collide with any static obstacle and to ensure that no two vehicles collide with each other. On top of this, vehicles may not go very close either to each other or to a static obstacle, which is a potential threat in driving. The traffic is assumed to possess large diversities in terms of the constituent vehicles. This means that vehicles vary in terms of their sizes ($l \times w$) and preferred driving speeds (vpref). There is no lower limit to the allowable speed, which means traffic may have extremely slow vehicles moving in it. Hence, the motion of the vehicle produced by the algorithm can only be regarded as desirable if any vehicle having a higher preferred speed is able to overtake a vehicle having lower preferred speed. Overtaking is preferred to take place on the right side, but this is not a mandatory condition. On wide roads a vehicle already lying to the left of a vehicle may proceed to overtake the vehicle on the left (with some cooperation from other vehicles) rather than having to go to the other side of the road to overtake. The vehicles need not arrange themselves laterally as per their preferred speeds (typical in speed-lane scenarios) which lead to overtaking mostly on the right.

The algorithm presented here is based upon the method of APF which is a widely used and studied algorithm for cases of both single and multiple mobile robot planning. In this work, however, it is preferred to model the algorithm from the perspective of the *thought process of a human driver* as if he/she was driving the automated vehicle. A conventional potential field design would demand using distance measures from surroundings, converting them into force vectors and moving the vehicle by the resultant force. Sonar sensors are found on a variety of robots which give the distance from obstacles directly as input and can easily be used for computation of the resultant force vector.

This methodology, however, does not enable modelling driving behaviours, and hence a modified scheme is used. The implemented methodology enables generation of travel plans which are more realistic, as well as mixing well in chaotic traffic comprising both autonomous and human-driven vehicles. Using this mechanism, it is intended to generate similar behaviours to those that are observed in countries where speed lanes are not followed. The task of planning may be easily broken down into lateral planning and longitudinal planning. Although the former deals with adjusting the steering, the latter deals with adjusting the speed.

The key task of the algorithm is to decide the *lateral position* of the vehicle which is done using LPs. The potentials may be positive, which force the vehicle to occupy a position with a larger value on the Y' axis, or negative, forcing the vehicle to go for a smaller value on the Y' axis. Potentials amalgamated from a few sources are used to decide the resultant lateral position of the vehicle. The different sources are summarized in Fig. 11.3. The key features of the approach are summarized in Box 11.2. The overall approach is also summarized in Box 11.3.

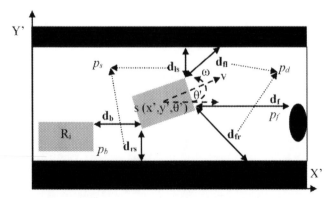

Figure 11.3 Sources of potential. d_f, d_{ls}, d_{rs}, d_{fl}, d_{fr} and d_b denote distance from ahead, left, right, forward-left, forward-right and back obstacles, respectively. p_f, p_s, p_d, p_b denote forward, side, diagonal and back potentials, respectively.

Box 11.2 Takeaways of Solution Using Lateral Potentials

- Modelling of *lateral potentials* suited for road scenarios to eliminate the known problems associated with the potential approaches.
- Modelling of potentials based on the principles of *time to collision* and *cooperation* apart from the distance measures for lateral planning of the vehicles.
- Use of obstacle and vehicle avoidance *strategy parameters* for higher-order planning.
- *Heuristic decision making* in deciding these strategy parameters for real-time planning.

Box 11.3 Key Aspects of Solution Using Lateral Potentials

- Lateral Planning (steering)
  - Vehicle is repelled from other vehicles, obstacles and road boundaries
  - *Forward Potential* using time to collision as metric. Steering direction is set using heuristics: Turn left if front obstacle is on the right, and vice versa.
  - *Side Potential* in opposite direction using distance as metric
  - *Diagonal Potential* in opposite direction using distance as metric
  - *Back Potential* using time to collision as metric, to facilitate cooperative overtaking of the vehicle at back. Steering direction is set opposite the overtaking direction.
- Longitudinal Planning (speed)
  - *Preferred speed* is the maximum permissible speed per the distance in the lateral and heading direction of the vehicle, limited to the maximum speed.
  - The *aggression factor* limits allowed acceleration/deceleration, based on which the actual speed is set.

11.4.1 Forward Potential

The first source of potential is from a vehicle or obstacle directly in *front* along the X' axis of the vehicle. Let the obstacle be at a distance of d_{f_i} units away from the vehicle when measured from a point f_i on the vehicle's front boundary (line C_1C_2). Assume that after a distance f_i longitudinally, there lies a static obstacle ($o = \text{obs}$), or another vehicle ($o = B$). The potential applied to the vehicle is given by Eq. [11.15].

$$
p_{f_i} = \begin{cases}
0 & o = B, v_B \geq \text{vpref} \quad \text{(a)} \\[2mm]
\text{sgn}(B)\dfrac{\text{vpref} - v_B}{d_{f_i}} & o = B, v_B < \text{vpref} \quad \text{(b)} \\[2mm]
\text{sgn}(o)\dfrac{\text{vpref}}{d_{f_i}} & o = \text{obs} \quad \text{(c)}
\end{cases}
\qquad [11.15]
$$

Here, v_b is the speed of the vehicle in front (B, if any). Eq. [11.15a] deals with the condition when the vehicle being planned (say A) is possibly *following the vehicle ahead*, vehicle B. There is no possibility that vehicle A may need to overtake vehicle B. As there is no other behaviour that vehicle A shows because of the presence of vehicle B, the potential is 0.

However, as per condition Eq. [11.15b], overtaking is possible if vehicle A accelerates. Hence, potential is applied in the direction of sgn(B) by vehicle B to vehicle A. sgn(B) may be $+1$ or -1, the value for which can be determined by considering whether overtaking should take place on the left or the right. In this algorithm both sides are possible; hence, the value is kept 1 if B lies at a *higher lateral position* to A or at the same lateral position (overtaking on the right preferred), and -1 otherwise. Eq. [11.15c] is the same scenario in which the potential is caused by a static obstacle in place of another vehicle. sgn(o) denotes the strategy to overcome the obstacle on the left (sgn(o) $= 1$) or the right (sgn(o) $= -1$).

Obstacle avoidance may be perceived as overtaking a static vehicle which accounts for the difference between Eq. [11.15b] and [11.15c]. Unlike conventional potential approaches, prospective *time to collision* is used as an indicator of potential rather than distance to collision. This accounts for the commonly observed driving phenomenon wherein manoeuvres are smaller on sighting a vehicle directly ahead which needs to be overtaken and larger if an obstacle is at the same distance. It may be noted that a sonar sensor may not be applicable for measuring this distance as it measures distance in heading angle of vehicle and not along the X' axis. However, knowing the positions and orientations of other vehicles and the position of vehicle R, this distance may be computed. The net value of potential due to front sources may be given by Eq. [11.16].

$$p_f = \text{sign}(\max\{\text{abs}(P_{f_i})\}) \cdot (\max\{\text{abs}(P_{f_i})\})^2, i \text{ lies on } C_1 C_2 \qquad [11.16]$$

This means that the *largest potential* measured along any point on the front boundary is used as the front potential. This potential gives the overtaking and obstacle avoidance behaviour of the vehicle. Conventional potential field modelling for a vehicle directly in front of another vehicle or obstacle would have pushed the vehicle A backwards instead, thereby disallowing any overtaking. On being marginally deviated in its lateral position, the lateral potential would have been too small to facilitate quick overtaking.

11.4.2 Side Potential

The next source of potential is an obstacle, another vehicle or road boundary to the *side* of the vehicle, with distances measured along the lateral direction or the Y' axis. Let the vehicle have a distance of d_{l_i} (or d_{r_j}) measured from a point l_i (or r_j) along Y' axis (or $-Y'$ axis) from a point l_i (or r_j) lying at the left (or right) boundary of the vehicle that is line C_1C_4 (or line C_2C_3). The resultant potential may be given by Eq. [11.17].

$$p_s = p_{l_i} + p_{r_i} = -\max\{(1/d_{l_i})\}^2 + \max\{(1/d_{r_j})\}^2 \qquad [11.17]$$

i lies on C_1C_4, j lies on C_2C_3.

Note that speed is not mentioned in Eq. [11.17] unlike Eq. [11.15]. The reason for this is that there is no concept of side speed which determines when the vehicle may collide with the sensed obstacle, road boundary or vehicle. In fact (unless the same obstacle or vehicle was sensed in Eq. [11.15], in which case it is governed by its dynamics), the vehicle may never collide with the obstacle, vehicle or road boundary end, because it does not lie directly in front and the vehicle mostly moves straight longitudinally.

11.4.3   Diagonal Potential

The next source of potential is the *forward diagonal* distance measured at points C_1 and C_2. Consider point C_1 (or C_2) which is used to measure distance d_{flC_1} (or d_{frC_2}) at an angle of 45 degrees (or −45 degrees) to X' axis. This potential may be given by Eq. [11.18].

$$p_d = p_{fl} + p_{fr} = -(1/d_{flC_1})^2 + (1/d_{frC_2})^2 \qquad [11.18]$$

The diagonal potential (p_d) acts as a forerunner to side potential (p_s). The lateral potential is recorded as a position which the vehicle would occupy in the future, if it does not make any lateral alterations. Diagonal potential enables the vehicle to make any corrections in advance.

11.4.4   Back Potential

The last source of potential is from a vehicle which may be to the *rear*. Let the distance be d_b in the $-X'$ axis and a vehicle B be behind at this distance. The resultant potential is given by Eq. [11.19].

$$p_{bi} = \begin{cases} 0 & B = \text{null} \lor v \geq \text{vpref}_B \\[2mm] \text{sgn}(B)\dfrac{\text{vpref}_B - v}{d_b} & v < \text{vpref}_B \end{cases} \qquad [11.19]$$

In case vehicle B has a higher (than A) preferential speed (vpref$_B$), it is possible that vehicle B overtakes vehicle A. Hence, vehicle A must drift towards the opposite side to which overtaking is being performed to *cooperate* towards the overtaking. In this algorithm sgn(A) has a value 1 if B lies at a higher lateral position to A or at the same lateral position (overtaking on the right preferred) and −1 otherwise. The resultant potential is given by Eq. [11.20].

$$p_b = \text{sign}(\max\{\text{abs}(p_{bi})\}) \cdot (\max\{\text{abs}(p_{bi})\})^2, i \text{ lies on } C_3C_4 \qquad [11.20]$$

11.4.5 Lateral Planning

Four sources of potential add up to the total potential given by Eq. [11.21]. However, the different potentials are at different scales and hence cannot be directly added up.

$$p = \text{sen}_{X'} \cdot p_f + \text{sen}_{Y'} \cdot p_s + \text{sen}_{X'Y'} \cdot p_d + \text{coop} \cdot p_b \qquad [11.21]$$

Here, $\text{sen}_{X'}$ is a factor that governs the *sensitivity* of the vehicle from an obstacle or another vehicle directly ahead. Higher values lead to early heavy steering to avoid the obstacle or another vehicle, even though it might be way ahead. Smaller values lead to small steering early until the vehicle reaches very close to the vehicle or obstacle when left lateral corrections take place. The factor $\text{sen}_{Y'}$ governs the lateral sensitivity of the vehicle. If the factor is high, the vehicle is prone to make too large steering changes for small behavioural changes. If the factor is small, the vehicle shows very slow lateral corrections and the majority of its journey is travelled in a straight line, until it reaches a state of potential collision in which case sharp steering is required.

The factor $\text{sen}_{X'Y'}$ governs sensitivity to forthcoming lateral corrections, which plays a role as a combination of the other two factors. The factor coop governs the degree to which the vehicle *cooperates* with another vehicle to the rear for potential overtaking. Small values are better for the vehicle being planned but painful for the overtaking vehicle, and vice versa.

Lateral control of the vehicle is done using the steering control which changes the orientation of the vehicle. The desired orientation of the vehicle θ'_{desired} is proportional to the lateral potential given by Eq. [11.22].

$$\theta'_{\text{desired}} = k \cdot p \qquad [11.22]$$

Here, k is a constant governing conversion of potential to orientation. In practice, it may not be possible to orient the vehicle to θ'_{desired} due to rotational speed restrictions ($-\omega_{\max} \leq \omega \leq \omega_{\max}$), in which case the maximum change possible is applied.

11.4.6 Longitudinal Planning

Longitudinally, the major decision to be taken is on the speed of travel. The lateral position of the vehicle or the steering is controlled by the lateral planner, and the longitudinal planner needs to only ensure that the vehicle keeps moving at the fastest speed possible. Hence, there are no longitudinal potentials used in this technique. The distance of the vehicle is measured on the X' axis.

Let the distance at any point be d_{f_i}. Let (after this distance) the vehicle meet an obstacle ($o = \text{obs}$) or another vehicle ($o = B$). The corresponding maximum speed possible because of an obstacle being found after d_{f_i} distance is given by Eq. [11.23].

$$v_{f_i} = \begin{cases} \text{vpref} & o = B, v_B \geq \text{vpref} \quad \text{(a)} \\ \min\left(v_B + \sqrt{2(\text{acc} \cdot \text{agg})d_{f_i}}, \text{vpref}\right) & o = B, v_B < \text{vpref} \quad \text{(b)} \\ \min\left(\sqrt{2(\text{acc} \cdot \text{agg})d_{f_i}}, \text{vpref}\right) & o = \text{obs} \quad \text{(c)} \end{cases} \qquad [11.23]$$

Eq. [11.23a] covers the case when there is no potential threat of a collision to the vehicle as no slower vehicle or static obstacle lies ahead, and hence it may attempt to travel at the fastest speed possible. Eq. [11.23b] is when there is a vehicle ahead in which case the relative motion of the two vehicles is studied to compute the desirable speed. agg ($0 < $ agg ≤ 1) is the *aggression factor*. More aggressive driving is marked by higher acceleration and decelerations. A higher value of this factor means that the vehicle continues to drive at fast speeds, even after seeing the obstacle or vehicle ahead, and sharply decelerates (if needed) to avoid the obstacle or vehicle. Lower values imply a slower deceleration scenario.

Sometimes it may be possible that no potential collision is visible in the lateral direction, but the vehicle is oriented at some angle θ' such that it is close to an obstacle, vehicle or boundary end. Hence, at the same point f_i calculations are repeated with distances measured along the current heading angle of the vehicle or θ', which gives another preferential driving speed indicator $v_{\theta'i}$. The resultant preferred driving speed is given by Eq. [11.24].

$$v_{\text{desired}} = \min\{v_{f_i}, v_{\theta'i}\}, i \text{ lies on } C_1C_2 \qquad [11.24]$$

This speed may not be obtainable due to acceleration limits ($-\text{accmax} \leq$ acceleration $\leq \text{accmax} \cdot \text{agg}$), in which case the maximum change allowed is made.

11.5   Results for Lateral Potentials

The algorithm was developed and tested using simulations in Matrix Laboratory (MATLAB). For computational reasons the various potentials were measured only at the corners of the vehicle, instead of measuring them at every point across the vehicle boundary and then taking the maximum. Unless a small vehicle or obstacle happens to lie strictly in between the vehicle corners, which would be the case with very small obstacles or vehicles, this approach holds good. A number of scenarios were generated to test the working of the algorithm with respect to its parameters.

11.5.1   Simulation Results

A variety of scenarios were created to assess the behaviour of the vehicle. First, the obstacle-avoidance capability of the vehicle was tested. A single vehicle was created on the road which needed to overcome two obstacles one after the other. The path followed by the vehicle is shown in Fig. 11.4A. The vehicle steered left to place itself to comfortably pass the first obstacle. Soon, the second obstacle was detected, and on being close enough, steering took place on the opposite side.

The next scenario was created to test the ability of the vehicle to overtake another vehicle. To make the scenario difficult, a static obstacle was added just after potential overtaking completion. The vehicle generated later was capable of high speeds. It overtook the slower vehicle ahead at a point p, and proceeded to pass the obstacle as shown in Fig. 11.4B, whereas the slower vehicle slowly moved on towards the

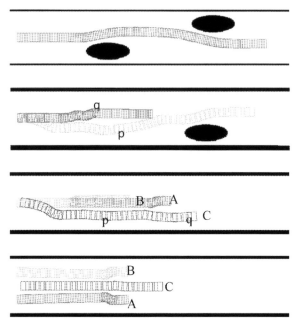

Figure 11.4 Simulation results of the algorithm. (A) Scenario 1, (B) Scenario 2, (C) Scenario 3, (D) Scenario 4.

obstacle. The slower vehicle showed cooperation and drifted left to allow the over-taking procedure as denoted by point q in Fig. 11.4B.

In the third scenario, the vehicle A is first made to enter, which travels straight. Then vehicle B is made to enter which simply follows the vehicle A, exhibiting vehicle-following behaviour. Then vehicle C entered, which was capable of high speeds. It succeeded to overtake first vehicle B (at point p) and then vehicle A (at point q). This scenario is shown in Fig. 11.4C.

In the last scenario, two vehicles (A and B) entered the map simultaneously, separated laterally by some distance. The vehicles continued to move parallel to each other, with the same speed. Lateral potentials from each other and road boundaries made them drift towards each other, to make lateral separations equal. Later, the vehicle C entered the scenario and proceeded to first push the two vehicles and then it succeeded in intercepting them. Finally, C overtook the two vehicles. The scenario is shown in Fig. 11.4D.

11.5.2 Algorithmic Parameters

Eq. [11.21] shows a number of parameters which govern the contributions of the various kinds of potential. One of the important parameters of the algorithm is $sen_{X'}$ which covers the sensitivity along the X' axis. This parameter was tested for the case of a single obstacle, in which the performance largely depends upon this

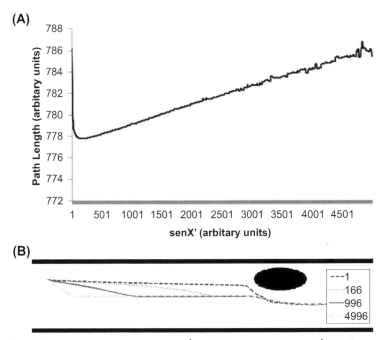

Figure 11.5 Effect of changing parameter senX'. (A) Path length versus senX', (B) Corresponding trajectories.

parameter. The effect of different values on the path length is shown in Fig. 11.5A. The paths corresponding to various values are shown in Fig. 11.5B. It is clear that low values lead to late steering, whereas high values cause immediate steering and early positioning to avoid an obstacle.

The other parameter of interest is coop, which governs the magnitude by which a vehicle cooperates with other vehicles. A simple scenario was created with a slow-moving vehicle ahead in road. A fast vehicle entering the scenario could simply overtake the slower vehicle. The magnitude by which the vehicle being overtaken cooperates with the faster vehicle is the magnitude by which it drifts on the road. This leads to less of a need for the overtaking vehicle to steer. The path length of the overtaking vehicle and the vehicle being overtaken for different values of the parameter coop are shown in Fig. 11.6A. The path corresponding to some of the values is shown in Fig. 11.6B. It can be seen that high values of this parameter are desirable for the overtaking vehicle and less desirable for the vehicle being overtaken.

The other important factor is $senY'$, which governs the sensitivity of the vehicle in the Y' axis. A simple scenario is taken with a single vehicle generated on the side of the road which would prefer to drift towards the centre of the road due to unequal lateral potential by the road boundaries. The behaviour of the vehicle for different values of this parameter is shown in Fig. 11.7. On further increasing, the value of $senY'$ the vehicle became highly sensitive and showed oscillations within the road.

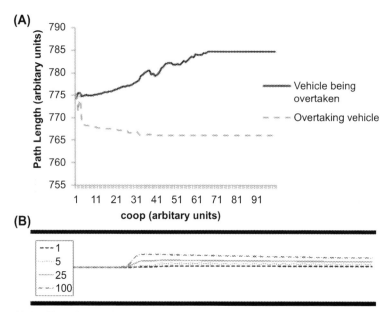

Figure 11.6 Effect of changing parameter coop. (A) Path length versus coop, (B) Corresponding trajectories.

Figure 11.7 Effect of changing parameter $sen_{\gamma'}$.

11.6   A Primer on Elastic Strip

Replanning is a very important problem in motion planning. Although the planning algorithms may take seconds to make a trajectory for the robot or vehicle to follow, the environment may have small changes in milliseconds. Calling the entire algorithm again is not an option and this necessitates the need of a good and fast replanning algorithm for quickly correcting any errors in the trajectory profile and making it be near optimal even if the environment changes. Replanning gives capabilities to the trajectory to *adapt* itself. *Elastic strip* (Brock and Khatib, 2000) is like a rubber band in the stretched form that models the trajectory of motion. Using this concept, the replanning algorithm is modelled.

11.6.1 Concept

Imagine a stretched rubber band. The rubber band has two very important properties because of which it is widely used in everyday tasks. First, that the rubber band can be stretched to take any desired shape, if one can characteristically hold the rubber band at the needed points. Imagine having infinite fingers, you could hold the rubber band to get any desired shape. Of course, practically, the rubber bands have a limit to which they can be stretched, after which they simply break away; here, it is assumed that the band can be stretched wide enough to get the desired shape. Second, once the rubber band is released from one or more positions of hold, it contracts back. Imagine holding the rubber band with four fingers, two of them being the extreme ends of the rubber band. On release of the four fingers, the rubber band contracts so that the shape is a straight line between every two points of hold. On releasing the third finger, the shape is simply a straight line between the two points of hold.

An elastic strip has *elastic properties* which gives it the property to imitate any shape on application of suitable force, and to contract back to the minimal length once the force is removed. Hence, the trajectory of a robot, being planned by any algorithm, can also be represented by an elastic strip. This strip has two forces acting upon it, external and internal.

11.6.2   External Forces

The *external force* is applied by the obstacles. Each obstacle applies a repulsive potential the derivative for which acts as a repulsive force to every point of the elastic strip. The external force usually attempts to expand the strip and attain a shape such that every point is as far away from the obstacles as possible. If only external forces are used, the elastic strip would attempt to lie on points nearly equidistant to obstacles, which is where the repulsive forces of the obstacles balance each other out. Although the external force is capable of maintaining very high clearance, this force, however, leads to irregularity and very large path length, both of which are undesirable.

11.6.3   Internal Forces

The *internal force* is the elastic force acting on the elastic strip which attempts to make the strip contract to the smallest possible size. The difference between the natural length of the strip and the expanded length gives rise to a tension which moves each of the points on the elastic strip towards each other to contract to the minimal length. In the absence of external forces, this force would make a straight-line connection between the source and the goal, which minimizes the path length. The resultant force, however, is a mixture of the two forces, internal and external, based on which every point on the elastic strip moves. Hence, a balance of the two forces keeps the trade-off between the path length, clearance and smoothness.

11.6.4   Working

The elastic strip continuously *adapts* itself as per the internal and external forces. The concept is summarized in Fig. 11.8. The only exception is that the source and the goal

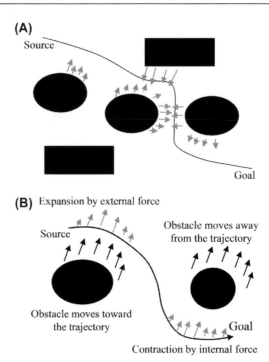

Figure 11.8 Working of an elastic strip: (A) external forces affecting the elastic strip, (B) trajectory adapting to changed obstacle positions.

are clamped and not allowed to move. It is assumed that no obstacle can lie on the trajectory or the elastic strip at any time, in which case the external forces are undefined. In all other cases, the external forces push all the points in the elastic strip away from them. The internal force counterbalances these forces to avoid the elastic strip from getting overly long shapes. As the obstacles move towards the strip, the repulsive forces become more dominant and hence the elastic strip expands. Upon expansion, the internal force also increases. The expansion stops when the two forces balance each other out. If an obstacle moves away from the obstacle, the repulsive external forces weaken and the balance between the internal and external force is broken, meaning the internal force dominates. The strip contracts, while doing so the external forces increase as the strip approaches the obstacle which balances the internal force. In this manner, the trajectory always adapts to the changes in the environment.

It is not possible to take the elastic strip as a collection of infinite points and make each of these points react to the internal and external forces. Hence, a continuous trajectory is discretized to the *sampled-out points*. The number of points should be high enough to resemble the trajectory and should be small enough to facilitate computations in small computation points. More points are taken near the obstacles, where the trajectory shows fine turns, and fewer points are taken away from the obstacles where the trajectory is largely straight. Each of these points adapts to the changes in the environment. When these points attain their equipotential region due to the opposing forces, a smooth trajectory is then carved from these sampled-out points. Any smoothening technique may be used for the purpose.

11.7 Problem Modelling With an Elastic Strip

Elastic strips have been used for the planning of a mobile robot. A number of homotopies may lead the robot from its source to the goal. The preferable homotopy is selected and represented as a strip which marks the robot's trajectory. For computational reasons, the strip is discretized to a number of waypoints. A change in the environment is marked by movement of these points, the addition of points or deletion of points such that the resulting trajectory is collision free. The intent is to have the fewest waypoints mostly near the obstacles. Each waypoint is acted on by a repulsive force from the obstacles which makes the trajectory lie far from the obstacles. An additional internal force pulls the waypoints towards each other, resulting in a shortening of the path. Given that two obstacles do not eventually intersect such that the robot is planned to travel in between them, the resulting travel plan is complete and reactive to real-time changes.

Given an initial optimal travel plan, the algorithm is near optimal to small changes in the environment. With some additional computational time, the algorithm is, as a result, better than the APF method and similar approaches which are neither complete nor optimal. The algorithm is in fact similar to the Elastic Roadmap (Yang and Brock, 2010; Quinlan and Khatib, 1993) and related approaches which maintain a roadmap in a dynamic environment.

The approach extends the approach described in Section 11.4 and works over two major limitations. (1) Potential methods are neither near complete nor near optimal. This approach solves both the problems. (2) The approach answers the questions of deciding on a strategy of avoiding obstacles: This was done earlier using heuristics which could be problematic in many situations. The key takeaways of the algorithm are summarized in Box 11.4. The overall algorithm is also summarized in Box 11.5.

A limited map of the road is assumed to be given. It is considered that the road does not contain any junctions or diversions. The road is characterized by its left and right

---

Box 11.4 Takeaways of Solution Using Elastic Strip

- Quick computation of the optimal *strategy* for obstacle and vehicle avoidance, and the associated trajectory.
- Real-time *optimization* of the trajectory in motion, making the resultant plan near optimal.
- Using *heuristics* to ensure the travel plan is near complete.
- *Cooperative* coordination strategy between vehicles.

Box 11.5 Key Aspects of Solution Using Elastic Strip

- **Objectives**:
  - Go as far as possible
  - Maximize lateral clearance
  - Minimize travel time
  - Maximize cooperation
  - Apply lateral-potential strategy heuristic
- **Feasibility**: In the projected travel, feasible plans allow the vehicle being planned to always
  - Slow down in time and avoid accidents with vehicles in front
  - Allow vehicles at back to slow down and avoid accidents with the vehicle being planned
  - Avoid collisions with obstacles, other vehicles and road boundaries
- **Terms**:
  - *Trajectory*: Path by which the vehicle is planned to move
  - *Obstacle-only Trajectory*: Path considering obstacles only and none of the other vehicles
  - *Strategy*: Specification of side (left or right) of avoiding every vehicle and obstacles
- **Planning**: Plan as vehicle moves; start with a null plan, then as the scenario changes:
  - Set speed to the maximum value per the current position
  - Delete infeasible part
  - Extend plan
  - Optimize plan
- **Modes of Operation**: *Mode 1: Trajectory ends in an obstacle*, and *Mode II: Trajectory does not end in an obstacle.*
- *Mode 1: Trajectory ends in an obstacle*
  - Plan extension can only happen using the *obstacle-only trajectory* strategy.
  - Acceleration is disallowed.
  - Adjust speed to make vehicle stand still at a distance d_{obs}.
- *Mode II: Trajectory does not end in an obstacle*
  - Normal operation, all possible subsequent plans explored.

---

boundaries. It is assumed that all obstacles (including vehicles) can be sensed with some degree of certainty. Let any such general vehicle R_i be located at position L_i (x'_i, y'_i, θ'_i), in which θ'_i denotes the heading direction and (x'_i, y'_i) correspond to the centre; and travelling with speed v_i. Let at any general time the vehicle being planned (Q) be at position L_q (x'_q, y'_q, θ'_q) with linear speed v_q ($\leq \text{vpref}_q$) and angular speed $\omega_q (\leq \omega_q^{max})$, in which vpref_q is the preferred (maximum) speed of travel and ω_q^{max} is the maximum angular speed. The linear acceleration is bounded by acc_q^{max}. All positions are denoted using the longitudinal (X') and lateral (Y') axis. Let the free workspace be given by $C_{\text{free}}^{\text{static}}$ which excludes any region with static obstacles or outside the road boundaries/road segment. Static obstacles and moving vehicles behave differently and are handled separately. Broken-down vehicles cannot move, hence these are taken as static obstacles.

11.7.1 Objectives

The purpose of the algorithm is to construct a travel plan τ. Let $\tau(t)$ $\{=(x'_q, y'_q, \theta'_q)\}$ be the planned position at time t and T denote the time up to which the vehicle is planned. The objectives in (nonstrictly) decreasing order of importance are:

1. vehicle should go *as far as possible*, or maximize $\tau(T)[X']$, the longitudinal position at the end of travel,
2. maximize the *minimum lateral clearance* of the trajectory, in which lateral clearance is the minimum distance of the vehicle from any obstacle measured on the lateral axis.
3. minimize T,
4. maximize *lateral cooperation*, the net lateral movement (measured in lateral axis) of a vehicle traversed solely with the intention of enabling some other vehicle to the rear to obtain a better plan.
5. In case overtaking an obstacle may be equally advantageous (as per the above-mentioned objectives) from both the left and right sides, the strategy used in Section 11.4 is regarded as better.

11.7.2 General Speed Bounds

Consider a point-sized object at state s moving with speed v_q ($\leq \text{vpref}_q$). Consider that an obstacle o lies ahead of it at a longitudinal distance d_f. Here, o may be a static obstacle ($o = \text{obs}$) or a vehicle ($o = B$) within a pool of vehicles P. The maximum speed (v^s_a) by which the object can move to avoid a collision with o is given by Eq. [11.25]. v^s_a should be low enough to allow the object to stop before a collision takes place, if there is no other alternative than following o; whereas it is assumed that o continues to travel with its current speed (v_B if vehicle, 0 if obstacle). This corresponds to slowing with the maximum uniform retardation of $\text{agg}_q \cdot \text{acc}^{\max}_q$.

$$
v^s_a = \begin{cases}
\text{vpref}_q & o = B, v_B \geq \text{vpref}_q \\[2mm]
\min\left(v_B + \sqrt{2 \cdot \text{acc}^{\max}_q \cdot \text{agg}_q \cdot d_f}, \text{vpref}_q\right) & o = B, v_B < \text{vpref}_q \\[2mm]
\min\left(\sqrt{2 \cdot \text{acc}^{\max}_q \cdot \text{agg}_q \cdot d_f}, \text{vpref}_q\right) & o = \text{obs}
\end{cases}
$$

$$[11.25]$$

Eq. [11.25] is similar to Eq. [11.23] with the major difference that the equation is not applied to the current position of the vehicle (as done in reactive planning) but to a projected position which the vehicle may or may not occupy (as done in deliberative planning).

Here, agg_q ($0 < \text{agg}_q \leq 1$) is the aggression factor which limits the planned acceleration. Lower values would indicate a more comfortable drive, whereas higher values sacrifice comfort for travel time. A minimum threshold distance of d_{unc} must always be maintained which is excluded from the measured d_f. This is employed to overcome uncertain speed changes of the vehicle in front or other uncertain environment changes.

Further, consider that an obstacle o lies behind a point-sized object at a longitudinal distance of d_b. Here, o may be a static obstacle ($o = $ obs) or any vehicle ($o = B$) within a pool of vehicles P. The minimum speed $\left(v_b^s\right)$ by which the object can move to avoid collision with i is given by Eq. [11.26]. Using the concepts from Eq. [11.25], v_b^s should be high enough to allow stopping o travelling with speed v_B in case of a vehicle ($o = B$) and 0 if case of an obstacle ($o = $ obs), before it collides with the point-sized object.

$$
v_b^s = \begin{cases} \max\left(v_B - \sqrt{2 \cdot \mathrm{acc}_i^{\max} \cdot \mathrm{agg}_i \cdot d_b}, 0\right) & o = B \\ 0 & o = \mathrm{obs} \end{cases}
$$
[11.26]

hence, for safe travel, $v_b^s \leq v_q^s \leq v_a^s$

11.7.3 Plan Feasibility

A driver only considers vehicles ahead of it while formulating his/her travel plan, as is the case in deciding the feasibility of a travel plan τ. The only exception is making a turn (lane change) when one might accidently drive in front of a vehicle which may not have enough time to slow down to avoid collision. Hence, the resulting pool of vehicles considered for feasibility consists of all vehicles R_i which either lie *completely ahead* of Q (or the longitudinal coverage for which is completely ahead of the longitudinal coverage of Q) or *do not lie in the same lane* as Q (or the lateral coverage for which is completely disjoint from the lateral coverage of Q). The set of vehicles is given by Eq. [11.27].

$$
P = \{R_i(L_i)[X'] > R_q(L_q)[X'] \lor R_i(L_i)[Y'] \cap R_q(L_q)[Y'] = \phi\}
$$
[11.27]

Here, $R_i(L_i)$ denotes the region in the workspace occupied when the vehicle i is at position L_i, the longitudinal occupancy for which is given by $R_i(L_i)[X']$ and lateral occupancy as $R_i(L_i)[Y']$.

A plan τ is called a feasible travel plan if

1. No collisions occur with static obstacles, road boundaries or Eq. [11.28]

$$
\tau(t) \in C_{free}^{static} \, \forall t \leq T
$$
[11.28]

2. No collisions occur with the projected motion of other vehicles, or Eq. [11.29]

$$
R_q(\tau(t)) \cap \bigcup_{i, i \in P} R_i((L_i \pm \Delta L_i) + t(v_i \pm \Delta v_i), \theta_i) = \phi, \, \forall t \leq T
$$
[11.29]

All vehicles are projected to travel straight (longitudinally), maintaining their current orientation, from the sensed initial position in the range $(L_i - \Delta L_i, L_i + \Delta L_i)$ with a constant sensed speed in the range $(v_i - \Delta v_i, v_i + \Delta v_i)$. At any time t the vehicle Q is at the state $\tau(t)$, occupying the region $R_q(\tau(t))$. A collision is said to have occurred if Q intersects with any projected position of the vehicle at any point in its trajectory. Here, ΔL_i and Δv_i denote the uncertainty in measurements of position and speed of the vehicle R_i. The projected region is hence $R_i((L_i - \Delta L_i, L_i + \Delta L_i) + \mathrm{t}(v_i - \Delta v_i, v_i + \Delta v_i), \theta_i)$.

3. At all times the speeds are within the desirable speed bounds, or Eq. [11.30]

$$v_b^s \leq v_q \leq v_a^s \; \forall \, s \in R_q(\tau(t)), t \leq T \tag{11.30}$$

In other words, for every point in the trajectory $\tau(t)$, and for every point (s) on the vehicle when placed at $\tau(t)$, the speed bounds for the point s must be satisfied. Checking speed bounds ensures that the trajectory being followed can be terminated at any instance (due to changed environment dynamics), and the vehicle can be made to drive straight ahead, giving enough time for itself and every other vehicle to adjust.

11.8 Solution With an Elastic Strip

Let the trajectory being followed at any time be τ. The algorithm additionally defines τ_{obs} to denote the trajectory constructed by *considering only static obstacles*. This term is only defined if the vehicle cannot compute a collision-free trajectory considering all the static obstacles and other moving vehicles. The static obstacle-only trajectory considers only static obstacles, and hence it should be possible to compute such a trajectory. Nonexistence of static obstacle-only trajectory (when a normal trajectory also does not exist) means that the route is blocked. This trajectory assures collision-free avoidance of static obstacles. The difference between the trajectory τ and obstacle-only trajectory is illustrated in Fig. 11.9.

τ_{strat} denotes the *operational strategy* and is a specification of direction (*left* or *right*) by which any obstacle needs to be overcome. This is a compact specification of the homotopy of the trajectory. The strategy variables may be referred for

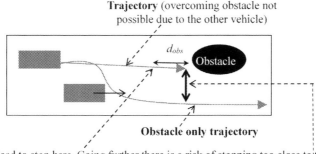

Trajectory (overcoming obstacle not possible due to the other vehicle)

d_{obs}

Obstacle

Obstacle only trajectory

Need to stop here. Going further there is a risk of stopping too close to the obstacle, preventing further motion even by the greatest steering.

Closer the trajectory to the obstacle only trajectory, more away is the final position of the vehicle from the obstacle, lesser is d_{obs}.

Figure 11.9 Concept of obstacle-only trajectory. The trajectory considers only static obstacles and no other vehicle, whereas a normal trajectory considers both.

multiple purposes in the working of the algorithm. As new obstacles and other vehicles are encountered, this term needs to be adequately updated. Initially, τ contains the immediate position only whereas τ_{obs} and τ_{strat} are both null.

11.8.1   Plan Extender

Consider a travel plan τ (known to be feasible as per the current traffic scenario) constructed using a strategy τ_{strat}. The task is to *extend* τ. The assumption is that the entire extended plan thereafter would be followed at a prespecified speed v_q. The extension is carried using LPs (Section 11.4) with the difference that:

1. The *speed of travel* is kept constant. The *computational frequency* is taken to be inversely proportional to the speed indicated by LP. A potential-driven approach operates in continuous mode; however, on computer systems it is implemented in a discrete mode wherein control actions are applied in cycles of some duration. This discretization denotes the resolution of operation or the computation frequency, and is attributed to the computational expense. A normal potential approach modifies the physical speed of the robot keeping this computation frequency constant. Here, a virtual vehicle is made to travel using LP only to get a trajectory. Hence, the speed can be kept constant whereas the computational frequency is altered as per need. The computational cycle time indicates the time span after which the steering action of LP is applied. When the (projected) vehicle is nearer to obstacles, the speed indicated by LP is smaller and hence frequent LP actions are applied, and vice versa. Every call to LP consumes computation. This methodology is hence adaptive to obstacle placements.

2. The *strategy parameters* appearing in Section 11.4 are looked up in τ_{strat}. For obstacles which have already been encountered during the generation of τ and a decision whether to overcome the obstacle from the left or right has already been taken, the same strategy is repeated. Because a part of τ is constructed while overcoming an obstacle from the right, the extended part of τ must also try to overcome the obstacle from the right and vice versa. For every new obstacle witnessed (which does not have an entry in τ_{strat}), all combinations of strategies are separately computed resulting in a variety of plans (eg, Fig. 11.10). Each plan is generated from a different strategy setting for the new obstacles. It is possible that some obstacle will go unnoticed in the process of generation of the bouquet of plans. It may never lie ahead of the vehicle at any point of time by any of the prospective plans. Out of all the plans constructed, the best one as per the performance criterion set in Section 11.7.1 is chosen.

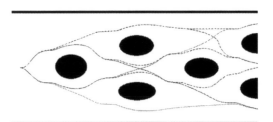

Figure 11.10 Plan extension by every strategy. *Blue (dashed) lines* (black in print versions) represent all possible strategy plans (only strategies experienced by the vehicle). *Red (continuous) line* (grey in print versions) shows the optimal plan.

Box 11.6 Plan Extension

- Use lateral potential to decide a unit move
- Extrapolate the motion of the other vehicles to create scenario at the next time step
  - Keep speed constant
  - Use lateral-potential speed indicator as granularity of motion generation
  - Granularity finer near the obstacle in front and coarser at a distance
- If a new vehicle or obstacle is encountered, try left and right sight-avoidance strategies separately
- Select best plan out of all plans formed by various strategies for various obstacles and vehicles
- If the plan ends with an obstacle
  - Compute additionally an obstacle-only trajectory
  - Re-generate main trajectory using the strategy that resulted in obstacle-only trajectory
  - Similarity results between an obstacle-only and the main trajectory

---

The general approach of extension is also summarized in Box 11.6. In such a system, while operating in unstructured environments in which the structural information about the obstacles may not be available, it is important to classify between the obstacles to decide whether two obstacles are the same or different. The obstacles are assumed to be convex, in which case if a straight line between two obstacles is always collision prone, they represent the same obstacle, and vice versa.

At every step, the extended plan needs to be feasible as per the criterion stated in Section 11.7.3. The plan extension may terminate for any one of three reasons: (1) End of road segment being planned is reached, (2) not possible to move any further due to feasibility criterion due to a static obstacle and (3) not possible to move any further due to the feasibility criterion of another vehicle.

The algorithm can be regarded as *complete* for the cases which end in (1). In the case of (3), the algorithm has an option to travel as per the computed plan, and then to start following the vehicle in front. Human drivers can be reliably followed, whereas following an automated vehicle can be regarded as near complete if the vehicle ahead is following a near-complete algorithm. Because the recurrence is applied to a forward vehicle, the planning algorithm due to termination at (3) can be regarded as complete.

In the case of termination due to (2), there needs to be an assurance of avoiding the static obstacle at a later time. Hence, an attempt is made to compute a plan τ_{obs} considering only static obstacles and all ways of avoiding them. Out of the range of different possibilities, the best plan is considered. Although completeness in construction of τ_{obs} cannot be guaranteed, all possible ways of avoiding static obstacles are considered, and no other vehicle is considered, the algorithm may therefore be stated as near complete. It may be assumed that the vehicles ahead would eventually clear making way for the vehicle to follow the plan τ_{obs}. In lieu of these points, the entire algorithm is *near complete* and (combined with Section 11.8.2) *near optimal*.

The extension is hence given by Algorithms 11.1 and 11.2. Algorithm 11.1 extends a plan consisting of τ and τ_{strat}. Every time a new obstacle is discovered, avoidance from both the left and the right are tried, and the best is retained. For obstacles encountered previously, the strategy indicated by τ_{strat} is used. Algorithm 11.2 computes τ_{obs}, in case τ as computed by Algorithm 11.1 ends in a static obstacle. $\tau_{stratObs}$ denotes the strategy to avoid static obstacles. *Extend1* is called again to make τ and τ_{strat} consistent with the strategy used in τ_{obs} that is $\tau_{stratObs}$. In implementation, all strategies may be stored in line 1 of *Extend()* and fetched in line 4 instead of a new function call.

Algorithm 11.1: Extend1(τ, τ_{strat},v_q)

```
while τ is feasible

  τ ← τ∪ LP(τ[end], τ_strat, v_q)

  if τ[end] encounters an obstacle i and τ_strat(i)=null

  < τᵃ,τᵃ_strat >← Extend1(τ, τ_strat ∪ <i, left>,v_q)
  < τᵇ,τᵇ_strat >← Extend1(τ, τ_strat ∪ <i, right>,v_q)

  < τ, τ_strat>←better of < τᵃ, τᵃ_strat> and < τᵇ, τᵇ_strat>
  break while
  end if
  end while
  return < τ, τ_strat>
```

Algorithm 11.2: Extend(τ, τ_{strat}, v_q)

```
< τ, τ_strat> ← Extend1(τ, τ_strat) with all obstacles
if τ ends with static obstacle
<τ_obs, τ_stratObs>← Extend1(τ, τ_strat) with only static obstacles
< τ, τ_strat> ← Extend1(τ, τ_stratObs) with all obstacles
end if
return <τ, τ_strat, τ_obs>
```

11.8.2 Plan Optimizer

Consider a travel plan τ which, as per the current traffic scenario, needs to be followed with a prespecified travel speed of v_q. The travel plan is *optimized* as the vehicle moves (and scenarios change). The plan τ is converted into a set of coarsely located waypoints (τ') vaguely representing the plan τ. The optimization of τ' is based on the analogy of an *elastic strip* with each waypoint τ'_i representing a virtual vehicle with a movable clamp. τ'_i is modelled as a clamp attached to the lateral axis (Y'). Hence, by the application of forces τ'_i can move along Y', but not along X'. The initial position τ'_0 is fixed. This constraint disallows two waypoints to come close to each other which may slow down the optimization process. τ_i' is influenced by four forces, which are:

1. *Lateral Force:* The force is applied by obstacles laterally left and laterally right in opposing directions, and the magnitude (F_l) is given by a methodology similar to the computation of the side potential in Eq. [11.17].

2. *Internal Force*: Each waypoint τ_i' is attracted by the waypoint ahead τ_{i+1}' (if any) and behind τ_{i-1}' (if any) with a force proportional to the extension given by Eq. [11.31]. Two points clamped to their lateral axis can have a minimal separation equal to their longitudinal separation. Any separation in excess is considered as an extension.

$$F_s = \left(\left\| \tau_{i+1}' - \tau_i' \right\| - \left(\tau_{i+1}'[X'] - \tau_i'[X'] \right) \right) \cdot u\left(\tau_{i+1}' - \tau_i' \right)$$
$$+ \left(\left\| \tau_{i-1}' - \tau_i' \right\| - \left(\tau_i'[X'] - \tau_{i-1}'[X'] \right) \right) \cdot u\left(\tau_{i-1}' - \tau_i' \right) \tag{11.31}$$

Here, $u(x)$ denotes the unit vector in the direction of x.

3. *Cooperation Force:* A plan τ may initially be made only considering the vehicles in the scenario. Additional vehicles may appear later at the rear, and they might then aim to overtake. Extension of τ does not account for cooperation in overtaking, and hence the same is modelled in optimization. The force (F_{coop}) is the same as is given by the back potential or Eq. [11.20]. The only difference is that this force is only applied when the intent of overtaking and the direction of overtaking is eminent.

4. τ_{obs} *Drift Force:* A nonnull value of τ_{obs} indicates that the current plan τ cannot overcome a static obstacle, whereas the plan τ_{obs} can overcome the static obstacle subjected to the absence of the other vehicles. Following τ would mean ending up close to the static obstacle and then having to steeply steer to avoid it, whenever feasible. Following τ_{obs} by waiting for the vehicles to clear and at every step computing the highest possible speed may mean excessive slowing down initially or for a large part of the journey. An attempt is made to induce advantages of both the techniques by *following τ, but slowly drifting towards τ_{obs}*. The force (F_o) is proportional to the distance between the closest waypoint in τ_{obs} (say τ_{obs}^i) applied in same direction. This is given by Eq. [11.32].

$$F_o = \left\| \tau_{obs}^i - \tau_i' \right\| \cdot u\left(\tau_{obs}^i - \tau_i' \right) \tag{11.32}$$

The total force is given by Eq. [11.33].

$$F_{total} = k_l \cdot F_l + k_s \cdot F_s + k_{coop} \cdot F_{coop} + k_o \cdot F_o \tag{11.33}$$

Here, k_l, k_s, k_{coop} and k_o are the associated weights of the different factors.

The *lateral component* of F_{total} is used for deviating τ_i'. Only changes resulting in a feasible plan are admitted. For a sample path, the optimization is shown in Fig. 11.11. The concepts are summarized in Box 11.7.

11.8.3 Complete Framework

The basic hypothesis behind the algorithm is simple. Compute the *highest speed* (v_q) which the vehicle can have as per the current scenario (step A); use v_q to *trim* (step B), *extend/construct* (step C) the plan such that the resulting plan is feasible; and *optimize*

Figure 11.11 Optimization of the plan. The blue line represents the initial plan, which after optimization is given by the red line. If optimized with the aim of maximizing average clearance, the plan is given by the *green (dashed) line*(light grey in print versions).

Box 11.7 Plan Optimization

- *Projected* positions of vehicles and road boundaries repel the trajectory modelled as an elastic strip
- Trajectory Drift
  - Used when trajectory ends in an obstacle (Mode I)
  - Normal trajectory takes the vehicle closer to the obstacle, but later requires steep steering to avoid the obstacle
  - Obstacle-only trajectory may initially excessively slow the vehicle to avoid collisions with other vehicles
  - Hence, trajectory *gravitates* towards the obstacle-only trajectory
- Forces
  - *Lateral Force*: Obstacles at side impel the waypoint in opposite direction
  - *Spring Extension Force*: Strip tries to self-straighten, whereas corresponding waypoints attract
  - *Cooperation Force*: A vehicle at back attempting overtake impels waypoint in the direction opposite to that of overtake
  - *Drift Force*: Main trajectory drifts towards obstacle-free trajectory, if any.

---

the plan (step D). The vehicle, at any time, may have two modes of operation which are Mode I: *travelling with a plan ending at some static obstacle* (τ_{obs} = null); and Mode II: *travelling with a plan not ending at some static obstacle* (τ_{obs} ≠ null). Each of the steps is applied for both modes (denoted *I A*, *I B*…*II D*), whereas switching between modes (denoted *I→II* and *II→I*) is monitored. These steps are applied as the vehicle moves and the scenarios change. Hence, at any time the vehicle may be seen to show initial signs of reacting to any new obstacle (or an obstacle the motion for which has changed a lot as per expectation), deliberating over later course of actions, adapting the plans to any changes in the scenario and optimizing any previously suboptimal plan. The algorithm is summarized by Fig. 11.12 and Algorithm 11.3.

First, each of the four steps (*A* to *D*) are discussed for each mode of operation (*I* and *II*), and these are assembled into Algorithm 11.3. Step *A* is different for both modes *I* and *II*. In *II A*, the speed v_q is the *highest speed possible* for the vehicle as per the feasibility plan considering its current position, taking into account the acceleration limits, aggression factor and deceleration limits. This update rule is formulated in realization of the fact that a fast-moving vehicle at a distance, on being unable to overtake a slow-moving vehicle in front, would be found to exhibit high

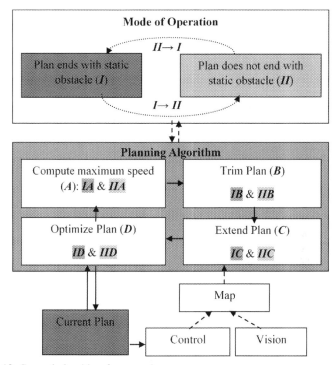

Figure 11.12 General algorithm framework.

speed up to the point when it comes close to the slower vehicle, after which a nonaggressive (as prespecified) deceleration takes place. Hence, at every step the vehicle displays maximum speed, which naturally plays a role in decreasing the total travel time.

Algorithm 11.3: Plan(τ_{obs}, τ, v_q)

```
I B         if (I) and τ_obs is not collision-free
                τ_obs=null, τ=null
II A        if (II), compute v_q using (8.52) with acceleration/
                aggression limits
I/II B      trim τ till it is feasible
            if (I)
I C             < τ, τ_strat > ← Extend1(τ, τ_strat, max(v_q, v_min))
I→II            if τ satisfies (II), τ_obs=null
            else
II C, II→I      < τ, τ_strat, τ_obs> ← Extend(τ, τ_strat, max(v_q, v_min))
            end if
I/II D      Optimize(τ, v_q)
I A         if τ_obs ≠ null
                v_q ← maximum speed to stop at ||τ||-d_obs
            return < τ_obs, τ, v_q>
```

An optimal travel plan consists of optimal trajectory computation as well as optimal speed settings. Working in the joint space of trajectory and speed is not possible due to computational and modelling constraints. However, by constant adjustment of speeds and trajectories separately, a near-optimal path can be obtained.

For $I\,A$, speeding up is disallowed, which arises from the commonly seen behaviour that drivers tend to slow down on seeing a static obstacle if a lane change is not allowed. It is expected that the vehicle would come to a standstill at a distance of d_{obs} before the static obstacle, in which d_{obs} is the minimum longitudinal distance needed to overcome the static obstacle given by Eq. [11.34] and accordingly the speed is iteratively reduced.

$$d_{\text{obs}} = d_{\text{obs}}^{\min} + \Delta\tau_{\text{obs}} \qquad\qquad [11.34]$$

Here, $d_{\text{obs}}^{\min}$ is a small fixed distance to be maintained from the static obstacle at all times, and $\Delta\tau_{\text{obs}}$ is the deviation between τ and τ_{obs} which is taken as the distance between last point in τ and the closest point in τ_{obs}. If the vehicle continues to follow τ it is expected that it would have to *stop* at the end due to no possible subsequent moves and *wait* for the other vehicles to clear. The distance d_{obs} gives enough scope for a vehicle to turn to avoid the static obstacle, in the absence of other vehicles. If the static obstacle is directly ahead of the vehicle, a larger steer would be required as compared to the case when only a small part of the vehicle is ahead of the static obstacle. A larger steer implies a requirement of a greater longitudinal distance which is modelled by $\Delta\tau_{\text{obs}}$. In case the vehicle actually stops, or almost stops, the subsequent extend operation (to move the vehicle when path is clear) is called with a small predefined speed $v_{\min}$.

Trim without ending at static obstacles or $II\,B$ is simply based on the feasibility of the plan τ as per the changed environment (if any) using the set feasibility constraints. For $I\,B$ this needs to be performed for both τ and τ_{obs}. An *infeasible* τ_{obs} (Algorithm 11.3, Line 1) implies emergence of a new static obstacle, in which case everything needs to be invalidated and recomputed to ensure a collision-free τ_{obs}, whereas τ follows the same concepts as $II\,B$. In conception, *extend* or step C is the same for both modes; however, they are called differently in the algorithm to eliminate recomputation of τ_{obs} which is already known. The *extend* algorithm also monitors for mode changes. Step D is same for both modes.

From Algorithm 11.3 it can be seen that the four steps are performed one after the other for mode II. Mode-checking conditions are introduced if a step is particular to a mode. For mode I, the algorithm starts with $I\,B$ and ends with $I\,A$ through $I\,D$. $I\,B$ (Algorithm 11.3, Line 1) is placed at the start because its action invalidates the plan; whereas $I\,A$ is kept at the end because in mode I speed is trajectory dependent, although once initiated, it does not matter which was the first step of the loop and which was the last. An exception to the algorithm is when the vehicle is initially to be found placed in an infeasible manner, or when not even a single step is feasible, in which case the immediate move is as directed by LP with maximum possible deceleration of $-\text{acc}_q^{\max}$.

11.9   Results With an Elastic Strip

The simulation setup is similar to the one employed in Section 11.5. The major difference is that although planning using LP was entirely reactive in nature, the planning using Elastic Strip is a little deliberative in nature. As per assumptions, all the vehicles operate on the same side, that is, all are either outbound or inbound.

11.9.1   Simulation Results

To test the algorithm, a number of diverse scenarios were constructed, each aimed at testing a different aspect of the algorithm. In all the experiments, vehicles are named in the order of their appearance. In all of the experiments, initially the road segment is empty with only the static obstacles (if any). The vehicles are made to emerge at the positions and emergence time. All vehicles are generated in the direction of the road segment. Nothing is known about a vehicle by any other vehicle before it is visible in the scenario. In other words, the vehicles suddenly appear in the scenario requiring other vehicles to adjust their plans accordingly.

The first experiment tests the ability of the algorithm to optimally drive a vehicle within a grid of complex obstacles (Fig. 11.13A). An initial expectation was that the vehicle would take the central route; however, the extra manoeuvres around the obstacles made the trajectory larger. The vehicle always maintained comfortable distances from all the obstacles, while travelling with the maximum permissible speed.

Fig. 11.13B shows a different scenario in which A could not overcome a static obstacle and hence had to slow down and wait for the other vehicles (B and C) to clear. At a later stage, it comfortably places itself before D. The scenario tests the algorithm's ability to restrain a vehicle from accidently pushing in front of another vehicle, for which it may not be prepared. Consider that D's entry is made much earlier, in which case A would wait for it as well. If A is synthetically made to move, D would obediently follow. If D is synthetically made to slow down, A would not wait any longer. Hence, individually the vehicles show expected behaviours; however, having already waited for two vehicles to pass through, A would be called too courteous to be waiting for D, which the algorithm currently simulates.

The next scenario tests the ability of the vehicle to decide the optimal direction for overtaking. The overtaking vehicles have infeasible entry conditions, and the plans are made in multiple extension operations. C overtaking the two vehicles (A and B) in front is visible in Fig. 11.13C, whereas it is likely to be overtaken itself by D. Fig. 11.13D shows how D overtakes C, whereas the other vehicles (A and B) themselves orient to enable the overtaking procedure to happen. All vehicles travel nearly at their maximum permissible speeds.

The last scenario is introduced to consider the traffic dynamics when overtaking is not initially possible and later still rather difficult due to competing vehicles. E entered the scenario before F and hence overtook the initial set of vehicles (C and D) earlier (Fig. 11.13E). However, it appeared that there was no room for overtaking

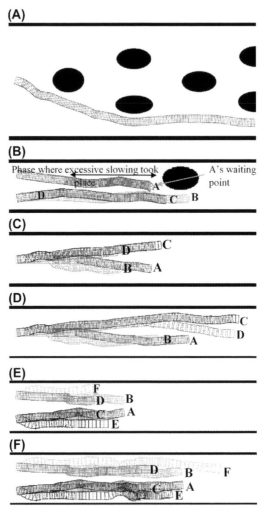

Figure 11.13 Simulation results: (A) computing the optimal path in a complex grid. (B) Slowing down to avoid a static obstacle while waiting for vehicles to clear. (C and D) Multiple overtaking, each from the optimal side. (E and F) Multiple overtakings, each after waiting for the correct time and in the right order.

the vehicle set ahead (*A* and *B*), and hence *E* had to rapidly slow down, to follow *A* ahead, whereas *F* succeeded in overtaking the initial set of vehicles (*C* and *D*). Subsequently, *F* occupied enough room and succeeded in overtaking the vehicle set ahead (*A* and *B*, Fig. 11.13F). Initially, *F* slowed while space was being created, but later accelerated during overtaking. Henceforth, the vehicles drifted to the other side to accommodate overtaking *E*. It must be emphasized that the scenario is largely driven by LP.

11.9.2   Parameters

The effects of the parameters of LP were demonstrated in Section 11.5. The settings used here, however, favour more-sensitive settings, which led to the disadvantages of introducing oscillations and steep turns, whereas the advantages included the capability to avoid obstacles, however far or near. The limitations are, however, eliminated by the *optimize* algorithm, whereas the advantages remain.

Trajectory planning always involves the problem of a trade-off between path length and average clearance. These factors are controlled by the parameters in the *optimize* algorithm. Experiments over a single-obstacle scenario were carried out, for which the participating parameters are k_l and k_s. Both a large clearance and a short path length cannot be simultaneously achieved, which is clear from Fig. 11.14. A large clearance leads to the vehicle quickly placing itself in the middle of the road and then between the obstacle and the road edge, whereas attempts to minimize the path length led to the vehicle travelling very close to the obstacle (Fig. 11.11). Only parameter settings leading to feasible results are plotted.

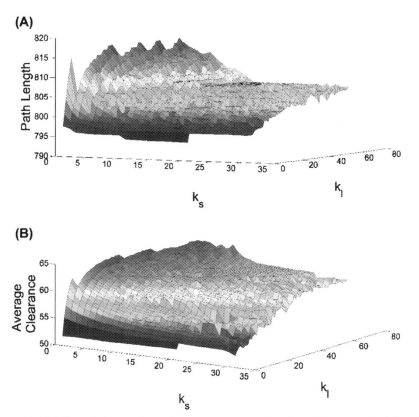

Figure 11.14 Effect of change of algorithm parameters on obstacle avoidance: (A) Path length, (B) Average clearance.

The worst-case complexity of the extend algorithm is $O(2^{(|O| + |R|) \cdot L})$ in which $|O|$ represents the number of static obstacles ahead, $|R|$ represents the number of vehicles ahead with a speed lower than the vehicle being planned, and L is the length of the segment being planned. In the presented scenarios, computing the trajectory for each vehicle takes less than a second, whereas the initial plan generation can take 2.5 s if τ_{obs} also needs to be computed. Scenario 1 takes 8 s to compute in total due to the computation of a single trajectory overcoming all static obstacles.

The high complexity is not a factor of concern considering that the length of segment planned can be adjusted to account for the number of obstacles. Further, for a high number of obstacles, the algorithm may fail to find a single trajectory which simultaneously avoids all the obstacles, forcing the algorithm to terminate.

The algorithm does not (in initial plan construction) have the liberty to reduce the speed. The uncertainties in projecting a vehicle's future motion increase with the vehicle's speed and with longer projections into the future. High uncertainties show as elongated projected positions of the vehicles which may be hard to surpass. In such a case, the algorithm would construct a trajectory to overcome the initial set of obstacles first, and as the vehicle moves, the trajectory would be modified to overcome later obstacles.

Avoiding closer obstacles is easier due to the high degree of certainty in position, speed and preferable speed settings as compared to the obstacles which are much farther away. The inability to overtake a vehicle may, however, result in vehicle following.

11.10   Summary

Optimality and completeness are always a requirement in the implementation of any motion-planning algorithm, which become extremely difficult criteria to sustain when planning in near real-time environments. The aim is then to make practical assumptions about the environment and deliver a solution that best caters to optimality and completeness without taking excessive computation time. Reactive methodologies perform very well in traffic scenarios and are a natural suggestion to make regarding the choice of algorithm, noting their real-time nature and the structured nature of the road environment. The chapter implemented Potential Field methods executing in a completely reactive framework for navigation of the vehicles; and Elastic Strip algorithm which used a little deliberation to add to the optimality and completeness. Both the approaches were restrictive in the sense that they did not allow the vehicles from both sides to share the same road. This was an important and practical assumption to make to meet the algorithmic criterions.

In the first approach, planning was done by using lateral potentials which used potentials from a few sources to decide the lateral position of the vehicle. Vehicles, obstacles and road boundaries on all sides of the vehicle contribute to the potential and ultimately the steering of the vehicle. Each source of lateral potential was carefully chosen to lead to an overall vehicle behaviour which is commonly found in chaotic traffic. The modelling was different for the different sources and represented some intuitive way of dealing with the steering action. The simulated results showed that a vehicle was able to avoid obstacles, navigate amidst other vehicles and overtake other vehicles.

The next approach extended the concept of lateral potential and used elastic strip to make a planning framework which is near optimal and near complete. The approach, however, was not purely reactive in nature. The algorithm at every step extended the plan and optimized it. The strategy variables were chosen by trying all the combinations, from which the best one was retained. The algorithm was near complete and near optimal, whereas the computational cost was only just larger than the reactive techniques. Using the algorithm, the ability to manoeuvre in a complex obstacle framework was presented, while showing complex overtaking, vehicle-following and waiting behaviours.

References

Baxter, J.L., Burke, E.K., Garibald, J.M., Normanb, M., 2009. Shared potential fields and their place in a multi-robot co-ordination taxonomy. Robotics and Autonomous Systems 57 (10), 1048−1055.

Baxter, J.L., Burke, E.K., Garibaldi, J.M., Norman, M., 2006. The effect of potential field sharing in multi-agent systems. In: Proceedings of the 3rd International Conference on Autonomous Robots and Agents, pp. 33−38.

Brock, O., Khatib, O., 2000. Elastic strips: a framework for integrated planning and execution. In: Experimental Robotics VI, Lecture Notes in Control and Information Sciences, vol. 250, pp. 329−338.

Choset, H., Lynch, K.M., Hutchinson, S., Kantor, G.A., Burgard, W., Kavraki, L.E., Thrun, S., 2005. Principles of Robot Motion: Theory, Algorithms, and Implementations. MIT Press, Cambridge, MA.

Fahimi, F., Nataraj, C., Ashrafiuon, H., 2009. Real-time obstacle avoidance for multiple mobile robots. Robotica 27 (2), 189−198.

Gayle, R., Moss, W., Lin, M.C., Manocha, D., 2009. Multi-robot coordination using generalized social potential fields. In: Proceedings of the 2009 IEEE International Conference on Robotics and Automation, pp. 106−113.

Ge, S.S., Cui, Y.J., 2002. Dynamic motion planning for mobile robots using potential field method. Autonomous Robots 13, 207−222.

Jaradat, M.A.K., Garibeh, M.H., Feilat, E.A., 2012. Autonomous mobile robot dynamic motion planning using hybrid fuzzy potential field. Soft Computing 16, 153−164.

Kala, R., Warwick, K., 2012. Planning autonomous vehicles in the absence of speed lanes using lateral potentials. In: Proceedings of the 2012 IEEE Intelligent Vehicles Symposium, Alcalá de Henares, Spain, pp. 597−602.

Kala, R., Warwick, K., 2013. Planning autonomous vehicles in the absence of speed lanes using an elastic strip. IEEE Transactions on Intelligent Transportation Systems 14 (4), 1743−1752.

Khatib, O., 1985. Real-time obstacle avoidance for manipulators and mobile robots. In: Proceedings of the 1985 IEEE International Conference on Robotics and Automation, St. Louis, Missouri, pp. 500−505.

Marchese, F.M., 2006. Multiple mobile robots path-planning with MCA. In: Proceedings of the 2006 IEEE International Conference on Autonomic and Autonomous Systems, Silicon Valley, California, USA, p. 56.

Parhi, D.R., Mohanta, J.C., 2011. Navigational control of several mobile robotic agents using Petri-potential-fuzzy hybrid controller. Applied Soft Computing 11 (4), 3546—3557.

Quinlan, S., Khatib, O., 1993. Elastic bands: connecting path planning and control. In: Proceedings of the 1993 IEEE International Conference on Robotics and Automation, IEEE, pp. 802—807.

Tiwari, R., Shukla, A., Kala, R., 2013. Intelligent Planning for Mobile Robotics: Algorithmic Approaches. IGI Global Publishers, Hershey, PA.

Tu, K.Y., Baltes, J., 2006. Fuzzy potential energy for a map approach to robot navigation. Robotics and Autonomous Systems 54 (7), 574—589.

Yang, Y., Brock, O., 2010. Elastic roadmaps-motion generation for autonomous mobile manipulation. Autonomous Robots 28 (1), 113—130.

Yin, L., Yin, Y., Lin, C.J., 2009. A new potential field method for mobile robot path planning in the dynamic environments. Asian Journal of Control 11 (2), 214—225.

Logic-Based Planning

12.1 Introduction

Considering no communication available between vehicles and the need to have short computation times, *reactive techniques* are a good choice as planning algorithms. Although optimality and completeness cannot be guaranteed with such approaches, the loss may not be practically significant. Chapters 'Fuzzy-based Planning and Potential-based Planning' discussed a set of reactive approaches which were mainly focussed upon selecting the immediate move of the vehicle based on the situational assessment. The approaches presented attempted to align the vehicle in the direction of the road, while simultaneously attempting to overcome obstacles and other vehicles as well as allowing for overtaking by other vehicles. Of course, at any time only one or two such behaviours may be significantly active. Chapters 'Optimization-based Planning, Sampling-based Planning, Graph Search-based Hierarchical Planning and Using Heuristics in Graph Search-based Planning' on the contrary, advocated the use of *deliberation* to generate efficient solutions. An effective balance between the approaches was the case in the use of an elastic strip presented in Section 11.7 in which deliberation was applied to some limited extent.

The working methodology behind all these chapters was to model the algorithm as per the problem requirements and to further embed knowledge of traffic systems and common driving habits as *heuristics* if the same can potentially result in a computational speedup. Such heuristics were mostly used for the purpose of *coordination* between the vehicles while attempting to maintain cooperative driving.

The aim of this chapter is to directly look into the manner in which humans drive and to map that into an algorithmic framework. A potential advantage of such an approach is to escape from the limitations of the base approaches which may be making autonomous driving different from the human counterpart. The key question is *whether human driving is deliberative or reactive*. As motivated in chapter 'Graph Search-based Hierarchical Planning', the planning seems to be highly reactive wherein humans tend not to think about actions much into the future, while driving their way out. General driving is marked by adjusting one's positions to the changing road curvature and looking for ways to avoid obstacles and other vehicles. However, if any specific driving trait is seen, eg, overtaking, there is some deliberation in the action. Deliberation is in the forms of deciding whether to initiate overtaking, how close to overtake, whether to continue overtaking or to cancel it etc. Such tasks cannot be performed without deliberating over all the vehicles concerned in the system. Hence, although the overall plan is reactive, the individual actions may be deliberative.

In practice, it is difficult to draw a line between *being deliberative* and *being reactive*. Situational assessment in reactive-planning techniques can be somewhat deliberative, whereas even the deliberative techniques do not deliberate much into the future and can hence be called reactive to a much broader sense of the environment. For these discussions, however, any attempt to draw even a part of the trajectory for planning is regarded as nonreactive.

The algorithm is modelled as a collection of *behaviours*. The different behaviours are studied from natural driving in countries with unorganized traffic systems. The objective behind the algorithm is to be able to move an autonomous vehicle flawlessly within unorganized traffic. The algorithm must not make the vehicle either wait a long time to move ahead (starve) or cause accidents. The algorithm must also allow the vehicle to move by means of the optimal route and speed, at the same time allowing the faster vehicles to overtake, and the vehicle to overtake slower vehicles whenever possible.

At any time instance, the vehicle assesses the current situation and selects the most appropriate behaviour. Behaviours have a *precondition* which must be true for the behaviour to be initiated. Each behaviour has an associated *priority*. At any instance, the highest-priority behaviour is called the precondition for which is valid. Unlike reactive methods, deliberation may be used to check the possibility of some behaviour or for construction of the behaviour.

Driver aggression has an important role to play in the dynamics of an organized traffic which is limited to accelerating or decelerating one's own vehicle, performing risk-prone cut-ins and speed lane changes, use of the horn etc. (McGarva and Steinerm, 2000; Lajunen et al., 1999; Shinar and Compton, 2004). In the absence of speed lanes, aggression plays an even greater role and governs critical decisions such as whether to overtake a vehicle in front.

Unorganized traffic has an apparent higher risk of accidents, and this needs to be a factor for consideration in design of a planning algorithm. A study of the accident causes for organized traffic (Clarke et al., 2005; Paulsson and Thomas, 2005; Wang et al., 2010) and unorganized traffic (Mohan and Bawa, 1985; Mohan, 2002) narrates the common reasons for accidents. For unorganized traffic, two wheelers are particularly prone to accidents (Clarke et al., 2004; Jain et al., 2009). Analysis reveals that most, if not all, of the head-on collisions on noncrossing scenarios are caused by factors absent in autonomous vehicles which have active intelligent sensing and actuation systems. However, this does place emphasis on keeping a *safe distance* from the vehicle in front to avoid accidents even if the vehicle in front stops suddenly. Accidents may however be possible when one vehicle crashes into the *side* of another vehicle, producing slight damage to both vehicles. Excessive steering is one reason for such a crash which, in a nonspeed lane system, may be produced due to an attempt to incorrectly fit into the available space or wrongly trying to give another vehicle space in the road.

It can be seen that the approach adopted in this chapter, as well as an *elastic strip*, seems to make use of both deliberative and reactive approaches for effective planning. The two approaches have different working philosophies. Although the approach used in this chapter attempts to model human driving as a set of behaviours which are implemented in an algorithm framework, the elastic strip has many human-driving behaviours arising naturally because of its modelling. Being slightly more deliberative, the elastic strip algorithm is near optimal and near complete, contrary to the approach used in this chapter. However, many times *comprehensibility* of results and extensibility of logic is an additional factor of concern to evaluate systems, which is higher in the current approach.

The basic concepts are summarized in Box 12.1. The key takeaways of the work are summarized in Box 12.2. Contrary to the approaches discussed in chapters 'Optimization-based Planning, Sampling-based Planning, Graph Search-based Hierarchical Planning, Using Heuristics in Graph Search-based Planning and Fuzzy-based Planning', the *single-lane overtaking* here is without intervehicle communication (unlike chapters: Optimization-based Planning, Sampling-based Planning, Graph Search-based Hierarchical Planning, Using Heuristics in Graph Search-based Planning and Fuzzy-based Planning) and applicable for all kinds of roads and vehicles (unlike chapter: Fuzzy-based Planning, because deliberative means are used for individual decision-making). As opposed to chapter 'Potential-based Planning', the modelling of driver aggression is on the basis of vehicle's planned position unlike maximum acceleration.

Box 12.1 Key Aspects of Logic-based Planning

Pros
- Direct interpretation of observed driving behaviours
- Each behaviour has a precondition of initiation and indicates the trajectory to be taken during implementation of the behaviour.
- Ties are resolved by priorities.

Issues
- Modelling of the individual behaviours

Concept
- Balance between being deliberative and being reactive

Assumptions
- *Straight Road*: Road will not suddenly narrow, thus making an initiated overtaking infeasible
- *Infinitely Long Road*: A vehicle may take long to move aside to enable an overtaking procedure; the road would not end while this happens, resulting in no overtaking.
- Vehicles projected to *move straight* with same speeds unless apparent with the pose

Aggression Factor
- Aggression leads to better travel for the aggressive driver at the cost of the other drivers and risk
- Modelled as minimum separation that must be kept with a vehicle/obstacle at side
- The maximum separation also under threshold
- Aggressive overtakings are closer to the vehicle's occupancy

---

Box 12.2 Key Takeaways of Logic-based Planning

- Vehicle behaviours in *unorganized traffic* are studied and identified.
- Each *behaviour* is modelled in an algorithm used for the motion of autonomous vehicles in unorganized traffic.
- In particular, the complex behaviour of *single-lane overtaking* is studied wherein a driver slips into the wrong lane to complete an overtaking procedure. The cases of initiation, cancellation and successful completion of the behaviour are studied.
- Driver *aggression* is studied and modelled as an algorithmic parameter in such traffic.

Segments of Sections 12.1—12.4 and 12.6—12.8 have been reprinted from Kala and Warwick (2013).

Segments of Sections 12.1, 12.5—12.8 have been reprinted from Kala and Warwick (2015).

First, a very brief overview of the relevant literature is given in Section 12.2. The problem and solution modelling aspects are highlighted in Section 12.3. Section 12.4 lists all the behaviours modelled to solve the problem. The list includes only those behaviours wherein a vehicle sticks to its own side of travel without going into the 'wrong side'. Section 12.5 illustrates a special behaviour known as the single-lane overtaking behaviour wherein a vehicle may go to the wrong side to complete an overtaking. Section 12.6 gives the entire algorithm framework. Section 12.7 gives the simulation results, whereas Section 12.8 provides the summary.

12.2 A Brief Overview of Literature

The closest related work is that of Paruchuri et al. (2002), who simulated vehicle behaviour on straight roads and crossings without traffic lights. They discussed vehicular behaviours in which one vehicle could follow another vehicle, overtake another vehicle or multiple vehicles could result in a traffic jam. The limitations include the assumption of intervehicle communication, no cooperation between vehicles, the algorithm seemed to be lane-prone and the overtaking decision module did not generalize to a high number of vehicles with unorganized patterns. Meanwhile, Leonard et al. (2008) chose Rapidly-exploring Random Trees (RRT) for planning a vehicle, although the algorithm was noncooperative. Furda and Vlacic (2011) used Automata by which they could show the behaviours of maintenance of position on the road, maintenance of a safe distance from other vehicles, collision avoidance etc. The selection of behaviour was through Multicriterion Decision-making. The assumption of speed lanes, however, makes the behaviour set small and simple.

A behaviour-based approach is common in the domain of multirobot path planning in which the problem is to move the robots from their sources to goals. Most approaches (Aguirre and Gonzalez, 2000; Beom and Cho, 1995) use sensors that sense the environment around, assess the situation, and use fuzzy-based methods to move the robot. Although mobile robotics often considers open spaces as the map, traffic scenarios have predefined roads which are long and narrow (in comparison) — this makes a big difference. In traffic scenarios, no particular point is regarded as the source or goal, and safety is an important concern. Hence, these methods are different from road-following behaviour in traffic scenarios, and they have different behaviours and dynamics.

In the work of Alvarez-Sanchez et al. (2010), every robot step was taken to maximize the distance from the obstacles. Here, a similar idea is used whereby a vehicle attempts to maximize its separation from other vehicles. The maximization effort is, however, under a threshold and only lateral separations are considered. This basic notion is further extended to develop complex behaviours and cooperative measures. Jolly et al. (2010) modelled the various behaviours observed in robotic soccer. The rule set was built with discrete field regions, which were later learnt using a fuzzy neural network. Whilst driving exhibits behaviours that are more complex, driving behaviours are clearly discrete in nature.

12.3   Problem and Solution Modelling

12.3.1   Problem Modelling

The basic problem considered here is to safely move a vehicle in a scenario amidst multiple robotic or manually driven vehicles. Hence, consider a road segment which is characterized by its two boundaries, say $Boundary_1$ and $Boundary_2$, lying on either side of the road. A vehicle R_i is assumed to be rectangular of length len_i and width wid_i. The speed of the vehicle is changed by acceleration (or retardation) over the interval $[-accMax_i, accMax_i]$. The maximum speed of any vehicle is fixed to a value of $vMax_i$.

Although the general planner developed here works for all types of roads having any kind of curves, this behaviour study is restricted in the first instance to a *straight road*. As a result, while initiating any behaviour a vehicle assumes the road ahead to be straight and *infinitely long*. The reason for this is that completion of a manoeuvre might involve cooperation of other vehicles about which the vehicle initiating the behaviour may never be sure. Because it will take a finite time for a manoeuvre to complete in cooperation with other vehicles, the vehicle needs to assume that the road is long enough for this. Further, the road is assumed straight to overcome decision-making on dynamics based on the curve. Vehicles travelling on the same curve would differ in their time of travel depending upon their lateral position on the curve.

Each vehicle R_i keeps a track of all the vehicles in its neighbourhood. Tracking, however, is error prone, and such errors can be especially high if the other vehicle is at a distance. According to the uncertainty model used, the position of vehicle R_j

as measured by vehicle R_i on the X' axis is given by Eq. [12.1] and the speed of vehicle R_j as measured by vehicle R_i is given by Eq. [12.2].

$$\Delta x'_{ij} = puMag \left(e^{puSpread.abs\left(x'_i - x'_j \right)} - 1 \right)$$

$$x'_{ij} = \left[x'_j - \Delta x'_{ij}, x'_j + \Delta x'_{ij} \right]$$

[12.1]

$$\Delta v'_{ij} = vuMag \left(e^{vuSpread.abs\left(x'_i - x'_j \right)} - 1 \right)$$

$$v_{ij} = \left[v_j - \Delta v'_{ij}, v_j + \Delta v'_{ij} \right]$$

[12.2]

Here, *puSpread* is the positional uncertainty spread and *puMag* is the positional uncertainty magnification factor. Similar factors for speed are *vuSpread* and *vuMag*. $X'Y'$ denotes the road coordinate axis system.

The *visibility* of any vehicle is restricted to a distance of $\Delta(v_i)$ along the length of the road. In the present implementation this factor is kept linearly proportional to the current speed of the vehicle v_i (or $\Delta(v_i) = k.v_i$). A vehicle would not be able to see any obstacle or another vehicle beyond this factor. In natural driving this factor is infinite for straight roads and depends upon the degree of curvature for nonstraight roads. However, the factor is restricted to the limits of the computational load of the obstacle discovery algorithm.

12.3.2 Single or Dual Carriageway

A common characteristic of traffic systems in many countries with narrow roads is that there is no physical barricade for inbound and outbound traffic on a single carriageway. The road may be divided by markers or drivers may assume that half of the road is for inbound traffic and the other half for outbound traffic. Hence, they stick to their own side, normally following the vehicle ahead. In an unorganized traffic scenario, there is, though, the chance of overtaking for motorcycles and smaller vehicles. On the contrary some countries have a dual carriageway, wherein the inbound and outbound traffic is divided by a physical barricade.

The problem is first solved with the assumption of *one-way roads only* or that the entire traffic flow is in one direction only (outbound or inbound) taking the dual carriageway model. With this assumption, an algorithm is designed to present interesting vehicular behaviours in complex scenarios. The assumption is not difficult for the case of a single carriageway, because a road can always be assumed to have a virtual boundary dividing the two directions of travel. Given this virtual boundary, for planning, each side can display its own behaviours, and the two sides do not need to interact with each other at any point of time.

The algorithm is then extended and generalized for performance when the road is *two-ways*, wherein both inbound and outbound traffic operate on the same road without any physical barricade in between. The focus here is to only study behaviours in which vehicles travelling in the opposite direction *interact* in some way or another. General travel, when the vehicles remain on their own side, is exactly described by the algorithm with the assumption of one-way roads.

This raises the important question: *should vehicles be allowed to slip across to the* 'wrong' *side* of a single carriageway with unbarricaded inbound and outbound traffic, for some time? In general, this is not regarded as safe, even for human drivers, because a driver slipping over to the wrong side may not be able to return to the correct side and might therefore cause an accident or a traffic jam. Hence, for nonautonomous traffic, even if it appears that for a vehicle to occupy some part of the road on the opposite side, leading to better traffic bandwidth and travel efficiency, it must be avoided at all costs due to the risks involved. Unfortunately, this eliminates much of the possible interesting behaviour involving the mixing of traffic from opposite directions.

However, it is common for a vehicle to slip onto the wrong side just to *overtake*. Overtaking is the key factor contributing to efficiency in diverse-speed unorganized traffic. Hence, every attempt is made to enable a faster vehicle to overtake a slower vehicle. If a faster vehicle considers it safe enough, it should therefore be allowed to slip onto the wrong side, complete the overtaking manoeuvre and return to the correct side whenever feasible. Such overtaking may thereby greatly enhance travel efficiency, although making it a little riskier for the traffic travelling on the opposite side of the road if the assessment of the overtaking vehicle is poor. Here, such overtaking is modelled as a *single-lane overtaking* behaviour.

The single-lane overtaking is largely inspired by narrow road traffic systems with one lane per side of travel. In such traffic systems, the addition of a slow vehicle can almost block a complete road. Hence, it is important for an autonomous vehicle to have some way of overtaking and allowing itself and the other vehicles to drive efficiently. Even in countries with organized traffic, human drivers tend to take any opportunity to overtake in such scenarios. On many occasions, all vehicles collectively decide a strategy. A lane-following law-abiding autonomous vehicle can be troublesome in such situations, if such a behaviour is not modelled. Another source of motivation is obtained from zones in traffic systems, which a vehicle may use for overtaking. Similarly, such behaviours are common in countries with unorganized traffic and are usually taken with great caution.

Overtaking, even in general, is regarded as a special behaviour and has been extensively studied in the literature, including overtaking assessment and actually performing overtaking. Overtaking involves a change of lane to the overtaking lane, driving ahead of the vehicle to be overtaken and then returning to the driving lane. Overtaking in organized traffic is easier to carry out as a set of lane changes, because the driving lane and the overtaking lane are on the same side and in the same direction of travel, so this minimizes the risks involved.

12.3.3   Aggression Factor

Aggression is a commonly used term to assess a driver's performance. Though there is no clearly defined metric to express how aggressive a driver is, aggression is qualitatively assessed with close cut-ins, too many lane changes, close overtakings, sharp speed changes and manoeuvres etc. The cons of aggressive driving include passenger discomfort, risk of accidents, and discomfort for other drivers who have to be more alert and reactive. The pros of aggressive driving include usually a shorter travel time for the aggressive driver. This advantage of the aggressive driver may trigger a corresponding disadvantage for a nonaggressive driver, who may have to take cautious steps to ensure safety causing their travel time to be longer. Traffic is an adversarial system, in which the different vehicles may not be able to simultaneously exhibit their optimal plans or, in other words, the different vehicles compete to be near optimal. Aggressive drivers are usually more inclined towards a win situation (with associated cons of aggressive driving) whereas the less aggressive drivers are more inclined towards the lose situation. As an example, consider a scenario in which a road narrows to allow only one of two competing vehicles to go forth. In such a scenario a more aggressive driver would normally make their way through. Of course, if both drivers are very aggressive, a collision could occur.

One of the important tasks in the algorithm is to model aggression as an algorithmic factor. In all displays of aggressive driving it is seen that the driver keeps very little separation from other vehicles or obstacles. The distance from the vehicle directly ahead is often enough to ensure no collision if the vehicle suddenly stops, and it barely contributes to aggression. The separation which constitutes aggression is the *separation that a vehicle plans to keep from the other vehicles and obstacles at its side*. In a lane-based system, this parameter is preset as the width of a lane minus a vehicle's width. In a nonlane-based system, this is a parameter.

The algorithm must always plan to keep a separation of more than $separMin_i$ at the side, in which $separMin_i$ is called the aggression factor. A larger separation would be preferred for increased safety, even if it be at the expense of the path length or travel time. However, after a separation of $separMax_i$, the path is regarded safe enough and any attempt to increase separation further would only be done if it brings an incentive in terms of the path length or travel time. This is studied here by an example. Consider a vehicle is driving behind another vehicle and wants to overtake it. If the vehicle behind cannot construct an overtaking trajectory, such that the separation of the projected motion is always greater than $separMin_i$ from the sides, the overtaking would not happen. However, if the road is wide enough, the overtaking would happen with the two vehicles separated by a lateral separation of approximately $separMax_i$. If the overtaking vehicle attempts to steer any farther, it would make its path worse while adding no safety. If the road is not wide enough, the lateral separation would be kept as large as possible.

12.3.4 Algorithm Modelling

The entire algorithm is modelled as a collection of *behaviours* that the vehicle must display. The different behaviours are interrelated. The overall behaviour to be displayed depends upon the surrounding vehicles. In this manner, the approach is similar to the use of a *rule-based system* for navigation of the vehicle. In a system with a speed lane, the rules can only produce a single output indicating the speed lane to travel along. However, here, speed lanes are not in play, hence many more behaviours must necessarily be exhibited. Although some of the behaviours may sound similar to a speed lane-based system, the ability of a vehicle to showcase similar behaviours in the absence of speed lanes adds to the design challenge.

Each behaviour has a set of *conditions* which must be true for the vehicle to display that particular behaviour. Even whilst one behaviour is being displayed, the conditions for other behaviours need to be checked. Each behaviour has its own *priority* which is used to decide which behaviour needs to be executed when a number are possible. The notion of behaviour here is simple. Consider driving on the road. In some situation a driver may decide to overtake a vehicle ahead. In another situation it may be necessary to overcome an obstacle. These actions constitute different behaviours. At any time instance, the decision taken results in the behaviour which is best for a vehicle in cooperation with the other vehicles. Here, two such behaviours are not simultaneously showcased; say overtaking a vehicle and avoiding an obstacle, an actual behaviour would only be one of them.

With the stated modelling scenario and assumptions, the set of behaviours is constructed with the following hypotheses:

1. No interesting behaviour may be initiated if it appears to have a separation less than $separMin_i$ in the future. On the other hand, if either currently or in the future it appears that the separations may drop below $separMin_i$, preventive measures need to be initiated.
2. The aim of the algorithm is to attempt to increase the separation as much as possible (upto $separMax_i$). Hence, obstacles and vehicles should be avoided by the maximum possible separations.
3. In case a separation of more than $separMax_i$ is available, the vehicle must attempt to minimize its *lateral movements* (amount of steering). Hence, obstacle avoidance and vehicle overtaking would happen from the sides which are closer to the vehicle. This also suggests that in general a vehicle would always attempt to travel parallel to the road.
4. For plans involving the same lateral moments, a plan which steers to the new lateral position in a *single attempt* is preferable to a plan which does the same in multiple steering events.
5. A vehicle must attempt any possible *safe overtaking* as well as *cooperate* to make a safe overtaking possible.
6. Every steering attempt must be as *smooth* as possible. However, smoothness may be put under a threshold if it improves any other performance metric. Hence, vehicles may be made to reasonably quickly align to overtake, to allow overtaking or to increase separations.
7. In the long term, a vehicle may attempt to maximize separations more than $separMax_i$ which means slowly *drifting to the centre* of the road on an empty road.
8. At every instance the attempt is to have the *maximum possible speed* considered safe as per the situation.
9. If a vehicle (by lane change) comes in front of another vehicle, it must give enough time for the other vehicle to *adjust its speed* to avoid any possible accident.

12.4   Behaviours

The algorithm is simply a set of behaviours which have their own priorities, preprecon-ditions and effects. This section models the behaviours of obstacle avoidance, centring, lane changing, overtaking, being overtaken, maintaining separation steer, slowing down, discovering conflicting interests and travelling straight.

12.4.1 Obstacle Avoidance

One of the most basic behaviours in the system is the ability of the vehicle to *avoid an obstacle*. Consider that the current position of the vehicle R_i is (x'_i, y'_i, θ_i). For trajectory generation, first, the best avoidance point is computed considering all the obstacles that lie within $\Delta(v_i)$.

Consider the Y' axis at a distance of p'_x along the X' axis. Segments of this line would lie in regions without obstacles and segments within obstacles. A vehicle may either avoid an obstacle on its left side or right side. The vehicle simply selects the *widest segment* for its navigation. Let the widest segment be defined by the set of points $[a, b] \subseteq [Boundary_2(p'_x), Boundary_1(p'_x)]$. Within this segment, the vehicle must attempt to move such that it has to change its relative position on the Y' axis by a minimal amount, and the obstacle is avoided by as large a separation (under a threshold of $separMax_i$) as possible on the Y' axis. By these objectives the point of avoidance $P\left(p'_x, p'_y\right)$ at a distance of p'_x is given by Eq. [12.3].

$$
p'_y = \begin{cases}
y' & a \le y'.rl - wid_i/2 - separMax_i \le y'.rl \\
 & + wid_i/2 + separMax_i \le b & \text{(i)} \\
(a+b)/2rl & b - a \le wid_i + 2separMax_i & \text{(ii)} \\
(a + separMax_i + wid_i/2)/rl & a > y'.rl - wid_i/2 - separMax_i & \text{(iii)} \\
(b - separMax_i - wid_i/2)/rl & b < y'.rl + wid_i/2 + separMax_i & \text{(iv)}
\end{cases}
$$

$$[12.3]$$

Here, wid_i is the width of the vehicle and rl is the width of the road given by $rl = \|Boundary_1(x') - Boundary_2(x')\|$.

Eq. [12.3(i)] covers the scenario in which no obstacle lies in the vehicle's normal path at a distance of $separMax_i$ from both sides. In some cases, the segment might however be too small for a vehicle to move without the desirable maximum separation, in which case it simply tries to maximize the separation by driving right in the middle as stated in Eq. [12.3(ii)]. In case the segment is wide enough to allow a separation of more than $separMax_i$ on both sides and the vehicle by its normal course does not enjoy this separation, planning is done so that the least amount of steering or change in relative position is required. If the point of avoidance lies towards the left side of the infeasible segment, the scenario is given in Eq. [12.3(iii)] and if it lies on the right side of the feasible segment, the scenario is given in Eq. [12.3(iv)]. The various cases are depicted in Fig. 12.1. The additional line shown is the planned trajectory which is not actually traced because an obstacle is encountered in the line of motion.

Out of all the competing points, the best obstacle avoidance point OA is chosen which is the *rightmost point* for a right-turn obstacle avoidance (or leftmost for a left-turn obstacle avoidance) along the considered road length (x'_i to l). Selection of the rightmost (or leftmost) point assures that no steering is required for obstacles in the road length beyond OA whereas a vehicle should be able to drive to OA with minimum separation. In other words, the intent is to *steer as much as possible and then drive straight.*

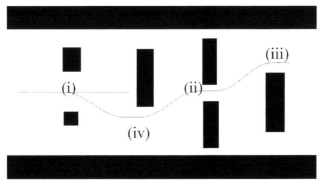

Figure 12.1 Computation of avoidance point in obstacle avoidance strategy using Eq. [12.3(i—iv)].

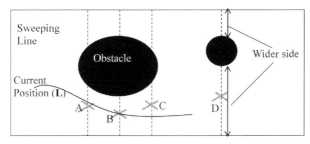

Figure 12.2 Obstacle avoidance point computation procedure. Best avoidance points are A, B, C and D based on widest segment and widest separation. Overall best is B. If A is selected, the vehicle will need two turns: L to A, A to B. If B is selected, the vehicle will need one turn: L to B. Vehicle cannot construct a collision-free trajectory from L to C or D.

The trajectory of a vehicle τ from its current position R_i to OA is constructed using spline curves. Every point $\tau(t)$ on the curve must be in the free configuration space ζ_i^{free} of the vehicle (which accounts for a minimal separation of *separMin$_i$*) for the trajectory to be accepted. The selection of the obstacle avoidance point is illustrated in Fig. 12.2. The procedure is given by Algorithm 12.1 and also simply summarized in Box 12.3.

Algorithm 12.1: ObstacleAvoidance(R$_i$, map)

```
Calculate P(pₓ', pᵧ') ∀ pₓ'∈[xᵢ', xᵢ'+Δ] as per Eq. [12.3]

l ← xᵢ'+Δ
while true
    OA ← P₁(p₁ₓ', p₁ᵧ'): abs(p₁ᵧ'-y') > abs(p₂ᵧ'-y') ∀ p₁ₓ', p₂ₓ'∈[xᵢ', l],
    P₁ ≠ P₂, P₁, P₂∈P
    τ ← curve(Rᵢ, OA)
    if τ(t)∈ζᵢᶠʳᵉᵉ ∀ t, return τ
    else l ← OA(x') - 1
end while
```

Box 12.3 Obstacle Avoidance Behaviour Procedure

- Traverse a sweeping line along the length of the road
- For every obstacle-free segment of line, find best avoidance point
  - Located on widest obstacle-free segment
  - Maximize separation
  - Minimum deviation from vehicle's lateral position on the road (minimum steering)
- Find overall **best avoidance point**
  - Rightmost or leftmost amongst all the candidate points (no steering should be required for subsequent part of obstacle)
  - Leads to a collision-free trajectory
  - Look at nearer obstacles if farther ones lead to collision
- Select the best obstacle avoidance point
- Join current position to avoidance point by a trajectory
- The trajectory is checked for feasibility

12.4.2   Centreing

Even though it might be comfortable for the vehicle to travel with its current relative position y' on the Y' axis, it may alternatively plan to *slowly drift towards the road centre*. Because the drift is slow, it does not require a large steering effort on the vehicle's part. Advantages of driving in the centre include easier adaptation to bends/curves and variations in road widths. To initiate this behaviour, slower vehicles or obstacles must not be ahead of the vehicle along the road length $\Delta(v_i)$. The vehicle must itself be travelling at a reasonable speed, which ensures that this behaviour is not invoked at low speeds, when it is not relevant. The vehicle chooses the point of aim at the centre of the road width or $CEN(x_i' + \Delta(v_i), 0.5)$. Spline curves are used for trajectory generation.

12.4.3   Lane Change

A *lane change* in itself is not behaviour but rather a precondition which must be true for a number of behaviours including obstacle avoidance and centreing. In other words, it may be seen as a superbehaviour which incorporates a number of behaviours in itself. The enabling condition of this superbehaviour leads to the enabling conditions of the subbehaviours. A possible lane change is queried when any vehicle needs to change its relative position y' by a reasonable amount.

Every vehicle, as a safety precaution, needs to ensure that it keeps a sufficient *safe distance* from the vehicle in front, considering its own acceleration limits. This means that even if the vehicle in front was to stop suddenly, no collision would occur. However, this does not account for the scenario wherein another vehicle might *suddenly cut in* from the side, thereby forcing the vehicle to recompute its speed and adjust accordingly. If the vehicle is travelling at a good speed and a vehicle suddenly comes in front from the side, it may not be able to decelerate fast enough to avoid a collision. Even if

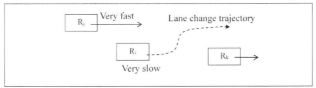

A very slow R_i will come ahead of a very fast R_j without collision, but R_j will not later have enough time to slow down and avoid collision with a very slow R_i.

Figure 12.3 Intuition behind lane change behaviour.

only a part of the vehicle cutting in lies ahead of the vehicle in question, a collision is possible. The concept is shown in Fig. 12.3.

Hence, any change in lane is allowed only if the vehicle can complete the entire trajectory of lane change without needing any vehicles in the rear to slow down. Consider that the vehicle R_i needs to change its speed lane as per the trajectory τ. In doing so it changes its relative position (or lane) from y_i' to $y^{t'}$. The time required by the vehicle to do so is given by Eq. [12.4].

$$t = \|\tau\|/v_i \qquad [12.4]$$

Here, $\|\tau\|$ indicates the total length of the curve τ and v_i is the current speed of the vehicle. It is assumed that no other vehicle lies within the speed lane occupancy of y_i' to $y^{t'}$ of R_i.

Vehicle R_i assumes that all other vehicles continue to move at the same speed with the same relative positions as per the uncertainty model till time t. Hence, the predicted path of a vehicle from the current time to time t may be given by Eq. [12.5] which is an extension of Eqs [12.1] and [12.2].

$$x_{ij}'(t) = \left[x_j' - \Delta x_{ij}' + \left(v_j - \Delta v_{ij}'\right).t, x_j' + \Delta x_{ij}' + \left(v_j + \Delta v_{ij}'\right).t\right] \qquad [12.5]$$

It is assumed here that while the vehicle R_i travels on its trajectory τ, other vehicles plan as per Eq. [12.5]. If this results in a potential collision, vehicle R_i will need to take an alternative behaviour. It is evident that only vehicles having any part of their vehicle within the lane occupancy of y_i' to $y^{t'}$ of R_i need to be checked. Further, if vehicle R_i has a smaller operating speed than any of the other vehicles, it might have to lower its speed to avoid a collision. However, it would have sufficient time to do this.

12.4.4 Overtaking

Because vehicles potentially have large differences in speed, it may not be 'fair' for a vehicle R_i with high-speed capability to follow a slow vehicle. Hence, if it is possible it tries to *overtake* the slower vehicle in front. The vehicle in front is first assessed. Any part of the vehicle should lie within the relative position of $[y_i' - (wid_i/2 + separMin_i)/rl, y_i' + (wid_i/2 + separMin_i)/rl]$ ahead of vehicle R_i.

The term denotes the speed lane occupancy of R_i with minimum thresholds at both ends. Here, rl is the current road width. Although multiple vehicles are potential candidates for the vehicle R_i to overtake, it is essential to select one such vehicle only. Hence, the vehicle which is *closest* to the vehicle R_i (in terms of its position on the Y' axis) is selected unless it lies far apart in terms of distance on the X' axis. In simple terms this means that a vehicle a long way ahead is not as significant as a vehicle a reasonable distance ahead, whilst a vehicle directly in front is preferred over a closer vehicle which is to the side. This means it is not only the vehicle directly in front which needs to be overtaken, but also any vehicle that is in front of the vehicle at a distance of *separMin$_i$* from both sides.

Consider that the vehicle in front is R_j. Let l_j be the distance available to the left of R_j measured along the Y' axis (beyond which some other vehicle or road boundary lies) and r_j be the distance available to the right of R_j. Note that if no vehicle is in front, the vehicle does not exhibit this behaviour. Two scenarios arise — *direct overtaking* and *assistive overtaking*. The overall concept of overtaking is also summarized in Box 12.4.

12.4.4.1 Direct Overtaking

Direct overtaking can take place by a vehicle R_i when there is sufficient distance on either the left- or right-hand side of another vehicle R_j so that R_i may simply steer and slip in. To slip in, R_i requires a minimum distance of $wid_i + 2.separMin_i$ considering the minimum separation that it must maintain from both ends. The first decision to be made, however, is the *side* that overtaking will take place — that is whether R_i overtakes R_j on the left or right. Mostly, overtaking on the right is what happens in countries which drive on the left and vice versa. However, in a situation with wide lanes and no speed lanes, overtaking on the right may not always be possible.

Box 12.4 Overtaking Behaviour Procedure

- Decide the **vehicle to overtake**
- Decide the **side to overtake**
  - If vehicle to be overtaken more towards left, overtake from right, and vice versa
- Decide the **point of overtaking**
  - **Laterally** attempt to maintain maximum separation (under threshold) and minimum deviation from the current lateral position
  - Point of overtaking is taken distant enough **along the length of the road** to allow the vehicle easily correct its lateral position
- Do other vehicles need to move?
  - **Direct Overtaking**: Sufficient separation available — movement of other vehicles not mandatory
  - **Assistive Overtaking**: Sufficient separation not available — movement of other vehicles mandatory
- Construct **overtaking trajectory**

The decision only needs to be made when sufficient distance from both sides is available, else only the side having sufficient distance can host the overtaking. A simple overtaking rule is defined here, which states that in the case when R_i is *more towards the left side* of vehicle R_j, overtaking takes place on the left and vice versa. If R_i is exactly behind R_j, overtaking takes place on the right.

The other task to be carried out is to decide on the position or the *point of overtaking* $P\left(p'_x, p'_y\right)$ to which the vehicle must travel. The trajectory τ is calculated until a point $P\left(p'_x, p'_y\right)$ on the road traversing which the overtaking would have happened, or the vehicles would be aligned, so that travelling straight would complete the overtaking. The decision is broken into the computation of p'_x and p'_y. The value p'_y denotes the relative position of the vehicle R_i on either side of the vehicle R_j as decided. The same concepts are used as noted in Eq. [12.3]. The value of p'_y is given by Eq. [12.6].

$$
p'_y = \begin{cases}
\left(y'_j.rl + wid_j/2 + separMax_i + wid_i/2\right)\big/rl & l_j \geq wid_i + 2separMax_i \wedge side = left & \text{(i)} \\[2mm]
\left(y'_j.rl + wid_j/2 - l_j/2\right)\big/rl & l_j < wid_i + 2separMax_i \wedge side = left & \text{(ii)} \\[2mm]
\left(y'_j.rl - wid_j/2 - separMax_i - wid_i/2\right)\big/rl & r_j \geq wid_i + 2separMax_i \wedge side = right & \text{(iii)} \\[2mm]
\left(y'_j.rl - wid_j/2 - r_j/2\right)\big/rl & r_j < wid_i + 2separMax_i \wedge side = right & \text{(iv)}
\end{cases}
$$

$$[12.6]$$

Eq. [12.6(i)] holds when a separation of more than $separMax_i$ is available on the left side of vehicle R_j and left is the overtaking side. If the same distance is not available and overtaking needs to take place on the left side, vehicle R_i simply attempts to maintain as large a distance as possible from other vehicles. Similarly, Eq. [12.6(iii)] is when separation $separMax_i$ is available and Eq. [12.6(iv)] when this separation is not available. Notations are shown in Fig. 12.4 for a random scenario for which one condition of Eq. [12.6] may be active.

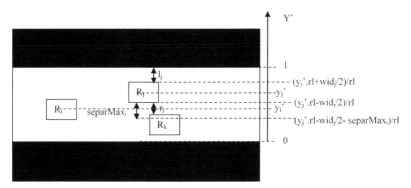

Figure 12.4 Notations used for a random scenario (all measurements on Y' axis).

The value of p'_x is the distance across the X' axis within which the vehicle should align itself and attain the value of p'_y along the Y' axis. Normally, the overtaking would not have been completed when the vehicle reaches the point $P\left(p'_x, p'_y\right)$, but this behaviour has more to do with *arranging the relative positions* of all vehicles and facilitating conditions such that R_i may travel straight ahead at high speed. Hence, enough distance is allowed along the length of the road for R_i to steer as per needs. A higher deviation in relative position from the current position along the Y' axis means a larger distance along the X' axis for alignment. Further, high speed may require large distances along X' axis for the same magnitude of steering. The value of p'_x may then be given by Eq. [12.7].

$$p'_x = c_1 + c_2.v_i + c_3.\text{abs}\left(y' - p'_y\right).rl \qquad\qquad [12.7]$$

Here, c_1, c_2, c_3 are constants. c_1 is the minimal distance across the X' axis needed to produce a smooth curve as per spline curves. Usually, this is set to be twice the length of the vehicle. c_2 and c_3 meanwhile denote contributions of the factors of speed and steering requirements.

12.4.4.2 Assistive Overtaking

The second scenario possible is *assistive overtaking* wherein an overtaking may still be possible but it requires the assistance of other vehicles. For the overtaking to be possible, the other vehicles must *rearrange themselves* by steering and changing their relative positions on the road, thereby allowing R_i adequate space to overtake. To check whether such a scenario is possible, both the left- (and right-) hand side of the vehicle in front R_j are iteratively traversed, at every iteration moving to the vehicle on its left (and right). Subtracting a basic constant amount equivalent to the presumed minimal width for every vehicle, a rough estimate of the maximum separation available can be made. If vehicle R_i itself comes into any iteration, it is not considered. If this space is more than $wid_i + 2.separMin_i$, the vehicle may decide to overtake. Whether an overtaking actually happens or not is uncertain as vehicles may have larger minimum thresholds and do not offer enough room for R_i to pass through. It is even possible that vehicles may have smaller thresholds, and hence a feasible overtaking is not attempted.

Further, multiple vehicles might simultaneously want to overtake with no one vehicle having enough separation of its own to overtake. A cooperative behaviour is not modelled here, in which all vehicles decide which one overtakes, but rather an *aggressive behaviour* is considered in which they all attempt to overtake. It is quite possible, therefore, that because of the aggression no vehicle is able to actually overtake at a specific time. The decision whether to initiate overtaking or not is also critical. If the vehicle decides not to overtake, it displays a simple behaviour such as following vehicles in front of it. However, if it tries to overtake, it constantly *pushes other vehicles* to give it more space to make its desired overtaking possible.

If an overtaking procedure is initiated, R_i must choose its relative position in terms of alignment on the Y' axis such that sometime later it might overtake easily if possible. Once more it might overtake a vehicle directly in front (R_j) either from the left or right side, decided purely on the basis of whether it is more towards the left or right of the vehicle. The *point of overtaking* $P\left(p'_x, p'_y\right)$ is computed for the case of assistive overtaking. The value of p'_x is kept the same as it was in Eq. [12.7]. Considering the minimum distance to be available, the value of p'_y is given by Eq. [12.8].

$$
p'_y = \begin{cases} \left(y'_j.rl + wid_j/2 + separMin_i + wid_i/2\right)\Big/rl & \text{side} = \text{left} \\[2mm] \left(y'_j.rl - wid_j/2 - separMin_i - wid_i/2\right)\Big/rl & \text{side} = \text{right} \end{cases} \qquad [12.8]
$$

Trajectory generation τ is done by the use of a spline curve. Once the trajectory τ is constructed, the task for the vehicle is to travel along it. Although the trajectory is planned and fixed, the only thing to be monitored and changed is the speed to avoid any collisions. The attempt is to make the vehicle travel with the *highest speed possible*. Hence, at any time while executing the trajectory, let the vehicle in front be at a distance of d units in front measured from the front of the vehicle R_i to the back of the vehicle in front. The minimum separation of $separMin_i$ needs to be maintained and further it needs to be assumed that the vehicle in front may stop suddenly. The maximum deceleration is taken to be $-accMax_i$. Hence, if v_i is the current speed of the vehicle, the changed speed at the next instant v'_i is given by Eq. [12.9]

$$
v_m = \sqrt{\left(2.accMax_i.\max(d - separMin_i, 0)\right)}
$$

$$
v'_i = \begin{cases} \min\left(v_i + accMax_i, v_i^{\max}\right) & v_i + accMax_i \leq v_m \\[2mm] \max\left(v_i - accMax_i, v_m\right) & v_i > v_m \\[2mm] v_m & \text{otherwise} \end{cases} \qquad [12.9]
$$

Here, v_m is the maximum speed as per the distance available.

Whichever is the scenario of overtaking, the other vehicles in the vicinity need to be made aware of the vehicle R_i attempting to overtake. This information helps them in their planning. In the case of an assistive overtaking, the other vehicles need to move and create space as shall be seen later. In the case of direct overtaking, vehicles need to be prevented from accidently moving in the opposite direction and reducing the available separation. For manual vehicles this may be in the form of a horn or similar gesture commonly used in traffic while overtaking. Autonomous vehicles may additionally have some alternative intervehicle communication. This is the only bounded communication expected/allowed in the simulation.

The overtaking gesture is *updated at every decision-making cycle*. Whenever a vehicle finds a situation not suitable for overtaking in any manner, the gesture is stopped. This would be, for example, when the vehicle has already *completed an over-taking* and vehicles ahead are now to the side, or when the vehicles in front have left enough room for the vehicle R_i to comfortably drive straight ahead, or the overtaking is later found *infeasible*. The infeasibility may be due to a reduction in road spacing or perhaps a new vehicle entering parallel to the stream of vehicles ahead.

Another prerequisite is that vehicles which are being overtaken must be travelling straight or on the opposite side (left or right) from which the overtaking is being initiated. This resolves any *conflict of interest* wherein a sufficient separation is seen and the vehicle initiates overtaking only to find that the vehicle in front continuously moves to the opposite side thereby reducing the available separation. If either of the scenarios does not hold or vehicles are not seen to be travelling straight or on the opposite side or a change of lane request is denied, no overtaking is possible.

12.4.5   Being Overtaken

A related behaviour along with overtaking is to *be overtaken*. In assistive overtaking a vehicle expects the vehicle in front to give it *more separation*. Hence, whenever possible the vehicle in front must cooperate and allow the vehicle behind to overtake. For this behaviour a vehicle R_i scans all the vehicles requiring to overtake behind it within a relative position of $\left[y'_i - (wid_i/2 + separMax_i)/rl, y'_i + (wid_i/2 + separMax_i)/rl \right]$. Note the use of $separMax_i$ in place of $separMin_j$. This is done for two reasons. Firstly, the vehicle behind, say, R_j may have a higher value of $separMin_j$, which means that even if the vehicle R_i sees the vehicle R_j having enough distance to go ahead, it might actually not proceed. To enable overtaking, the vehicle R_i must further give more space for vehicle R_j to overtake. Secondly, giving more space than the minimal amount is always considered better as per the modelling problem. Out of all the available vehicles, the vehicle R_i selects the vehicle R_j which is closest to it in terms of separation on the X' axis. R_i can easily 'guess' the side which R_j wants to overtake by the same rules.

Now R_i must give the necessary separation to R_j for overtaking, or at least the maximum that it can give. In case it does not have adequate separation, further space may be given by a vehicle to the side of R_i or various vehicles may steer themselves and move in due course of time (as shall be seen in the later behaviours) giving more separation to R_i to allow space for R_j. Let l_i be the separation available with R_i on its left and r_i be the separation available on its right. Assuming that the separation thresholds (minimum and maximum) of the two vehicles are the same, the separation needed by R_j may be anywhere in the range $[wid_j + 2.separMin_i, wid_j + 2.separMax_i]$, which needs to be created within the separation available with R_i which is $(l_i + r_i)$, also considering the minimum separation requirements of R_i.

Say the vehicle R_j needs to overtake R_i on the right, which means R_i must drift leftwards to facilitate the overtaking. Because R_i drifts leftwards, the space r_i is already available for R_j to overtake, making the requirements within the range $[a, b] = [wid_j + 2.separMin_i - r_i, wid_j + 2.separMax_i - r_i]$. The preferred relative position of R_i to account for the overtaking of R_j is given by Eq. [12.10].

$$
p'_y = \begin{cases}
\text{NIL} & b \leq 0 & \text{(i)} \\[2mm]
(y'.rl + l_i - separMin_i)/rl & a > l_i - separMin_i & \text{(ii)} \\[2mm]
(y'.rl - r_i + 2.fs + wid_j)/rl & b > l_i - separMin_i & \text{(iii)} \\[2mm]
(y'.rl - b)/rl & \text{otherwise} & \text{(iv)}
\end{cases}
\qquad [12.10]
$$

Here, fs is the free space after both R_i and R_j fit in side by side. It is given by $fs = l_i + r_i - wid_j$.

Eq. [12.10(i)] is the case when no assistance from R_i is required and there is sufficient space (with $separMax_i$ distance additional at both sides) available for R_j on the right side of R_i. Eq. [12.10(ii)] deals with the case when R_i is unable to give sufficient space, keeping the minimum separation reserved for itself, and hence it can give the maximum separation possible to R_j. Eq. [12.10(iii)] is the case when the separation available is not large enough so that R_j can enjoy a separation of $separMax_i$ from both their ends. Hence, the available separation (fs) is equally divided between R_i and R_j. The case in Eq. [12.10(iv)] is when R_i can offer a separation of b to R_j which is the maximum that it needs.

A similar equation is used when R_j overtakes R_i on the left side. The process of selecting an equivalent p'_x and using this as the space creation point is similar to the method discussed earlier. Similarly, the trajectory is generated and the vehicle is moved. Again, caution needs to be taken that distances are only measured if vehicles are travelling straight or drifting to the side needed by R_i. Lane-change behaviour is checked for every vehicle other than R_j which is overtaking. The process is summarized in Box 12.5.

12.4.6 Maintain Separation Steer

This behaviour attempts to *maximize a vehicle's separation* from both sides, under a threshold of $separMax_i$, in cases in which steering may increase the separation. Say the vehicle R_i has a separation of l_i units on the left and r_i units on the right. Separations are

---

Box 12.5 Being Overtaken Behaviour Procedure

- Select the vehicle attempting overtaking
  - Closest overtaking vehicle behind
- Guess the **side of overtaking** and plan to move to the opposite side
- Decide the **magnitude to move**
  - Separation needed for overtaking
  - Any excess separation is shared (under threshold)
  - If required separation unavailable maximum that the vehicle can offer is offered
- Construct the **be overtaken trajectory**

regularized by selecting a regularization point $P\left(p'_x, p'_y\right)$ the p'_y value for which is given by Eq. [12.11]. Other principles are the same as those considered with the lane change superbehaviour.

$$
p'_y = \begin{cases}
\text{NIL} & l_i + r_i < 2separMin_i \vee l_i > separMax_i \wedge r_i > separMax_i & \text{(i)} \\
(y'.rl + l_i - separMax_i)/rl & l_i + r_i \geq 2separMax_i \wedge l_i < separMax_i & \text{(ii)} \\
(y'.rl - r_i + separMax_i)/rl & l_i + r_i \geq 2separMax_i \wedge r_i < separMax_i & \text{(iii)} \\
(y'.rl + (l_i - r_i)/2)/rl & l_i + r_i < 2separMax_i & \text{(iv)}
\end{cases}
$$

$$[12.11]$$

Eq. [12.11(i)] is operative when the behaviour is not applicable; that is when the vehicle enjoys a separation of more than $separMax_i$ on both sides or when it cannot enjoy a separation of $separMin_i$ on both sides. Eq. [12.11(ii, iii)] satisfy the condition when a vehicle enjoys a separation of more than $separMax_i$ on one side and less separation on the other side, whereas the total separation is such that the vehicle can enjoy its threshold value $separMax_i$ on both sides. Eq. [12.11(iv)] attempts to centre the vehicle by maintaining equal separation on both sides.

12.4.7 Slow Down

If a vehicle, by any steering mechanism, cannot enjoy a distance of more than $separ Min_i$ on both sides (that is $l_i + r_i < 2.separMin_i$), there is a potential threat of collisions due to uncertainties, possible control errors etc. This situation marks the state of the vehicle in which it is not 'comfortable' driving due to the very low separation available. Whenever this situation is encountered, vehicle R_i decreases its speed by a maximum amount making the new speed $v' = \max(v_i - accMax_i, 0)$. Because the dynamics of the road, in terms of the relative positions of vehicles, is constantly changing, it is possible later that the separation needed will be available and the vehicle can accelerate again. Further, as there is a possibility that a vehicle going ahead may create more space so that other vehicles can adjust and create space for the vehicle R_i. In case the separation is continuously low, R_i could instead join the stream of vehicles behind it, or follow the stream of vehicles in front. Note that a slowing down of vehicle R_i would mean its separation would be eventually distributed across the stream of other vehicles.

12.4.8 Discover Conflicting Interests

Planning without formal communication makes it very likely that two vehicles showcase behaviours which, when executed simultaneously, lead to collision. Consider the scenario in Fig. 12.5, the two vehicles compute and aim to occupy the same space in front, and hence their trajectories, when executed, would lead to a collision. Such a situation rarely happens in daily life that two vehicles simultaneously make opposing plans. If one of the vehicles had been a little late in decision-making, the other vehicle

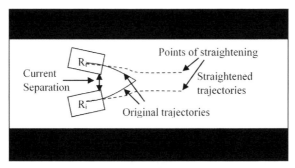

Figure 12.5 Discover conflicting-interests behaviour.

would have displayed its intention to turn. This intention was a consideration in all the behaviours involving separations.

The *smallest separation* between any two vehicles is always monitored while one vehicle traverses its planned trajectory τ. If this distance happens to be such that after unit move of both vehicles the separation would be less than the minimum threshold, vehicle trajectories are *straightened*. The point of straightening is computed the value for which on the Y' axis is found by measuring the least subsequent motion on the Y' axis before the vehicle can straighten itself which is the projection of the vehicle's length on the current Y' axis.

12.4.9 Travel Straight

The last behaviour to be considered is in fact the simplest behaviour as this deals with *travelling straight* on the road. There are no prerequisites for this behaviour, it can be invoked at any time and is usually invoked when there are no other behaviours to be exhibited. A small curve is generated for a unit move of the vehicle from the current position, keeping the relative position (y') of the vehicle constant. The vehicle attempts to move by the highest speed possible given by Eq. [12.9].

12.5 Single-Lane Overtaking

For this section, it is assumed that the road is marked in the middle by a real or virtual boundary separating the outbound and inbound directions of travel. The behaviours introduced in Section 12.4 are all used, and their computations are done using this boundary. As a result, while driving using any of the behaviours, the vehicle does not slip onto the wrong side of the road at all, even though there may be no physical barrier separating the sides. The *single-lane overtaking* behaviour is modelled in addition to these behaviours. The computation of the single-lane behaviour, however, uses the actual road boundaries. Various aspects of the behaviour are detailed as follows.

12.5.1 Single-lane Overtaking Initiation

Consider that the vehicle being planned R_i cannot overtake a slower vehicle in front by normal means. It may attempt to see if single-lane overtaking is possible. It is required that the vehicle is currently on the correct side of the road and not already attempting single-lane overtaking. As per the general methodology of overtaking trajectory computation, it first needs to be decided whether the overtaking will happen on the *left or right side* of the vehicle ahead and then select the *preferred lateral position* for overtaking. Here, it is assumed that the traffic operates with a 'drive on the left' rule, and hence, single-lane overtaking can only happen on the right side.

Consider that the vehicle in front is R_j with a separation of r_j available on its right (without considering the virtual boundary). Furthermore, consider that the vehicle R_i has a separation r_i on its right side. The availability of these separations may be partly on the correct (outbound) carriageway and partly on the wrong (inbound) carriageway. The most important precondition for the overtaking to be initiated is that both of these separations must be larger than the minimum separation required by R_i for it to overtake, which is its width (wid_i) and a safety distance ($separMin_i$) on both sides, totalling $wid_i + 2.separMin_i$. The notations are shown in Fig. 12.6A.

On availability of the separation, the next task is computing a point $P\left(p'_x, p'_y\right)$ to which the vehicle may travel for overtaking to occur. Usually, this overtaking point would lie on the wrong side of the road. On reaching $P\left(p'_x, p'_y\right)$, the vehicle may travel almost straight ahead to move in front of the vehicle being overtaken and subsequently may attempt to return to the original lane. p'_y denotes the lateral position that the vehicle attempts to achieve during the overtaking procedure. The basic requirement

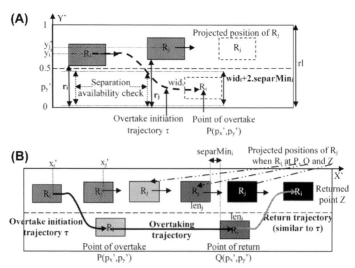

Figure 12.6 Notations used in single-lane overtaking. (A) Overtaking trajectory; (B) return trajectory.

is separation maximization, and hence, this point must be far enough from the vehicle being overtaken, as well as from any other vehicle or road boundary that may lie towards the right. The maximization is under a threshold of $separMax_i$ which disallows the vehicle from going too far on a wide road that would require large steering movements. Computation of p'_y is given by Eq. [12.12].

$$
p'_y = \begin{cases}
\left(y'_j.rl - wid_j/2 - separMax_i - wid_i/2\right)\big/rl & r_j \geq wid_i + 2separMax_i \quad \text{(i)} \\[2ex]
\left(y'_j.rl - wid_j/2 - r_j/2\right)\big/rl & r_j < wid_i + 2separMax_i \quad \text{(ii)}
\end{cases}
$$

$$[12.12]$$

Here, rl is the current width of the road and y'_j is the position of R_j on the Y' axis. Eq. [12.12(i)] denotes the condition when the separation available on the right of R_j is wide enough for the vehicle R_i to enjoy a maximum separation of $separMax_i$ on both sides. Eq. [12.12(ii)] denotes the condition when the largest possible separation is not available, and hence, R_i simply attempts to drive in between the separation available.

The value of p'_x is the same as per the general guidelines, given by Eq. [12.7]. Spline curves are used to compute the overtaking initiation trajectory τ from the current vehicle position R_i to $P\left(p'_x, p'_y\right)$, such that the vehicle is parallel to the road at $P\left(p'_x, p'_y\right)$.

The time required for the vehicle to place itself in the overtaking zone is given by $\|\tau\|/v_i$, in which $\|\tau\|$ is the length of the trajectory and v_i is the current vehicle speed. The vehicle may then be seen to accelerate from its original speed v_i to the maximum speed $vMax_i$ to be well ahead of R_j (with an additional safety distance of $separMin_i$) before its return to the correct side can take place. In normal circumstances, the trajectory for the vehicle to return to the correct side of the road is similar (flipped) to the overtaking trajectory. However, its speed will have increased to $vMax_i$, giving a return time of $\|\tau\|/vMax_i$, which will be shorter than $\|\tau\|/v_i$. Overall, the total overtaking period can be computed with the corresponding expected straight trajectory, called the overtaking trajectory. The notations are shown in Fig. 12.6B. The total time for overtaking may hence be approximated by Eq. [12.13].

$$
T = \frac{\|\tau\|}{v_i} + \frac{x'_j - x'_i + \max(len_i, len_j)/2 + separMin_i}{\frac{v_i + vMax_i}{2} - v_j} + \frac{\|\tau\|}{vMax_i} \qquad [12.13]
$$

Here, x'_i and x'_j denote the current position of R_i and R_j, respectively, whereas len_i and len_j denote the corresponding lengths. v_j is the current speed of R_j.

For single-lane overtaking to be initiated, the preconditions of the general overtaking behaviour (Section 12.4) are included, that is the lane change behaviour for trajectory τ must be true, as well as the condition that R_j must not be drifting rightwards.

Further, no collisions should occur between the vehicle R_i and any other vehicle (as per their projected travel) on the wrong side of the road, if R_i occupies the computed lateral position for the time interval T. This means, ideally, if the vehicles stick to their current speeds and lateral positions, the vehicle R_i should be able to complete the overtaking procedure within time T without causing any vehicles to slow down.

12.5.2 General Travel

In case a vehicle is travelling on the wrong side of the road and wishes to attempt single-lane overtaking, it must always test *whether such overtaking is feasible* or *whether it has been completed*, which are modelled as separate behaviours and discussed in Sections 12.5.3 and 12.5.4. In all other cases, the vehicle must travel in the wrong lane using the same behaviour set as any general vehicle (travelling on the correct side of the road), with the exception that no overtaking can be performed. This would mean that mostly either the vehicle travels straight ahead or adjusts its lateral position to maximize its separation from all other vehicles.

The speed needs to be set for every vehicle in the planning scenario whilst it travels during single-lane overtaking or any other behaviour. As per the heuristics used, each vehicle must always attempt to travel within its maximum safe speed. This notion is, however, different for the different classes of vehicles possible.

The first type of vehicle (i) is a normal vehicle, which is on its *correct side of travel, and no other vehicle attempting single-lane overtaking is to be found ahead* (Fig. 12.7A). For such a vehicle, it is assumed that, in the worst case, the vehicle driving in front might brake suddenly, and hence, the speed should correct to enable the vehicle to react instantaneously and avoid a collision with a safe distance of *separ-Min$_i$*. The preferred speed is given by Eq. [12.16(i)].

The second category of vehicle (ii) is one which is *on the wrong side of the road* in the middle of attempting single-lane overtaking and finds a vehicle R_j directly ahead of

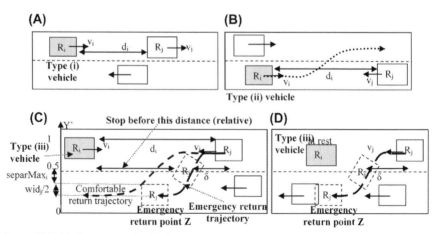

Figure 12.7 Various types of vehicles and notations for their speed computation. (A) Type (i) vehicle, (B) type (ii) vehicle, (C) type (iii) vehicle and (D) type (iii) vehicle.

it (Fig. 12.7B). Hence, R_i and R_j are travelling towards each other with speeds v_i and v_j. The attempt here is to stop R_i, which is the vehicle being planned, before a possible collision. The speed must always be such to enable R_i to stop with the maximum possible deceleration. The speed setting is hence the same as for a normal vehicle, with the resultant safe speed interpreted as the resultant safe relative speed. The preferred speed is given by Eq. [12.16(ii)].

The last category (iii) is if the vehicle R_i is *on the correct side of the road with vehicle R_j travelling towards it on the wrong side*, attempting single-lane overtaking (Fig. 12.7C). Vehicle R_i is aware of the fact that R_j needs to travel back to its correct side, and hence, mere collision prevention is not enough. If R_i and R_j stand almost touching each other, neither can move until one of the vehicles backs up. R_j needs some distance to return to its correct side. In an extreme case, R_j should have additional space available to steer left and return to its correct lane (Fig. 12.7D) using an *emergency return trajectory*, while still avoiding collision with R_i. R_j would need to stop and wait for the other vehicles to clear on the correct side and slowly trace the emergency return trajectory. In extreme cases, both the vehicles R_i and R_j should slow down sufficiently for R_j to return. This additional distance was not maintained by the earlier category (or R_j itself does not contribute much in slowing down), as it is important for R_j to get close to R_i while tracing the emergency return trajectory. R_j must return to the correct side and not stop because of the presence of R_i. Otherwise, both vehicles may wait indefinitely.

As perceived by R_i, if the vehicle R_j immediately needs to return to its original lane, it would need to choose an emergency point of return $Z\left(z'_x, z'_y\right)$. Here, z'_y is the chosen lateral position to which R_j would return. Assuming sufficient distance would be available with R_j, its lateral position at the return can be given by Eq. [12.14]. $Y' = 0.5$ is the virtual boundary from which the maximum separation distance is desired.

$$z'_y = 0.5 + (wid_j/2 + separMax_i)/rl \qquad [12.14]$$

The distance required (δ) along the road to change the lateral position as computed in Eq. [12.14] is given by Eq. [12.15], the notations for which are the same as those in Eq. [12.7]. It is assumed here that half of this distance would be required on the wrong side of the road, whereas the other half would be on the correct side of the road, and hence, only the first half is in question. This is shown clearly in Fig. 12.7.

$$\delta = \left(c_1 + c_2.v_i + c_3.abs\left(y' - z'_y\right).rl\right)\Big/2 \qquad [12.15]$$

The vehicle R_i attempts to change its speed such that this distance can be assured even if R_i has to stop to make this distance available. The relative speed is computed, which is used to compute the actual speed given by Eq. [12.16(iii)].

The preferred speeds for various cases are given by Eq. [12.16], whereas the actual speeds considering acceleration limits are given by Eq. [12.17]. Here, d_i is the distance of R_i from the vehicle or obstacle in front.

$$v_m = \begin{cases} \sqrt{2.accMax_i.\max(d_i - separMin_i, 0)} & \text{(i)} \\[2ex] \max\left(\sqrt{2.accMax_i.\max(d_i - separMin_i, 0)} - v_j, 0\right) & \text{(ii)} \\[2ex] \max\left(\sqrt{2.accMax_i.\max(d_i - separMin_i - \delta, 0)} - v_j, 0\right) & \text{(iii)} \end{cases}$$

$$[12.16]$$

$$v_i' = \begin{cases} \min\left(v_i + accMax_i, v_i^{\max}\right) & v_i + accMax_i \leq v_m \\[2ex] \max(v_i - accMax_i, v_m) & v_i > v_m \\[2ex] v_m & \text{otherwise} \end{cases} \qquad [12.17]$$

12.5.3 Cancelling Single-Lane Overtaking

Although a vehicle R_i is attempting single-lane overtaking, factors not considered in its initiation may stop the overtaking from being successfully completed, and the vehicle may need to return to its original side as quickly as possible. Driving on the wrong side is dangerous and creates a problem for the entire traffic system. Hence, the *feasibility of overtaking* must be continuously tracked during the whole overtaking process. At any time, overtaking may be regarded as infeasible if there is some vehicle R_j in front, driving towards R_i and the distance between the two vehicles is insufficient (or below) to meet the minimal distance required by R_i to execute the emergency return trajectory. This *minimum distance* (δ) is given by Eq. [12.15]. If the current distance is just equal or less than the required amount or R_i cannot decelerate fast enough to stop the distance (δ) from reducing below this amount, single-lane overtaking is regarded as being cancelled on that occasion.

The vehicle now needs to actually *return to its original side of travel*. In this stage, the actual *point of emergency return* $Z\left(z_x', z_y'\right)$ is computed by an analysis of the vehicles on the correct side. Suppose l_i is the separation to the left that is currently available. This distance does not account for the fact that after R_i returns to its lane, the virtual boundary would become the boundary to keep away from, and hence, the effective distance available to the left is that beyond the virtual boundary given by Eq. [12.18]. The notations are shown in Fig. 12.8.

$$l^{eff}_i = l_i - \left(rl/2 - y_i'.rl - wid_i/2\right) \qquad [12.18]$$

Based on the guidelines of separation maximization under a threshold of *separ Max_i*, the desired lateral position of the vehicle is given by Eq. [12.19]. In case the minimum separation is not available, R_i needs to attempt to return back to the correct side whilst maintaining at least a minimum separation.

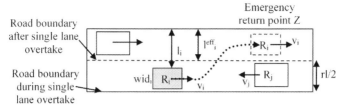

Figure 12.8 Return point calculation.

$$
Z'_y = \begin{cases}
0.5 + (wid_j/2 + separMin_i)/rl & l_i^{\text{eff}} \leq wid_i + 2.separMin_i \\[2mm]
0.5 + (wid_j/2 + separMax_i)/rl & l_i^{\text{eff}} \geq wid_i + 2.separMax_i \\[2mm]
0.5 + l_i^{\text{eff}} \big/ 2.rl & wid_i + 2.separMax_i < l_i^{\text{eff}} < wid_i + 2.separMax_i
\end{cases}
$$

$$[12.19]$$

The (emergency) return trajectory τ is computed using spline curves joining the current point to the point of return Z. The trajectory is traversed at all times with the highest possible speeds as calculated in Eq. [12.17]. Because the return trajectory might not be clear, it is evident that the vehicle may have to slow down drastically and even stop and wait for some vehicles to clear. However, it would eventually return to its original side.

12.5.4 Completing Single-Lane Overtaking

Once R_i, which was attempting single-lane overtaking, *is ahead of the vehicle being overtaken*, it may aim to return to its original side of travel. This behaviour is in fact the same as cancelling the single-lane overtaking behaviour with the exception that it is invoked post-overtaking and only when the minimum separation is available. In the absence of minimum separation, the vehicle continues to traverse on the wrong side of the road. It may return when the minimum separation is available or may be forced to return by cancellation of the single-lane overtaking behaviour when a vehicle in front, travelling in the opposite direction, is too near. Implementation details are then the same as for the cancelling single-lane overtaking behaviour. The overall procedure of single-lane overtaking is given by Fig. 12.9.

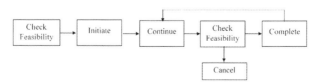

Figure 12.9 Single-lane overtaking procedure.

12.6   Complete Algorithm

The various behaviours discussed here are summarized in Table 12.1. Note that many of the behaviours are specific in terms of their initiation and hence a change of their

Table 12.1 **Summary of Vehicle Behaviours**

S. No.	Behaviour	Precondition	Description	In-behaviour specifications	Priority
1.	Obstacle avoidance	Obstacle discovery, lane change true	Strategy to avoid obstacle	Check for collisions with vehicle in front, obstacle avoidance	1
2.	Centring	No vehicle, no obstacle, vehicle travelling at high speed	Put vehicle in the centre of the road	NIL	2
3.	Lane change	Called by other behaviours	Whether possible to steer	NIL	NA
4.	Overtaking	Slower vehicle ahead, sufficient separation available assuming the cooperation of all vehicles ahead, lane change true, not undergoing single-lane overtaking	Strategy to initiate overtaking, ask other vehicles to move, and eventually align so that travelling straight completes overtaking	Discover conflicting interests, check for collisions with vehicles in front	3
5.	Single-lane overtaking	Slower vehicle ahead, sufficient separation available, lane change true, no collisions with a vehicle on the wrong side for the expected time of completion of overtaking assuming no cooperation	Attempt to initiate single-lane overtaking by placing the vehicle on the wrong side	Check for cancellation of single-lane overtaking, check for collision with vehicle in front	4
6.	Cancel single-lane overtaking	Performing single-lane overtaking, sufficient separation not available or not expected to be available in the future even with the largest deceleration to return to the original lane because of a vehicle ahead on the wrong side	Attempt to place the vehicle on the correct side and not allow any subsequent motion on the wrong side	Check for collision with the vehicle in front and if clear move ahead	NA

7.	Complete single-lane overtaking	Performing single-lane overtaking, sufficient separation available on the correct side, lane change true	After completing overtaking, return to the correct side	Check for collision with the vehicle in front	5
8.	Be overtaken	Vehicle to the rear shows the need for overtaking, separation available to offer, lane change true, not undergoing single-lane overtaking	Align so that the vehicle to the rear needing to overtake has more overtaking separation	Discover conflicting interests, check for collisions with the vehicle in front	6
9.	Maintain separation steer	Maximum separation possible not available at both ends whereas steering in some manner can increase current lowest separation, lane change true	Steer to maintain as high separation as possible (not more than threshold) from both ends	Discover conflicting interests, check for collision with vehicle in front	7
10.	Slow down	No adjustment of steering capable of generating minimal separation at both ends	Reduce speed	NIL	8
11.	Discover conflicting interests	A neighbouring vehicle steering towards vehicle being planned found too close while the vehicle being planned was steering towards it, vehicle following a nonstraight trajectory	Straighten trajectory being followed	Check for collision with the vehicle in front	NA
12.	Travel straight	No preconditions	Take a unit step forwards as per the road's current orientation	Check for collisions with the vehicle in front	9

priority would not affect the algorithm. Priorities are useful only when multiple behaviours may simultaneously meet their preconditions.

The algorithm as a pseudocode is given by Algorithm 12.2. Here, the various behaviours have been listed one after the other. Lane-change behaviour is marked by *lane_ change*(τ) which returns true if a change is possible. This characteristic also appears in most other behaviours. If behaviour results in the generation of a trajectory τ, the same is returned in the plan, which becomes the input for the next time instance. A non-null value of τ signifies a behaviour being followed and the same behaviour should be followed unless it has been completed. In all cases in which τ is null, a new behaviour is searched for. Behaviours which complete in a unit move (such as moving straight) return a null value of τ, signifying a behaviour was selected and was completed in the same step.

Algorithm 12.2: Plan(Vehicle R_i, Map, Previous Plan τ)

```
If new obstacle found

        Compute τ for obstacle avoidance using Algorithm 12.1

        If lane_change(τ), return τ
        Else, vᵢ ← max(vᵢ—accMaxᵢ,0), Rᵢ ← move a unit step, return null
If no slower vehicle and obstacle ahead in vicinity ∧ vᵢ close to
vᵢᵐᵃˣ ∧ τ = null
        CEN ← (xᵢ'+Δ(vᵢ), 0.5)
        τ ← curve(Rᵢ,CEN)
        If τ(t)∈ζᵢᶠʳᵉᵉ ∀ t, return τ
If τ ≠ null

        vᵢ ← Safe speed as per Eq. [12.17]

        If Conflicting Interests ∧ τ is nonstraight
                τ ← straighten(τ), return τ
        Else If performing single-lane overtake

                Calculate δ using Eq. [12.15]

If distance δ not available or not expected to be available in the
future even with largest deceleration

                        Compute Z for return using Eq. [12.19]

                        τ ← curve(Rᵢ,Z)
                        return τ
                Else
                        Rᵢ ← Move a unit step by τ
                        If τ is over, return null, else return τ
Else If performing single-lane overtake

        Calculate δ using Eq. [12.15]

        If distance δ not available or not expected to be available in
        the future even with largest deceleration
```

Compute Z for return using Eq. [12.19]

$\tau \leftarrow$ curve(R_i,Z)

return τ

If slow vehicle in front $\land$ sufficient separation exists for overtaking assuming cooperation $\land$ not undergoing single-lane overtake

$R_j \leftarrow$ Vehicle to overtake, side $\leftarrow$ side of overtaking

Compute P for overtake using Eqs [12.6]–[12.8]

$\tau \leftarrow$ curve(R_i,P)

If $\tau(t) \in \zeta_i^{free} \; \forall \, t \land$ lane_change(τ) $\land R_j$ not steering towards side, return τ

If Slower vehicle in front $\land$ sufficient separation available $\land$ no collisions with vehicle in wrong side for expected time of completion of overtake assuming no cooperation

$R_j \leftarrow$ Vehicle to overtake

Compute P for single-lane overtake using Eq. [12.12]

$\tau \leftarrow$ curve(R_i,P)

Calculate T by Eq. [12.13]

If $\tau(t) \in \zeta_i^{free} \; \forall \, t \land$ lane_change(τ) $\land$ no collisions with vehicles in wrong side till T assuming no cooperation $\land R_j$ not steering towards side

return τ

If performing single-lane overtake $\land$ sufficient separation available at correct side

Compute Z for return using Eq. [12.19]

$\tau \leftarrow$ curve(R_i,Z)

If $\tau(t) \in \zeta_i^{free} \; \forall \, t \land$ lane_change(τ), return τ

If vehicle overtaking at back $\land$ separation available to offer $\land$ not undergoing single-lane overtake

$R_j \leftarrow$ Vehicle to allow overtake, side $\leftarrow$ side of being overtaking

Compute P for being overtaken using Eq. [12.10]

$\tau \leftarrow$ curve(R_i,P)

If $\tau(t) \in \zeta_i^{free} \; \forall \, t \land$ lane_change(τ), return τ

If $l_i + r_i \geq 2.\min\_separ_i \land (l_i < \max\_separ_i \lor r_i < \max\_separ_i)$

Compute P for separation maintenance using Eq. [12.11]

$\tau \leftarrow$ curve(R_i,P)

If $\tau(t) \in \zeta_i^{free} \; \forall \, t \land$ lane_change(τ), return τ

If $l_i + r_i < 2.\min\_separ_i$

$v_i \leftarrow \max(v_i - accMax_i, 0)$, $R_i \leftarrow$ move a unit step, return null

$v_i \leftarrow$ Safe speed as per Eq. [12.17]

$R_i \leftarrow$ Move a unit step parallel to the road

return null

12.7   Results

The simulation tool was developed in Matrix Laboratory (MATLAB). Base modules carried the tasks of boundary determination, vehicle tracking, graphical display and distance measurements. Different behaviours were coded as separate modules. Scenarios could be generated by specifying entry times, initial speeds, maximum speeds, maximum accelerations and sizes of vehicles. The map was supplied as an image file. The algorithm was tested for various scenarios which are discussed in subsequent sections.

12.7.1   General Traversal

The first task associated with the testing of the algorithm was assessing the ability of a vehicle to travel along any general road. A number of maps were given to a single vehicle for its traversal. All these maps involved some challenges as to the method of taking a turn or the smoothness of the overall path. The results in three of the scenarios, each having a different road orientation, are presented in Fig. 12.10. Because only one vehicle was travelling in the scenario, encouragement was given to centring.

Fig. 12.10A shows a gradual turn in which the vehicle was supposed to turn smoothly before encountering a major straightening of the road. Clearly, the straightening was sudden and nonsmooth, and the vehicle was able to handle this by its planning, making the entire curve relatively smooth to follow with high speeds. Fig. 12.10B depicts quite the opposite behaviour in which the road is almost straight with a sudden curve right at the end. Such roads are a common phenomenon especially in crossing-like situations. Fig. 12.10B shows the path of the vehicle which was as smooth as possible in making the turn. Fig. 12.10C, meanwhile, shows a road in which the complete road segment may be broken down into further segments. The challenge was not so much to travel within segments but rather to steer between segments. Although the vehicle makes a smooth steer in the first change, the manner of handling the second change is unique and no major steering is visible.

12.7.2   Obstacle Avoidance

The next major challenge associated with autonomy in vehicles is to be able to avoid any static obstacles, while still maintaining a smooth trajectory. The challenge becomes even more difficult when the vehicle is to avoid an obstacle in the presence of a curved road or another vehicle. Various possibilities have been studied via simulations, the results of which are presented in Fig. 12.11. Fig. 12.11A shows a road with a general curve and an obstacle. The vehicle was supposed to plan its smooth trajectory such that it is centred at the road end and the obstacle is avoided comfortably. The same analysis for a straight road and single vehicle is shown in Fig. 12.11B. The same scenario was again used for an additional vehicle and the trajectory of the two vehicles is shown in Fig. 12.11C and D. Note that the vehicle which originated lower in the scenario was generated later. This vehicle needed to drive straight, keeping its relative distance constant to avoid the obstacle. Hence, here a straight-driving behaviour was applicable rather than obstacle avoidance. The first vehicle, generated

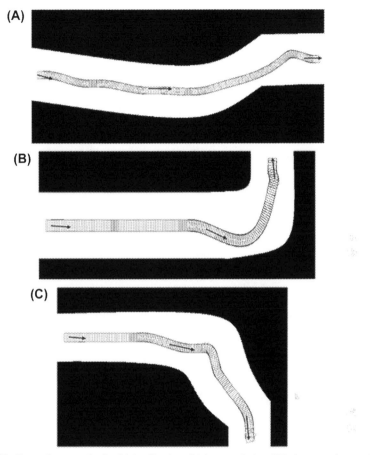

Figure 12.10 General traversal of vehicle. Single vehicle negotiating (A) sharp road curve left to right, (B) right-angle left turn, (C) irregular right turn.

in the middle of the road, saw the obstacle ahead but was not allowed to change lanes to overtake it, as is shown in Fig. 12.11C. Hence, instead it slowed down, waiting for the other vehicle to cross. It then passed the obstacle, as shown in Fig. 12.11D.

12.7.3 Overtaking

The next behaviour explicitly studied was overtaking. The interest was primarily in vehicular behaviours which are best visible on a straight road but which are also common scenarios in travelling. The simplest scenario is when a vehicle comes from behind and attempts to overtake the vehicle in front with a lot of space to do so. The corresponding trajectory is shown in Fig. 12.12A. The overtaking vehicle keeps the maximum possible distance possible as per its set value of *separMax$_i$* while overtaking. It quickly uses a trajectory of centring due to seeing no other vehicle in the vicinity, apart from the vehicle it has already overtaken. The same scenario was repeated with the overtaking

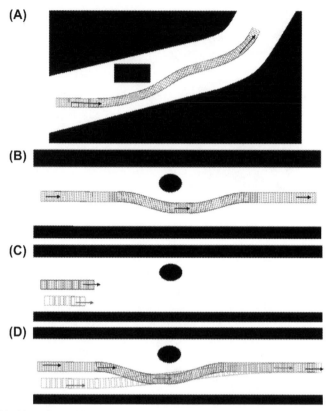

Figure 12.11 Obstacle avoidance. (A, B) Single vehicle avoiding single obstacle, (C, D) two vehicles avoiding single obstacle.

vehicle being more aggressive than the vehicle being overtaken (in terms of its *separMax$_i$* value). The trajectory in such a case is shown in Fig. 12.12B. Here, the vehicle being overtaken feels it is possible to increase separation and be safer for the brief period of time the vehicle is overtaking. This is because of its less aggressive nature, more cautious driving, indicated by a larger value of *separMax$_i$*.

The last case dealt with was when the overtaking vehicle did not have enough separation to overtake. The vehicle being overtaken was hence expected to move while the overtaking vehicle aligned itself. As soon as the vehicle attempting to overtake obtained enough separation, overtaking was initiated. The corresponding trajectory is given in Fig. 12.12C. Fig. 12.12C shows the importance of cooperation by the vehicle being overtaken to the overtaking vehicle. Overtaking is important to allow the two vehicles to drive at their maximum speeds as it might be 'painful' for a fast vehicle to follow a slow vehicle. Cooperation of the vehicle being overtaken first makes it possible to make overtaking possible and allows the vehicle to benefit from the same. Secondly, it eliminates the need of the overtaking vehicle to steer by very large amounts which would make overtaking a cumbersome job. Cooperation of the vehicle being overtaken

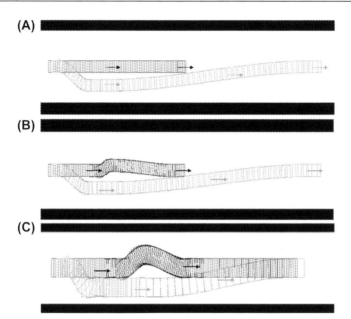

Figure 12.12 Overtaking behaviour. (A) Without aggression, (B) with aggression, (C) with cooperation.

by drifting leftwards by some amount does not make its travel any worse; on the other hand, it allows the overtaking vehicle to overtake easily and smoothly.

12.7.4 Complex Formulation

Based on the experimental results presented in the previous subsections, basic vehicular behaviours are well understood. Hence, many vehicles may be simulated over a road segment in which each vehicle displays its individual behaviour at every time instance. One of the unique behaviours was identified and used for discussion. The complete scenario is shown in Fig. 12.13. In this simulation, all vehicles have different values of aggression factor, indicated by $separMin_i$.

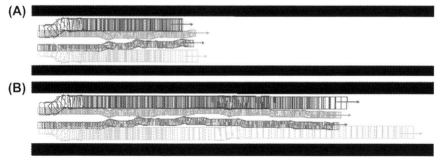

Figure 12.13 Complex formulation. (A) With successful overtaking, (B) without successful overtaking.

The first two (smaller) vehicles were generated simultaneously and travelled parallel to each other on the road. After some time two more (larger) vehicles were generated simultaneously. The newly generated vehicles were capable of higher speeds and attempted to overtake the slower vehicles. The faster and larger vehicles travelled on either side of the road, forcing the two smaller vehicles to create some separation for them. Each of the vehicles computed the total available space and believed that overtaking was possible, which might actually not be the scenario as two vehicles are simultaneously trying to overtake, each blocking the chances of the other vehicle. The smaller vehicles attempted to create some separation and moved to the opposite side, each unaware of the contradictory plans of the other vehicle. The vehicles soon displayed conflicting interests and straightened their trajectories.

The motion henceforth was purely driven by aggression factors. The top smaller vehicle was less aggressive (having a higher value of $separMin_i$) and attempted to maintain a larger separation (equal in magnitude to its $separMin_i$ from the bottom small vehicle. Hence, it steered slightly to the top/left when the desired magnitude of separation was not available. However, the bottom small vehicle was more aggressive with a smaller value of $separMin_i$. It saw that the separation created a possibility to create more separation for the bottom bigger vehicle. When the bottom vehicle moved towards the top, the top vehicle again saw that the separation available was less than the magnitude desirable. Therefore, it steered more towards the top. The bottom vehicle hence kept pushing the top vehicle, till the required separation was available for the bottom large vehicle to overtake, accounting for its additional minimum separation value.

Once the required separation was available, the bottom large vehicle overtook as shown in Fig. 12.13A. The centring behaviour was disabled and the vehicle travelled straight. Upon completion of overtaking there was ample separation available for the smaller bottom vehicle and it easily created a separation of $separMax_i$ for itself on both sides. However, the small top vehicle was still struggling to create necessary separation for the larger vehicle behind it. It steered towards the bottom thereby reducing the separation of $separMax_i$ for the bottom small vehicle. The bottom small vehicle further steered towards the bottom to maximize the distance. In this manner reverse pushing occurred as the top vehicle pushed the bottom vehicle to create separation for the large vehicle at the rear. As soon as the required separation along with the minimum separation threshold was available for the larger vehicle, overtaking took place as is shown in Fig. 12.13B. Finally, the two smaller vehicles travelled straight, parallel to each other.

The same experiment was repeated, but this time the value of the aggression factors remained constant. This scenario is naturally symmetric and hence no vehicle was capable of pushing another vehicle as was witnessed in Fig. 12.13. The result was that the smaller vehicles kept looking for an opportunity to give the larger vehicles more separation, but they were unable to do so for the entire journey. The scenario is shown in Fig. 12.14.

12.7.5 Aggression

To judge the working and effectiveness of the aggression factor another scenario was considered. A narrowing road was constructed with three vehicles travelling on it. It is

Figure 12.14 Complex formulation with constant aggression factors.

evident that if all three vehicles travelled parallel to each other, none would be able to reach the end of the segment. However, if aggression factors are different for the vehicles, it would be expected that vehicles that are more aggressive would reach the end, whilst less-aggressive vehicles would fear danger at some time and would give way to other vehicles, allowing them to go ahead. To maximize the effect, all vehicles were given generally high values of *separMin$_i$*. One case is discussed here and shown in Fig. 12.15. The bottom vehicle was most aggressive followed by the central and top vehicle. Hence, reasonably early in the journey, both the central and top vehicles waited for the bottom vehicle to go ahead.

12.7.6 Single-Lane Overtaking

The next important set of behaviours concerned single-lane overtaking. The road considered was wide enough for only two vehicles to be accommodated comfortably. Half the width was reserved for traffic travelling in one (outbound) direction and the other half for traffic travelling in the other (inbound) direction. This is the most typical and the only likely scenario in which such overtaking would take place. Indeed, wider roads may not necessitate that traffic from either direction mixes with the other. Different scenarios involving single-lane overtaking were tested.

Being a very risky behaviour, most of the time single-lane overtaking would not be attempted. This can be practically seen in everyday life, as such overtaking is rarely seen and, then, only too when the conditions are the most favourable or idealistic. The visual display of all single-lane overtaking behaviours was the same. The most interesting or the closest overtaking cases are discussed, as the other results are much easier to obtain.

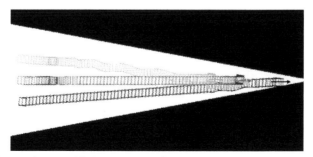

Figure 12.15 Experiments with the aggression factor.

In the first scenario, a slower vehicle (*A*) was generated in the centre of the road, and this vehicle naturally simply drifted towards the correct side of travel. A faster vehicle (*B*) was generated later. The faster vehicle (*B*) judged the presence of the slower vehicle (*A*) ahead and the feasibility of single-lane overtaking. By the projected motion of *B*, it was clear that single-lane overtaking could easily happen by the vehicle going to the wrong side, and hence, the vehicle decided to perform single-lane overtaking. *B* jumps to the wrong lane or overtaking lane by the initiation of the single-lane overtaking behaviour, surpasses *A* by general travel during the single-lane overtaking behaviour and then decides to return to the original lane by the completion of the single-lane overtaking behaviour. There was no change at all in the motion of *A*, and overtaking was easily completed as shown in Fig. 12.16A.

The second scenario was created to test the ability of a vehicle when such overtaking is not possible. In this scenario, an additional vehicle (*C*) was made to appear on the other side, travelling in the opposite direction. When the faster vehicle (*B*) emerged, it computed the feasibility of overtaking the slower vehicle (*A*) by considering the motions of *A* and *C*. *C* made single-lane overtaking seem risky, and hence, *B* did not attempt to overtake *A*. Instead, it followed *A*. When *C* had passed both *A* and *B*, again, the feasibility of single-lane overtaking was judged. This time, there was no vehicle that could make overtaking seem infeasible or risky. Single-lane overtaking by *B* was now judged to be feasible. It was initiated and subsequently completed in a manner similar to Scenario 1. The trajectories of the three vehicles and the scenario are shown in Fig. 12.16B.

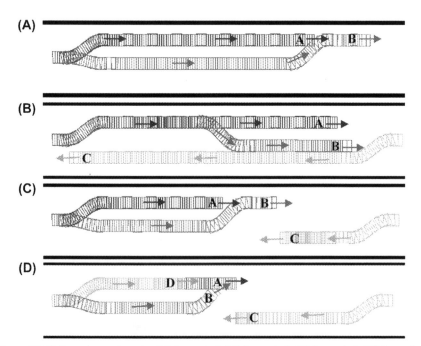

Figure 12.16 Experimental results. (A) Scenario 1, (B) Scenario 2, (C) Scenario 3, (D) Scenario 4.

In the third scenario, vehicle B judged single-lane overtaking to be feasible and initiated the same. There was, however, an oncoming vehicle (C) travelling on the other side of the road. The emergence and speed of C was just enough for overtaking to be feasible and just possible without cooperation. A slower speed for C would have made overtaking comfortable and, thus, not challenging, whereas a higher speed would have made overtaking infeasible. By fixing the speed of C in such a manner, the narrow overtaking behaviour was studied. On committing to overtaking, B quickly placed itself on the wrong side of the road, went ahead of the vehicle being overtaken (A) and returned in front of A in the correct lane. However, C was driving towards B during single-lane overtaking, which could have been a threat. C assessed a possible threat in advance by computing the relative speeds of the vehicles as a part of its normal driving behaviour and showed cooperation to some extent by slowing down. This gave A some extra space to return back to the correct side of the road and make overtaking comfortable and risk-free. The results are shown in Fig. 12.16C.

The aim in the last scenario was to put the overtaking vehicle (B) in an awkward scenario, which is the worst that such a vehicle can enter. The vehicle was made to initiate overtaking. In doing so, it did not account for A ahead, which would not allow B to return back to the correct lane easily on completion of single-lane overtaking. Overtaking was initiated, and the vehicle did go ahead of the slower vehicle D, with C coming from the other direction. After a little general driving on the wrong side of the road, it was clear that overtaking was not complete, because of the presence of A, and general driving could not continue, because C was approaching on the other

Table 12.2 Summary of the Results

Scenario	Vehicle	Time to destination	Distance travelled till termination	Maximum allowed speed	Average speed
1.	A	147	716.2713	5	4.8726
	B	75	726.2893	10	9.6839
2.	A	130	632.2713	5	4.8636
	B	85	690.4683	10	8.1232
	C	74	716.1901	10	9.6782
3.	A	137	667.8519	5	4.8748
	B	76	730.2858	10	9.6090
	C	99	721.2544	10	7.2854
4.	A	147	716.2713	5	4.8726
	B	99	746.8892	10	7.5443
	C	103	714.8350	10	6.9401
	D	138	645.2713	5	4.6759

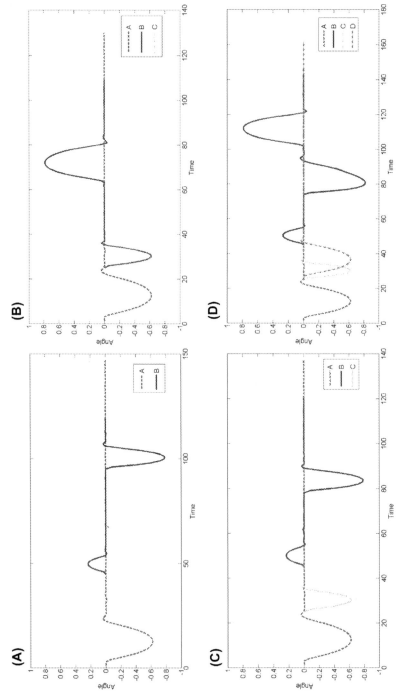

Figure 12.17 Angle profiles of different vehicles for different scenarios. (A) Scenario 1, (B) Scenario 2, (C) Scenario 3, (D) Scenario 4.

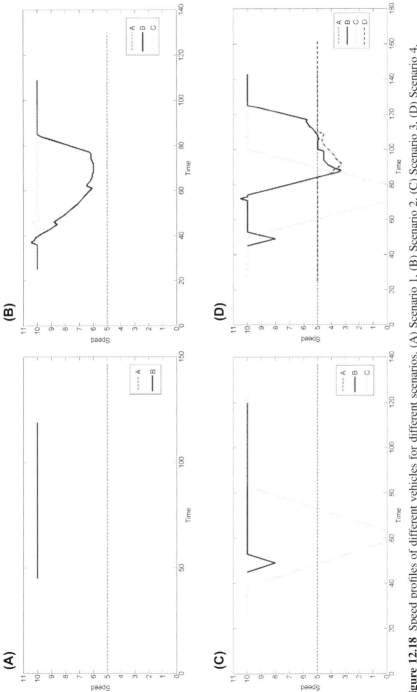

Figure 12.18 Speed profiles of different vehicles for different scenarios. (A) Scenario 1, (B) Scenario 2, (C) Scenario 3, (D) Scenario 4.

side of the road. Hence, the infeasibility of overtaking was discovered, and an emergency return trajectory was computed. However, the emergency return trajectory was infeasible, due to a possible collision with A. C detected the infeasibility of B during its general driving speed assignment and slowed down to give enough space for B to eventually return back to its original side. B therefore placed itself in between the two slower vehicles (D and A). D saw B trying to accommodate during its general driving, and when B was able to slip in, D did well to give B accommodated in between. The scenarios are shown in Fig. 12.16D.

The statistical results relating to different scenarios are given in Table 12.2. Further, the angle and speed profiles for different vehicles are given in Figs 12.17 and 12.18. In Fig. 12.17, both B and C turn towards their left simultaneously on entering the scenario, using similar dynamics. Hence, their angle profiles intersect. The units of distance and time are arbitrary and depend on the simulator. These can be mapped to real-world units by appropriate constants.

12.8 Summary

The objective of the research described here was to construct a planning algorithm for an autonomous vehicle amidst other nonautonomous vehicles and obstacles. The motivation was to solve the problem of planning along straight roads for autonomous vehicles driving in countries without speed lanes. Further effort was to advocate the absence of speed lanes for a diverse traffic system which autonomous vehicles may bring in the future. The absence of speed lanes brought about a rich set of behaviours which were identified and modelled.

At any time, the vehicle could display any of the behaviours depending upon its state and the conditions that govern the execution of the behaviour. The overtaking behaviour is particularly exciting, wherein it is difficult to decide which set of vehicles facilitates the overtaking of which other vehicles. The chapter also looked in depth at the important overtaking procedure wherein a vehicle for some time occupies a part of the wrong side of the road to accomplish its overtaking activity. The chapter looked at aspects of assessing the feasibility of such overtaking and formulating an initiation trajectory, the mechanisms involved with driving on the wrong side, judging the infeasibility of such overtaking and cancelling and successfully completing an overtaking procedure. As in natural driving, aggression is another important factor which affects traffic dynamics.

Experiments were performed for a variety of scenarios. Vehicles were tested for their ability to avoid obstacles, go around curved roads, overtake, help other vehicles overtake and perform complex formulations. The single-lane overtaking behaviour was also tested. In all cases, the vehicle either successfully completed the overtaking action or was able to cancel doing the same. No collisions were recorded, and the vehicles apparently behaved in a similar fashion to actual vehicles on the road.

References

Aguirre, E., Gonzalez, A., 2000. Fuzzy behaviors for mobile robot navigation: design, coordination and fusion. International Journal of Approximate Reasoning 25 (3), 255−289.

Alvarez-Sanchez, J.R., de la Paz Lopez, F., Troncoso, J.M.C., de Santos Sierra, D., 2010. Reactive navigation in real environments using partial center of area method. Robotics and Autonomous Systems 58 (12), 1231−1237.

Beom, H.R., Cho, H.S., 1995. A sensor-based navigation for a mobile robot using fuzzy logic and reinforcement learning. IEEE Transactions on Systems, Man and Cybernetics 25 (3), 464−477.

Clarke, D.D., Ward, P., Bartle, C., Truman, W., 2004. In-depth Study of Motorcycle Accidents. Road Safety Research Report No. 54. Department of Transport, London.

Clarke, D.D., Ward, P., Bartle, C., Truman, W., 2005. An In-depth Study of Work Related Road Traffic Accidents. Road Safety Research Report No. 58. Department of Transport, London.

Furda, A., Vlacic, L., 2011. Enabling safe autonomous driving in real-world city traffic using multiple criteria decision making. IEEE Intelligent Transportation Systems Magazine 3 (1), 4−17.

Jain, A., Menezes, R.G., Kanchan, T., Gagan, S., Jain, R., 2009. Two wheeler accidents on Indian roads − a study from Mangalore, India. Journal of Forensic and Legal Medicine 16 (3), 130−133.

Jolly, K.G., Kumar, R.S., Vijayakumar, R., 2010. Intelligent task planning and action selection of a mobile robot in a multi-agent system through a fuzzy neural network approach. Engineering Applications of Artificial Intelligence 23 (6), 923−933.

Kala, R., Warwick, K., 2013. Motion planning of autonomous vehicles in a non-autonomous vehicle environment without speed lanes. Engineering Applications of Artificial Intelligence 26 (5−6), 1588−1601.

Kala, R., Warwick, K., 2015. Motion planning of autonomous vehicles on a dual carriageway without speed lanes. Electronics 4 (1), 59−81.

Lajunen, T., Parker, D., Summala, H., 1999. Does traffic congestion increase driver aggression? Transportation Research Part F: Traffic Psychology and Behaviour 2 (4), 225−236.

Leonard, J.J., How, J.P., Teller, S.J., Berger, M., Campbell, S., Fiore, G.A., Fletcher, L., Frazzoli, E., Huang, A., Karaman, S., Koch, O., Kuwata, Y., Moore, D., Olson, E., Peters, S., Teo, J., Truax, R., Walter, M., Barrett, D., Epstein, A., Maheloni, K., Moyer, K., Jones, T., Buckley, R., Antone, M.E., Galejs, R., Krishnamurthy, S., Williams, J., 2008. A perception driven autonomous urban vehicle. Journal of Field Robotics 25 (10), 727−774.

McGarva, A.R., Steinerm, M., 2000. Provoked driver aggression and status: a field study. Transportation Research Part F: Traffic Psychology and Behaviour 3 (3), 167−179.

Mohan, D., 2002. Traffic safety and health in Indian cities. Journal of Transport and Infrastructure 9 (1), 79−94.

Mohan, D., Bawa, P.S., 1985. An analysis of road traffic fatalities in Delhi, India. Accident Analysis & Prevention 17 (1), 33−45.

Paruchuri, P., Pullalarevu, A.R., Karlapalem, K., 2002. Multi agent simulation of unorganized traffic. In: Proceedings of the First International Joint Conference on Autonomous Agents and Multiagent Systems: Part 1, pp. 176−183.

Paulsson, R., Thomas, P., 2005. Deliverable 5.2: In-depth Accident Causation Data Study Methodology Development Report, Project: Safety Net, Building the European Road Safety Observatory Integrated Project, Thematic Priority 6.2 "Sustainable Surface Transport". Vehicle Safety Research Centre.

Shinar, D., Compton, R., 2004. Aggressive driving: an observational study of driver, vehicle, and situational variables. Accident Analysis & Prevention 36 (3), 429−437.

Wang, J., Xu, W., Gong, Y., 2010. Real-time driving danger-level prediction. Engineering Applications of Artificial Intelligence 23 (8), 1247−1254.

Basics of Intelligent Transportation Systems

13.1 Introduction

So far, the book has presented an in-depth analysis into ways to move a vehicle in different traffic scenarios, specifically the problem of trajectory planning of autonomous vehicles. The vehicle is just one entity in the entire transportation system. The transportation system comprises traffic signals, intersections, merging roads, special vehicles like public transport vehicles and emergency vehicles etc. This chapter (and chapters: Intelligent Transportation Systems With Diverse Vehicles and Reaching Destination Before Deadline With Intelligent Transportation Systems) is devoted towards a broader study of intelligent transportation systems. Although the first part of the book focussed upon everything that the vehicle could have done, with or without communication, for efficient travel through a lower level of planning; this part of the book focusses upon the advantages of an efficient higher-level planning.

The transportation system does the tedious task of transporting a large number of vehicles from their source to their destination. Traffic is highly unbalanced in the sense that the traffic for a specific direction for some time of the day and for some routes may be under heavy demand, whereas the demand for some other roads may be very light for most times of the day. This leads to slow traffic and congestion, questioning the researchers to formulate mechanisms to avoid congestion and maximize the flow of traffic. To devise mechanisms to facilitate fast and efficient flow of vehicles, it is necessary to understand the science of traffic, which also guides the driving primitives of autonomous vehicles. This chapter looks at the broader perspective of understanding the entire transportation network and its operation. In the process, the chapter talks about the utility of all the components of the transportation network like intersections, traffic lights, mergers, diversions etc. and their impact on the overall traffic. Section 13.2 discusses the basics of *traffic-flow theory* useful in the understanding of traffic.

Based on the understanding of the operation of traffic, there is a natural quest to imitate traffic on computer systems to study the effects of every traffic policy, addition of traffic infrastructure and making traffic rules without experimenting on the real traffic. The *traffic simulation systems* are used for the process. Although a large number of modules need to be designed, and the development of the overall traffic simulator is a mammoth task, in this book the modules related to different behaviours of the vehicle and other transportation system constituents are noted in Section 13.3.

The last part of the chapter carries forwards the basic concepts to the level of intelligence. It is interesting to note how each of the components of the transportation system operates and the mechanisms by which they can be made intelligent. This creates the foundation of an *Intelligent Transportation System* (ITS), wherein all the entities like vehicles, transportation management centres, traffic lights, intersection managers, merging managers, roadside units etc. can talk to each other to aid the navigation of the vehicles. This intelligence enables the vehicles to make more informed decisions about their route, navigation profile, lane changes, overtakings and driving speeds. Similarly, this intelligence enables the transportation system to move the vehicles such that congestion is minimized whereas the efficiency of the transportation system is maximized. Intelligence would be a central theme in the discussion of the transportation systems. This is done here in Section 13.4. The notion of intelligence will be carried forwards to chapters 'Intelligent Transportation Systems With Diverse Vehicles' and 'Reaching Destination Before Deadline With Intelligent Transportation Systems'.

13.2 Traffic Systems and Traffic Flow

Traffic is extremely dynamic, hosting a large number of vehicles passing through the road infrastructure at every point in time. It is possible to find congestions, deadlocks, generally slow-moving traffic etc. Considering that traffic impacts a large number of people, there have been extensive attempts to model and study traffic, thereby attempting to make operational policies which maximize the resource utilization and lead to efficient travel. It is painful to wait prolonged periods in a traffic jam, which motivates the need to have mechanisms that prevent such situations. This section studies the basic principles of traffic flow, thereby leading to the development of intelligent transportation systems. For a more detailed discussion, please refer to Immers and Logghe (2002), Lighthill and Whitham (1955a,b), Maerivoet and Moorm (2005) and Newell (1993).

13.2.1 Traffic Flow

The study of *traffic flow* deals with assessing the behaviour of a pool of vehicles constituting the traffic in different conditions. Traffic flow has been studied by numerous people coming from diverse domains. It has been extensively studied using the theory of fluid dynamics, wherein the stream of vehicles with all intersections and diversions is modelled as a fluid flow. Physicists have also modelled traffic using Particle Flow theories, assuming every vehicle as a particle and tracing its behaviour in a stream of similar particles. The control engineers model traffic as a control problem, to control the traffic network parameters to minimize congestion using control laws.

To understand the traffic flow, let us first study the simplest scenario of operation, which is a straight road with no intersections, mergers, traffic lights etc. This is also called the *uninterrupted flow of traffic*. Consider one such vehicle which is following another vehicle in front. The two vehicles are separated by a *space gap*, which is the distance between the two vehicles, from the back bumper of the vehicle in front to the front bumper of the vehicle in back. The same notion can also be defined in terms of time. A *time gap* between the vehicles is the time that the vehicle at back will take to touch the current occupancy of the vehicle in front while operating at the current speed. The two terms can also be extended to the case of multiple lanes, in which case the time gap is different for different vehicles in different lanes. The gap with the immediately left- and right-lane vehicles is important as it helps the vehicle to make decisions regarding lane changes, in some cases leading to overtaking.

Consider any particular point on the road at a particular instant of time. One could keep time constant by taking a snapshot of the current traffic and noting the vehicles ahead and behind in space. Alternatively, one could keep space constant by standing at the current position only, while noting the vehicles passing through at different time steps. This makes the *space—time graph* of the transportation system, with every vehicle denoting a trajectory in this graph showing its occupancy in space and time, shown in Fig. 13.1. The space—time graphs are widely used to assess traffic. For most readings, let us study the traffic on a road of length K till time duration of T. All vehicles falling in this space—time window are recorded for computations. Fig. 13.1 shows only two vehicles in a small observation window, whereas in most dense traffic, there will be numerous such vehicles with their trajectories displayed in the observation window.

This modelling can be used to define some macroscopic variables used for defining the macroscopic behaviour of the transportation system. The *average speed* of a vehicle is a good indicator of traffic, affecting most of the decisions. The speed can be averaged in space or in time; here averaging across space for an observed time duration is more important. The space-averaged speed $(\bar{v})$ is defined as the total distance travelled by all vehicles in the observation window, over the total time of travel of all the vehicles. This is given by Eq. [13.1].

$$\bar{v} = \frac{\sum_i x_i}{\sum_i T_i} = \frac{\sum_i v_i dt}{\sum_i dt} = \frac{\sum_i v_i}{N} \qquad [13.1]$$

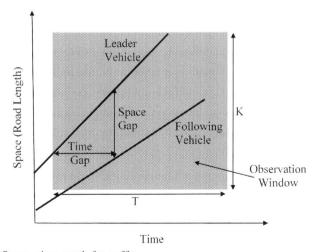

Figure 13.1 Space—time graph for traffic.

Here, x_i is the distance travelled by the vehicle i, T_i is the time of travel of vehicle i. v_i is the average speed of vehicle i. The second term of the equation is for an observation window of time dt.

Density (k) is the number of vehicles per unit length. It is measured by the number of vehicles (N) in a small observation window, over the length of the small observation window (K). If the road has multiple lanes and N_i is the number of vehicles in lane i, the density (k) is given by Eq. [13.2]

$$k = \frac{N}{K} = \frac{\sum_i N_i}{K} \qquad [13.2]$$

The above density is based on space, wherein we keep a constant snapshot of time and measure the density for some region of space (road). The density can also be defined in terms of time, wherein the density is the total time spent by all vehicles in a small region, over the area of the observation region. This formulation is given in Eq. [13.3].

$$k = \frac{\sum_i T_i}{Tdx} = \frac{\sum_i \frac{dx}{v_i}}{Tdx} = \frac{1}{T} \sum_i \frac{1}{v_i} \qquad [13.3]$$

Here, T_i is the time spent by vehicle i, v_i is the average speed of vehicle i. The equation is for an observation made in a space of length dx for T units of time.

Flow is the number of vehicles passing an observation point per unit time. The flow (q) is given by Eq. [13.4].

$$q = \frac{N}{T} \qquad [13.4]$$

The flow can also be defined as the distance travelled by the vehicles in an observation window, over the area of the observation window. The relation is given by Eq. [13.5].

$$q = \frac{\sum_i x_i}{Kdt} = \frac{\sum_i v_i \cdot dt}{Kdt} = \frac{\sum_i v_i}{K} \qquad [13.5]$$

Here, x_i is the distance travelled by vehicle i. The calculation is done for a road of length K and for a time duration of dt. The density and flow may be added over the number of lanes, and time averaged over an observation duration.

The traffic-flow theory is based upon a *fundamental relation* between the three entities, mean speed, flow and density. This enables conversion from one to the other to understand the traffic flow. The relation is given by Eq. [13.6].

$$q = k\bar{v} \qquad [13.6]$$

The traffic is assumed to be homogeneous over the road network and stationary for some time. This can happen when the conditions are stationary and the traffic shows no change in trend in terms of density, flow and speed. When the vehicles leaving become fewer than the vehicles entering, there is a mismatch and queues start to form. Alternatively, the queues can clear when the flow of vehicles outside is more than the flow entering.

Floating cars are widely used to record the real traffic data in a transportation network. These are special vehicles which are deputed to move around and record the essential parameters of traffic, based on which the condition of traffic can be assessed.

13.2.2 Fundamental Diagrams

Traffic operates in multiple phases and states, the properties for which change with time depending upon the demand dynamics. To understand traffic flow, let us start with an empty road with no vehicles such that the density is zero. Let us slowly add vehicles such that the density increases. Even though the number of vehicles and thus the density increases with time, because the road in such a state is so sparsely occupied that the vehicles are hardly affected by each other. This stage is called the *free-flow traffic*, and is characterized by small density and high flow, whereas the mean speed is close to the preferred speed of the drivers. On increasing the density, the flow increases as the number of vehicles increases with time, whereas there is a very small drop in the mean speeds. On further increasing the density, the network reaches a threshold flow, which is the maximum flow that the network can sustain and is called the *capacity flow*. The mean speed reduces in this region.

Suppose the traffic density is increased even further by packing in more vehicles. This results in smaller distances maintained between the vehicles. As a result, the drivers have to slow down. Thus, further increase in traffic leads to a severe drop in speeds and a situation of *congestion*. Severely increasing the number of vehicles can cause jammed traffic, wherein the distances between the vehicles is very small and the vehicles are nearly always in a state of rest.

Consider that the density at a region of road is suddenly increased, which will make the vehicles of the region drop their speeds. Common examples are traffic lights, mergers and blockages. This produces commonly seen *shock waves*. As a result, the vehicles behind will also drop their speeds. In the worst cases, the speed will become zero, in which case the vehicles will staring queueing one after the other. This is a shock wave which travels from the front to back. If a pool of vehicles continues to move a small distance, the wave will propagate backwards and the vehicles will start moving in a wave-like manner, creating a stop-and-go wave. On the contrary, consider that the traffic suddenly clears and now it is possible for a pool of vehicles to have free-flow travel as the vehicles ahead have suddenly cleared or a blockage is cleared. In this case the vehicles start moving and clearing the way for other vehicles which further start moving. Going forwards in space, one sees the speed continuously increase. This cases a kinematic wave to travel forwards with vehicles possessing high speeds. In traffic systems, such upstream and downstream waves are continuously produced and propagated. The waves from different sources may merge and cancel each other's effect, or add up and traverse. The waves maintain their own speeds and effects.

Based on the previous discussion, it is intriguing to study the behaviour between mean speed, flow and density under different conditions. A plot between these quantities is called a *fundamental diagram* and represents the basic fundamental principles behind the operation of traffic. The first fundamental diagram is between traffic density and mean speed. To plot the graph, start with a very small density, in which case the speed will be very high. As one starts to pack more vehicles, the intervehicle separation starts to decrease, resulting in the vehicles having to reduce their speed. In this duration the vehicle makes a transition from free flow to a capacity flow and finally to a congested flow wherein the speeds are nearly zero. A generic figure showing the relation is given in Fig. 13.2A.

The next diagram is between traffic flow and traffic density. Initially, when the density is very small, the vehicles move with their free-flow speeds. However, because the number of vehicles is small, the flow is small. An increase in the traffic density increases the flow, till the network starts exhibiting the maximum capacity flow. Any further increase in traffic density causes the traffic flow to reduce, till it reaches the congested level, wherein the vehicles stop moving and correspondingly the flow is zero. The diagram is shown in Fig. 13.2B.

The last diagram is between the traffic flow and mean speed. This diagram is interesting. A very low traffic flow is an indicative of a very congested traffic scenario

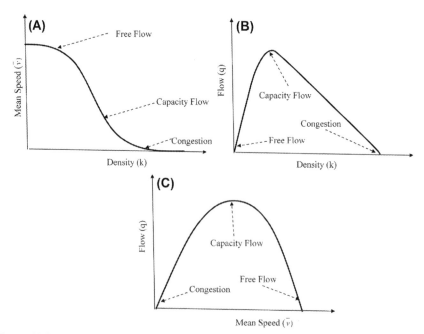

Figure 13.2 Fundamental diagrams: (A) mean speed versus density, (B) flow versus density, (C) flow versus mean speed.

wherein the speeds of the vehicles are nearly zero and the density is alarmingly high. It is also possible to get very small traffic flow when the vehicles are operating at high speeds, but the density is far too low making the flow small. The former is the case of a congested phase, whereas the latter is a case of free-flow phase. As the vehicle goes from the free-flow region to a capacity-flow region, the speed reduces due to the insertion of more vehicles. However, the flow constantly increases till it reaches the capacity flow. Any further increase in density will cause vehicles to show signs of congestion and they will start slowing down, also reducing the flow. This happens till a jam occurs and all vehicles nearly stop, creating zero traffic flow. An indicative diagram is shown in Fig. 13.2C.

13.2.3 Interrupted Traffic Flow

The traffic system discussed so far was based on a single road, with all vehicles travelling on it. This is an uninterrupted flow of traffic. In the real world, the normal traffic flow is *interrupted* by conditions such as traffic lights, mergers, diversions, intersections etc. The case of mergers and diversions is particularly interesting, which results in either a new source of vehicles to feed in or an alternative way for vehicles to feed out. Consider the case wherein two roads merge into one, meaning the traffic from both roads is expected to merge and make a single flow. Consider a main road on which a ramp road merges and adds traffic. Let k be the density and q be the flow of the main road. Let the ramp road feed in vehicles at the rate of z per unit length and per unit time. We know that the number of vehicles in the main road after merger will be the vehicles given by the main road before merger and the ones given by the ramp road, also accounting for any change in flows in the process. Note that here z is expressed in per unit time and space both, hence the units. Formally, this law is given by Eq. [13.7] and is the *fundamental law of traffic with external sources*.

$$\frac{\partial k}{\partial t} + \frac{\partial q}{\partial x} = z \qquad [13.7]$$

The conditions of traffic lights are interesting. The stoppage due to traffic lights causes the traffic of a particular road to stop, sending a shock wave downstream. As a result, the vehicles start to queue. The green light results in clearing up of the traffic, sending another kinematic wave upstream wherein the vehicles start to clear up.

13.2.4 Congestion

Congestion (Güner et al., 2012; Skabardonis et al., 2003) is commonly seen in traffic, wherein the traffic barely moves. Congestion can be very troublesome for people. Congestion arises when there are too many vehicles entering the road network, whereas the same outflow cannot be accomplished. This results in queueing in the road network, thereafter leading to more congestion. The congestions may be classified as recurrent and nonrecurrent. *Recurrent congestions* are the ones which are repetitive and occur in patterns. Say a place has a large number of people going to the office in the morning in the same direction, which dramatically increases the number of vehicles on the road causing congestion. This pattern is observed every day. The recurrent congestions happen due to prior known causes and can be predicted. People are normally aware of the recurrent congestions and account for the same before making travel decisions.

NonRecurrent Congestion arises due to reasons unaccounted for a priori and because such congestion shows no trends, it cannot be easily computed in advance. The nonrecurrent trends are caused due to factors such as accidents on a road, blockage of a road due to various reasons, public demonstrations, public events, adverse weather conditions etc. Such congestions suddenly arise and cause a great deal of discomfort to people. As the entire road network is connected, congestion at one place can severely affect the traffic flow at other places as well.

13.3 Traffic Simulation

Traffic simulation studies are widely used to imitate traffic on natural and synthetic scenarios and to experiment with different operational policies to understand the effect of those policies (Jaume, 2010). The traffic may be studied from a microscopic or macroscopic perspective. The *microscopic* study of traffic models each vehicle as an independent entity, which interacts with the other vehicles in the vicinity. The motion is affected by conditions such as intersections, traffic lights, mergers and diversions. On a straight road, the intention is largely to model the vehicle flow based on the separation from the vehicle in front. This is the simplest car-following behaviour, wherein the immediate speed of the vehicle depends upon the speeds of the two vehicles and their distances. The other behaviours can be similarly modelled. Microscopic traffic simulation gives the ability to study the individual vehicle behaviours based on traffic studies and to imitate the same in a computational framework.

The *macroscopic* traffic simulation models the vehicles as a dense flow, dealing with the average speed and density of vehicles at every intersection. It concerns the interaction, building up and clearing of the vehicle density due to different scenarios. The flow, overall, is modelled, rather than a collection of individual vehicles. The *mesoscopic* traffic simulation models combine the two methodologies of microscopic and macroscopic simulation and enable controlling the individual vehicle behaviours while operating as a flow of vehicles.

Traffic simulation requires a number of modules to be developed, which knit together enable simulation of the entire road-network graph for a long duration of time. One of the most important modules is the car-following module, showing the dynamics of a vehicle when it follows another vehicle. The basic module is supplemented with the modules of lane change, merging, traffic lights, intersection management, pedestrian handling, routing etc. The important modules and concepts are illustrated in the following subsections. The intelligence in Intelligent Transportation Systems is added on top of these modules. The intelligent aspects of some of these constituents are discussed in Section 13.4.

13.3.1 Basic Concepts

The traffic simulation systems take as input the road-network graph of the place. The data for the same can be obtained from multiple places like Openstreetmap (2015) and are also publically available for a large number of cities. The roads may have information about the lanes, speed limits, whether it is a one-way, whether overtaking is permitted etc. The road-network data are supplemented with the specification of the controlled and uncontrolled intersections, pedestrian crossings, car parks etc. The data may be noisy, and therefore the simulation is expected to filter out the data to remove errors like roads which are not connected to any other road, estimation of the presence of traffic lights, identification of intersections from raw files etc.

To simulate the network, we need to generate some traffic. The data about the traffic is specified by an *Origin−Destination* (OD) *matrix*. The matrix contains the number of vehicles going from every origin to every destination on the map for every unit time. The matrix is loaded onto the simulation, which generates such a source stream of vehicles for the specified time. The destination consumes the vehicles as the vehicles disappear from the network on reaching it. A road map of Reading, United Kingdom, and a simulated OD matrix is shown in Fig. 13.3.

The first problem associated with simulation is to convert the OD matrix into flows for different regions of the network at different points in time. This problem is called

(A)

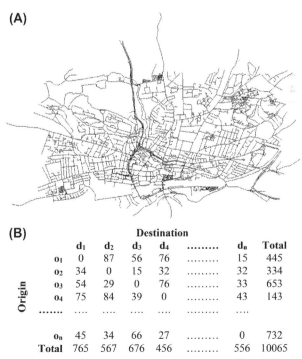

(B)

		Destination					
		d_1	d_2	d_3	d_4	.... d_n	Total
Origin	o_1	0	87	56	76	 15	445
	o_2	34	0	15	32	 32	334
	o_3	54	29	0	76	 33	653
	o_4	75	84	39	0	 43	143
	......	....	....	....	....	....	
	o_n	45	34	66	27	 0	732
	Total	765	567	676	456	 556	10065

Figure 13.3 Inputs for traffic simulations: (A) road network graph and (B) origin destination matrix.

the *traffic assignment* problem. To make their travel every vehicle must select the route to travel through, which is called the problem of *routing*. The person may prefer to take not only the shortest routes, but also routes which do not have much traffic, hence the congestion is low, and the operating speeds are high. So the routing decisions are based on the network model and the demand model of the system. The route information can be used for loading the network with the vehicles as per the given OD matrix and choice of route.

Consider a case wherein there is a short road connecting some origin and destination, which is a route normally under heavy demand and therefore congested. In such a case, the speeds of the vehicles on the route will be low and the time of journey will be long. Now suppose a vehicle decides to take a longer route, and reaches the destination earlier due to the less-congested state of the longer route. This encourages more vehicles to take the longer route in pursuit of reaching the destination early. This will finally end in an equilibrium, wherein the longer route will be as good an option to take as the shorter route considering the combined effect of journey time, convenience, fuel cost etc. In general, for a transportation network, for every origin and destination in the OD matrix, if the utility of two vehicles departing at the same time is equal and highest across all options offered by the network, the traffic flow is said to be in a state of *Dynamic User Equilibrium* (DUE). The traffic assignment must hence take care that such an equilibrium is maintained. Any change in the equilibrium must be dynamically managed by readjusting the traffic at the alternative routes.

13.3.2   Intelligent Driver Model

Based on the fundamentals of traffic flow, numerous models have been developed over the years. These models enable the simulation of traffic. The most dominant and hard to model behaviour is the vehicle-following model. The *Intelligent Driver Model* (Treiber et al., 2000) is one such model widely used for traffic simulation and analysis. The model computes the motion of a vehicle in the state when it is following another vehicle in front. The model can be naively extended to the situation of travelling on a free road, wherein there is no vehicle in front. The model accounts for the general equation for vehicle driving which states that once the driver perceives some stimulus from the environment, there is a lag of reaction time from the driver during which time no actions are made. After the reaction time, the driver responds to the stimulus by making an appropriate action, based on the driver's sensitivity. This is given by Eq. [13.8].

$$\text{Response}(t + t_r) = \text{Sensitivity} \times \text{Stimulus}(t) \qquad [13.8]$$

Here t is time and t_r is the reaction time.

The Intelligent Driver Model uses the concept of a safe distance, which is the distance ($s*$) that the vehicles prefer to maintain. The safe distance is given by Eq. [13.9].

$$s* = s_0 + s_1\sqrt{\frac{v}{v_0}} + Tv + \frac{v\Delta v}{2\sqrt{ab}} \qquad [13.9]$$

Here, s_0, s_1, v_0, T, a and b are constants. Δv is the relative speed of the vehicle with respect to the vehicle in front and is given by the speed of the following vehicle minus the speed of the vehicle in front. It is also called the approaching rate. The first term s_0 is the minimum safe distance that the vehicle always maintains with the vehicle in front, even if both the vehicles are stationary in a jam situation. The second term $s_1\sqrt{\frac{v}{v_0}}$ is not very intuitive and suggests that higher speeds require more safe distance. The third term Tv is the distance that the vehicle will travel with the current speed till a moderate time T, which must be maintained. This accounts for the reaction time. The last term accounts for a vehicle moving with a general acceleration given by the maximum acceleration (a) and braking (b) terms, while the vehicle is travelling with a general speed given by the approaching speed Δv and the current speed v.

The vehicle may not be maintaining the preferred distances and therefore will have a tendency to brake or accelerate in response to the situation. The vehicles have a tendency to accelerate if the road is free with a value given by Eq. [13.10].

$$\text{acc} = a\left(1 - \frac{v}{v_0}\right)^{\delta} \qquad [13.10]$$

Here, δ is a constant. The equation states that, everything kept aside, a higher velocity means a smaller magnitude of acceleration.

The vehicles also have a tendency to brake if the current distance is much smaller than the desired distance given by Eq. [13.11]

$$\text{brake} = -a\left(\frac{s^*}{s}\right)^{2} \qquad [13.11]$$

The equation suggests that the smaller the distance in contrast to the preferred distance, the larger is the retardation or braking. The resultant acceleration model displayed by the vehicle by the application of brake and throttle is thus given by Eq. [13.12].

$$\dot{v} = \text{acc} + \text{brake} = a\left[\left(1 - \frac{v}{v_0}\right)^{\delta} - \left(\frac{s^*}{s}\right)^{2}\right] \qquad [13.12]$$

13.3.3 Other Modules

The other modules are well illustrated either as already discussed in the book or in the following sections/chapters. Therefore, the modules are very briefly summarized here. *Lane change* is one of the dominant modules which plays a major role in everyday traffic. Lane-change behaviour involves decisions whether to change lanes, which lane to change to and thereafter the trajectory of lane change. The decisions are based on time to collision, time and space gaps, density of vehicles, length of queues etc. *Overtaking* is another module dealing with whether to overtake and thereafter the overtaking trajectory. The *merging* module handles the vehicle behaviour when two roads merge into one. The mergers may be controlled or uncontrolled. The decisions involved are whether the main vehicle goes next or the ramp vehicle, and thereafter the trajectory for each of these. Similarly, *diversions* deal with the situations when the road diverges. *Intersections* may similarly be controlled or uncontrolled. In case of controlled intersections, the operation of traffic lights is very important. The *traffic light* policies are also simulated by the simulator.

The problem of *routing* deals with deciding the fastest, shortest and most economical (in terms of tolls or fuel) route for the vehicle considering the predicted traffic data. The modelling of *pedestrians* is another important feature as at many places pedestrian traffic severely affects the main traffic. Ways of dealing with *unexpected situations* like road blockages, special incidents, emergency vehicles and public buses on public bus-only lanes are other important aspects of traffic simulation.

13.3.4 Empirical Studies

So far, many constants and parameters have been introduced. Some of these parameters are controlled, the values for which depend upon the operational traffic policy, whereas some others are uncontrolled. The aim is to make a traffic simulation system using policies such that the performance is maximized for the controlled variables. Here, the performance may be given by many factors including the efficiency of the system, safety, economy etc. The uncontrolled variables cannot be controlled by the transportation centres, and the network is operated such that the expected performance is maximized against all values of these variables.

All the constants depend upon the vehicle. One may assume that the transportation system has all vehicles of the same nature and therefore assume the constants to be the same for all vehicles. This facilitates easier study of the network, not having to compute a large number of parameters. To fix the parameters, traffic data are collected over a number of roads and a number of scenarios. The data will require filtering to remove noise. The traffic data are used to compute the values of the parameters. This process is called calibration. *Calibration* may be different for each module in complex models. The techniques give the best fit of the parameters which make the assumed model behave closely with the real-world model recoded by the data.

13.3.5 Common Traffic Simulators

Numerous research groups worldwide have developed a number of traffic simulation tools, which can be used straightaway for carrying traffic simulation. These tools are extensively used to assess the utility of changing the transportation operation policies and rules, new infrastructure in terms of roads and public vehicles, and further enabling study of the effect of addition of intelligent constituents in the otherwise nonintelligent transportation system. One of the popular open-source traffic simulators is Simulation of Urban Mobility (SUMO) (Krajzewicz et al., 2012) from Germany, which allows all features mentioned while using microscopic simulation. Other popular simulators include Dynamic Route Assignment Combining User Learning and microsimulAtion (DRACULA) (Liu, 2005), Microscopic Traffic

Simulator (MITSIM) (Ben-Akiva et al., 2010), Traffic in Cities Simulation Model (VISSIM), PARAllel MICroscopic Simulation of Road Traffic (PARAMICS), Advanced Interactive Microscopic Simulator for Urban and non-urban Networks (AIMSUN), DYNAMic EQuilibrium (DYNAMEQ) etc.

13.4 Intelligent Constituents of the Transportation System

This section specifically focusses on the intelligence part of the transportation system. Although the theme of intelligence of transportation systems will continue to chapters 'Intelligent Transportation Systems With Diverse Vehicles' and 'Reaching Destination Before Deadline With Intelligent Transportation Systems', some of the notable and basic concepts are highlighted here for motivation. Section 13.3 mentions numerous modules necessary for traffic simulation, which are ultimately obtained from the operation of real-life traffic. This section and the next two chapters focus on specific constituents and investigate the application of intelligent concepts that make the overall transportation system intelligent.

13.4.1 Intelligent Traffic Lights

Traffic lights play a major role in management of the traffic. Without traffic lights, dense traffic would come from all sides of the road and would block the intersection without letting any vehicle pass through. The traffic lights allow and disallow traffic, based on time, to pass through the intersection. The duration and order of change of the traffic light is very important. The problem with the current orthodox traffic light systems is that the traffic light period and order are predefined. The order may be made to change at different times of the day. However, intelligent traffic lights can look at the traffic density at the different roads around and itself decide the order and duration of the traffic light. These traffic lights are facilitated by live inputs like the queue length at different roads. This can be accomplished by a variety of ways including the use of loop detectors at different roads, video-processing techniques etc.

There are two methodologies to work with intelligent traffic lights. One could consider a microscopic approach wherein every traffic light is dealt with independently and each traffic light decides its own operation policy based on the current input of traffic. Alternatively, macroscopic systems consider a set of traffic lights as one system and simultaneously generate an operational policy for all of them, such that the performance of the overall system is maximized. This methodology enables one traffic light to avoid sending too many vehicles too early to another traffic light, which will not be easily able to consume all vehicles at such a large pace. These traffic-light—traffic-light interactions can be modelled by the approach.

The problem of deciding the traffic light operation time and order is a decision-making problem. In this problem, the different parameters affecting the decision like queue length, priority of vehicles, traffic density, number of lanes, predicted traffic etc. are given as inputs to an intelligent system, which needs to assess these inputs and accordingly decide the output, which is the green-light time and the order of operation. One of the simple ways to solve the problem is by using Fuzzy Inference Systems (Collotta et al., 2015; Nair and Cai, 2007; Trabia et al., 1999), which use rules that can be handcrafted and tuned to get a good performance over different mixes of traffic densities. The tuning of rules is done to maximize the performance measure, which can be conveniently taken as the average and the maximum waiting time for a vehicle. Traffic simulators are largely used for the purpose. One gets the ability to generate different types of traffic at different times of the day, and study the performance of the handcrafted rules. Looking at the performance, the human designer or, more preferably, an automated system can learn the rules and membership functions.

The problem can also be modelled as an optimization problem (Garcia-Nieto et al., 2011; Kwasnicka and Stanek, 2006; Teo et al., 2010), wherein the aim is to reduce the average waiting time of the vehicles. The waiting time of the vehicles is a function of the traffic signal green time and the operational order. So using simulations for any traffic scenario, the performance for any given traffic-light policy can be computed. This serves as an objective function for the techniques, to be optimized by changing the traffic-light operation time and order. Any optimization technique, like Genetic Algorithm, Particle Swarm Optimization etc. can be used for the optimization.

13.4.2 Intelligent Intersection Management

Waiting at intersections is a common phenomenon which accounts for much of the time of the journey. Intersections and traffic lights lead to properly regulating the flow of traffic. Besides straight roads, intersections form a dominant scenario of operation through which an autonomous vehicle must be able to navigate. Intersections may be controlled or uncontrolled. The *controlled intersections* have traffic lights located at each of the roads and the vehicles can only move when the traffic light is green. This demands the vehicles to have a vision system which can detect and read traffic lights. If the light is green, the conventional trajectory planning mechanisms and algorithms work, although overtaking is rarely conducted. If the light is red, the autonomous vehicle will have to graciously stop with enough distance from the vehicle in front. Intersections with roundabouts mostly have smooth turns for the vehicles to navigate. Sometimes sharp steering may be required at small intersections.

On the contrary, *uncontrolled intersections* do not have any traffic lights to regulate traffic. Going through an intersection is subjected to one's own decision. Hence, an autonomous vehicle must be able to sense the environment, look at all the vehicles occupying and desirous to occupy the intersection, judge the safety of going through the intersection. Then it must decide whether to go through the intersection or wait for some other vehicle, and, once decided to go through the intersection, make a suitable trajectory and trace that trajectory avoiding all other vehicles passing through the intersection.

To focus upon the perspectives of planning, a simple *intersection management* example is considered. An intersection manager frames policies and plans the flow of vehicles at different roads of the intersection. Here, the work of Dresner and Stone (2004) is discussed. The authors study the *reservation-based intersection management*, wherein a vehicle may reserve its journey at the intersection. The authors first discuss a First-Come-First-Served reservation policy, wherein a vehicle coming near the intersection first attempts to make a reservation. The vehicle specifies the time to reach the intersection, the speed, direction of reaching and the destination direction. The reservation manager, an intelligent agent, simulates if the vehicle can travel without colliding or disturbing any other vehicle passing through the intersection. If the journey is possible, the reservation manager makes the reservation and allows the vehicle to pass through if it reaches at the mentioned time and speed. Otherwise, the reservation is not made and the vehicle may have to wait. If the vehicle fails to reach at the quoted time, it may request cancellation of the previous reservation and make a new reservation. The authors also study uncontrolled intersection management and traffic light-based intersection management.

In another work, Dresner and Stone (2006, 2007) allow for human-driven vehicles which may not have computational devices to make automated reservations at the intersections. Here, the intersection operates using traffic lights. The traffic lights may be operated via different policies, typically allowing all vehicles from one direction simultaneously or allowing vehicles between every source—destination direction pair. The human vehicles may only move when the traffic light is green. However, the automated vehicles are allowed to move even if the traffic light is red, provided they have a valid reservation. Reservations may be requested to the reservation agent in advance. The reservation agent checks whether all kinds of human vehicles with all from—to direction pairs can pass collision free if the reservation is allowed. In such a case the reservation is made and the vehicle is allowed to pass. If some vehicle may potentially have a collision, while operating with the traffic-light policy, the reservation is not made. The resultant bandwidth is hence the normal bandwidth allowed by the traffic light and additional bandwidth allowed by reservations. The reservation policy prohibits reservations which affect the normal functioning of the traffic.

13.4.3   Traffic Merging Management

The overall efficiency of the transportation system stresses the elimination of congestion for which the merging areas play a dominant role. *Merging* denotes the scenarios wherein (normally) two roads merge into one. This is shown in Fig. 13.4. Vehicles from both roads cannot simultaneously enter the merged road unless the bandwidth of the merged road is greater than the sum of the bandwidths of the individual roads, which is rarely the case. Many approaches have been made in the past to effectively

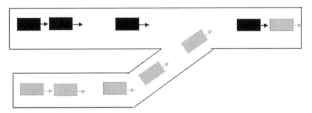

Figure 13.4 Merging scenario.

manage traffic in the merging area. Ramp metering is a popular method to control merging which imposes restrictions on the on-ramp vehicles for entering into the mainline roads (Papageorgiou et al., 1997). The use of the microscopic approach to manage the merging traffic based on the intelligence at the individual vehicle level is also common (Antoniotti et al., 2007). Most of the work related to traffic merging is based on the assumption that the vehicles are automated and equipped with a communication system. Studies have also been carried out in the area of lane distribution for effective implementation of ramp metering. Variable speed limits might change the lane distribution and in turn affect the merging (Carlson et al., 2010).

Merging can be categorized as merging of two highways or freeways, merging of on-ramps into the freeway mainstream, merging within a toll plaza infrastructure and merging of freeway mainstream lanes due to road works (Papageorgiou et al., 2008). For designing any merge-assistance system, there are certain basic questions which are to be addressed like which vehicle of the two traffic streams in the case of two-lane merging, should merge first, what action should be performed by the mainline vehicles and the ramp vehicles to merge at the right time and place etc. The answer depends mainly on the scheduling criterion to be followed between the vehicles of the two streams. Many researchers have been trying to formulate the best type of merging criterion for merging two vehicles efficiently, but the result depends on the effectiveness of the scheduling mechanism employed. The type of merge-assistance system employed depends largely on the types of vehicles involved ie, whether the vehicles are automated and equipped with an effective communication system, semiautomated (limited technological usage) or manual (no automation).

Merging is studied by a small example presented in the work of Raravi et al. (2007), with some adaptations. Consider an intersection of two roads. It is assumed that the number of vehicles at the intersection, the distance of each vehicle from the intersection, the speeds of the vehicles and the maximum speeds and accelerations of the vehicles are known. The vehicles need to move so that a minimum safe distance between all vehicles is always available, whereas the safe distance depends upon the speeds of the vehicles. It is assumed that the vehicles can communicate with the merging manager, which makes the decisions regarding the merging. The algorithm attempts to frame an optimal merging sequence for the vehicles at both streams in the merging scenario.

At any instance of time a decision needs to be made which of the two competing vehicles should go on to the merged road. The scheduling algorithm works on the simple hypothesis that a vehicle which in its normal course of action (without considering the stream of vehicles in the other stream) is projected to reach the intersection first, should be allowed to merge. Them decision regarding which vehicle should be allowed to merge needs to be made as soon as possible to admit a vehicle in the intersection area. The vehicle which takes less time to reach the intersection region is allowed to enter into the merge region.

The time that the vehicle at the head of the lane takes to reach the intersection is calculated from the kinematic parameters of the vehicles. Based on the positions, speeds and accelerations of the vehicles and their current distance from the merged road, the time to reach the intersection can be computed. The vehicle with a lesser time drives through, whereas the other may have to wait for its turn, if the merged road is not clear till it reaches the merging point. In case of any conflict with reference to the time taken to reach the intersection region, the cost to merge associated with the vehicle is taken into account which acts as a secondary factor considered for decision-making. The cost here is the distance of each vehicle from the intersection region at the time of decision-making. The vehicle having the least distance is assigned the least cost and is allowed to enter the intersection region. It is based on the notion that the nearer is the vehicle, the greater the chances it has of adapting itself to any changes. If the distance from the intersection also happens to be the same, then any vehicle may be randomly picked for entrance to the intersection region.

All other vehicles are assumed to be operating on a simple vehicle-following mode, wherein they drive to always maintain a safe distance from the vehicle in front, while trying to drive at high speeds as long as possible. No overtaking may be attempted. Safe distance is computed from knowledge of the relative speeds, vehicle acceleration and current distance. The distance between any two vehicles must be larger than the given safe distance. Speed changes account for acceleration limits.

13.5 Summary

This chapter looked at the complete transportation system, rather than a single vehicle alone, and took the task of understanding the complete system to maximize its performance. The first dominant issue in such a case is the modelling of such a complex system. Traffic-flow theory was explained which relies upon the basic principles of traffic flow, traffic density, mean speed and the relations amongst these. The effect of interrupted traffic in the form of mergers or traffic lights was also studied. Fundamentally, an increase in traffic density reduces the speed of the vehicles, from free-flow speed to a nearly zero congested speed. The flow initially increases to a capacity value and thereafter reduces to a nearly zero value.

These fundamentals were used to make a traffic simulator, which simulates given traffic over a given road network. For this, the traffic simulator should be able to imitate all behaviours of the vehicles including route-choice behaviour, merging behaviour, lane-change behaviour, overtaking behaviour etc. The most important behaviour is the vehicle-following behaviour. The Intelligent Driver Model was discussed for the same.

With this background the task is to make all the components of the transportation system intelligent. The chapter presented the ways of making the traffic lights intelligent to minimize the wait and the queue length on the roads. The intelligent systems try to do so by changing the green-light time and the order of change of traffic lights. The popular methods include fuzzy logic, optimization techniques and reinforcement learning. Similarly, the intersection can be made intelligent by allowing the autonomous vehicles, capable of better control, to pass through the intersection if the same do not conflict with the normal flow of vehicles for the particular state of the traffic lights. The merging can itself be made intelligent by using metrics to decide whether it will be better to allow the ramp vehicle to pass through next or the main-road vehicle.

References

Antoniotti, M., Deshpande, A., Girault, A., 2007. Microsimulation analysis of multiple merge junctions under autonomous AHS operation. In: Proceedings of the IEEE Conference on Intelligent Transportation Systems, Boston, USA, pp. 147−152.

Ben-Akiva, M., Koutsopoulos, H., Toledo, T., Yang, Q., Choudhury, C., Antoniou, C., Balakrishna, R., 2010. Traffic simulation with MITSIMLab. In: Barcelo, J. (Ed.), Fundamentals of Traffic Simulation. Springer, New York, pp. 233−268.

Carlson, R.C., Papamichail, I., Papageorgiou, M., Messmer, A., 2010. Optimal motorway traffic flow control involving variable speed limits and ramp metering. Transportation Science 44 (2), 238−253.

Collotta, M., Bello, L.L., Pau, G., 2015. A novel approach for dynamic traffic lights management based on wireless sensor networks and multiple fuzzy logic controllers. Expert Systems with Applications 42, 5403−5415.

Dresner, K., Stone, P., 2004. Multiagent traffic management: a reservation-based intersection control mechanism. In: Proceedings of the Third International Joint Conference on Autonomous Agents and Multiagent Systems, NY, USA, pp. 530−537.

Dresner, K., Stone, P., 2006. Multiagent traffic management: opportunities for multi-agent learning. In: Lecture Notes in Artificial Intelligence, vol. 3898. Springer Verlag, Berlin, pp. 129−138.

Dresner, K., Stone, P., 2007. Sharing the road: autonomous vehicles meet human drivers. In: Proceedings of the 20th International Joint Conference on Artificial Intelligence, pp. 1263−1268 (Hyderabad, India).

Garcia-Nieto, J., Alba, E., Olivera, A.C., 2011. Enhancing the urban road traffic with swarm intelligence: a case study of Córdoba city downtown. In: Proceedings of the 2011 11th International Conference on Intelligent Systems Design and Applications, pp. 368−373.

Güner, A.R., Murat, A., Chinnam, R.B., 2012. Dynamic routing under recurrent and non-recurrent congestion using real-time ITS information. Computers & Operations Research 39 (2), 358−373.

Immers, L.H., Logghe, S., 2002. Traffic Flow Theory. Lecture Notes, Faculty of Engineering, Department of Civil Engineering, Section Traffic and Infrastructure, Belgium.

Jaume, B., 2010. Fundamentals of Traffic Simulation. Springer, New York.

Krajzewicz, D., Erdmann, J., Behrisch, M., Bieker, L., 2012. Recent development and applications of SUMO - simulation of urban mobility. International Journal on Advances in Systems and Measurements 5 (3&4), 128−138.

Kwasnicka, H., Stanek, M., 2006. Genetic approach to optimize traffic flow by timing plan manipulation. In: Proceedings of the 2006 Sixth International Conference on Intelligent Systems Design and Applications, vol. 2, pp. 1171−1176.

Lighthill, M.J., Whitham, G.B., 1955a. On kinematic waves. I. Flood movement in long rivers. In: Proceedings of the Royal Society of London A: Mathematical, Physical and Engineering Sciences, vol. 229. No. 1178. The Royal Society.

Lighthill, M.J., Whitham, G.B., 1955b. On kinematic waves. II. A theory of traffic flow on long crowded roads. In: Proceedings of the Royal Society of London A: Mathematical, Physical and Engineering Sciences, vol. 229. No. 1178. The Royal Society.

Liu, R., 2005. The DRACULA dynamic network microsimulation model. In: Kitamura, R., Kuwahara, M. (Eds.), Simulation Approaches in Transportation Analysis: Recent Advances and Challenges. Springer, pp. 23−56.

Maerivoet, S., Moorm, B.D., 2005. Traffic Flow Theory, Technical Report 05-154. Katholieke Universiteit Leuven, Department of Electrical Engineering ESAT-SCD (SISTA).

Nair, B.M., Cai, J., 2007. A fuzzy logic controller for isolated signalized intersection with traffic abnormality considered. In: Proceedings of the 2007 IEEE Intelligent Vehicles Symposium, pp. 1229−1233.

Newell, G.F., 1993. A simplified theory of kinematic waves in highway traffic, part I: general theory. Transportation Research Part B: Methodological 27 (4), 281−287.

Openstreetmap, 2015. Available online: https://www.openstreetmap.org (accessed October 2015).

Papageorgiou, M., Hadj-Salem, H., Blosseville, J.-M., 1997. ALINEA local ramp metering: summary of field results. Transportation Research Record 1603, 90−98.

Papageorgiou, M., Papamichail, I., Spiliopoulou, A.D., Lentzakis, A.F., 2008. Real-time merging traffic control with applications to toll plaza and work zone management. Transportation Research Part C 16, 535−553.

Raravi, G., Shingde, V., Ramamritham, K., Bharadia, J., 2007. Merge algorithms for intelligent vehicles. In: Next Generation Design and Verification Methodologies for Distributed Embedded Control Systems, pp. 51−65.

Skabardonis, A., Varaiya, P., Petty, K., 2003. Measuring recurrent and nonrecurrent traffic congestion. Transportation Research Record 1856, 118−124.

Teo, K.T.K., Kow, W.Y., Chin, Y.K., 2010. Optimization of traffic flow within an urban traffic light intersection with genetic algorithm. In: Proceedings of the 2010 Second International Conference on Computational Intelligence, Modelling and Simulation, pp. 172−177.

Trabia, M.B., Kaseko, M.S., Ande, M., 1999. A two-stage fuzzy logic controller for traffic signals. Transportation Research Part C 7, 353−367.

Treiber, M., Hennecke, A., Helbing, D., 2000. Congested traffic states in empirical observations and microscopic simulations. Physical Review E 62 (2), 1805−1824.

Intelligent Transportation Systems With Diverse Vehicles

14

14.1 Introduction

Transportation systems are amalgamations of numerous subsystems which operate harmoniously with each other for overall efficiency and ease of travel. It is hence encouraging to intelligently operate each of these subsystems for efficient travel and to further exchange information between these subsystems for higher efficiency. Many factors play a major role in regulating traffic and hence need to be intelligent and adaptive towards changing traffic trends. This includes the mechanisms to handle traffic lights, intersection management, overtaking policies, speed-lane policies, reservation policies etc. Each of these management systems and policies largely depends upon the kind of traffic that is operating in a region. Increasing autonomy in the vehicles and the transportation infrastructure facilitates engineering innovative solutions that make the best out of available information and enable sharing information amongst transportation subsystems.

The core interest behind the book is to study a traffic system that operates without speed lanes. It was advocated that such a traffic system is largely caused due to diversity between vehicles. A macroscopic study of the entire transportation system operating without lanes is currently not possible. However, it is still intriguing to study the transportation system when operated with diverse traffic, which is the theme of this chapter. Although it would be incorrect to state that the concepts and results presented here would be applicable for transportation systems without lanes even though both systems are triggered by diversity of traffic, some indications can always be drawn. More importantly, the focus of the chapter is towards the design of algorithms at the transportation level which perform well in conditions of high diversity in vehicles.

Section 14.2 discusses some very interesting works in the intelligent management of transportation systems, particularly from the point of view of routing. Section 14.3 assumes the vehicles have some communication with a central transportation authority or other vehicle infrastructure system. This is similar to the assumption of communication placed in chapters 'Optimization-based Planning', 'Sampling-based Planning', 'Graph Search-based Hierarchical Planning' and 'Using Heuristics in Graph Search-based Planning'; however, the assumption is softer here. Such an assumption can lead to a variety of ways in which the traffic can be managed which would not be possible in the case in which the vehicles are not connected. The chapter discusses numerous such possibilities through simulations.

Section 14.4 does not assume any communication which means any number of human-driven vehicles can be on the road. This makes many of the possibilities introduced in Section 14.3 impossible to implement. The most important problem in such a context for a macroscopic study is vehicle routing. Effective routing can avoid road congestions and hence inefficiency in the transportation system. The aim is to study an urban kind of transportation system, for which the design of an effective routing technique is displayed.

Segments of the chapter have been reprinted from R. Kala and K. Warwick (2015a). Congestion Avoidance in City Traffic. Journal of Advanced Transportation 49(4), 581−595 Copyright © 2015 John Wiley & Sons, Ltd. with permission. Segments of the chapter have also been derived in part from an article R. Kala and K. Warwick (2015b) published in International Journal of Computational Intelligence Systems, 2015, Vol. 8, No. 5, pp. 886−899, Intelligent Transportation System with Diverse Semi-Autonomous Vehicles, Available online: http://www.tandfonline.com/10.1080/18756891.2015.1084710.

14.2 A Brief Overview of Literature

Here, some of the notable works in the intelligent management of the transportation system are very briefly discussed. Diversity is an important aspect which most of these works do not particularly address. These works make different elements of the transportation system intelligent and thus result in a better traffic efficiency. Routing and related decisions are particularly important for effective traffic operation.

The complete road network graph can be viewed as a Markov network with the route as a Markovian process. Kim et al. (2005) presented their results to routing. The model could account for dynamic traffic, and hence congestion could be monitored. Wahle et al. (2000) used Cellular Automata and defined various traffic behaviours like braking, accelerating, avoiding obstacles etc. as rules. Based on this approach the authors showed a routing strategy such that the entire traffic was more distributed. In another related approach, Furda and Vlacic (2011) also use Automata for system modelling. The authors exhibited vehicular behaviours including maintenance of position on the road, maintenance of a safe distance from other vehicles, collision avoidance etc. Current behaviour was selected using Multicriterion Decision-making.

Inspired by the ant algorithms, digital pheromone is another popular mechanism by which traffic dynamics in a road network may be modelled and decisions may be made. Ando et al. (2006) represented various driving actions by a digital pheromone distribution. These pheromones gave an indication of traffic congestion. The same information was used for routing decisions. Narzt et al. (2007) also used the notion of digital pheromones. Their model used micro simulations with a decentralized routing strategy for vehicular motion.

In terms of route planning in a static sense, when the given road network graph becomes too complex, it is viable to use some hierarchical planning. Song and Wang (2011) employed heuristics to divide the entire road network into hierarchical communities. Each community marked a highly connected region. Li et al. (2009) represented a graph in a multilayered approach with edges between the layers. Voronoi diagrams were used as the basis for hierarchical separation of the road network map. Tatomir and Rothkrantz (2006) presented another hierarchical approach. Their algorithm divided a road network into zones with identified road links connecting the zones. The authors used an ant-based swarm algorithm to find the shortest route. In addition, the authors displayed how the time of journey may be computed when rerouting in accident situations.

Fawcett and Robinson (2000) displayed a system which monitored live data with relevance to the available road infrastructure and the mechanism by which these data may be made available to route planning of the vehicles. Kesting et al. (2008) looked at the problem of traffic congestion and proposed a model wherein vehicles could adapt their driving model parameters based on the available information of traffic flow.

In a situation in which the number of vehicles is too large and the road infrastructure is limited, reservation seems a viable alternative. Congestion may be avoided by a careful reservation strategy. Dresner and Stone (2004) studied a subset of this problem of intersection management and proposed a scheme in which the autonomous vehicles could navigate by reservation irrespective of the traffic signal states. The authors further extended the model to incorporate learning behaviour and market economics of reservation (Dresner and Stone, 2006, 2007). Vasirani and Ossowski (2009) continued this work by presenting a market economy model for reservation. For roads, an approach similar to reservation was used by Reveliotis and Roszkowska (2011) who modelled the entire road infrastructure as resources with a judicious resource allocation algorithm.

14.3 Semiautonomous Intelligent Transportation System for Diverse Vehicles

For reasons of safety, driving efficiency and sometimes driving comfort, much research is now being done in the domain of driving assistance systems and autonomous vehicles. Autonomous vehicles are technologically more advanced and are capable of driving on their own without any human input, making all driving decisions on their own. *Semiautonomous vehicles*, however, provide limited capabilities, restricted to either or all of automated parking, overtaking, lane following etc. State-of-the art research in these domains showcases a promising future in which vehicles increasingly become more advanced, to the extent that most vehicles on the road will be semiautonomous with communication abilities, access to advanced travel information and dynamic route guidance systems, amongst other safety and decision-making systems. In reality though, nonintelligent vehicles may still exist in small numbers for a very long time. A fully autonomous scenario, with only autonomous vehicles is also a possibility, although this may take a much longer time to materialize.

Vehicles with different levels of autonomy have different capabilities related to vision, control and reaction time, all of which lead to capabilities to drive at different maximum speeds. Autonomous vehicles may especially be designed for different commercial or business requirements due to which they may vary in their type, make and performance, leading to different driving speeds. For semiautonomous vehicles, driving speeds also depend upon driver preference, passenger preference, purpose of travel, social stature etc. All of which leads to different vehicles driving at different speeds. In fact, at present traffic showcases limited *diversity in driving speeds*, wherein different drivers prefer to drive at different speeds, leading to lane changing, overtaking and a distribution of traffic across lanes roughly as per the preferential driving speeds. Increasing levels of autonomy are though very likely to increase this speed band, making traffic more diverse in terms of speed capabilities.

Intervehicle communication (Sen and Matolak, 2008) systems help a group of vehicles to talk to each other and share information which provides advantages including collision warning, obstacle alert, cooperative obstacle avoidance etc. *Vehicle—infrastructure communication* (Ma et al., 2009) meanwhile enables a vehicle to communicate with a transportation infrastructure found on the road. Such a system may be useful to communicate traffic or road conditions to the vehicle. Communication helps vehicles to make optimal decisions at the local and global levels, while reducing any uncertainties. Having extra information via communication, unknown to the driver's or the vehicle's normal vision, is always helpful in decision-making. Here, the focus is not only on making a vehicle's personal plan better, but in enabling vehicles to collaborate to make the overall transportation system better, even if it is at the cost of one's own personal plan.

Traffic systems play a major role in regulating the movement of vehicles in any country. In most scenarios, static traffic rules lead to reasonable traffic management for most general driving. Traffic may further be managed by making rule changes for certain days and times (eg, heavy vehicles at night only) or to cope with certain scenarios (eg, possibly a large number of vehicles before/after a concert or sporting event). These rules need to be effective as they impact a large number of vehicles. However, in the case of semiautonomous vehicles, this allows for a transportation system-wide communication between all entities, thereby enabling a *central transportation authority* to *dynamically* regulate traffic as per the available information or traffic policies. Alternatively, an intelligent system may be placed to constantly monitor traffic, anticipate traffic conditions and make traffic regulating decisions, which may be communicated to the vehicles for them to follow.

Traffic simulation allows the study of various ways or rules by which traffic can be regulated. Traffic simulation systems are classified into microscopic systems, macroscopic systems and mesoscopic systems (Helbing et al., 2002). The approach towards this domain is microscopic in nature, in which the individual vehicle behaviour is considered, when planned amongst a group of vehicles in a scenario.

This work continues the discussion on motion planning for autonomous vehicles in a traffic environment with diverse speed capability vehicles and operating in unorganized traffic. The solutions were largely restricted to a straight road. It was observed that diversity leads to interesting driving behaviours. It is hence encouraging to study the effect of diverse unorganized traffic on the overall transportation system. This requires the need for a traffic simulator working with unorganized traffic and diverse vehicles. Creating a complete simulation system for unorganized traffic requires solving complicated subproblems. Here, a simulation system for diverse and organized traffic is presented.

The simulation system described here was built by taking a futuristic view in which *most of the vehicles would be semiautonomous* within an *intelligent transportation infrastructure*. A semiautonomous vehicle may be defined as one that can be networked, with the ability to take basic driving instructions which may be implemented by a human driver or the autonomous vehicle itself. The aim is to exploit all possibilities with such traffic by closely observing each and every transportation entity. This notion opens a pool of new possibilities and issues, some of which are presented. Although vehicles being semiautonomous may not necessarily be a requirement in all cases, it will certainly benefit the system in making dynamic and fast changes, which cannot be done in the present traffic system.

With this simulation, the approach is to introduce a number of *concepts* in the present traffic system and to study the behaviour of vehicles under these concepts by a simulation. By this, the effectiveness of the introduced concepts could be measured against the rules presently in the system for a diverse vehicle scenario. At the same time, the attempt was to make a traffic simulation system that accounts for the ability to statically (at the start) or dynamically specify the applicability of all these concepts on various roads. The simulator also provides a base for current and future research to make traffic systems more sophisticated, with the presented assumptions.

In this simulator, it is assumed that the task is to enable a large number of vehicles to reach their destination from their source. The road network map is already available within the system. Each vehicle starts from its own source and attempts to reach its destination in the shortest time possible, in *cooperation* with the other vehicles. The vehicles emerge from their source at a predefined time and leave the map on reaching their destination. At any time during the simulation process, the position of all the vehicles is assumed to be known. Further, the vehicles can communicate with the *central information system*, which helps them in decision-making. The key takeaways of the algorithm are summarized in Box 14.1. The major concepts are also summarized in Box 14.2.

Box 14.1 Key Takeaways of the Semiautonomous Intelligent Transportation System Approach

- The approach presents an *integrated study* of an intelligent transportation system covering all the various concepts which are separately studied in the literature.
- The study proposes *architecture of the transportation systems* of the future covering both decentralized vehicle control and centralized management control.
- The approach is designed for *diverse semiautonomous vehicles* operating in a scalable environment, which is the likely future of the transportation system.
- The approach is a positive step towards creation of a *traffic simulation tool* for diverse and unorganized traffic.

Box 14.2 Key Concepts of the Semiautonomous Intelligent Transportation System Approach

- **Assumptions**
  - All vehicles are semiautonomous, or all can communicate
  - All vehicles can be tracked
  - There might still be some human-driven vehicles
  - Vehicles have very diverse speeds
- **Key idea**
  - Explore all the possibilities with such assumptions
  - Enable vehicles to cooperatively reach their destinations in the best way
  - Make transportation rules dynamic
- **Systems**
  - *Traffic Light System*, to immediately change light when a queue gets empty, and to subsequently allow traffic from the road with the most-waiting vehicle.
  - *Speed Lanes*, to dynamically vary the speed limits of lanes based on the diversity of the vehicle speeds.
  - *Route Planning*, to avoid building up of congestion on any road.
  - *Reservations*, to give reserved rights to some vehicles for a lane or for the entire road.

Traffic lights are an important aspect of a traffic management system. They ensure that vehicles reach their destinations on time, at the same time avoiding congestion. Both the order and duration of operation are important. The manner of handling traffic lights is discussed in Section 14.3.1. Similarly, *speed lanes* play a major part in the distribution of traffic. This especially becomes important in the case of vehicles with diverse speed capabilities. Deciding the speed limits for individual lanes is important, which in this system is discussed in Section 14.3.2. *Route planning* deals with deciding on the roads to use to reach the destination. *Continuous replanning* enables escaping from densely crowded roads, traffic regularization and the avoidance of traffic jams. Section 14.3.3 presents the route-planning algorithm used in this system. Increased traffic density, slow traffic and a wide diversity in speed capabilities, especially at some times of the day and for some roads, necessitate the need to use roads as a *reserved* infrastructure. In Section 14.3.4 the mechanism by which a road or a lane may be made available only by means of making a reservation is discussed. This enables important vehicles to reach their destination as early as possible, which includes emergency vehicles. Section 14.3.5 presents the general architecture of the system.

14.3.1 Traffic Lights System

Traffic lights constitute one of the most important aspects of a traffic system. The waiting time for traffic lights to turn green may constitute a significant proportion of the time of journey. Efficient operation of traffic lights can lead to overall traffic efficiency.

14.3.1.1 Concept

Presenty employed traffic light systems allow traffic going from multiple sources to travel within a specified time. Usually this time is prespecified to a threshold value. Common problems include having to wait for one's turn to cross when no other vehicle coming from the other direction, an equal waiting time for traffic on high-density roads and low-density roads, fixed traffic light operation times during the day, nonadaptability to changing traffic trends, having to wait for too many traffic light changes whilst travelling on a high traffic-density road etc. It is evident that intelligent traffic lights operating within an intelligent transportation system are capable of overcoming these problems.

The proposed traffic light system considers both the number of vehicles as well as the operation time. It is assumed that the number of vehicles waiting to cross at each crossing is known in advance. In the case of networked semiautonomous vehicles, this task is trivial as the position of the vehicles is reasonably well known by Global Positioning System (GPS) or a local mapping algorithm. Additionally, sensors at the crossing region may help record the same data. Assume a crossing c with intersecting roads $R = \{R_1, R_2, R_3, R_4, ... R_n\}$. Vehicle information is noted in a data structure Q_c, which maintains a queue of all vehicles waiting at the crossing region. An entry comprises the triplet $<V_i, R_i, t_i>$ denoting the vehicle V_i entered the crossing scenario (when it is ready to go over the crossing) at time t_i from road $R_i \in R$. The data structure is updated for every vehicle (V_i) entering (Eq. [14.1]) or leaving (Eq. [14.2]) the crossing scenario:

$$Q_c = Q_c \cup \langle V_i, R_a, t_i \rangle \tag{14.1}$$

$$Q_c = Q_c - Q_c[V_i] \tag{14.2}$$

in which $Q_c[V_i]$ is the entry of V_i in Q_c.

The traffic lights operate in a manner such that at any time only one road has the green signal, whereas the signal for all other roads is red. All vehicles on the road in possession of the green signal can travel, irrespective of their exits. Lateral time is employed to clear the crossing region between light changes. The objective of the traffic lights is to reduce the maximum waiting time for any vehicle. Hence, the order of traffic light changes is such that each change allows all traffic from road R_a in possession of the *longest waiting vehicle* at crossing c to cross. Let D_c denote the road with the green traffic signal, which changes as per (Eq. [14.3]).

$$D_c = R_i, i = \operatorname{argmin}_{i, \langle V_i, R_i, t_i \rangle \in Q_c} t_i \tag{14.3}$$

Traffic lights change as per the stated order if a traffic light change event (say E_c) occurs. After every change, the traffic lights stay in the same mode until a maximum of

η vehicles have crossed or for a maximum of T units of time. Here, η is taken as the minimum of a threshold value (η_{th}) and the present queue size ($|Q_c[D_c]|$) for vehicles originating from the road which is currently green (D_c). T is the time threshold which is a constant. If no vehicles are left on the road which is currently green, the traffic lights change. The change may however only occur if currently no other vehicle is in the crossing area. Leaving this lateral time ensures that no deadlock exists in the crossing region. The prerequisite for a change to occur is given by Eq. [14.4].

$$E_c = T_c \geq T \vee \eta_c > \eta \vee |Q_c[D_c]| = \phi \qquad [14.4]$$

Here, T_c is the time elapsed since the last change and η_c is the number of vehicles that passed since the last change.

Limiting both the *maximum time* and the *maximum number of vehicles* enables much better control of traffic when it contains diverse vehicles. Because the vehicles are semiautonomous, this is realizable. An important criterion here is to keep η_c as the minimum of the current queue size and a threshold. The effect of this is to discourage a newly entered vehicle from passing over the crossing region without waiting. This is because such a crossing might be at the cost of another waiting vehicle. If there are no other waiting vehicles, any change would automatically be in favour of the newly entered vehicle, which would be allowed to move. In this way, the heuristic of minimizing the waiting time can be realized.

14.3.1.2   Simulations

The purpose behind the simulations was to test the working of the proposed system under diverse traffic conditions. A map was employed with a crossing at the centre. A random number of vehicles were generated, each with its own speed capability, and these vehicles were made to travel from one road to the other over the crossing. The emergence time of the vehicles was randomly fixed. This system of traffic light operation was compared to a system in which the lights were changed at regular intervals of T units of time, which is the method frequently encountered in practice. In such a strategy, a lot of time could be wasted if the light is green yet no vehicles need to cross. The simulation system generated vehicles within random small time intervals that are uniformly distributed.

Another simpler traffic light change system was also studied. The change condition was kept largely as in Eq. [14.4], with the only difference being that changes were produced in a *cyclic order*, rather than that proposed in Eq. [14.3]. In all the simulations, the average time of travel was used as a metric. These three algorithms were simulated for varying numbers of vehicles. A smaller number of vehicles in a scenario meant less densely occupied roads and vice versa. The resultant graph between the number of vehicles in the scenario and the average time of travel is shown in Fig. 14.1A. Every point on the graph represents a scenario with random entry time, exit time and speed. Hence, the randomness has been smoothed by plotting a trend line produced by the moving-average method. The simulation used arbitrary units of

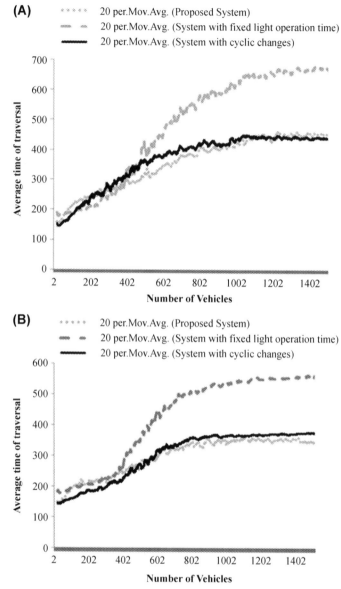

Figure 14.1 Comparative analysis of traffic-light system for different numbers of vehicles: (A) traffic from all sides and (B) traffic from one side blocked.

distance and time which are specific to the simulation tool and can relate to real-world units proportionately.

The general trend in all the curves was an increase in the average time of traversal with increasing numbers of vehicles, until this time became a constant. The increase was due to increasing traffic density. On reaching a congested (saturated) traffic density, any subsequent increase in the number of vehicles had no effect. The curve with fixed traffic-light operation time was though much more inefficient when compared to the other two approaches which showed almost the same trend. That said, for middle-density traffic the proposed system did exceed the system with cyclic crossing changes in terms of average traversal time. Small changes in average traversal time could though be regarded as insignificant, considering the random nature of the generation of vehicles. The significance of this was further highlighted by disallowing traffic generation from one side of the road, while the traffic lights operated in the same order. The resulting graph is shown in Fig. 14.1B. It is clear that the difference between the approaches is magnified, which shows the clear limitation of having fixed traffic-light operation times.

Two important parameters in the approach govern the algorithmic performance. These are the time threshold (T) and the threshold number of vehicles to cross (η_{th}). Both these parameters are analysed. In the first experiment, T was varied and the average traversal time for the different situations was compared. The factor η_{th} was set to infinity, so that it had no effect on algorithmic performance. The study was broken down into a densely packed road (with 2000 vehicles) and a lightly packed road (with 250 vehicles). The resulting graph produced is shown in Fig. 14.2A for a densely packed road and Fig. 14.2B for a lightly packed road.

Fig. 14.2A shows that the general trend was an initial decrease in the average traversal time as the time threshold was increased, which then became a steady value. Increasing this factor causes fewer traffic light changes, which as a result reduces the lateral switching time. The decrease reduces further as a very small proportion of vehicles are affected by the lateral time and the effect is balanced by the gain in reduced traversal time of the vehicles on the other roads at the crossing. This certainly suggests that making relatively few changes in traffic light signals could actually be beneficial overall. In a practical sense, it would be sensible for no vehicle to stop at a crossing for more than two traffic light changes, which is practically observed mainly due to lower traffic density. However, it may be noted that too few changes might drastically increase the wait time for some vehicles, while lowering the time for some other vehicles. Hence, someone may be lucky enough to queue when the lights were about to change to green, whereas another person might arrive at the front of the queue when the traffic lights have just changed to red. Maximum wait time is not a factor studied here, but in reality may be a value for a parameter setting.

The traversal time for fixed light operations is much higher than the other two cases, which seem to nearly follow each other. The entire scenario has start and end stages in which the density is fairly low and a central stage of high density of vehicles. In the start and end stages the fixed light operation time algorithm performs very poorly as compared to the other two. Spending excessive time on a road with no vehicle in the queue is the major reason for this. Further, in the central stage, the other two approaches put stress on allowing the motion of vehicles which were in the queue at the time of a change (a rule used in light operation), whereas the fixed light operation time algorithm allows motion of any newly entrant vehicles as well. This decreases its effectiveness.

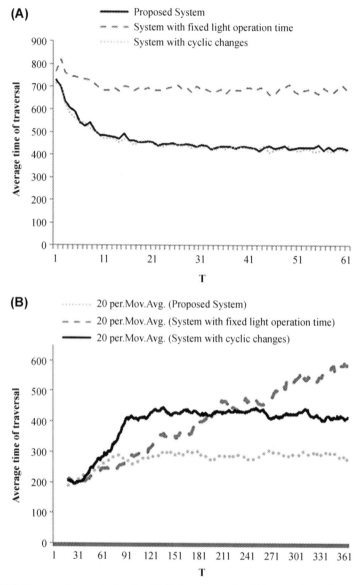

Figure 14.2 Comparative analysis of traffic lights system for different values of T: (A) densely occupied scenario and (B) lightly occupied scenario.

The same experiments performed with a lightly packed scenario show a different trend. The resulting graph is shown in Fig. 14.2B. The fixed light change algorithm showed a general increase in traversal time in line with increasing T. This was because of the waiting time when the queue was empty or was predominantly occupied by vehicles that had just arrived into the scenario and the traffic light was not changed. This waiting time increased with an increase in T making the algorithm consistently inefficient. The other two systems however showed similar trends, with the proposed system being better as the value of T was increased. The increase of this factor in this case had the effect of increasing the average traversal time, which soon settled around the same value. The increase was due to the fact that the road was lightly occupied and the increased time meant fewer changes which, as a result, increased the waiting time for vehicles. After some increase, the changes were caused only due to the vehicles in the queue being cleared rather than the time threshold. Hence, this factor, on further increase, was not used. It may be observed that increasing T increases the wait time for vehicles in a queue (unpreferred) but at the same time decreases the overheads of excessive crossing changes (preferred).

The other parameter of study was η_{th}. It is natural that this factor plays no role in the system with fixed light operating times, and hence this study did not consider that system. This factor is discussed here separately for low- and high-density roads. For high-density roads, there was a little decrease in traversal time on initial increase of this factor, but this soon became constant. The increase of this factor leads to fewer traffic-light changes and hence less overhead. The corresponding graph is plotted in Fig. 14.3. The irregular trends are because the actual points, rather than a trend line, are plotted. Experiments with low-density roads showed that the vehicles continued to travel without unnecessary queues. Because traffic lights did not wait in the case that the initial queue had cleared, the simulation continued. Hence, this parameter had no effect on the performance and showed constant results for any value of set parameter.

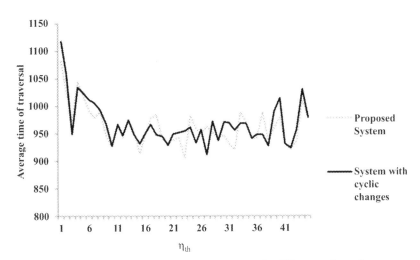

Figure 14.3 Comparative analysis of traffic-light system for different values of η_{th}.

14.3.2   Speed Lanes

The idea of assigning different speed limits to different lanes is an important concept that especially comes into play if the traffic on a road has a high diversity in vehicles ranging from those with low speeds to those with high speeds. Having slower vehicles migrate onto all lanes clearly would make the traffic slow on all lanes, which is not good for higher preferential speed vehicles. Similar to the case of fixed traffic light operational times, the optimality of fixed speed limits for different speed lanes is questionable. The intention here therefore is to *dynamically adjust the speed limits* of the various lanes. The central system vigilantly changes the speed limits depending upon the speed capability of the set of vehicles on the road, as they arrive and exit. The optimal speed limits of the lanes at any time may be a complex function depending upon the preferential speed distribution of the road. The concept is however simplified to some extent by assuming a uniform distribution. Let $V = \{V_i\}$ be the set of vehicles on a road R_a. Let s_i denote the preferential speed of a vehicle V_i. Let $\min_i(s_i)$ and $\max_i(s_i)$, respectively, be the minimum and maximum preferential speeds exhibited by any vehicle on the road. Let the road have b speed lanes. The speed limits of the various speed zones (in increasing order of limits, measured from the left to right) are allocated by Eq. [14.5]:

$$L_j = (1 - w_j)\min_i(s_i) + w_j\max_i(s_i) \qquad [14.5]$$

in which: $w_1 < w_2 < w_3... < w_b$; $w_1 = 0$, $0 < w_2$, $w_3,...w_b \le 1$. w_j is the weight associated with the lane j.

It is important to realize that unlike the current traffic system, here the speed limit implies a lower bound, which is the least preferential speed that a vehicle must possess to drive in a particular lane. The upper bound is, however, set as infinity for every lane. Semiautonomous vehicles may have their preferred speed set as per their capabilities as it would not be appropriate to force them drive subject to lower limits. This concept by itself results in *overtaking capabilities* for a vehicle with a reasonably high preferred speed. A vehicle on seeing another vehicle moving with a lower speed may opt to change its speed lane and drift rightwards (the driving rule is assumed be on the left, which means overtaking on the right is preferred). After some time the overtaking vehicle would be ahead of the vehicle being overtaken. At this time, if the vehicle finds another slower vehicle in front of it and no higher-speed lane is available, it can always drift leftwards and return to its original speed lane. This is because the upper speed-limit bound was set to infinity, and it is quite valid for a high-speed vehicle to drive on a speed lane with a lower bound. This would then complete the overtaking procedure. It is evident though that driving with a higher speed in a low-speed lane may not be optimal, and hence eventually the vehicle would seek a chance to drift to the higher speed lane.

The system was studied via simulations. The map given was a simple straight road on which different vehicles were generated at different times with their own speeds. The road had two lanes (for each side of traffic − outbound and inbound). This made the lower bounds of the speed limits of the two lanes as 0 and $(1 - w)\cdot$

$\min_i(s_i) + w \cdot \max_i(s_i)$ by Eq. [14.5]. First, the parameter w is studied. The study was performed separately under low and high-density conditions. The system was compared to a system that had no speed lanes and hence any vehicle could drive on any lane (even though the parameter w plays no role for a system without speed lanes).

The graph produced on a densely packed road (with 2000 vehicles in the scenario) is shown in Fig. 14.4A. As the factor w was increased, so did the speed limit for the high-speed lane. As a result, fewer vehicles were allowed to move to it. Although this decreased the time of travel for the vehicles using the high-speed lane, at the same time it led to the high-speed lane being underutilized. As a result, whenever the speed limit was increased, the decrease in travel time for the vehicles in the high-speed lane was averaged out by an increase in travel time for the vehicles in the low-speed lane. Conversely, keeping this factor close to 0 makes the system equivalent to one with no speed lanes.

The same patterns may be observed in Fig. 14.4A. Lightly packed roads ideally do not require speed lanes as the various vehicles can easily pass each other without resulting in obstruction. In such a scenario, high-speed vehicles lose out as they have to spend time overtaking slower vehicles. However, this lost time is very small in comparison with the time that the slower vehicles gain by having access to the

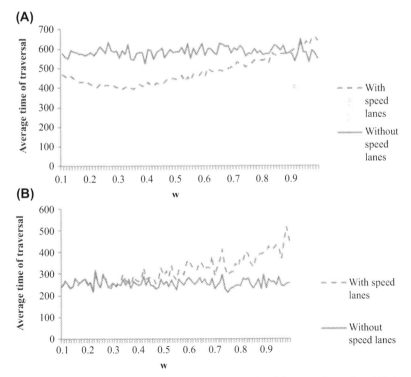

Figure 14.4 Comparative analysis of speed-lane system for different values of w: (A) densely occupied scenario and (B) lightly occupied scenario.

high-speed lane which increases the traffic bandwidth. The results of lightly packed road are shown in Fig. 14.4B.

Diversity of traffic is a major factor that plays an important role in the speed-lane system. To better test the system, the effect of a varying diversity of vehicle speeds on the algorithm performance is also studied. The simulation tool produced vehicles the speeds for which lay within a specified upper and lower bound. The first experiments were done by changing the upper bound (at the same time keeping the lower bound fixed to a value of 0.2 unit distance per unit time, arbitrary units) and then subsequently experiments were done by changing the lower bound (keeping the upper bound fixed to a value of 1 unit distance per unit time, arbitrary units). The corresponding graphs are shown in Fig. 14.5. Here, the time of traversal was compared with that of a fixed speed-lane system, in which the speed limit (lower bound) of the high-speed lane was fixed to a value of 0.5 unit distance per unit time. The general decrease in time in both cases is due to the increased average speed of the vehicle. It may be easily seen that the variable speed-limit system nicely adapts the speed limits for enhanced performance.

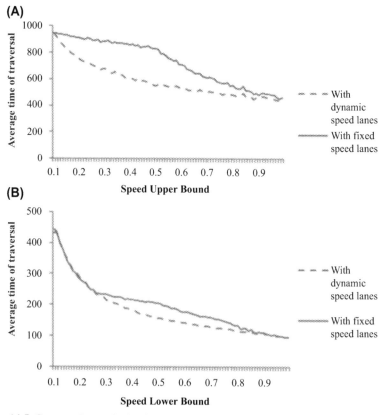

Figure 14.5 Comparative analysis of speed-lane system for different diversities of vehicles: (A) variable speed upper bound and (B) variable speed lower bound.

14.3.3 Route Planning

Routing plays a major role in distributing traffic across the road network, enabling every vehicle to reach its destination in the shortest possible time in cooperation with the other vehicles. In reality, it is common to have many vehicles using a popular road which enables quick access to a particular destination. This, however, leads to increased *congestion* and lower driving speeds for all, thereby resulting in reduced performance. Not considering other vehicles while planning one's own route may hence lead to poor results. The solution is to *distribute traffic* wisely on the roads, exploiting the entire transportation infrastructure for collective travel. An alternative longer road may be used, if it appears to have lower traffic density as compared to the main road. However, if the alternative road is too long, the choice may not be beneficial.

A *Uniform Cost Search* algorithm is used to route plan every vehicle in this approach. The aim of the Uniform Cost Search algorithm is to minimize the time of travel of all vehicles. Any road being selected for travelling is added a penalty which is proportional to the traffic density of the road. The central information system does know the current traffic density of the roads, but not the expected traffic density at the expected time of arrival of the vehicle on the roads. Hence, the traffic density is predicted using the historical information of traffic flow. For roads near the current position of the vehicle, the current density is of a higher relevance, as it would not change much till the vehicle arrives at that road. However, for roads far away from the current position of the vehicle, the prediction from historical data is more important as the current traffic scenario could change drastically. Hence, the expected density at road R_a at time t may be given by Eq. [14.6].

$$\rho(R_a, t) = \begin{cases} \rho_{\text{current}}(R_a) & t \leq \beta \\ \rho_{\text{historic}}(R_a, t) & t > \beta \end{cases} \qquad [14.6]$$

Here, $\rho_{\text{current}}(R_a)$ is the current traffic density and $\rho_{\text{historical}}(R_a, t)$ is the predicted traffic density. β is the time until which the current traffic density is reliable. The total cost computed for a node c_1 when expanding from a node c_2 connected by a road R_a for vehicle V_i may hence be given by Eqs [14.7] and [14.8].

$$t(c_1) = t(c_2) + \frac{|R_a|}{s_i} \qquad [14.7]$$

$$f(c_1) = f(c_2) + \alpha \cdot \rho(R_a, t(c_1)) \qquad [14.8]$$

Here, α is the penalty constant, $t(c_1)$ is the time of arrival at c_1, $|R_a|$ is the length of road R_a.

The route, as calculated by the Uniform Cost Search algorithm, is based on the current and predicted traffic, which changes with time. Vehicles may add in the desired route, vehicles may clear from the desired route and irregular trends may make the actual traffic much different from the predicted traffic. The route needs to be constantly adapted against these changing trends. *Continuous adaptation* by all vehicles in the traffic scenario enables vehicles to collaboratively make an efficient travel plan. This adaptation is done by replanning the route. Once the vehicle is on a road, it is considered that it will not turn back, even if turning back leads to a better route. Hence, maximum adaptation corresponds to replanning on reaching every crossing. The replanned route then reflects any changed traffic trends.

To test the working of this approach, a synthetic scenario was generated with a straight road with source/destination at the extreme ends. Initially, a very high traffic density was given to the road. Two alternative roads of unequal length were however also available to be taken and these finally merged again with the main road. Vehicles were continuously generated on both sides of the road. The vehicles first drove along the straight road. After some time vehicles were seen on the smaller alternative, whereas some kept going on the straight road to maintain the density. Still later, the larger alternative came into play and the vehicles used this for traversal as well.

An important factor of the algorithm is the parameter α which plays an important role in regulating traffic. The effect of changing this factor on the system performance with a large number of vehicles was studied. The algorithm was compared to a distance minimization Uniform Cost Search algorithm in which all the vehicles followed the same straight road and the other alternatives remained unused. The corresponding graph is given in Fig. 14.6. Increasing this factor encouraged vehicles to balance traffic densities across the different road options. An initial decrease points to the vehicles preferring the alternative roads when the main road is dense, which is the right strategy to follow. However, further increase of the factor encourages vehicles to take alternatives even though traffic on the main road is not very dense. This therefore increased the average time of traversal.

14.3.4 Reservations

Traffic systems may differentiate between important and nonimportant vehicles owing to their social importance. Roads may alternatively be seen as a business resource with possibilities of reserving a road or lane for a time exclusively or on shared basis. This is

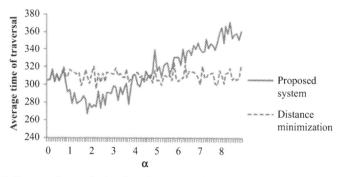

Figure 14.6 Comparative analysis of routing system for different values of α.

highly beneficial as *reservation* ensures reasonable travel speeds even though a general road may be heavily packed. This is especially important for roads in which the density of traffic is usually high or is expected to be very high because of some event. In such a case, it may be useful to have a road that can be reserved per a pricing model. At the same time, other general traffic can still move on other available roads. An important factor here is the *number of vehicles that may be reserved*. Reserving too few vehicles would make the road underutilized. Reserving too many vehicles, on the other hand, could make the traffic on the reserved road slower than the traffic on the nonreserved roads, giving no incentive for reservation. Assuming a high density of overall traffic in the system, it is further assumed that p percent of these vehicles were reserved.

A map was generated for simulation that had a straight road from the source to destination and a rather long and highly curved alternative road. The straight road was made a reserved resource. Hence, the reserved vehicles could travel straight on the road, whereas others would need to travel along the alternative, meandering, road. Unlike the map presented in Section 14.3.3, the long length of the alternative road in this scenario made the use of the straight road highly beneficial. The percentage of reserved vehicles p was varied and its effect on the average travel times of the reserved and the nonreserved vehicles was studied. The graph is shown in Fig. 14.7. Fig. 14.7 also shows cases with all vehicles reserved, in which they all used the straight road; and no vehicle reserved, in which they all used the diversion, and the main road was left unused. The graph shows the increase in travel time of the reserved vehicles as more and more of them travelled on the reserved road within the duration of simulation time. At the same time, the average travel time for nonreserved vehicles decreases as their number is reduced. This may help in determining the number of vehicles to be reserved, keeping the trade-offs matched.

A similar case may happen if *only a lane on a road is made a reserved resource* rather than a complete road. This makes it possible to use the existent road infrastructure for reservation as separate roads may not always be available giving reasonably alternate access. Reserved vehicles in such a case may be free to use general lanes and to *overtake* vehicles on the reserved lane but general vehicles may never be

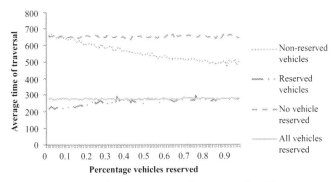

Figure 14.7 Comparative analysis of road reservation system for different percentages of vehicles reserved.

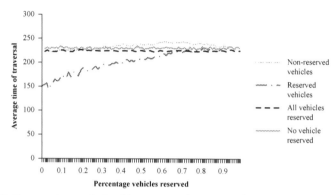

Figure 14.8 Comparative analysis of lane reservation system for different percentages of vehicles reserved.

allowed to move to the reserved lanes. Another manner in which the problem can be seen is in the case of emergencies. It would be viable to reserve a lane for an emergency service vehicle rather than for the vehicle to wait for other vehicles to give way at the time of travel. The concepts of lane reservation are the same as road reservation. A graph was plotted between the reserved percentage p and the time of travel for reserved and nonreserved vehicles. All reserved and nonreserved cases were also taken into account. The resultant graph is shown in Fig. 14.8. Here, the average time of traversal of vehicles increased as the number of vehicles reserved was increased. However, it should be noted that in this case there was no difference between all vehicles reserved or no vehicles reserved. The graph shows the same trend.

14.3.5 General Architecture

The general architecture of the simulator is given in Fig. 14.9. The architecture clearly shows the four modules discussed separately. Each one of them links to the other for information passing. Central information is maintained by the central information system. The initial settings or scenario may be given to this system as an initial specification file. The central information system is queried by all the other modules for information. The general architecture may be separately studied for vehicle subsystem, reservation subsystem, traffic lights subsystem and speed lane subsystem.

The vehicle subsystem has a route-planning algorithm that uses traffic information for deciding on the path. The lower-level planner uses traffic information to decide the motion of the vehicles. This may be to follow the vehicle in front (or simply to drive straight ahead), to change lane, to stop at a crossing, to start from a crossing or to turn at a crossing. Updated positions are always monitored and communicated to the central information system. For a lane change, the vehicle must be informed of the speed limits of the lane. Additionally, it must not be a reserved resource. Speed limits are always computed by the speed lane subsystem which gets all the traffic information from the central system for computation. The route planning subsystem must assess

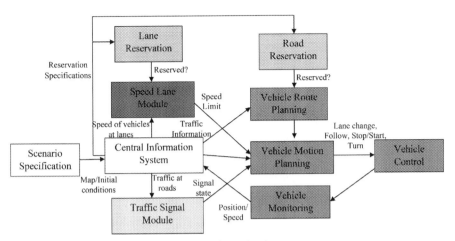

Figure 14.9 General planning architecture of the simulator.

if the road is a reserved resource or not. The reservation subsystem gives directions to the lane subsystem and the vehicle's route-planning module whether to use or not use the resource. Reservation may itself be separately handled by a system which is ultimately reflected in the central information system. The traffic lights subsystem operates the traffic lights based on the traffic behaviour on the various roads. The traffic-light state is communicated to the vehicles on approaching a crossing.

14.3.6 Simulations

So far, the mechanisms of the working of intelligent traffic lights, speed lanes, route planning and road/lane reservations were separately studied. All these concepts showcased benefits when compared to the current traffic management system. For traffic involving all these concepts at the same time, it can hence be expected that the individual system gains would all contribute to overall transportation system performance. The traffic simulation system is aimed at running a large number of diverse vehicles using all the modules stated earlier. The testing methodology in this section is to invoke all the modules and test the performance in complicated traffic scenarios.

To make the testing easy, a utility is created which can take an image representation of the environment and parse it to produce the road network map. It then becomes easy to make maps and use them for testing. The only parameter to be given to the simulator is the demand or the number of vehicles generated per unit time. This is needed for control of the induced congestion which the algorithm tries to eliminate. The locations of the vehicles, initial speeds, emergence time etc. are all set randomly. The initial and final positions are always kept at the extremity of the map so that the vehicles travel the maximum distance.

A number of maps were generated from the drawing utility tool and parsed to create a road network map. Large maps ensured that planning was a complex task. The

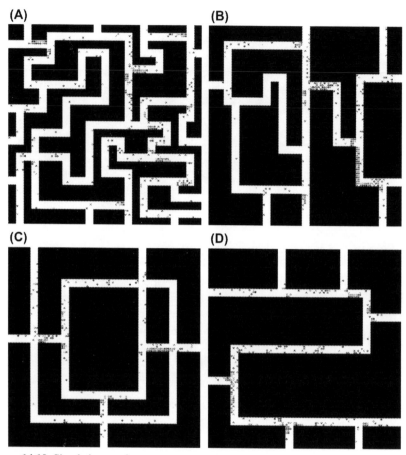

Figure 14.10 Simulation results.

densities were usually kept high as the benefits of the proposed system are particularly apparent in high-density circumstances. To better assess the working of the algorithm, the vehicles were plotted on this map with different colors. Screenshots of some of the simulations for test scenarios are shown in Fig. 14.10. For each simulation, it was observed that the vehicles were able to easily reach their goals while avoiding congestions and excessive waiting times. In a congested setting with too many vehicles asking to go though some central regions, queues were formed for some time when a red-light signal was apparent, but the queue cleared very quickly and the vehicles spread out to avoid congested areas.

The simulator measures a number of metrics which are indicators of the performance of the system. The primary performance indicator is the average travel time of the vehicles, which was the main metric used to assess the performance of the system. The other important metrics include the average travel distance and average travel speeds. To save space only one scenario is used for further analysis. The metrics for a different number of vehicles with all modules activated are shown in Fig. 14.11.

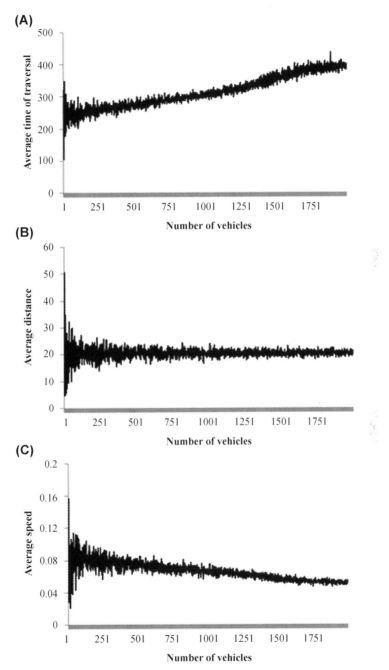

Figure 14.11 Analysis for integrated scenario. (A) Average time of traversal versus number of vehicles, (B) average distance versus number of vehicles, (C) average speed versus number of vehicles.

All units are arbitrary and specific to the simulation tool. The metrics can be mapped to real-world units by multiplying by suitable constants.

Scenario specifications such as the sources and goals of vehicles were randomly generated. This randomness contributes to the oscillations seen in Fig. 14.11. Based on a large number of simulations, it was observed that all the vehicles reached their destinations in acceptable times. Hence, it can be ascertained that the simulation tool developed for the purpose was able to cooperatively plan the paths of the different vehicles involved.

14.4   Congestion Avoidance in City Traffic

Large numbers of vehicles within a road network commonly give rise to congestion which is marked by a large drop in the average speed of the moving vehicles. As a result, every vehicle takes a considerable time to reach its final destination. On a particular road, congestion may be *recurrent* or *nonrecurrent* (Gordon, 2009). Although regular drivers are normally prone to adjust their departure times and routes for recurrent traffic, nonrecurrent traffic congestion is hard to predict and adjust to. Nonrecurrent congestion is caused by unusually high demand (like a sporting event) or suddenly low capacity (like an accident or a road closure).

An increasing amount of autonomy in vehicles and transportation management systems has given impetus to the possibilities of congestion avoidance. Although it is possible to locate, track and measure traffic density on various roads by intelligent agents concerned with road infrastructure, intelligent devices in vehicles are capable of collecting live data and using the same for planning purposes. This makes it possible to enable a vehicle to avoid roads in which congestion is likely to occur and to use alternative routes.

Traffic congestion (Verhoef, 1999; Maniccam, 2006) may be avoided largely by routing techniques, which tell the vehicle the route they need to travel. Presently installed devices and systems like Satellite Navigation only take static data. Unfortunately, this results in multiple vehicles using the same roads, which leads to congestion.

Here, first, the true state of the traffic system is analysed with an eye on possible future developments, and a traffic scenario prone to traffic congestion for everyday travel is assumed. Henceforth referred to as the *city traffic scenario*, this traffic scenario is analysed and a traffic-routing strategy is made for the guidance of vehicles. Experimental results are performed on the city (town as per local terminology) of Reading, United Kingdom. The key takeaways of the algorithm are summarized in Box 14.3. Box 14.4 summarizes the main concepts.

14.4.1   Problem Formulation and Scenario of Operation

The problem of study is to move a number of vehicles on a map such that congestion is avoided. The vehicles must not violate any traffic rules. Every vehicle may emerge on any part of the map at any time. The origin, destination or movement plan of any

---

Box 14.3 Key Takeaways of the Congestion Avoidance in City Traffic Approach

- Proposing *city traffic* as a scenario to study traffic congestion.
- Proposing the importance of considering *traffic lights* in decision-making regarding routes.
- Proposing a simple *routing algorithm* that eliminates the high density of traffic and hence minimizes congestion.
- Stressing *frequent short-term replanning* of the vehicle in place of long-term (complete) infrequent replanning.

Box 14.4 Key Concepts of the Congestion Avoidance in City Traffic Approach

- **Assumptions**
  - Vehicles have very diverse speeds
  - Nonrecurrent traffic (does not follow historical traffic patterns)
  - *City* traffic scenario
- **City Traffic Scenario**
  - *Infrastructure*: Many short-length roads (alternative roads) intercept each other in city traffic, meaning very computationally expensive routing; unlike highway traffic with fewer roads and longer roads.
  - *Vehicle Emergence*: Many entry/exit points are located at road ends/between roads in city traffic, because of which new vehicles constantly enter and anticipation not possible; unlike highway traffic which has distant entry/exit points and new vehicles do not much invalidate anticipated plans.
  - *Planning Frequency*: Low anticipation invalidates long-term plans in city traffic, while high anticipation favors long-term planning in highway traffic.
- **Hypothesis: Make frequent effective short-term plans**
  - *Frequent*: Constantly adapt to changes, replan at every crossing
  - *Effective*: Minimize (1) expected travel time, (2) expected traffic density, (3) expected time to wait at crossings
  - *Short Term*: Limit computational requirement. Plan for a threshold distance from the source, assume it is possible to reach the goal from the planned state. Like human drivers, always see the *current* traffic and take the best route *towards* the goal, assuming no dead ends.

---

vehicle is not known by any other vehicle. This means there is a provision for manual vehicles in the traffic scenario. U-turns can only be taken from a traffic crossing and not in the middle of the road. The efficiency of this routing system is judged by the average time of travel of the vehicles. This metric is considered to reflect the magnitude of congestion that a vehicle faces during its travel. The algorithm is motivated by the characteristic scenario on which it operates, which is explained in detail in the following subsections.

14.4.1.1 City Scenario

The scenarios of moving within a *highway map* and a *city map* are clearly different. Both, however, place stress on judiciously selecting the roads to travel on and forecasting the scenarios well in advance to avoid traffic congestion. However, the former scenario has long highways which, if entered, need to be followed for a significantly long time before an alternative path may be available, whereas the latter has numerous alternative roads from which a vehicle may diverge and reconnect through any other close cut-in. The other point of difference lies in traffic emergence. Highway scenarios have distant entry and exits points, whereas in city traffic any vehicle can enter or leave from any road. Thus, within a city, anticipation may not always help as it accounts for only recurrent traffic (in forecasting-based systems) (Dia, 2001; Kirby et al., 1997) or intelligent vehicles which are on the road and the travel plan for which is known (for anticipatory routing systems) (Weyns et al., 2007; Kaufman et al., 1991). In reality the vehicles may emerge from car parks (or homes) located at any point along any road and in doing so affect the entire network plan.

The difference between highway traffic and city traffic emphasizes the fact that while in highway scenarios it may be advisable to make *long-term plans*, the same are not so useful for city traffic. In highway scenarios the vehicles can be expected to stick to their anticipated plans, thereby indicating which highway to follow. In city traffic, on the contrary, vehicles may make *very frequent changes* in travel plan due to the variety of options in terms of the roads to take to reach their destination. Because the number of vehicles is large, the total changes may be too large for any system to monitor and every change will affect all vehicles, which makes the system too dynamic to handle. Present approaches (eg, Claes et al., 2011) limit the changes and only accept the changes which result in a significant improvement and hence control the highly dynamic nature of the problem in this way. Here, a part of the road map of Reading, United Kingdom, is taken as the city map given in Fig. 14.12.

Routing may be classified into *centralized approaches* (Kuwahara et al., 2010) and decentralized approaches (Pavlis and Papageorgiou, 1999). *Decentralized approaches* consider every vehicle separately during plan generation and are hence able to generate

Figure 14.12 Map of Reading, United Kingdom, used for experimentation.

a travel plan in a short time. During planning of each vehicle, decentralized techniques may prefer (1) not to account for another vehicle's motion, (2) to predict the motion of other vehicles or (3) use traffic forecasting information from the historical data (Taniguchi and Shimato, 2004). Method (1) leads to high traffic congestion and method (3) does not account for nonrecurrent traffic. *Microsimulation* is a common tool for method (2) wherein it is assumed that the travelling information of all the vehicles is available and the system operates by simulating the different possible plans. The method has limitations including the fact that it is computationally difficult to simulate a large number of vehicles for every replan of a vehicle's trajectory or in the case of any new vehicle entering. If a vehicle replans, the plans of some other vehicles may get affected and it may sometimes take a long time for vehicles to obtain their best plans. All vehicles on the road need to be intelligent and the assumption is usually that they have similar driving speeds. In addition, simulation uncertainties can become very large with time. These uncertainties are especially large when accounting for overtaking and traffic signals. All this puts an emphasis on long-term plans being of less use for city traffic.

Although a high number of roads or high connectivity leads to a significant variety of travel options, it further makes the problem computationally expensive. Cities are normally large. Most studies are restricted to traffic over only part of the overall city map.

14.4.1.2   Inferred Hypothesis

Understanding the stated points, it must be noted here that it is important to make *frequent effective short-term plans*, rather than making plans which are too long term, investing in heavy computation and hence limiting the planning frequency. From a simulation tool perspective, the computing infrastructure that simulates, renders and moves every vehicle is limited, and it has to give simulation results within limited times. As a result, researchers usually have to limit the frequency of replanning for each vehicle, and this has a considerable impact on the study.

From the perspective of a physical system, every vehicle has its own computing infrastructure which interacts with the other computing infrastructures to get information. In a scenario in which the static map is itself complex, loading the vehicle with excessive information on the motion of the other vehicles makes the computing even more difficult. This is of less use when considering that long-term plans are uncertain and hence the real/actual information is likely to change.

14.4.1.3   Other Scenario Specifics

Traffic systems in most countries consist of vehicles which travel with nearly the same preferred speeds. However, the study is aimed at scenarios in which the vehicles may greatly differ in their preferred speeds. The difference in present-day traffic mainly reflects the urgency, driving capability and experience of the drivers. Technology has led to modern day vehicles to be classified as autonomous, semiautonomous or manual vehicles. The first may typically see vehicles that differ in speeds as per size, price, features, sensing and control algorithms.

Considering city-based traffic, it is further considered that a major proportion of the roads are two lanes with one lane each for inbound and outbound traffic. With diverse speeds, it is naturally unpleasant for a high-speed vehicle to be following behind a low-speed vehicle for a large part of the journey, in which there is no multilane to overtake by lane changes. Hence, it is allowed for vehicles to travel on the 'wrong' side of the road for some time to complete an overtaking operation.

14.4.1.4 Decentralized Anticipatory Routing

A significant attempt is made to use a decentralized anticipatory approach to vehicle routing. In a related work, Claes et al. (2011) presented a system wherein every vehicle considers all possible routes before selecting a 'best' route for its journey. The authors realized a formula to convert the anticipated traffic density into an average travel speed. This extends the work of Weyns et al. (2007), who used traffic microsimulation to compute the anticipated traffic speed. *Replanning* is done after some time steps for every vehicle. In light of the discussions, the limitations of the approach are too infrequent replans, the impossibility of computing every possible route in real time for large maps, the assumption that every vehicle is intelligent, no consideration given for traffic lights or overtaking and finally all vehicles are assumed to have the same preferred travel speed which makes conversion of traffic density to predicted speed possible. Most of these limitations might however hold if the complete map was itself small.

The attempt is to present a fairly simple system not making assumptions which may not hold true in the real world. In Section 14.4.3 it is shown how this may lead to a better performance in city traffic. In fact, the complete system may be implemented by the adoption of a simple changeable message sign (CMS) at every road end, along with some detectors (such as an array of loop detectors, a counter for vehicle entry/exit etc.) to measure traffic density.

14.4.2 System Working

This section describes the entire system used to route and move the vehicles. The system is designed such that congestion is avoided as much as possible, while it is assumed that the density of vehicles at every road can be sensed.

14.4.2.1 Traffic Simulation

Vehicle motion is done using an *Intelligent Driver Model* (Treiber et al., 2000). The model states the manner by which one vehicle follows another vehicle depending upon the preferred driving speeds, operational speeds and available separation distances. The preferred speed of the vehicle V travelling along road R_{ij} connecting nodes v_i and v_j is taken to be $v\text{pref}_{ij} = \min(s, \text{speed\_limit}_{ij})$, in which s is the preferred speed of the vehicle V. speed_limit_{ij} is the maximum allowable speed on the road R_{ij}. Because the road traffic is diverse, the vehicles vary in their maximum speed s.

Considering a high diversity, a lot of *overtaking* is possible solely by lane-change mechanisms. Hence, a vehicle must always be ready to overtake a slow vehicle in front and to be itself overtaken by an even faster vehicle to the rear. For decision-making regarding lane changes, time to collision is used as a metric (if the vehicles continue to travel at the same speed).

If the vehicle is travelling close to its maximum speed limit $vpref_{ij}$ it may further attempt to stay in the left hand lane (to facilitate others to easily overtake it, in left-side driving countries — United Kingdom/Japan style considered) unless it sees a slower vehicle ahead of it in the left-hand lane. In such a situation naturally, no question of overtaking arises whilst a vehicle to the rear has already requested to overtake. Overtaking on the right is preferred as compared to overtaking on the left in case both options are available and likely.

For simulation purposes, *traffic lights* are assumed intelligent in that they know the number of vehicles at each entry point and their time of arrival. The traffic lights do not change in a cyclic manner, but allow traffic to flow from the direction of the vehicle with the longest waiting time. The switching frequency of the traffic lights is taken as a maximum of mtim. If no vehicles are waiting to cross the road, the traffic light for which is currently green, and some other road has vehicles waiting to pass through, the traffic lights would change before the normal scheduled time. mtim is a constant and is set using simulations, such that it allows enough vehicles to pass through in a heavily congested traffic scenario. The factor is only of importance when far too many vehicles are waiting to cross an intersection; as in other cases, the queue clears well-before this factor comes into play.

14.4.2.2   Single-Lane Overtaking

In case the vehicle is travelling on a road on which a single lane exists for each side of the traffic, it may be undesirable to follow a slower vehicle in front. Hence, the vehicle is allowed to move onto the wrong side of the road to overtake the slower vehicle and to then return to the left-hand lane. Such an overtaking is termed as *single-lane overtaking*.

For overtaking, it is essential that the vehicle being overtaken is travelling almost at its preferred speed which is slower than the preferred speed of the overtaking vehicle. In addition, the vehicle being overtaken must itself not currently be overtaking another vehicle, and any other vehicle on the wrong side of the road (if any) must not be overtaking. An overtaking vehicle is projected to be accelerating till it reaches its preferred speed (or until overtaking completes) while overtaking. Separations from the vehicle being overtaken, to any vehicle on the wrong side of the road (if any), and the vehicle ahead of the one being overtaken (if any) are checked at every instance. All these must be greater than the preferred separations (as per the intelligent driver model) at all times during the overtaking procedure. Overtaking cannot happen if (by projections) it cannot be completed before the end of the road. If all the conditions hold good, overtaking then takes place. A simulated overtaking scenario is shown in Fig. 14.13.

It may be interesting to observe that the applied overtaking mechanism assumes *no cooperation* with other vehicles. In simple words, the other vehicles may assume that

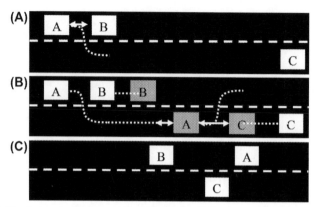

Figure 14.13 Single-lane overtaking. (A) *A* checks feasibility to overtake *B* while *C* is coming from opposite end. (B) Projected positions of vehicles when *A* is expected to lie comfortably ahead of *B*. (C) Completion of overtake. Arrows indicate separation checks. Because *A* and *C* are moving in opposite directions, needed separation is much larger.

the overtaking vehicle is absent and move normally thereby still making the required separation with all the vehicles. The only exception is that other vehicles may not accelerate. In the real world the oncoming vehicle or the vehicle being overtaken may slow down as an act of cooperation. The important decision of whether overtaking is to take place is done solely by the overtaking vehicle without assuming cooperation, and even if an error is made the oncoming vehicle and the vehicle being overtaken compulsorily slow down to facilitate the overtake.

14.4.2.3   Vehicle Routing

Routing deals with the route selection of the vehicles. The *frequency of planning* is a key aspect which, per the hypothesis, needs to be as large as possible for efficient congestion avoidance. Considering that taking a U-turn in the middle of the road is not allowed, the earliest a vehicle can react to any change of plan is before a crossing. The planning should be done well before reaching the crossing so that the required lane changes are made, traffic lights are read and suitable indicators are given before making the required turn at the crossing. Hence, the maximum magnitude of replanning corresponds to planning the vehicle before every crossing.

The basic planning algorithm employed is A*, which is a search algorithm that finds a path from a given source to a given goal depending on a cost function supplied in the solution design. The A* algorithm finds a solution by constantly expanding nodes with the best expected cost from the source to the goal. The historic cost $\text{hist}(v_j)$ of a node v_j refers to the actual cost from the source to reach that node per the designed cost function. The heuristic cost $\text{heuristic}(v_j)$, on the other hand, estimates the cost from the current node v_j to the goal. The algorithm searches by constantly expanding nodes based on these costs.

Per assumptions, it is computationally very expensive to plan the entire route. An inspiration is taken from the manner in which human drivers plan their route. Drivers can reach their destination by a simple attempt to select the roads that make the vehicle head *towards the goal*. In case multiple such roads are possible, *short-term planning* may be done to reach some point by the best manner, beyond which an approximate travel cost may be assumed. However, it is important to be assured that the selected point is actually connected to the destination, without having the vehicle turn back or go by a long route. Although doing so the travel plan is made suboptimal, it is a compromise to the computational cost.

Hence, the A* algorithm stops if the historical cost is more than *maxHistorical* and the current node (best in the open list) is termed as the goal, in which the factor *maxHistorical* controls the computational cost. A low value of this factor makes the routing algorithm largely heuristic, in which heuristic estimates determine the route, whereas a large value may be too computationally expensive. The factor is given the highest value as per the available computation. In the preliminary version of the algorithm, a heuristic search (which is nonoptimal but very fast) was used to ensure that the subsequent motion from the node does reach the goal without having the vehicle move backwards. However, experimental results showed that such a path was always possible in the experimented scenarios and hence the check was removed, thereby saving on the computational cost. Having high connectivity, it is natural that from any point the vehicle would be able to reach the destination by travelling towards it.

Let the historic cost of node v_j be given by hist(v_j) and let e_{ij} be the average length of the road R_{ij} from node v_i to node v_j. The historic cost is given by Eq. [14.9]. As the cost minimizes both the density of the road network as well as the number of traffic lights that the vehicle may encounter, the method is called *density-based routing with traffic lights*.

$$\text{hist}(v_j) = \text{hist}(v_i) + \frac{e_{ij}}{S\big(\text{vpref}_{ij}, n(R_{ij})\big)} + \eta(R_{ij}) \cdot \text{mtim}(R_{ij}) \qquad [14.9]$$

Here, $S(\text{vpref}_{ij}, n(R_{ij}))$ is a function that predicts the average speed of a vehicle per the current traffic scenario at the road R_{ij} having a current number of vehicles $n(R_{ij})$. In the present approach this is given by Eq. [14.10].

$$S\big(\text{vpref}_{ij}, n(R_{ij})\big) = \begin{cases} \text{vpref}_{ij} & n(R_{ij}) \le n_{th} \\ \dfrac{\text{vpref}_{ij}}{n(R_{ij})/k \cdot n_{th}} & n(R_{ij}) > n_{th} \end{cases} \qquad [14.10]$$

Here, it is assumed that the operating speed is inversely proportional to the density. $\eta(R_{ij})$ is the fraction of traffic-light changes that the vehicles at road R_{ij} wait for before getting a chance to get through the traffic lights. mtim(R_{ij}) is the average time of wait at the traffic light for a single change. mtim(R_{ij}) = 0 if R_{ij} does not end at a traffic light. The factor n_{th} accounts for the number of vehicles that may leave within the

traffic-light change as per the factor $\eta(R_{ij})$, whereas the factor k relates density with the driving speed in dense traffic.
The heuristic cost is given by Eq. [14.11].

$$\text{heuristic}(v_j) = \frac{\|v_j - \text{Goal}\|}{\min\left(s, \max_{i,j}\left(\text{speed\_limit}_{ij}\right)\right)} \qquad [14.11]$$

Here, $\|v_j - \text{Goal}\|$ denotes the distance between the node v_j and the Goal measured using the coordinates of the two places on the road map. s is the preferred driving speed of the vehicle. The term $\max_{i,j}(\text{speed\_limit}_{ij})$ is the maximum speed possible on any road, which would point to the maximum allowable speed in the transportation network. Note that the heuristic function is admissible and assures optimality of the A* algorithm. However, the expansion of the A* algorithm is proportional to the difference between the actual cost and the heuristic estimates, which is high for the presented approach. This results in a significant number of nodes being expanded. The output of the A* algorithm is a *route* consisting of the roads that the vehicle must follow. A simulated planning procedure showing the iterations of repetition is shown in Fig. 14.14.

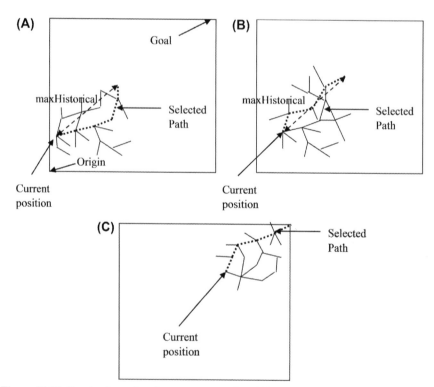

Figure 14.14 Routing by replanning at every crossing. (A) From current position, vehicle plans towards goal and after *maxHistorical* cost stops the current search and moves by best path. (B) After reaching next crossing, change of plan takes place as per new information available. (C) Vehicle finally reaches a point from where the goal is near.

14.4.3   Experimental Results

14.4.3.1   Initialization

Experiments were done on a traffic simulator based on an intelligent driver model and other features as discussed in Section 14.4.2. Experiments were done on a part of a road map of Reading, United Kingdom, which is shown in Fig. 14.12. The map was obtained from (Openstreetmap, 2011). A Depth First Search algorithm was used to eliminate isolated nodes. The processed map had a total of 7765 road nodes. Major roads were all assumed double lanes, whereas the general roads were all considered single lanes. Speed limits were fixed to 30 miles per hour or 40 miles per hour. The left-side driving rule was followed.

Traffic is produced in the road network by randomly generating an origin–destination matrix. The number of vehicles per second that enter the traffic scenario is taken as a human input using which vehicles are generated continuously for 10 min. Henceforth, the generation of vehicles stops but the simulation runs till all vehicles reach their destinations. The origins and destinations are preferred to be on the opposite sides of the map separated by a displacement of more than the radius of the map. The origin is selected by using a Gaussian distribution with the mean centred outside the map's central point by a magnitude of half the radius. The angle of origin to the map's centre (θ) is chosen randomly. The destination is also chosen from a Gaussian distribution with its mean at half the map's radius. The angle of destination to the map's centre is chosen from a Gaussian distribution with mean located at $\pi + \theta$. The speed limit of the individual vehicles is selected from a uniform distribution varying from 20 miles per hour to 40 miles per hour.

14.4.3.2   Alternative Methods

The strict constraint in the choice of the alternative methods for comparison was that no communication must exist between the vehicles or between all the vehicles and a central transportation system. Most research work on microsimulations, replanning etc. hence gets eliminated. Further, because the approach is for nonrecurrent traffic, most of the learning-based systems get eliminated. Based on these assumptions, a limited choice of methods is available, which were experimented on. However, further issues relate to the diversity of the vehicles in terms of travel speeds which make many of the alternative methods unacceptable.

Comparisons of the technique have been carried out with a variety of other methods. Each basic method has two modes of operation, a static case wherein the route is planned initially and the same is followed unaltered till the goal is reached; and a dynamic case wherein the routing takes place at every intersection.

The first method employed is the *optimistic fastest routing strategy* used for static planning. The strategy computes a route by minimizing the expected time of completion of the journey, which is given by Eq. [14.12].

$$\text{cost (optimistic)} = \sum_{e_{ij} \in \text{Route}} \frac{e_{ij}}{v\text{pref}_{ij}} \qquad [14.12]$$

The second strategy used for comparison is the *pessimistic fastest routing strategy* which is similar to the optimistic fastest routing strategy with the difference that the cost to be minimized is given by Eq. [14.13]. The attempt is to prefer roads which have a higher number of lanes. Here, $w_{ij} = 1/\text{lanes}(R_{ij})$, $\text{lanes}(R_{ij})$ is the number of lanes in road R_{ij}.

$$\text{cost (pressimistic)} = \sum_{e_{ij} \in \text{Route}} \frac{w_{ij} \cdot e_{ij}}{v\text{pref}_{ij}} \qquad [14.13]$$

The next strategy used is the *Traffic Messaging Channel* (TMC, Davies, 1989). In this strategy, every vehicle, on reaching the road segment end, informs the system about the average speed at the particular segment and this is used for planning other vehicles. Considering the simulation scenario, the cost minimized by this planning is given by Eq. [14.14].

$$\text{cost (TMC)} = \sum_{e_{ij} \in \text{Route}} \frac{e_{ij}}{\min\left(\text{TMC}_{ij}, v\text{pref}_{ij}\right)} \qquad [14.14]$$

TMC_{ij} is the average speed as known by the TMC system. The update is done as per Eq. [14.15].

$$\text{TMC}_{ij}(t) = \begin{cases} (1 - \text{lr}) \cdot \text{TMC}_{ij}(t-1) + \text{lr} \cdot v_{ij}^{\text{avg}} & v_{ij}^{\text{avg}} < v\text{pref}_{ij} - \varepsilon \\ \text{TMC}_{ij}(t-1) & \text{otherwise} \end{cases} \qquad [14.15]$$

v_{ij}^{avg} is the average speed of the vehicle at road R_{ij}, ε is a small number. Eq. [14.15] avoids vehicles with lower preferable speed to slow the TMC known average values, if the actual traffic is moving reasonably fast. lr is the learning rate.

These three strategies find the route from source to goal which does defy the assumption that the map is too complex for timely computing the route from the source to goal. This was, however, done only for comparative purposes. The time was large enough to disallow continuous replanning.

The next set of alternative methods belongs to the *dynamic domain* in which vehicles are replanned at every crossing. Considering the computation time, in each case replanning is done for short durations as explained in Section 14.4.2.3. The four methods are experimented, namely, TMC, density, TMC with traffic lights and density with traffic lights. The last method is the proposed method as discussed in Section 14.4.2. Density-based planning is same except for the fact that it disregards the traffic-light factor. The TMC method is the dynamic equivalent of the static TMC method. TMC with traffic lights has an additional cost on encountering traffic lights. The cost functions for each of these methods are given by Eqs [14.14], [14.16]–[14.18].

$$\text{cost (density)} = \sum_{e_{ij} \in \text{Route}} \frac{e_{ij}}{S\left(\text{vpref}_{ij}, n(R_{ij})\right)} \qquad [14.16]$$

$$\text{cost (TMC with traffic lights)} = \sum_{e_{ij} \in \text{Route}} \frac{e_{ij}}{\min\left(\text{TMC}_{ij}, \text{vpref}_{ij}\right)}$$
$$+ \eta(R_{ij}) \cdot \text{mtim}(R_{ij}) \qquad [14.17]$$

$$\text{cost (density with traffic lights)} = \sum_{e_{ij} \in \text{Route}} \frac{e_{ij}}{S\left(\text{vpref}_{ij}, n(R_{ij})\right)}$$
$$+ \eta(R_{ij}) \cdot \text{mtim}(R_{ij}) \qquad [14.18]$$

14.4.3.3 Comparisons

The metric that judges the effectiveness of the algorithm is the average time to destination. The results for increasing demands for various routing strategies are shown in Fig. 14.15. The algorithm was tested for a maximum of 45 vehicles per second which meant that there were a total of 27,000 vehicles. The general trend expected was an increase in the average time of completion of the journey per vehicle, which is visible in the graph barring a few regions. The difference in trend is due to the fact that for every demand, different origins, destinations and speeds were selected.

From Fig. 14.15, it can be easily seen that the proposed method performs best for all demands. The trend is closely followed by the TMC method with traffic lights. Clearly, considering traffic lights was beneficial as density and TMC methods with the traffic-light factor proved to perform better. An anomaly is the static TMC performing better than that the dynamic TMC. However, although the static TMC invested heavily

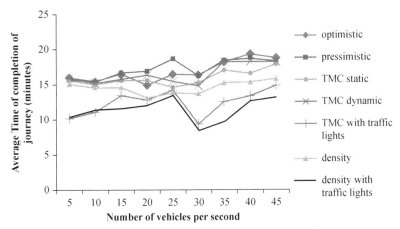

Figure 14.15 Average time of completion of journey for various algorithms and demands.

on computation at the start to find the best route from source to goal as per the set metric, the dynamic TMC plans only up to a point ahead. Hence, restricting the search for computational betterment has a payoff for the algorithm when taking longer routes. Further, at higher demands it takes a little time for the traffic level to rise on the popular roads. Later vehicles prefer alternative routes, thereby keeping the congestion level the same or balanced. Because part of the city map was simulated, the static congestion level was enough information as it did not change much.

Although the time of completion of journey was the sole metric of use which was optimized, it is also seen how the different routing strategies behaved in terms of total distance of travel. The graph showing the distance of travel for different demands is given as Fig. 14.16. For large demands, the proposed method had the shortest distance, whereas the largest distances were recorded by the optimistic and the pessimistic strategies. The distances for these strategies were largest due to the fact that faster strategies assumed roads with a large number of lanes and higher speed limits would lead to the shortest travel time. These roads usually have a high degree of congestion, and hence the assumption is incorrect.

The reason for longer time of travel for the dynamic TMC is visible in the distance graph which took longer routes. The TMC group of algorithms though had a better view of the applicable traffic speed. Density-based methods taking shorter distances indicate the cooperative measure of vehicles in the front for vehicles behind, in case the main route important for the vehicles behind is congested. The algorithms including the traffic-light factor show low distance indicating that it was precomputed that the majority of the time would be wasted at traffic lights.

The travelling speed is simply distance upon time, and hence it is simple to understand. The trend for various algorithms is given in Fig. 14.17. Considering that a significant portion of the time was wasted while waiting at crossings, the actual travel speeds were much higher. The optimistic and pessimistic algorithms emphasized taking wide roads which were longer, and hence the average speed was sufficiently fast even though it was reasonably less than the allowed speed limit. Other strategies emphasized taking shorter routes which were less congested, and hence a decent travel speed could be maintained.

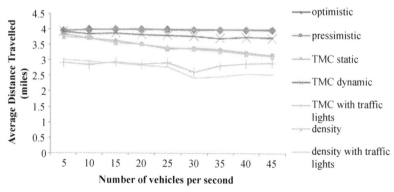

Figure 14.16 Average distance travelled for various algorithms and demands.

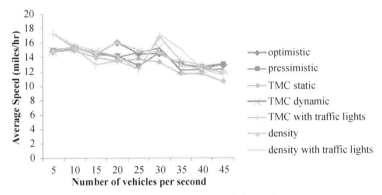

Figure 14.17 Average speed for various algorithms and demands.

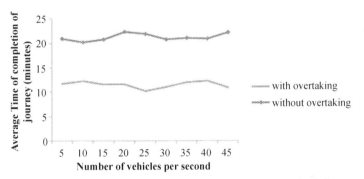

Figure 14.18 Average time of completion of journey with and without single-lane overtaking.

14.4.3.4   Analysis of Single-Lane Overtaking

An important feature of the algorithm was the ability of having overtaking in single-lane roads. Even if a single slow-moving vehicle is somewhere ahead in the lane, the entire lane traffic could suffer even in low-congestion areas. Most roads being single lane make overtaking impossible without this feature. The average time of journey is shown in Fig. 14.18. It can be clearly seen that single-lane overtaking resulted in a great boost to travel time. Without the feature, the main traffic scenario was primarily that all vehicles followed a slow vehicle ahead.

14.5   Summary

The chapter addressed the problem of managing traffic in a transportation system with diverse speed vehicles. Traffic density is constantly increasing, and this puts a lot of stress on the present transportation infrastructure. Semiautonomous vehicles with the option of communicating with other vehicles, road infrastructure and transportation management units are capable of efficiently planning themselves resulting in higher transportation efficiency. In the first part of the chapter, some of the various possibilities have been addressed, whereas the aim is to build on the existing traffic management system by making each of its components intelligent and efficient. The system was broken down into four main modules: traffic-light management, route planning, speed-lane management and reservation. Each of these concepts resulted in better management of traffic, reducing the average traversal time of vehicles. The resulting system is a dynamically managed traffic system which attempts to make traffic flow as efficient as possible.

Diversity in vehicle speeds makes systems behave differently from general expectations. Here an attempt was made to investigate the effect of increased diversity of vehicular speeds on overall transportation performance. Slow vehicles can lead all traffic in a lane to be slow, resulting in reduced efficiency, and this has to be managed by the transportation system. In the simulations, it was seen that slow vehicles do affect the travel efficiency of the fast vehicles and the effect is unavoidable; however, the effect is much lower in lower-density traffic wherein fast vehicles have the option to overtake by lane changes. Hence, there is an advantage in eliminating high traffic density on roads, which was a key objective of the routing system. Alternatively, on high-density roads, speed limits need to be intelligently adapted so as not to punish high-speed vehicles too much by making them drive in low-speed lanes while also not forcing vehicles to follow a slow vehicle ahead. Reservation signifies social diversity of vehicles. Expected dense traffic with diverse vehicles cannot be guaranteed a reasonable performance unless social diversity of vehicles is exploited as an additional factor.

In the second part of the chapter a method was presented to solve the traffic congestion problem, accounting for the factors of traffic density and traffic lights for a city transportation infrastructure. The solution attempted to make frequent short-term plans for each vehicle. The decentralized nature of the algorithm enabled its scalability. With this, the highly uncertain nature of long-term plans was also stressed based on which no decision-making can be done. The algorithm performance was reasonably better when the vehicles were allowed to overtake in a single lane. Experiments showed that the traffic lights played a vital role in planning.

Any routing algorithm for vehicles has a strict decision point regarding preferring shortest path to goal, fastest roads or reducing the waiting time at crossings. Considering the nature of the problem in which additional vehicles may pop up anytime and anywhere and known vehicles may change their plans without notice, it is impossible to predict these metrics for all roads. High diversity in terms of vehicular preferred speeds makes the choices even more difficult. Experiments show that the proposed algorithm is the best trade-off between these selections. Frequent replanning ensures plans are adaptive to changing traffic.

References

Ando, Y., Fukazawa, Y., Masutani, O., Iwasaki, H., Honiden, S., 2006. Performance of pheromone model for predicting traffic congestion. In: Proceedings of the Fifth International Joint Conference on Autonomous Agents and Multiagent Systems, New York, pp. 73–80.

Claes, R., Holvoet, T., Weyns, D., 2011. A decentralized approach for anticipatory vehicle routing using delegate multiagent systems. IEEE Transactions on Intelligent Transportation Systems 12 (2), 64−373.

Davies, P., 1989. The radio system-traffic channel. In: Proceedings of the Vehicle Navigation and Information Systems Conference, pp. A44−A48.

Dia, H., 2001. An object-oriented neural network approach to short-term traffic forecasting. European Journal of Operational Research 131 (2), 253−261.

Dresner, K., Stone, P., 2004. Multiagent traffic management: a reservation-based intersection control mechanism. In: Proceedings of the Third International Joint Conference on Autonomous Agents and Multiagent Systems, pp. 530−537. NY, USA.

Dresner, K., Stone, P., 2006. Multiagent traffic management: opportunities for multi-agent learning. In: Lecture Notes in Artificial Intelligence, vol. 3898. Springer Verlag, Berlin, pp. 129−138.

Dresner, K., Stone, P., 2007. Sharing the road: autonomous vehicles meet human drivers. In: Proceedings of the 20th International Joint Conference on Artificial Intelligence, pp. 1263−1268. Hyderabad, India.

Fawcett, J., Robinson, P., 2000. Adaptive routing for road traffic. IEEE Computer Graphics and Applications 20 (3), 46−53.

Furda, A., Vlacic, L., 2011. Enabling safe autonomous driving in real-world city traffic using multiple criteria decision making. IEEE Intelligent Transportation Systems Magazine 3 (1), 4−17.

Gordon, R., 2009. Functional its design issues. In: Intelligent Freeway Transportation Systems. Springer, NY, pp. 17−40. Chapter 3.

Helbing, D., Hennecke, A., Shvetsov, V., Treiber, M., 2002. Micro-and macro-simulation of freeway traffic. Mathematical and Computer Modelling 35 (5−6), 517−547.

Kala, R., Warwick, K., 2015a. Congestion avoidance in city traffic. Journal of Advanced Transportation 49 (4), 581−595.

Kala, R., Warwick, K., 2015b. Intelligent transportation system with diverse semi-autonomous vehicles. International Journal of Computational Intelligence Systems 8 (5), 886−899.

Kaufman, D.E., Smith, R.L., Wunderlich, K.E., 1991. An iterative routing/assignment method for anticipatory real-time route guidance. In: Vehicle Navigation and Information Systems Conference, vol. 2, pp. 693−700.

Kesting, A., Treiber, M., Schönhof, M., Helbing, D., 2008. Adaptive cruise control design for active congestion avoidance. Transportation Research Part C: Emerging Technologies 16 (6), 668−683.

Kim, S., Lewis, M.E., White III, C.C., 2005. Optimal vehicle routing with real-time traffic information. IEEE Transactions on Intelligent Transportation Systems 6 (2), 178−188.

Kirby, H.R., Watson, S.M., Dougherty, M.S., 1997. Should we use neural networks or statistical models for short-term motorway traffic forecasting? International Journal of Forecasting 13 (1), 43−50.

Kuwahara, M., Horiguchi, R., Hanabusa, H., 2010. Traffic simulation with AVENUE. In: Fundamentals of Traffic Simulation. Springer, New York, pp. 95−129.

Li, Q., Zeng, Z., Yang, B., 2009. Hierarchical model of road network for route planning in vehicle navigation systems. IEEE Intelligent Transportation Systems Magazine 1 (2), 20−24.

Ma, Y., Chowdhury, M., Sadek, A., Jeihani, M., 2009. Real-time highway traffic condition assessment framework using vehicle−infrastructure integration (VII) with artificial intelligence (AI). IEEE Transactions on Intelligent Transportation Systems 10 (4), 615−627.

Maniccam, S., 2006. Adaptive decentralized congestion avoidance in two-dimensional traffic. Physica A: Statistical Mechanics and Its Applications 363 (2), 512−526.

Narzt, W., Pomberger, G., Wilflingseder, U., Seimel, O., Kolb, D., Wieghardt, J., Hortner, H., Haring, R., 2007. Self-organization in traffic networks by digital pheromones. In: Proceedings of the IEEE Intelligent Transportation Systems Conference, pp. 490−495.

Openstreetmap, 2011. Openstreetmap: The Free Wiki World Map [Online] Available at: openstreetmap.org (accessed 02.07.11.).

Pavlis, Y., Papageorgiou, M., 1999. Simple decentralized feedback strategies for route guidance in traffic networks. Transportation Science 33 (3), 264−278.

Reveliotis, S.A., Roszkowska, E., 2011. Conflict resolution in free-ranging multivehicle systems: a resource allocation paradigm. IEEE Transactions on Robotics 27 (2), 283−296.

Sen, I., Matolak, D.W., 2008. Vehicle−vehicle channel models for the 5-GHz band. IEEE Transactions on Intelligent Transportation Systems 9 (2), 235−245.

Song, Q., Wang, X., 2011. Efficient routing on large road networks using hierarchical communities. IEEE Transactions on Intelligent Transportation Systems 12 (2), 132−140.

Taniguchi, E., Shimamoto, H., 2004. Intelligent transportation system based dynamic vehicle routing and scheduling with variable travel times. Transportation Research Part C: Emerging Technologies 12 (3−4), 235−250.

Tatomir, B., Rothkrantz, L., 2006. Hierarchical routing in traffic using swarm-intelligence. In: Proceedings of the 2006 Intelligent Transportation Systems Conference, pp. 230−235.

Treiber, M., Hennecke, A., Helbing, D., 2000. Congested traffic states in empirical observations and microscopic simulations. Physical Review E 62 (2), 1805−1824.

Vasirani, M., Ossowski, S., 2009. A market-inspired approach to reservation-based urban road traffic management. In: Proceedings of the 8th International Conference on Autonomous Agents and Multiagent Systems, pp. 617−624.

Verhoef, E.T., 1999. Time, speeds, flows and densities in static models of road traffic congestion and congestion pricing. Regional Science and Urban Economics 29 (3), 341−369.

Wahle, J., Bazzan, A.L.C., Klügl, F., Schreckenberg, M., 2000. Decision dynamics in a traffic scenario. Physica A: Statistical Mechanics and Its Applications 287 (3−4), 669−681.

Weyns, D., Holvoet, T., Helleboogh, A., 2007. Anticipatory vehicle routing using delegate multi-agent systems. In: Proceedings of the IEEE Intelligent Transportation Systems Conference, pp. 87−93.

Reaching Destination Before Deadline With Intelligent Transportation Systems

15.1 Introduction

The focus of the book so far has been upon reaching the destination as early as possible, be in a manner of driving a road without speed lanes, or driving within a transportation framework. The attempt has been to maximize the available transportation infrastructure. However, availability of transportation infrastructure is always a limitation which is not in control. A transportation system can only deliver vehicles to their destinations under a strict threshold of maximum bandwidth.

Hence, another task associated with making the transportation system efficient is to *distribute the traffic at different times of the day*, to maximize efficiency or minimize congestion, which is the focus of this chapter. This aspect of the problem talks about the ways to *schedule* different vehicles to enter the transportation system at different times, apart from the regular problem of enabling vehicles drive efficiently within the transportation system.

Chapter 'Intelligent Transportation Systems With Diverse Vehicles' assumed the traffic nonrecurrent, wherein the traffic trends do not follow a historical pattern. This enabled vehicles to escape from unpredictable congestions and to collaboratively prevent congestion from happening. However, historical traffic trends carry rich information and are largely the basis by which humans make their everyday travel decisions. This chapter exploits the *recurrent trends* of traffic to make the transportation system efficient. A disadvantage is, however, that the system might fail in cases in which unpredictable congestion occurs.

For scheduling vehicles within the day, an important classification is whether the journey plans of different vehicles are known beforehand. Practically, humans make their plans suddenly or, even if the plan is made a priori, it is not entered into a system. Hence, no assumption of knowing vehicle plans beforehand can be made. Further, the system does not assume communication between the vehicles, although the assumption that the vehicles can be monitored is made.

Some of the related works are discussed in Section 15.2. The first part of the chapter (Sections 15.3 and 15.4) deals with the problem of *start time prediction* and *route selection*. The aim is to *learn* the recurrent trends of traffic at different times of the day. These trends are used to report to the user the best time to start a journey and the route to take to reach the destination at the predecided time. Broadly, the system attempts to schedule a vehicle such that the person can start late, drive fast and reach just before the predecided time, at the same time ensuring that the probability of arriving on or before the time is high. When repeated for a large number of vehicles over time, the system has the capability of intelligently managing the transportation system such that the different vehicles are *scheduled* to avoid congestion and increase efficiency. The problem is primarily solved by scheduling vehicles in the time domain, unlike the approaches of routing, lane management etc. used in chapter 'Intelligent Transportation Systems With Diverse Vehicles'.

Diversity and *cooperation* have been the two central and related themes of the book. In different problems and algorithms, diversity led to a need for cooperation. Diversity was usually on the basis of travel speeds. In the second part of the chapter (Sections 15.5 and 15.6), the same concept of cooperation and diversity was introduced. However, here diversity is taken to be the degree by which a vehicle is *arriving late* and has to reach the destination on time.

A common picture of the traffic would be a bunch of vehicles *running very late* within a large pool of vehicles which may not mind being late, or may not actually be running late. Currently, the vehicles running late may not be allowed to speed up, and may have no option to enable them to arrive at the destination on time, no matter what the emergency may be. Sections 15.5 and 15.6 talk about the mechanisms by which (with the assumption of communication) the other vehicles and traffic lights can be made *cooperative* to *favour* such vehicles and enable them to reach their destination by the due time at some cost to the other vehicles which may be acceptable to them. A likely use of the system is that it would enable vehicles to depart a little late with an assurance that if they arrive somewhat late due to unaccounted traffic, it should still be possible to cover up with the cooperation of the transportation system.

Segments of the chapter have been reprinted from R. Kala and K. Warwick (2014), Computing Journey Start Times with Recurrent Traffic Conditions. IET Intelligent Transportation Systems 8(8), pp. 681−687, © 2014 IET, with permission. Segments of the chapter have also been reprinted from R. Kala and K. Warwick (2015), Reaching destination on time with cooperative intelligent transportation systems. Journal of Advanced Transportation http://dx.doi.org/10.1002/atr.1352. Copyright © 2015 John Wiley & Sons, Ltd. with permission.

15.2 A Brief Overview of Literature

Although the problem, as described in Section 15.1, has high relevance, it has not been appreciably studied in the literature in its direct form. The closest work is that of Kim et al. (2005). Here, the authors assumed that vehicles have a driver whose presence costs a salary. The authors used the Markov decision process for their modelling. Based on this model, the authors addressed three issues, namely driver attendance time, vehicle departure time and routing policy. Searching for the optimal policy in a time-varying Markov decision process is time-consuming. This must be done in the Kim algorithm by every vehicle attempting to compute their start time. The presented approach meanwhile is based on learning the historical and current data and summarizing them into speed information. This is done centrally by the intelligent transportation system infrastructure. The vehicles simply act as agents in using this information.

Here, the twin related problems of finding the route of the vehicle as well as computing the best start time have been solved. A number of algorithms are prevalent in the literature for vehicle route finding. In general, approaches are either centralized (Kuwahara et al., 2010) or decentralized (Pavlis and Papageorgiou, 1999). For a large number of vehicles the decentralized approach is computationally inexpensive and is clearly the choice in which every vehicle is an individual entity and is dealt with separately.

Claes et al. (2011) used a decentralized routing strategy in which every vehicle considered all possible routes. For every route, the vehicle queried the road infrastructure agent which gave an estimate of the total travel time. The estimate was made on the basis of the vehicle density expected at the particular time that a road segment was traversed. Hence, the same approach may be reversed to compute the latest start times as well as the optimal route. A similar method was presented by Weyns et al. (2007) who used traffic microsimulation to estimate both travel times and travel speeds. Based on these factors, the best route for the vehicle was decided. However, the problem with these approaches is that all vehicles need to be intelligent so that the overall traffic agent knows well in advance the expected number of vehicles. Further, as time progresses, more vehicles start their journeys which increases the expected density and hence the travel time. In both Claes et al.'s and Weyns et al.'s work, it was however possible to replan and get the best route. The objective was, though, simply to minimize the travel time. In this work, the discovery of new vehicles starting up might mean that no route could subsequently be guaranteed for a vehicle to reach its destination on time. Further, both these approaches are computationally expensive for maps with many intersections and roads.

Some algorithms have been designed specifically to solve the problem of traffic congestion (Maniccam, 2006; Verhoef, 1999) which is usually the source of excessive delays. People may initially allow sufficient time for normal travel; however, in congestion scenarios their speed of travel may be reduced by a large amount. Another method of avoiding routes with congestion is by using digital pheromones (Ando et al., 2006; Narzt et al., 2007). Here, the vehicles deposited pheromones on the routes they followed, which evaporated with time. The deposition depended on the travel speed. Later vehicles avoided routes having a high-pheromone content which symbolizes a high number of vehicles and low travel speeds. Again, a limitation of this method is that all the vehicles need to be intelligent.

Another related problem is that of travel-time prediction. The problem is a challenging one due to the stochastic nature of travel time. van Hinsbergen et al. (2011) predicted the travel time of a vehicle to travel to a node using Bayesian Neural Networks. The authors assumed that the historic information about traffic flows was given as an input signal. The signal was broken down into high-frequency and low-frequency signals. Denoising was done by using Wavelet Transforms under the assumption of Gaussian white noise. The denoised signal was then used by a neural network for prediction. Other neural network approaches include Dia (2001), Innamaa (2005), Kirby et al. (1997) and van Lint (2006). Neural network approaches can be a good technique with which to learn from historical data and compute expected travel times. However, uncertainties associated with the prediction using these approaches become very high if the planning is being done too far in advance. In other words, the uncertainties increase with time if no new data are available. In this application, planning the start time is done reasonably well in advance. The problem considered is when the expected travel time and expected arrival time must be well known, with only small uncertainties, before starting the journey. The approach exploits the recurrent nature of the traffic flow for learning and is hence more reliable for long-term planning.

Routing in a real-traffic sense is a stochastic and time-varying problem. For the same reasons a standard shortest-path graph search may not give the optimal solution, considering a known probability distribution of travel times with respect to the time of the day. Miller-Hooks and Mahmassani (2003) considered the problem of finding the best paths in such a model. The authors employed methods to compare two probabilistic paths and selected a path better than the other only if it dominated in all probability realizations. The authors computed all possible paths and build a pareto front of nondominant solutions. Similar work exists in Opasanon and Miller-Hooks (2006). The first problem with these approaches is that they do not state how the probability distribution of travel times is determined. Further, the approaches are computationally very expensive in which the entire combination of paths needs to be computed. This limitation exists in all the approaches with such models. In addition, the approaches may work for small maps; however, modern city maps are rather extensive with a high number of nodes and edges. A large amount of research has been carried out in hierarchical planning (eg, Song and Wang, 2011; Tatomir and Rothkrantz, 2006; Li et al., 2009) in which the researchers believe a shortest-path search algorithm may itself not be able to search for an optimal path in very large cities.

Here, the problem is converted into a deterministic equivalent, in which the required certainty (as defined by the user) is used to get the best time estimate. This enables solving the problem within short computation times. The actual travel of the vehicle is stochastic which may not follow the time-stamps computed by the algorithm, but the best trade-off is sought between ensuring safety and running late as opposed to not reaching the goal too early. The other problem with these approaches is that the final solution is returned based on the best-expected utility. However, for this problem the certainty of reaching the destination on time is an important factor which depends on the application. The proposed modelling of path selection is hence much more problem specific.

15.3 Computing Journey Start Times

The problem of *start time prediction* deals with deciding the time a person should leave from the source to reach their destination by a given time. The problem is commonly seen in everyday life. A person may have to meet some distant relatives, catch a train or flight, watch a movie show, consult a doctor etc. For all these scenarios there is usually a preagreed time at which the person should arrive at their destination. The problem of going to the office is however not necessarily included in this category as this task is performed every day, and the person has the scope to experiment with different start times and choose the best one, considering the safety of reaching their goal on time.

Box 15.1 Takeaways of the Approach

- *Decentralized agents* at the intersections are presented which record the traffic speeds and variations along with time. The use of centralized agents (or single-agent systems) for such an approach is common, which is, however, not a scalable approach. The use of decentralized agents for traffic speeds is also common. Here, recording an extra *deviation factor* helps in answering the user query.
- A new problem of *start time prediction* is studied, in which a user may adapt the algorithm based on the *penalty of late arrival*. A single factor governs the performance. Guidelines enable a user to set the parameter.
- Using the existent notion of advanced driver information system, the twin problems of *start time prediction* and *routing* are solved.
- A *graph search* method is used to compute the route and the start time for the vehicle. The algorithm attempts to select a route which is the shortest in length, has a high reliability and gives the latest starting time.

The difference between the two categories of problems is that in the former the person is either unfamiliar with the route and the traffic that trends on it, is not updated about any change in the traffic trend or has all the available information but has to make a *guess* regarding the start time based on his/her experience. It is possible that the trends might change the next time the same place is visited, travel may be at a different time when the trends are different or the person may vaguely remember the time it took previously. In the office case, the person *tunes* the start time every day and is generally aware of any changing trends, how close the arrival time was last time and whether there exists a possibility to safely make the actual arrival time closer to their desired arrival time. The key takeaways of the algorithm are summarized in Box 15.1. The problem is also summarized in Box 15.2.

15.3.1 Problem Characteristics

Traffic trends, as seen on the road, may be recurrent or nonrecurrent. *Recurrent trends* are repetitive which means that every day, at the same time, a similar density of traffic may be observed. Weekends and sometimes Fridays may, however, show different trends than weekdays. Recurrent trends may also change with time such as a steadily increasing number of vehicles on the road every year and changes in trends along with the seasons.

Nonrecurrent trends are those that occur once and do not repeat. Some of these may be known a priori like a rise in the volume of traffic due to a festival, football match, road repairs etc. Some, however, may not be known a priori such as an accident on the road, an increase in the number of vehicles due to unknown causes, weather conditions affecting travel etc. In making a decision regarding the journey start time, only recurrent trends tend to be considered. The start time decision needs to be made before the journey starts, when the information regarding nonrecurrent trends is unavailable or is

---

Box 15.2 Key Aspects of the Problem

Assumptions
- All roads can get very congested
- There may be no alternative roads
- Recurrent traffic (historic traffic trends are repeated)
- No communication

Concept
- Distribute traffic in different times of the day

Approach
- Learn historic traffic trends
- Calculate *start time* to reach destination at the prespecified time
- Calculate *route* to reach destination at the prespecified time

Applicability
- Human judgement is better if the human has performed the journey routinely with enough experimenting and thus knows everything by experience
- The approach is better than assuming average speeds, which are subjected to variations, and do not use the uncertainty to select safe routes.

---

largely uncertain. That said, a transportation authority may occasionally advertise the possibility of slow traffic on some roads due to preplanned reasons, in which case more spare journey time may be allowed for. In other cases depending upon the urgency to reach the destination on time, some spare time may be taken into account for any nonrecurrent traffic or problems which might provide delays on a trip.

In this problem it is assumed that the cost of a journey is the entire time spent by a person in driving to reach their destination as well as the time that the person would need to wait in case of an early arrival. Hence, it is assumed that the task for which the person travels starts at the scheduled time. In the case of arrival at a time which is later than the desired time there is a *penalty* which entirely depends upon the *purpose of travel*. For example, in the case of catching a train or plane the person may actually miss their connection and may have to cancel their entire plan. In such cases the penalty of late arrival is possibly infinite. However, in the case of meeting friends and relatives the penalty of late arrival is possibly negligible. Meanwhile in other cases, such as going to a movie or lecture, the penalty may increase with time. It often occurs though that late arrivals amounting to merely a few minutes are acceptable.

A popular method employed to solve this problem is to assume *average travel speeds* and compute the overall travel time based on distance. The distance can be estimated quite accurately by various routing algorithms which include SatNav, Google-Maps etc. Many of these also give the expected travel times along with distance information. However, on some roads at some times the actual travel speed may be far lower or a little higher than the assumed average, which increases the uncertainty of travel. The contradictory aims of maximizing (in the sense of it being as late as possible) the start time and maximizing the certainty of reaching the destination on time demand that computations are as accurate as possible with minimum uncertainty. Hence, such assistance systems cannot be relied upon as they stand.

An important factor to consider is the *scalability* issues associated with an approach to tackle the problem. The examples given earlier primarily targeted urban traffic (though the approach must also hold for freeway/motorway traffic and a combination of both as well). Modern cities may have a large number of nodes (junctions) and intersections making a very complex road network. In such cases, the approach needs to be computationally efficient from all perspectives for the system to be sustainable. Further, in a fully operative system the number of vehicles each computing their individual start times may be large, which is proportional to the total traffic. Because traffic density is generally increasing with time, it may be expected that the total number of users of the service also will grow considerably with time.

15.3.2   Problem Statement

Assume that a person needs to travel via a transport network to reach a location L at the latest by a *scheduled time T*, whereas the journey needs to start from a location S. The road network graph G of the city is assumed to be known. The first part of the problem is to compute the *start time* T_s for the person. It is assumed that the start time denotes the time of departure from the source. Hence, the person would time himself/herself such that there is no delay and the vehicle starts at time T_s. The second part of the problem deals with computing the *route R*. Here, R denotes a set of vertices $<S, V_1, V_2,...L>$ starting from the source S and ending at the destination L. The algorithm needs to ensure that if the person starts at T_s from S and uses the route R the certainty of reaching L on or before T is as high as desired by the person.

Let the *duration of the journey* be denoted by T_t and time of reaching the destination by T_f (which gives $T_s + T_t = T_f$). For a given T_s, the values T_t and T_f are *stochastic* in nature as different runs of the same vehicle may differ in travel durations and finish times due to the presence of other vehicles, traffic lights etc. Although T denotes the scheduled time that the vehicle aims to reach the destination, the vehicle may or may not be able to do so because of traffic uncertainties. The actual time at which the vehicle reaches the destination is denoted by T_f. Let $P(t \leq T \mid T_s, R)$ denote the probability that given the start time T_s and the route R, the vehicle reaches destination L at a time on or before the desired time T. Here, $P(t = T \mid T_s, R)$ is a probability distribution, whereas T_f and T_t are unit samples from the related distributions.

The first objective of the algorithm is that the *start time must be as late as possible* (maximize T_s). The second objective of the algorithm is that the route R must be the fastest way to reach the destination or in other words the *travel time needs to be the shortest* possible (minimize T_t). The third objective of the algorithm is that the route R should be *as reliable as possible* ensuring the person reaches their destination on or before the scheduled time T (maximize $P(t \leq T \mid T_s, R)$). If for any reason the person reaches the destination before the scheduled time T, the fourth objective states that this *waiting time should be as little as possible* (*if* $T_f \leq T$, *minimize* $T - T_f$, *high penalty otherwise*).

15.4 Algorithm for Computing Journey Start Times

The algorithm uses intelligent agents to enable computation of the start time and the route. These agents are assumed to exist at every intersection and monitor the traffic. The agents record the *mean travel speeds* and the *variance* at the different times of day and on different days. This information is used by a graph search algorithm which employs a shortest-path search to compute the route and hence the start time. The graph search algorithm converts the stochastic travel speeds into deterministic values for its computations. Full details are given in the subsequent sections. The general algorithm framework is given in Fig. 15.1.

15.4.1 Learning Travel Speeds

To solve the problem the search algorithm needs to know the *average expected travel speeds* at particular times of the day. The assumption behind the algorithm is that the traffic is recurrent. Hence, it is assumed that traffic flow and density observed at a *particular time of the day* would be similar to that observed at the same time on a *similar day*. From everyday driving as well it may be realized that most traffic trends are recurrent. On any day, people travelling may guess likely roads for high- and low-speed driving. In such a case, one may simply record the traffic trend of a particular day (or better still take an average over a number of such days) and use it as information input to the search algorithm for similar days. *Similar days* means the days of the week when the traffic is expected to be similar. Traffic on Mondays and Sundays clearly has different trends, whereas the trends may be only slightly different for Mondays and Tuesdays. However, traffic in general shows a trend of a gradual increase along with time, seasonal variations and some minor variations with time, such as noise and unexpected trends. Hence, to make the process adaptive to such variations, an element of *learning* is introduced. The learning stage is summarized in Box 15.3.

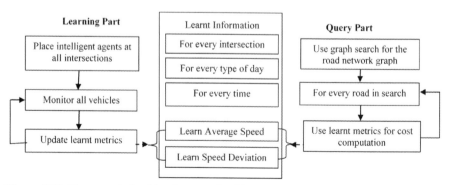

Figure 15.1 The general algorithm framework without cooperation.

Box 15.3 Key Aspects of the Learning Stage

- Intelligent agents at every intersection learn *mean* and *variance* of *travel speeds*
- Speed assumed similar on some *similar days* and in *pockets of 10 min*
  - Too small an interval creates too many parameters to learn, which is difficult and uncertain
  - Too large an interval results in high deviation of speeds within the time interval.
- *Learning rate*
  - At small rate the algorithm behaves passively and does not capture any changing trend.
  - At high rate the algorithm may treat any delay due to immediate uncertainties as a change in trend
- Decentralized architecture (an agent at every intersection)
  - Query time requires multiple connections to all intersections of competitive routes (unlike one in centralized approach)
  - Computation spread across agents and hence manageable (unlike the central server taking too many requests from all vehicles in centralized approach)
- Dealing with immediate *nonrecurrent traffic*
  - If observed speed too different from the current average, immediate nonrecurrent traffic is detected, learning pauses
  - If the same continues in the future, a new trend is assumed and learning continues.
  - If nonrecurrent traffic is due to preknown events, the learning may be manually paused.

It is assumed that the road network graph G has a total of $|V|$ vertices for which every vertex is an intersection. Two intersections V_1 and V_2 are said to be connected by an edge $<V_1,V_2>$ if there exists a road from V_1 to V_2 with travel from V_1 to V_2 permitted (in the case of one way). Each of these V intersections is occupied by an intelligent agent. The agent may be implemented using Vehicle Infrastructure Integration systems. Hence, the total map has $|V|$ intelligent agents in which $|.|$ denotes the set size. The purpose of each agent is to monitor the vehicles and record the speed information. The agent at intersection V_2 monitors all vehicles that come from any intersection V_1 and reach intersection V_2, in which V_1 is any intersection such that (V_1, V_2) is an edge.

Let speed(V_1, V_2, t, d) denote the average speed in going from intersection V_1 to intersection V_2 at time t of the day and at a particular day-type d of the week. Here, time t is broken down into buckets of 10 min. The assumption is that the average speeds in real-life traffic do not change much in an interval of 10 min. Hence, t between 10:00 am to 10:10 am would point to the same value of speed(V_1, V_2, t, d). Making the time interval too small results in too many parameters to learn, which may hence be difficult to compute and uncertain due to less data. Too large a time interval may show a high deviation of speeds within the time interval as any change in trend within the time interval cannot be captured.

Suppose a vehicle A left the intersection V_1 at time t_1 as observed by the agent at V_1. The agents may be sophisticated to track and identify the vehicle or intelligent vehicles (Bishop, 2000; Reichardt et al., 2002) may themselves communicate their identity.

The same vehicle is seen by the agent at intersection V_2. Suppose it leaves the intersection V_2 at time t_2. The agent at V_1 communicates time t_1 and the vehicle A to the agent at intersection V_2. The agent at intersection V_2 hence observes the average speed of the vehicle A in moving from V_1 to V_2 as given by Eq. [15.1].

$$\text{speed}(A) = \frac{||V_1 - V_2||}{t_2 - t_1} \qquad [15.1]$$

Here, $||V_2 - V_1||$ denotes the distance between intersections V_1 and V_2 which is assumed known. The average speed includes any time spent in *waiting for traffic lights* at V_2 (if any). If normally the traffic lights take a long time to change, the average speed of the vehicle may therefore be expected to be low.

Learning the factor speed(V_1, V_2, t, d) by the agent at V_2 is done using Eq. [15.2]. This equation constantly adapts the speed to the changing traffic trends. The updated speed estimate speed(V_1, V_2, t_1, d)^{new} is taken partly from the actual speed of the vehicle A and partly from the old speed estimate speed(V_1, V_2, t_1, d)^{old}. The old speed estimate may be initialized based on observations of a few initial vehicles or to the road's speed limit. Algorithmically, the value of speed(V_1, V_2, t_1, d) is constantly changed in consideration of the newly recorded speed of the vehicle A. As more and more vehicles pass by, their recorded speeds are used to correct the overall speed estimate. Only a fraction of the estimate, equalling learning rate, is taken from the vehicle A to eliminate any noise or slow-driving preference of the vehicle A.

$$\text{speed}(V_1, V_2, t_1, d)^{\text{new}} = (1 - \text{lr})\text{speed}(V_1, V_2, t_1, d)^{\text{old}} + \text{lr} \cdot \text{speed}(A) \qquad [15.2]$$

Here, $\text{lr}(0 < \text{lr} \leq 1)$ is the *learning rate*. A small value of learning rate implies that the algorithm is passive and does not capture any rapidly changing trends. A high value, meanwhile, denotes that the agent may treat any delay which is due to the personal preferences of the driver, or similarly small delays due to weather conditions, as a change in trend for all drivers and hence store the wrong information.

The agent further measures the *standard deviation* $\sigma(V_1, V_2, t, d)$ in a similar fashion to speed, given by Eq. [15.3]. Although the speed is learnt using a learning rate for each vehicle as it passes by, the deviation is measured for all the vehicles that passed by in the previous δ similar days. Here, N denotes the number of vehicles considered for computing the deviation. Too small a value of δ might mean too few vehicles are considered for the computation, which in turn would mean uncertainty in the recorded deviation. Taking too high a value, however, might cause undue effects of historical data which may have changed with time. For most high-density roads, a small value would suffice.

$$\sigma(V_1, V_2, t_1, d)$$

$$= \sqrt{\frac{\displaystyle\sum_{\text{previous similar } \delta \text{ days}} \left(\text{speed}(V_1, V_2, t_1, d)^{\text{new}} - \text{speed}(V_1, V_2, t_1, d)^{\text{old}}\right)^2}{N}} \qquad [15.3]$$

Unlike speed, the deviation is taken for all the vehicles for previous δ similar days only. For implementation purposes, the agents must hence have sufficient memory to store the differences needed for the calculation of deviation.

One of the key aspects of the algorithm is that a *decentralized architecture* of the agents is proposed wherein every agent stores some information, rather than proposing a centralized architecture with all the information being held at one place. In a practical system with centralized approach, a large number of users would be attempting to compute the start time and route. They would all query the central server and occupy it for a long time till the graph search algorithm finds a route. All the queries from the city would go to the same server. If there are other users willing to travel into the city, there would be even more queries. It may not be possible to simultaneously handle so many users, whereas the updates are being continuously made. The centralized approach, however, makes the algorithm require a single connection, and hence the speed of the algorithm may be faster for the case of a single user.

In a decentralized approach the computation is spread across the agents at intersections. The graph search algorithm considers only competing routes, intersections corresponding to which are queried. An algorithm hence queries a small number of agents out of all those available. This implies that every agent has limited demands despite the total number of queries throughout the system being high. Hence, agents may swiftly reply to the queries. However, this forces the search algorithm to make a large number of connections.

On many occasions, *nonrecurrent trends* appear in traffic in which the traffic flow is temporarily altered from the expected trends, and such an alteration will most likely not be seen in the future. Such trends must be identified and neglected. Hence, at every intersection the speeds of the vehicles are assessed. If the assessed speeds are very different from the expected averages, learning is not carried out for that vehicle considering it as a nonrecurrent trend. If such irregular trends continue in the future, the learning framework interprets it as some new trend which might have suddenly started and hence the learning continues. Many times such trends may be known a priori, for example a football match, public event, demonstration etc. In such cases the transportation authorities may ask the algorithm to neglect such cases by pausing the learning. Routing and start time computation in the presence of such trends known a priori may be done by humans themselves under the guidance of the transportation authorities.

15.4.2 Routing

The problem is to enable a vehicle to decide its starting time and route of travel. The expectation is to have the vehicle at the destination L at the predecided scheduled time T. Let $T(V_i)$ denote the latest time by which the vehicle must be at the intersection V_i so that it can hope to reach the destination L at a time T with a high probability. The algorithm proceeds by computing $T(V_i)$ for all the nodes. The value of $T(V_i)$ at the source is hence the starting time, whereas the route may be conveniently given by selecting the roads in terms of the shortest time to goal. The routing is summarized in Box 15.4.

The problem is modelled as a *graph search* which goes *from the goal towards the source*, which is an inverted version of a regular graph search problem. In a regular

---

Box 15.4 Key Aspects of the Routing Stage

- Objectives
  - Start as late as possible
  - Select fastest route
  - Probability of arriving on time should be highest
  - In case of early arrival, wait time should be least
- Cost function
  - *Latest time to reach a node to expect to reach the goal at the prespecified time*
  - is same as maximizing start time (for source)
  - is same as minimizing delay in case of an early arrival
  - is same as minimizing travel time
  - is *opposite* to maximizing probability of arriving on time (the earlier, the better)
- Combining the costs, making it deterministic, speed chosen = mean speed − α deviation.
  - α is a user-chosen parameter per task (maintains trade-off between contradictory objectives)
  - More importance of arriving on time implies more resistance to risk, or higher α, and vice versa
  - High α implies more resistance to risk, or earlier start time, or high probability of arriving, and vice versa
  - High deviation means vehicles in that road vary largely in speed, meaning the road is less reliable and should be avoided, and vice versa
  - A minimum threshold is fixed to avoid roads for which the data are too few.
- Inverted graph search
  - Cost is known for the goal, needs to be computed for the source
  - From goal, go in pursuit of the source
- α and probability of reaching on time
  - Simulate the system for various α for a region
  - For each simulation, compute the percentage of vehicles arriving late
  - Draw a graph between α and the other metrics
  - User can read this graph and decide the value of α

---

graph search, starting from the source at a time 0, the intent is to reach the goal with the shortest time (or any other metric) using any of the available nodes or edges. In this problem it is known that $T(L) = T$, or that the vehicle should be at the destination at the latest by time T, whereas the same needs to be computed for the other nodes, especially the source.

The objective of the problem was to simultaneously maximize start time T_s, minimize the travel time T_t, maximize the probability of reaching before the predetermined time $P(t \leq T \mid T_s, R)$ and minimize the waiting time as much as possible (if any) or *if $T_f < T$, minimize $T_f - T$*. These objectives need to be *fused into a single objective* that the graph search might optimize. For a deterministic approach the objective to minimize the travel time T_t and minimize the delay are the same as the objective to maximize the start time T_s. However, the factor $P(t \leq T \mid T_s, R)$ is contrary to all of the above. To be assured of reaching the destination on or before the desired time, one has to reserve spare time to overcome any unexpected delay which increases the start time. A related problem is that the time variables mentioned are stochastic in nature which pressurizes the need to work with probability distributions to optimize the objectives. However, considering the computationally expensive nature of the problem for a road network of large size, it may not be possible to go with such an approach.

For these reasons the approach followed in this work is to *fix the probability* $P(t \leq T \mid T_s, V_i)$ which would normally be required by the user, and *go forth with maximizing* T_s. Considering contradictory objectives, there has to be some way of making a tradeoff between them. Considering the nature of the problem, only a user can decide whether he/she is travelling for a task for which being late is not allowed or as a task for which being a bit late may not matter. This further converts the problem into a discrete equivalent, in which a probability distribution may be replaced by the best value of the metric which lies above the performance threshold.

The graph search takes place from the destination with $T(L) = T$ and proceeds in pursuit of the source. Consider expansion of an intersection V_1. The node V_1 is said to be connected to all nodes V_2 such that $<V_2, V_1>$ is an edge. Note that the edge definition is opposite due to the inverted nature of the problem. The vehicle needs to be at V_1 on or before $T(V_1)$ to reach the destination on time. The time taken to travel from V_2 to V_1, denoted by $T_m(V_1, V_2)$, may be given by Eq. [15.4]. $T(V_2)$ may hence be given by Eq. [15.5]

$$T_m(V_1, V_2) = ||V_1 - V_2||/(\text{speed}(V_2, V_1, T(V_1), d) - \alpha \cdot \sigma(V_2, V_1, T(V_1), d))$$
[15.4]

$$T(V_2) = T(V_1) - T_m(V_1, V_2)$$
[15.5]

Note that speed($V_2, V_1, T(V_1), d$) denotes the expected speed at time $T(V_1)$ whereas the requirement is to measure the speed at time $T(V_2)$. $T(V_2)$ is an unknown quantity (which is being computed). In reality the two times would be very similar (and may in fact lie in the same 10 min window), and hence the speeds would be very similar.

The expected speed of travel from V_2 to V_1 is speed($V_2, V_1, T(V_1), d$). However, the actual speed may be different from the general expectation, and hence a penalty is added which is proportional to the learnt deviation. A low deviation means that all vehicles on the road travel with almost the same speed and hence the learnt speed is *reliable*, whereas a high deviation means that the learnt speed is not reliable and hence an extra safety time has to be included. The factor α establishes a tradeoff between the two opposing factors of the probability of reaching the destination on time and the latest arrival, and is set by the user in understanding his/her requirements. The realization of the speed factor is shown in Fig. 15.2. If the vehicle takes more than $T_m(V_1, V_2)$ for travel to V_2, it cannot be said that a late arrival will happen. Spare time taken at all the edges is transferred from one edge to the other. Hence, in case the vehicle reaches a node later than expected, it may take a shorter than expected time to travel the next edge and hence a late arrival at the final destination may be avoided.

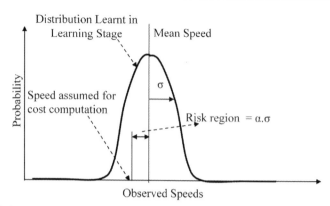

Figure 15.2 Selection of speed for cost computation.

It is possible at times that some roads, maybe for some parts of the day, are rather under used. Hence, there may *not be enough data* with which to learn the speed on a road section, and the recorded deviation (σ) may therefore itself be unreliable. High deviation implies high unreliability, and such roads are seldom used by drivers. In this algorithm, such roads have a high penalty and are hence unlikely used by the algorithm. Hence, high deviation due to lack of data produces very limited undesirable effects. However, lower deviation values due to lack of data indicate reliable roads which is not really the case. The algorithm may choose such a road due to its low penalty, whereas the vehicle may actually take longer.

Consider the case when a road is unused and only a single vehicle passed in a particular time frame, based on which the variance computations were made. Consider that this vehicle was driving fast and further did not have to wait for any traffic signal. However, now taking this road might not be that attractive. The problem is not having enough data for learning. Hence, a *minimum threshold σ_{min}* of deviation has been set. The effective deviation used in the calculations would hence be the larger of σ_{min} and the learnt σ. σ_{min} ensures that no road is regarded as too reliable (or possesses a small deviation) which may actually be due to the small amount of data. A higher value of σ_{min} is too pessimistic an approach, wherein the algorithm does not trust the recorded deviations and prefers to take large margins on all the roads. A small value, on the other hand, has the risk of the road being selected despite an uncertain value due to a small amount of data.

The vehicle is projected to be arriving at V_2 at a time $T(V_2)$. In reality it may arrive a lot earlier, particularly if the user was very conscious of arriving late and set a high value of α. The actual speed and the speed at the projected time can hence be very different. The risk is whether the vehicle would still be able to reach V_1 on or before $T(V_1)$. If the vehicle arrives at V_1 later than $T(V_1)$, of course, the probability of reaching the destination on time is dependent on the spare time available for the rest of the journey along with the accuracy of the learnt information. Consider, as per the A* algorithm, a vehicle is expected to reach a particular node at 10:00 am and the next node using the available road at 10:10 am. These computations kept some slack time at every node. So it is possible that the vehicle reaches the node by 9:30 am instead of the scheduled 10:00 am. It is also possible that traffic at 9:30 am is highly congested due to office crowding, which clears at 10:00 am. Due to this the expectation at the node by the A* algorithm was a clear road; however, because the vehicle reached the node early it saw congested roads. It needs to be proved that it can still reach the subsequent node at the latest by 10:10 am. Otherwise, the algorithm does not hold.

Suppose two vehicles A and B reach V_2 at times t_{2A} and t_{2B}. Suppose they reach V_2 at times t_{1A} and t_{1B}. If vehicle A reaches V_2 earlier, that is $t_{2A} < t_{2B}$, it can be ascertained that it also reaches V_1 earlier, that is $t_{1A} \leq t_{1B}$, provided that they travel by the same preferred speed and no single lane is reserved for high-speed traffic which one of the vehicles can utilize whereas the other cannot. In a single-lane case, there is no way that B would come from behind and overtake A. In a multilane case, the vehicles arrange themselves so that the speed on all the lanes becomes equal, and hence there is no way that B would come from behind and overtake a vehicle nearly moving parallel to A. In case the lanes are with different speed limits or the road is not densely occupied, the speeds of the lanes may be different. If A is running lane, it would certainly change to a high-speed lane and the proof for a single-lane case holds. Hence, if the vehicle reaches V_2 at $t_{i1} < T(V_2)$ it should be able to reach V_1 at $t_{f1} \leq T(V_2)$.

The approach taken here uses the A* algorithm from the goal, at each instance expanding nodes as per the priority queue. Every node V_2 with parent V_1 is associated with time $T(V_2)$ given by Eq. [15.5], time to destination $T_d(V_2)$ which is the expected journey time from V_2 to the destination given by Eq. [15.6], heuristic time to source or $T_s(V_2)$ given by Eq. [15.7] and the total expected duration of the journey $T_j(V_2)$ given by Eq. [15.8]. $velmax$ is the maximum preferred speed of the vehicle. The nodes are sorted as per T_j values in the priority queue and the smallest value is taken for expansion. The parent node is maintained to formulate the route after the A* algorithm has found the source.

$$T_d(V_2) = T_d(V_1) + T_m(V_1, V_2) \qquad [15.6]$$

$$T_s(V_2) = ||V_2 - S||/velmax \qquad [15.7]$$

$$T_j(V_2) = T_d(V_2) + T_s(V_2) \qquad [15.8]$$

Once the A* algorithm terminates at the source. $T(S)$ is the start time and the parent information from S to L is used to compute the route to take. $T_g(S)$ is the expected journey time.

15.4.3 Probability of Reaching the Destination on Time

In Section 15.4.2, it was stated that α controls the tradeoff between the *probability of reaching the goal on time* and the other objective of *maximizing the start time*. The factor, however, cannot be used to compute the precise probability of reaching the destination on time. In fact, computing the probability distribution over time for the journey is a rather hard and computationally expensive task. Attempting to set α by seeing a particular edge might lead to keeping the value too high which may cause the start time to be very early and vice versa. Here, it is proposed that the relation of α to the probability of reaching the destination on time can be conveniently studied by experimentation for every region. Considering that the duration of the journey does not vary alarmingly, the generality of the conversion of α to the probability of reaching the destination on time is high. Such a study can be performed for different regions. The study can be presented to the user, who may then be able to fix a value of α depending upon the risk of late arrival he/she is ready to take or the purpose of the journey.

15.5 Cooperative Transportation Systems

The problem with the approach in Sections 15.3 and 15.4 is that the system discourages a user from leaving too early and waiting at the destination for long times, as it is clearly suboptimal, whereas such a decision may be important accounting for sudden and unpredictable nonrecurrent trends. The parameter α can be used to reserve lengthy spare time; however, considering that most of the time such trends would not be observed, the prolonged time spent in waiting is unjustifiable. The work gives a start time and a route for the user to start. However, if the user is running late, there is nothing that he/she can do. Imagine leaving your home well on time to reach an important meeting or to catch a flight, only to realize that somehow traffic seems a little slower. A common reaction is to marginally exceed the speed limits which is neither legal nor advisable.

This section extends the concept and enables the user to *ask for cooperation* from the other vehicles. The cooperation helps the user to reach on time, despite running late. The aim is to select a modest start time and route, judiciously keeping some spare time to combat traffic uncertainties using the approach in Sections 15.3 and 15.4. Once the user starts the journey, the approach discussed in this section enables him/her to combat any large uncertainties while driving. This eliminates the need to reserve much spare time. The general architecture of the overall system, including the aspects of cooperation, is given in Fig. 15.3.

General traffic is mostly under the occupancy of vehicles going for leisure, catering to appointments which are somewhat flexible, meeting every-day schedules in which being a little late does not mean a huge loss, or on an important appointment being comfortably on time. In a queue of people, it may be acceptable for a person running very late and for an important task to ask for cooperation from the other people. This would lead to the others being at a little loss, but would be of a great help to someone running very late. It is acceptable as long as the number of people running late form a small part of the entire queue.

The transportation system currently has no mechanism by which such a priority could be given. This approach discusses the mechanisms by which other vehicles and traffic lights can be made *cooperative*. An example of priority is taken from the

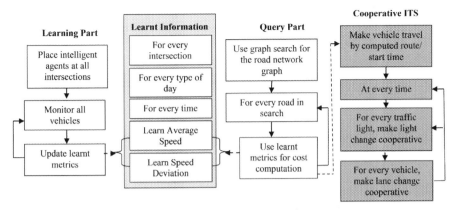

Figure 15.3 The general algorithm framework with cooperation.

emergency vehicles. Other vehicles make every attempt to allow emergency vehicles to overtake. Emergency vehicles can cross a traffic crossing irrespective of the traffic-light state, and assume priority in intersections and pedestrian zones. Although a simple implementation of the concept would be to allow a vehicle running late to flash some emergency light, such a solution is not acceptable because the cooperation of the other vehicles for emergency vehicles usually comes with some breach of traffic rules which badly affects the other vehicles. The key takeaways of the algorithm are summarized in Box 15.5, whereas the problem is summarized in Box 15.6.

It is assumed that the traffic system has a number of vehicles which need to arrive at the destination on or before the predecided time T. The current problem definition assumes that these vehicles are on a *real-time operation* and hence arriving at the destination after T is regarded as worthless. For the complete transportation system, the sole objective is to minimize the number of vehicles arriving late, given whatever time they actually start.

Box 15.5 Takeaways of the Cooperative Approach

- The notion of *cooperative traffic lights* is used which is biased to allow vehicles which are running late to pass through.
- The concept of *cooperative lane changes* is introduced by which a lane change attempts to minimize some vehicle from running late.
- Different *states of a vehicle* which desires to be on time are designed which include being comfortable on time, running a little late (may still arrive even without cooperation), running very late (difficult to arrive without cooperation) and impossible to arrive (given up).
- A *cost metric* is designed which maps the different states of a vehicle running late to a consciousness of being late, used for decision-making regarding all cooperative measures.

Box 15.6 Key Aspects of the Problem

Problem
- A bunch of vehicles are on real-time task and have to arrive on time
- These vehicles are running late, enable them to arrive on time
- No alternative routes may be available

Concept
- Make vehicles running late take priority
- Make the transportation system cooperative to these vehicles

Vehicle Travel States
- Comfortable, Late, Very Late and Give Up

Vehicle Travel Times
- Expected time with parameter α
  - High α implies more resistance to risk, or earlier start time, or high probability of arriving, and vice versa
- Late time with parameter β
  - High α and β imply too much of cooperation too early, or higher probability of arriving on time, which is bad for other vehicles, and vice versa
- Cancel time with parameter γ
  - High γ implies giving up too early or a possibility of arriving on time may be denied
  - Low γ implies giving up too late which is bad for the other vehicles

---

The transportation system attempts to make the vehicles cooperative such that the different vehicles enable each other to go through as per their needs. The first task is to measure whether the vehicles are running late and, if they are, by what time (Sections 15.5.1 and 15.5.2). The second task is making the system cooperative. This is done using two methods: *traffic lights* (Section 15.5.3) wherein the changing order of the traffic lights is altered to allow the vehicles running too late to travel first; and *lane changes* (Section 15.5.4), wherein a vehicle may make a lane change to enable some other vehicle running late to travel faster or overtake to get more comfortable with time. The cooperative constituents are also summarized in Box 15.7.

15.5.1 Vehicle Travel State

A traffic system at various times has different vehicles running late by varying degrees, depending upon which the cooperative measures are applied. Consider that a vehicle is moving using a precomputed route R which consists of a set of intersections. Say that the vehicle crossed an intersection V_i at time t. Three times are associated with the intersection V_i.

Box 15.7 Cooperative Constituents of the Transportation System

Cooperative Traffic Lights
* Change lights with a preference to the vehicles running late
* Primary criterion: *Total lateness*
* Secondary criterion: *Earliest Emergence*

Cooperative Lane Changes
* Every lane change of any vehicle is to aid a vehicle running (more) late
* Standard criterion:
    * Choose lane with maximum *Time to Collision* (if any)
    * Stay on the leftmost lane (if currently close to maximum speed).
    * This allows other vehicles to overtake (from the right)
* Added constraints for cooperation
    * Always change lanes on encountering a vehicle running late behind
    * Change lanes unless all options of lane change have vehicles running more late behind
    * Never change a lane to come in front of a vehicle running more late

---

The first time (*expected time* denoted by $T(V_i)$) is the latest time by which the vehicle should reach the intersection V_i to reach the destination on time with the pre-decided probability. This time accounts for the average speeds along with the safety factors measured from the knowledge of the deviation. The second time (*late time* denoted as $T_L(V_i)$) is the latest time by which the vehicle should reach the intersection V_i, after which it would become very difficult to reach the destination unless the traffic clears unexpectedly or some cooperation is offered. The third time (*cancel time* denoted by $T_C(V_i)$) is the latest time by which the vehicle should reach the intersection V_i at any cost, after which the vehicle may not expect to reach the destination, no matter whatsoever cooperation be offered. The times are explained in Fig. 15.4.

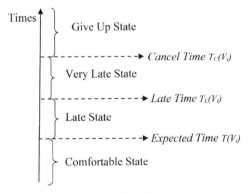

Figure 15.4 Times and states associated with the algorithm.

This gives rise to various *states* that a vehicle may be associated with at any time. The first state is a *comfortable state* ($t \leq T(V_i)$), which denotes that the vehicle is being navigated as expected and should reach the destination on time. The second state is a *little late* state ($T(V_i) < t \leq T_L(V_i)$), which denotes a need to worry as the vehicle is running late; but it should be possible to reach the destination either without cooperation with the expected situations but with a slightly lesser probability, or with some cooperation from the transportation system or with improved conditions. The third state is a *very late state* ($T_L(V_i) < t \leq T_C(V_i)$), which denotes that the vehicle is running very late and cooperation is very badly required to arrive on time. Alternatively, the vehicle may reach its destination if the subsequent traffic is a lot better than expected. The last state is a *give up* state ($t > T_C(V_i)$) which denotes that the vehicle cannot reach its goal on time, no matter how much cooperation is offered.

The different states are easy to understand from a human-driving experience perspective. Imagine leaving home for a meeting and for a long time thereafter one is in a *comfortable state*. If one encounters a traffic signal, which did not change for a long time due to congestion, one may begin to worry a little, but the driver may not show any change in driving behaviour which marks a *late state*. Subsequently, if there is high traffic congestion which eventually gets cleared, drivers tend to drive fast or make rapid lane changes, which indicate a *very late state*. A *give up* state is rather hard to observe from a human perspective, wherein even if one is running very late, there is always a hope that some way things would get better. From a more practical perspective, this state is modelled.

Computation of different times is done while computing the route and the start time. The graph search happens in the same way with the same costs involved. Computation of different times associated with an intersection V_2 with parent V_1 may hence be given by Eqs [15.9]–[15.12]. The equations have the same form as the Eqs [15.4] and [15.5].

$$T_{m_L}(V_1, V_2) = \|V_1 - V_2\|/(\text{speed}(V_2, V_1, T_L(V_1), d) - \beta \cdot \sigma(V_2, V_1, T_L(V_1), d))$$
[15.9]

$$T_{m_C}(V_1, V_2) = \|V_1 - V_2\|/(\text{speed}(V_2, V_1, T_C(V_1), d) - \gamma \cdot \sigma(V_2, V_1, T_C(V_1), d))$$
[15.10]

$$T_L(V_2) = T_L(V_1) - T_{m_L}(V_1, V_2)$$
[15.11]

$$T_C(V_2) = T_C(V_1) - T_{m_C}(V_1, V_2)$$
[15.12]

Although α controlled the trade-off between the two opposing factors of early start and arriving probability, the factors β and γ do the same for cooperation. High values of α and β lead the vehicle to *ask for too much cooperation* too early (apart from the role of α in computation of the start time) which may not be good for the other vehicles, at the same time leading to a better chance of the vehicle actually reaching its destination on time. Smaller values of these parameters may lead the vehicles to ask for too little cooperation too late. A high value of γ may lead the vehicle to *give up too early* whereas small values may make the vehicle give up too late. The vehicle giving up too late may be disadvantageous for the other vehicles which may be uselessly cooperating. Typically, $\alpha > \beta > 0 > \gamma$.

An important aspect is that three different times are maintained instead of using a single time (say $T(V_i)$) and computing the other times from $T(V_i)$ using fixed-size time windows. The implemented approach is better for planning long journeys, wherein computing the latest times by α, β and γ may lead to very different times at which the predicted speeds may be very different (speed(V_2, V_1, t, d)). Hence, monitoring time t for the different strategies is needed.

15.5.2 Lateness Consciousness Cost

The transportation system makes a vehicle running late preferred over the other vehicles. In practice, however, a number of vehicles may be running late and may be found in different groups. Hence, it becomes an important question to designate a cost with every vehicle which indicates the *magnitude by which a vehicle is conscious of running late* and hence the transportation system must prioritize it. The cost is given by Eq. [15.13].

$$\text{Late}(t, V_i) = \begin{cases} 0 & \text{if vehicle is not for a real time task} \\ \varepsilon & \text{if } t \leq T(V_i) \wedge \text{vehicle is for a real time task} \\ \varepsilon + (t - T(V_i)) & \text{if } T(V_i) < t \leq T_L(V_i) \wedge \text{vehicle is for a real time task} \\ \varepsilon + (t - T(V_i)) + (t - T_L(V_i))^2 & \text{if } T_L(V_i) < t \leq T_C(V_i) \wedge \text{vehicle is for a real time task} \\ 0 & \text{if } t > T_C(V_i) \wedge \text{vehicle is for a real time task} \end{cases}$$

[15.13]

The cost is only valid for the vehicles which need to compulsorily reach their destination on time. An important aspect is that a small cost ε is given even if the vehicle is moving comfortably. This gives it some extra time for anything uncertain in the future at the cost of the other vehicles which do not mind running late. During a vehicle's travel, a rough span of time may come in the future which makes the vehicle reasonably late. It is more risky to make the vehicle late and then solve the problem with cooperation. Rather, the algorithm does a little to give it enough time well in advance.

15.5.3 Cooperative Traffic Lights

The first manner of making the transportation system cooperative is by the installation of *cooperative traffic lights*. Consider a traffic light operating at an intersection V_i with a queue of vehicles at every road j which leads to the intersection. For every road two quantities are recorded. These are the total lateness (Late$_j$) and the earliest emergence (emer$_j$). *Total lateness* (Late$_j$) is the importance of preferring a road j due to the lateness of all the vehicles combined. The quantity is given by Eq. [15.14] in which Late$_j^k$ denotes the lateness of a vehicle k in the queue of vehicles formed at road j. Emergence of a vehicle k is denoted by t_j^k and measures the arrival time of the vehicles prior to waiting. *Earliest emergence* is the time associated with the earliest entering and consequently the most waiting vehicle, given by Eq. [15.15].

$$\text{Late}_j = \sum_k \text{Late}_j^k \qquad\qquad\qquad\qquad [15.14]$$

$$\text{emer}_j = \min_k t_j^k \qquad\qquad\qquad\qquad\qquad [15.15]$$

The traffic light changes to green for the road j with the largest lateness. If two roads have the same lateness, emergence is taken as the secondary criterion with a preference given to an earlier emergence. This means preference is given to the road with the vehicle which has been waiting for the largest time (for the light to change to green) in a First-Come-First-Served mechanism. The traffic lights are changed after a threshold number of vehicles, or when the queue gets empty and there are vehicles on the other road waiting to pass through.

15.5.4   Cooperative Lane Changes

The other manner of making the transportation system cooperative is by the use of *cooperative lane changes*. The speed of a vehicle depends both on the speed of the vehicle it is following and the distance between the vehicles. If a vehicle is running late, it may hence be useful to have the vehicle in front change lanes. Further, it would be painful to have another vehicle change lanes and come right in front, even if it is acceptable from a safety perspective. These factors have to be accounted for while making any lane change.

The basic mechanism for a lane change is assumed an *overtaking-based lane change* as discussed in chapter 'Intelligent Transportation Systems With Diverse Vehicles'. It is assumed that the traffic operates using a 'keep-left' rule and overtaking takes place on the right. In low traffic density, a vehicle (on seeing a slow vehicle ahead) moves to the lane on its right to aim an overtaking, and later returns to the left to allow other vehicles to overtake if interested. If jumping to the right lane is not possible or not more efficient, the overtaking is attempted from the left side. The clause handles the exceptional cases wherein a slow-moving vehicle is on the right overtaking lane, prohibiting overtaking. In all cases, *time to collision* is used as a metric to check whether it would be better to change lanes or stay in the current lane. Safe distances from the vehicles must be available for the lane changes to happen.

Cooperative behaviour is added on top of this mechanism. A vehicle has to make a lane change if it finds a vehicle running late behind (or a vehicle running late by an amount larger than itself). The lane change enables the vehicle running late to speed up which hence improves its chances of reaching the destination on time. The time-to-collision metric is used to decide which side the lane change must take place, if both the sides are possible. Further, any lane change computed by the previous strategies (either due to the general lane change rule or while cooperating with another vehicle) can only happen if the vehicle by its action of lane change does not happen to come in front of another vehicle which is running late by a larger amount. These rules make the transportation system cooperative in which a lesser late vehicle (or a nonreal-time task vehicle) clears the way for a vehicle running late.

The lane change rule can be easily implemented if all the vehicles are connected by a communication framework. However, general traffic would be filled with both semi-autonomous and human-driven vehicles. Hence, a relaxed format of this rule is proposed, wherein a vehicle running late may *flash its state* for information to the other vehicles. If the vehicle has communication facilities, it may communicate the lateness cost as well. The lateness cost largely depends upon the state, and hence the human-driven vehicles may easily estimate the cost from the state. In any case it is not mandatory for any vehicle, with or without communication, to cooperate.

A problem associated with the entire approach is that there is an incentive to falsely mark one as running late by the highest degree to get priority over all the other vehicles. People may do this to arrive as early as possible. Because the vehicles are being monitored, it should be easy for the transportation authorities to set some *global thresholds* on the maximum monthly/yearly lateness or a *monetary cost* may be incurred. Hence, it should be possible to control cheating or charge a price for priority.

15.6 Results

The approach was tested using reasonably realistic simulations. The simulation tool was developed in JAVA. The first major task was specifying the roadmap. For the experiments the Map of Reading, United Kingdom, was used which was obtained from Openstreetmap (2012) and loaded into the simulator as an Extensible Markup Language (XML) file. Fig. 15.5 shows a view of the map so obtained. The data were parsed to mine out the roads and intersections. The simulator assumed all intersections of three or more roads a traffic crossing. The data can consist of isolated nodes which were eliminated using a Depth First Search. The resultant roadmap was hence such that all pairs of nodes can be traversed. The map had a total of 7765 road nodes.

The simulator used a primitive traffic-light system which allowed traffic from a particular side to flow for a specific duration of time or up to the time when there was no vehicle left to cross. The side of travel was cycled in a clockwise manner. All major roads were assumed double lanes, whereas all other roads were assumed single. The speed limit of all roads was specified as 40 miles/hour. The traffic obeyed to-the-left rule in which the vehicles are supposed to travel on the left side of the road, as is the case in the UK.

The motion of the vehicles was made using the Intelligent Driver Model (Treiber et al., 2000) using traffic microsimulation. The method states the manner in which a vehicle follows another vehicle and hence the concept can be extended to describe the motion of each vehicle in the transportation system. The parameters of the model are set the same as in Treiber et al. (2000) with the exception that the maximum speed of the vehicle was taken to be 40 miles/hour.

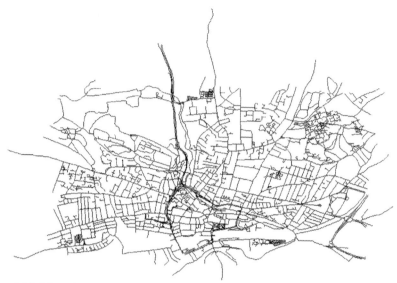

Figure 15.5 Map of Reading, United Kingdom, used for experimentation.

The traffic simulation uses an origin–destination matrix specifying the vehicles along with the time of emergence. For all vehicles, the origin was selected using a Gaussian distribution with mean centred outside the map's central point by a magnitude of half the radius. The angle of origin to the map's centre (θ) was chosen randomly. The destination was also chosen from a Gaussian distribution with mean at half the map's radius. The angle of destination to the map's centre was chosen from a Gaussian distribution with mean located at $\pi + \theta$. The closest node in the map was used as the origin and destination for the vehicle. The time of emergence was made random.

The first task associated with the algorithm was to learn the travel speeds for the recurrent vehicles. The generation of vehicles was done for a total of 12 h. For half the time the rate of vehicles uniformly decreased from five vehicles per second to zero vehicles per second, and for the other half the rate was uniformly increased from zero vehicles per second to five vehicles per second. This signified peak times in the mornings and evenings, with less congestion during the middle of the day. At each iteration of learning, the emergence time was shifted by an amount taken from a Gaussian distribution of mean 0 and with a deviation of 10 min. Further, an additional 10% of vehicles were added to introduce some nonrecurrent nature in the traffic system. These vehicles created uncertainties in the travel times of the vehicles marking recurrent traffic. Learning with these recurrent vehicles with shifted emergence time and nonrecurrent vehicles was carried out for a number of iterations. Each vehicle used the A* algorithm to compute its route with an objective to minimize the travel distance, whereas the actual travel happened with the Intelligent Driver Model. Learning was carried out for a total of 10 iterations with a learning rate of 0.4. For the simulations the value of σ_{min} was fixed to 7 miles/hour.

For testing, some additional vehicles were used which brought in a nonrecurrent aspect. Now the time to reach the goal was specified. The vehicles attempted to find the best route as well as the start time using the learnt speeds and deviations. The simulations were repeated for a number of vehicles over a number of scenarios. The experiments were further repeated for a number of values of α. The values of β and γ were fixed to $\alpha/4$ and $-\alpha/4$. It is reasonable to require that these values scale in proportion to α.

Each vehicle had its own start time, ideal finish time, time the vehicle actually finished its journey and travel time which were specific to its source, goal and ideal finishing time. These metrics cannot be compared between the vehicles. Hence, for the purpose of study the deviation of the vehicle arrival time at its goal was used. The ideal value is zero which would mean that a vehicle arrived at its destination at the time it was supposed to. A positive value specifies that the vehicle was late and the magnitude specifies the amount of time by which the vehicle was late. A negative value specifies that the vehicle arrived early and the magnitude specifies the duration by which the person would have to wait at the destination due to an early arrival.

For every value of α, a histogram showing the percentage of vehicles for the duration is given by Fig. 15.6 for the system with cooperation and Fig. 15.7 for the system

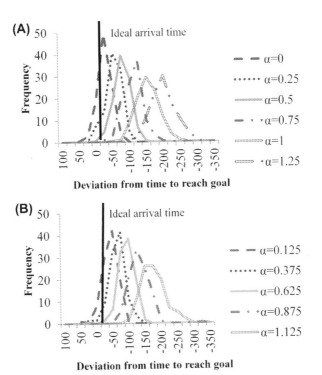

Figure 15.6 Histogram for deviation from time to reach goal for the transportation system with cooperation. (A) Interleaving half of the experimented values, (B) the other interleaving half of the experimented values.

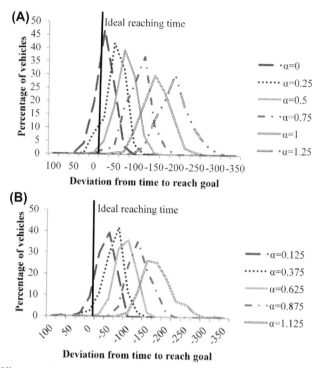

Figure 15.7 Histogram for deviation from time to reach the goal for the transportation system without cooperation. (A) Interleaving half of the experimented values, (B) the other interleaving half of the experimented values.

without cooperation. The histogram is produced in pockets of 50 s. The average travel time for any vehicle was on the order of 15 min for the considered map. The positive region of the graph shows late arrival which is undesirable. As expected, the higher values of α shift the histogram towards the negative region signifying that the vehicle reaches its destination much earlier than the expected time. This though causes a reduction in the percentage of vehicles arriving late.

Fig. 15.8 specifically shows the percentage of vehicles arriving late for different values of α. The percentage is high for very low values of α, whereas it dies off

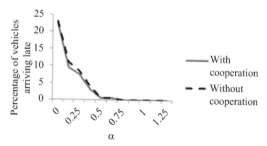

Figure 15.8 Percentage of vehicles arriving late.

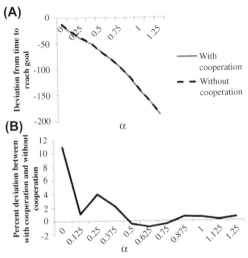

Figure 15.9 Deviation from time to reach goal: (A) measured value and (B) percentage with respect to noncooperative system.

very quickly as α increases. The figure further shows that making the system cooperative results in a lower percentage of vehicles running late. The mean deviation from time to reach the goal for different values of α is shown in Fig. 15.9A. Higher values of α make an average vehicle arrival time very early, compared to low values of α in which the average vehicle more or less reaches its destination at the specified time. It is difficult to observe the difference between cooperative and noncooperative traffic systems and hence the difference as a percentage to noncooperative systems is shown in Fig. 15.9B. In general, the cooperative transportation system makes the vehicles arrive earlier, whereas the degree of early arrival is higher for smaller values of α.

Another metric is used to study the effect of adding cooperation to the system. The metric of the percentage of vehicles running late does not account for whether the lateness was by a few seconds or by a few minutes. Although a good system would minimize the number of vehicles running late, the vehicles arriving late would be late by a small time only. Hence, a cost metric was used, which was kept as zero if the vehicle arrived on time and given a penalty equal to the time by which the vehicle was late otherwise. The metric was averaged over the number of vehicles and the resultant graph is shown in Fig. 15.10. Clearly, cooperation results in fewer vehicles running late, whereas the vehicles which were counted as late were actually late by small amounts.

From the results it can be seen that a lower value of α is preferable for most times when it is not very important to reach the destination either on or before the specified time. Mostly, one can take a change of 10−15%, and for an odd time when one does reach the destination late, it may be socially acceptable. Taking such a risk gives a big boon in terms of departure time which can be reasonably late. However, on occasions the 10−15% risk might be very high. One may not always be ready to miss an important flight or arrive late at a very important interview and hence may not be willing to

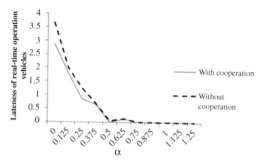

Figure 15.10 Net amounts by which real-time operation vehicles arrive late.

take a risk as high as 10−15%. It is perhaps best to take moderate values of α in such situations using which the risk may drop to as low as 5%. The departure time is affected as a result, but the rise is justifiable, given the risk. The need to be assured of reaching the destination on time is (usually) a rare requirement, in which case one must be willing to leave very early. It can also be seen that there is a large region in which the risk stays on the order of 1−5%, corresponding to which there is a significant increase in the deviation from the time to reach the destination, implying that the typical result will be a very high waiting time at the goal.

This is in fact an experimental verification of a naturally observed phenomenon of how humans decide their start time and route under known traffic conditions. Arriving just before a regular meeting is common wherein at times one may even turn up a little late. Similarly, waiting for a long time for a flight is common when one does not want to miss out. What has been done here, however, is to automate such procedures, at all times keeping the necessities and individual requirements of travellers to the fore. The proposed system is certainly better than using a system which gives some estimate of travel time with the user required to further guess the additional time as a precaution to traffic uncertainties. The guess is based on principles which are unreliable.

15.7 Summary

This chapter first introduced the mechanism of computing the journey start time and route. The objective was to enable vehicles leave as late as possible while still being assured of reaching the destination within the decided time. The algorithm attempted to make both the journey time short as well as reducing the waiting time in case of an early arrival. The algorithm attempted to solve the problem by exploiting the recurrent nature of traffic. The algorithm assumed the presence of intelligent agents at all intersection points and proposed a mechanism by which these agents could be used to learn the expected speed of travel at the various times of day. This information was used in computing the latest start time for a vehicle to reach its destination on time, at the same time ensuring that the probability of reaching the goal on time was as specified by the user.

An inverted graph search was performed from the goal to the source, whilst at each expansion speed and deviation information was used to compute the latest time of departure from each node to reach the goal on time. Experimentally, the tradeoff between the probability of reaching the destination on time and having to wait on arrival were shown. It was observed that one is likely not to need to wait for very long on average, provided he/she is ready to accept a little risk in arriving late. In case the user needs to guarantee that they reach their destination on time, waiting times at the goal can be significantly large.

The second part of the chapter talked about the ways by which the transportation system could favour a vehicle arriving late. The approach aimed at making the transportation system cooperative to favour the vehicles on a real-time task and running late. The attempt was to ensure that the vehicles can reach within the fixed time, even if some initially unaccounted-for delay may have happened on the way. Such a vehicle requiring to reach within time may be in a state of being comfortable with time, a little late, very late, or having given up. For implementation, the traffic lights as well as the lane changes were made cooperative to prioritize the vehicles arriving late. By such an approach it was seen that the vehicles could reach within the decided time, even though they may be departing late.

This explains many practical phenomena such as timing oneself to reach a destination just in time under normal circumstances, having to wait at the goal for a very long time for critical appointments and reaching the destination somewhat early when being a little late may be acceptable. In general life, one may time oneself, but knowing the risk involved and the benefits thereafter is important for decision-making. Further, a mechanism is needed to achieve authentic information for decision-making, which the users may themselves not have. This chapter made use of an intelligent road infrastructure system to solve the problem.

References

Ando, Y., Fukazawa, Y., Masutani, O., Iwasaki, H., Honiden, S., 2006. Performance of pheromone model for predicting traffic congestion. In: Proceedings of the Fifth International Joint Conference on Autonomous Agents and Multiagent Systems, New York, pp. 73–80.

Bishop, R., 2000. Intelligent vehicle applications worldwide. IEEE Intelligent Systems and Applications 15 (1), 78–81.

Claes, R., Holvoet, T., Weyns, D., 2011. A decentralized approach for anticipatory vehicle routing using delegate multiagent systems. IEEE Transactions on Intelligent Transportation Systems 12 (2), 64–373.

Dia, H., 2001. An object-oriented neural network approach to short-term traffic forecasting. European Journal of Operational Research 131 (2), 253–261.

van Hinsbergen, C.P.I.J., Hegyi, A., van Lint, J.W.C., van Zuylen, H.J., 2011. Bayesian neural networks for the prediction of stochastic travel times in urban networks. IET Intelligent Transportation Systems 5 (4), 259–265.

Innamaa, S., 2005. Short-term prediction of travel time using neural networks on an interurban highway. Transportation 32, 649–669.

Kala, R., Warwick, K., 2014. Computing journey start times with recurrent traffic conditions. IET Intelligent Transportation Systems 8 (8), 681—687.

Kala, R., Warwick, K., 2015. Reaching destination on time with cooperative intelligent transportation systems. Journal of Advanced Transportation. http://dx.doi.org/10.1002/atr.1352.

Kim, S., Lewis, M.E., White III, C.C., 2005. Optimal vehicle routing with real-time traffic information. IEEE Transactions on Intelligent Transportation Systems 6 (2), 178—188.

Kirby, H.R., Watson, S.M., Dougherty, M.S., 1997. Should we use neural networks or statistical models for short-term motorway traffic forecasting? International Journal of Forecasting 13 (1), 43—50.

Kuwahara, M., Horiguchi, R., Hanabusa, H., 2010. Traffic simulation with AVENUE'. In: Fundamentals of Traffic Simulation. Springer, New York, pp. 95—129.

Li, Q., Zeng, Z., Yang, B., 2009. Hierarchical model of road network for route planning in vehicle navigation systems. IEEE Intelligent Transportation Systems Magazine 1 (2), 20—24.

van Lint, J., 2006. Incremental and online learning through extended Kalman filtering with constraint weights for freeway travel time prediction. In: In Proceedings of the Intelligent Transportation Systems Conference, Toronto, pp. 1041—1046.

Maniccam, S., 2006. Adaptive decentralized congestion avoidance in two-dimensional traffic. Physica A: Statistical Mechanics and Its Applications 363 (2), 512—526.

Miller-Hooks, E., Mahmassani, H., 2003. Path comparisons for a priori and time-adaptive decisions in stochastic, time-varying networks. European Journal of Operational Research 146 (1), 67—82.

Narzt, W., Pomberger, G., Wilflingseder, U., Seimel, O., Kolb, D., Wieghardt, J., Hortner, H., Haring, R., 2007. Self-organization in traffic networks by digital pheromones. In: Proceedings of the IEEE Intelligent Transportation Systems Conference, Seattle, WA, pp. 490—495.

Opasanon, S., Miller-Hooks, E., 2006. Multicriteria adaptive paths in stochastic, time-varying networks. European Journal of Operational Research 173 (1), 72—91.

Openstreetmap, Available at: openstreetmap.org, (accessed August, 2012.).

Pavlis, Y., Papageorgiou, M., 1999. Simple decentralized feedback strategies for route guidance in traffic networks. Transportation Science 33 (3), 264—278.

Reichardt, D., Miglietta, M., Moretti, L., Morsink, P., Schulz, W., 2002. CarTALK 2000: safe and comfortable driving based upon inter-vehicle-communication. In: Proceedings of the 2002 IEEE Intelligent Vehicle Symposium, vol. 2, pp. 545—550.

Song, Q., Wang, X., 2011. Efficient routing on large road networks using hierarchical communities. IEEE Transactions on Intelligent Transportation Systems 12 (2), 132—140.

Tatomir, B., Rothkrantz, L., 2006. Hierarchical routing in traffic using swarm-intelligence. In: Proceedings of the 2006 Intelligent Transportation Systems Conference, Toronto, pp. 230—235.

Treiber, M., Hennecke, A., Helbing, D., 2000. Congested traffic states in empirical observations and microscopic simulations. Physical Review E 62 (2), 1805—1824.

Verhoef, E.T., 1999. Time, speeds, flows and densities in static models of road traffic congestion and congestion pricing. Regional Science and Urban Economics 29 (3), 341—369.

Weyns, D., Holvoet, T., Helleboogh, A., 2007. Anticipatory vehicle routing using delegate multi-agent systems. In: Proceedings of the IEEE Intelligent Transportation Systems Conference, Seattle, WA, pp. 87—93.

Conclusions

16

16.1 Conclusions

The book is devoted to an extensive study into the technology behind intelligent vehicles. Throughout the book, the intention was to make next-generation vehicles ones which drive all by themselves, while interacting with other vehicles, transportation management centres, road-side units, traffic lights etc. This promises a fascinating future wherein the transportation system will comprise intelligent agents which work in close cooperation with each other. Imagine the ability to have all of the information about current and future traffic, and the ability to talk to all the vehicles round, while travelling in the most sophisticated transportation system. At one end, this enables a vehicle to make informed navigation plans and decisions, whereas at the other end this enables the transportation system to most efficiently manage the movements of all the vehicles. Fundamentally, there are two related problems which were critically discussed in the book, the navigation of a vehicle amidst a pool of vehicles and the intelligent management of the transportation system.

The first problem was that of *navigation of an autonomous vehicle*. To navigate the vehicle, discussions started with the physical setup of the vehicle, basic vehicle capabilities, the technology behind the vehicle's ability to look around, and the ability to interpret and implement driving decisions. A basic primer of decision-making and executing action in the real world was given under the hood of the Advanced Driver Assistance System. Thereafter, the discussion was on the specific problem of trajectory planning of autonomous vehicles.

The second problem attempted to look at the transportation system at large, and devise traffic policies and intelligent concepts for the working of the entire system, given information in the form of sensors embedded at different places on the road which record and give information about the traffic density at different roads and areas. The aim was *intelligent management of the transportation system*. Here, the discussions started with a basic understanding of traffic and its components. Thereafter, the intention was to make each of the components behave intelligently for an efficient overall performance of the transportation network.

So far, the entire domain may appear extremely broad and scary, having a number of models, methods, and results. This chapter very sweetly puts everything into perspective. The focus is on a multitude of approaches in the form of algorithms to solve the same problem. This chapter attempts to knit together the different approaches with respect to the current and future states of autonomous vehicles and intelligent transportation systems. The aim is to present how the different approaches are related, when they become similar, how they differ, and which approach finds application at what places.

16.2 Autonomous Vehicles

The first part of the book was devoted towards the problem of navigation of autonomous vehicles. Autonomous vehicles contain a number of sensors which enable the vehicle to perceive the physical environment. The raw sensor readings are given to the processing units to process the sensor values and generate meaningful decisions, which are used to navigate the vehicle. Vision modules are responsible for converting the sensor readings into information about the navigable areas, nonnavigable areas, the type and location of obstacles etc. The information is given to a mapping server to continuously make a map of the world, while a localization module locates the vehicle in this world. The motion planning module makes all the vehicle decisions, which are given to a control system for the physical navigation of the vehicle.

Perception by the vehicle is a complicated task, making use of a number of sensors to redundantly see the same thing, and making inferences based on the same. The most dominant sensors used for decision-making are video cameras and light and distance detection (lidar). The aim is to make a three-dimensional (3D) map of the world around the vehicle for making navigation decisions. Vision using video cameras is useful for identifying types of nearby objects such as pedestrians, vehicles, road and nonroad regions, lane markings, sign boards,obstacles etc. The input image may be denoised and used for segmentation of different objects and regions of interest, further worked upon by feature extraction techniques to define features, and finally used by a classifier for identifying the object. Each of these features has some uniqueness that can be exploited by the recognition algorithms.

As a motivation towards autonomous driving, the book first presented Advanced Driver Assistance Systems. Inattention alert systems were used to detect driver inattention from different modalities. Similarly, Adaptive Cruise Control and Vehicle Following systems could be used to drive the vehicles and relieve the human from much of the driving. Similarly, the overtaking-assist and the parking-assist systems perform difficult manoeuvres. Multiple vehicles can also be connected by a communication system, to make more effective driving decisions. The vehicles can further be connected by the road-side units and transportation management centres for effective two-way broadcast of information.

The specific problem studied in the book was the trajectory planning of autonomous vehicles for which a number of algorithms were used. Readers may look back at the summary sections of the various chapters to recall the methods. For a general overview of the algorithms, please refer to Tiwari et al. (2013). Here, the algorithms are first assessed with respect to each other, and this knowledge is projected into a practical context to understand what actually the different algorithms offer or what they do not. The other possibilities are also commented upon.

16.2.1 Algorithm Analysis

For trajectory planning of a single vehicle the book presented a number of algorithms. From the perspectives of a single vehicle, the different algorithms are conveniently summarized by Table 16.1.

Table 16.1 **Summary of the Different Algorithms for the Planning of a Single Vehicle**

S. No.	Algorithm	Optimality	Completeness	Computation Time	Scalability	Iterative (If Deliberative)
1.	Genetic algorithm	Probabilistically optimal. More exploitative version was implemented, which meant less global optimality	Probabilistically complete for a reasonable number of obstacles	Little high	Reasonably scalable	Yes
2.	Rapidly-exploring random trees (RRT)	No	Near-complete	Fair	Largely scalable	No
3.	RRT-Connect	Locally optimal, globally optimal for simple cases	Near-complete	Fair	Largely scalable	No
4.	Multilevel planning	Near optimal. Can miss overtakes with very fine turns	Near-complete	Little high	Poorly scalable	No
5.	Planning using dynamic distributed lanes	Near optimal. Can miss overtakes with very fine turns	Near-complete	Somewhat high	Poorly scalable	No
6.	Fuzzy logic	No	No	Very low	Completely scalable	N/A
7.	Lateral potentials	No	No	Very low	Completely scalable	N/A
8.	Elastic strip	Near optimal. Can miss very fine turns	Near-complete (less than 2, 3, 4 and 5)	Medium	Very scalable (more than 2 and 3)	N/A
9.	Logic-based planning	Locally near-optimal. (Less than 3)	No (more than 6 and 7)	Low	Almost scalable	N/A

In terms of *optimality*, the order of algorithms is (more to less): Planning using Dynamic Distributed Lanes, Multilevel Planning, Genetic Algorithm (GA), Elastic Strip, Rapidly-exploring Random Trees (RRT)-Connect, RRT, Logic-based Planning, Lateral Potentials and Fuzzy Logic.

In terms of *completeness*, the order is (more to less): GA, RRT-Connect, RRT, Multilevel Planning, Planning using Dynamic Distributed Lanes, Elastic Strip, Logic-based Planning, Lateral Potentials and Fuzzy Logic.

In terms of *computational time*, the order is (least to highest): Fuzzy Logic, Lateral Potentials, Logic-based Planning, Elastic Strip, Multilevel Planning, Planning using Dynamic Distributed Lanes, RRT-Connect, RRT and GA. Theoretically, the graph search-based approaches scale poorly compared to the GA, which has an additional advantage of being iterative. However, practically the number of other vehicles or obstacles would never be very large. Further, it is practically difficult to produce scenarios wherein every vehicle needs to consider every other vehicle and obstacle in all ways of interactions.

The coordination handling of the different algorithms was somewhat different. The coordination mechanism is summarized by Table 16.2. It is evident that no decentralized coordination strategy can be globally optimal, whereas for the particular problem any complete algorithm (the ability to find a solution, if one exists) would make the resultant approach complete whatsoever the coordination strategy (and vice versa). In the worst cases, the vehicles start following each other. It is hard to comment upon the different strategies as they are all applied with different assumptions. The graph search-based approaches have coordination tightly bound with the planning algorithm. For both these approaches, the coordination is hence somewhat cooperative as these approaches can deliberate and schedule each vehicle correspondingly, which is something absent in the other approaches in which the coordination is entirely based on simple overtaking and vehicle-following heuristics.

For the other approaches, in place of studying the coordination per algorithm, it would be viable to broadly list the different types of coordination strategies followed. Two types of strategies are used, *deliberative* and *reactive*. The first type of coordination strategy is used with the deliberative (or near-deliberative) algorithms and the other with the reactive algorithms. Reactive coordination techniques simply treat the other vehicles as obstacles for deciding the immediate move. Another classification applicable for the deliberative strategies is on the basis of *communication*. With the presence of communication, the positions of the other vehicles in the future are precise, whereas with no communication the positions are guessed.

Further, there are *cooperative* and *noncooperative strategies* depending upon whether they ask the other vehicles to shift to some particular side. The next factor to consider is the *overtaking mechanism*. Some strategies can compute whether overtaking would be possible and hence push the vehicle in front for overtaking only if it is feasible. The other vehicles cannot compute the feasibility of overtaking and always attempt to overtake. The last factor to consider is *speed determination* in which some coordination strategies try hard to optimize the travel speeds, whereas others simply set the speed to overtake or follow the other vehicle.

Table 16.2 **Summary of the Different Algorithms for the Coordination of Vehicles**

S. No.	Algorithm	Coordination	Communication, Assumptions	Optimality	Computational Complexity
1.	Genetic algorithm	Traffic inspired heuristics for path/speed, Prioritization	Yes, vehicles stay on their left sides mostly	Suboptimal, global knowledge makes it more desirable	Somewhat high to continuously alter speed and check overtake feasibility. Computation is distributed as the vehicle travels
2.	Rapidly-Exploring random Trees (RRT)	Prioritization, Attempt to maintain maximum collision-free speed	Yes	Noncooperative, suboptimal	A Little high due to multiple attempts to compute speed
3.	RRT-Connect	Prioritization, vehicle following/overtake-based speed determination	Yes	Noncooperative, suboptimal	Small time needed to decide between overtaking and vehicle following
4.	Multilevel planning	Layered Prioritization, each layer uses separation maximization heuristic, vehicle following/ overtaking-based speed determination	Yes	Largely optimal	High due to a large number of replanning of different vehicles at different levels
5.	Planning using dynamic distributed lanes	Pseudocentralized, each state expansion uses separation maximization heuristic, vehicle following/ overtaking-based speed determination	Yes	Largely optimal, cooperation can be slow	High as part trajectories of a number of vehicles need to be continuously altered

Continued

Table 16.2 Summary of the Different Algorithms for the Coordination of Vehicles—cont'd

S. No.	Algorithm	Coordination	Communication, Assumptions	Optimality	Computational Complexity
6.	Fuzzy logic	Vehicles treated as obstacles, distances assessed for overtaking decision-making, speed controlled by fuzzy rules	No, vehicles stay on their left sides mostly; roads notwide enough to accommodate multiple vehicles per side of travel	Suboptimal, not accounting for global knowledge makes it undesirable	Nil
7.	Lateral potentials	Vehicles treated as obstacles, always overtaking strategy, distance from front used for deciding speed	No, one way only	Suboptimal, not accounting for global knowledge makes it undesirable	Nil
8.	Elastic strip	Vehicles treated as moving obstacles, always overtaking strategy, distance from front used for deciding speed	No, one way only	Suboptimal, not accounting for global knowledge makes it undesirable	Very short time needed to extrapolate vehicle motion
9.	Logic-based planning	Vehicles treated as moving obstacles, Lateral distances measured for overtaking decision-making, distance from front used for deciding speed	No, vehicles stay on their left sides mostly	Suboptimal, cooperation can be slow, not accounting for global knowledge makes it undesirable	Very short time needed to extrapolate vehicle motion

In terms of *computational time*, the deliberative strategies would need some time to query or extrapolate the vehicle's position into the future. Further computation would be needed to determine the overtake mechanism if it is a requirement. Heuristic-based speed determination is faster than iteratively tuning it.

In terms of *optimality*, the cooperative strategies are always better. Reactive and near-deliberative planners do not use global knowledge of the environment and hence can be more suboptimal. The logic-based planner further uses the notion in which the vehicles check the situation, make a lateral movement, again check the situation, and so on. Each lateral motion is a small trajectory. Hence, the cooperation can be slow. Similar is the problem with planning using dynamic distributed lanes in which the vehicle's path can only be altered at the pathway segment start. This induces suboptimality.

16.2.2 Practical Scenarios

Based on Section 16.2.1, it is clear that optimality and completeness come with computational costs or with an assumption of communication, and hence there may not be a perfect algorithm. This section talks about the mechanism of selecting the best algorithm for practical traffic scenarios, at the same time practically explaining Tables 16.1 and 16.2.

Especially for the early models of autonomous vehicles, it appears nice to recommend a fuzzy logic-based planner which does not require a computationally heavy system which at the same time works without the need of communication in all types of roads and traffic. In fact, many autonomous vehicles do use fuzzy controllers for lane keeping, lane changing and overtaking (eg, Jin-ying et al., 2008; Naranjo et al., 2008; Onieva et al., 2011). One may currently not mind some overtakings being missed or a slightly longer path being taken. The risk is, however, what happens if a car breaks down near the centre of the road. Considering a two-lane road, depending on whether the breakdown took place more on the right or more on the left of the road, it can be decided whether the left or right side of the road is blocked. This is what the fuzzy logic cannot make out. The same argument holds true for lateral potentials. The additional limitation of that approach is that it is not allowed to move to the wrong side of the road, even though it is a typical practice to do so for obstacle avoidance. So if the car broke down directly in front, the vehicle would actually stop as the next lane is reserved for the traffic from the other side. These algorithms cannot hence be used if there is no human inside to solve for such deadlocks.

Hence, if the vehicle is really to be operated on a fully autonomous mode without a human driver inside, it might be more advisable to go for logic-based planning. For the presented scenario, it would have been able to compute both of the possibilities and decide which way to go. The only additional disadvantage would be a little higher computational time. In the above problem, say a bicycle on the road broke down instead of a car, such that the vehicle could navigate on eitherside. If one does mind optimality, then obstacle avoidance from the left would be unacceptable if another bicyclist is riding on the left a little ahead. If, however, a little loss of optimality is acceptable, having to follow the bicyclist for a little while before overtaking should not be a problem. However, consider that two bicycles were broken down, one at the centre and the other a little more ahead towards the left. Now when the vehicle comes and takes the left to avoid the first bicycle, it would be considered dim-witted as it could not foresee that there was insufficient distance to overtake the second. To make matters worse, consider that both the broken bicycles had been previously autonomously driven with no human inside any of the three vehicles. This would imply that the system would be in a deadlock until help arrives from somewhere.

Elastic strip, of course, would have been able to detect the situation and act accordingly, which would have come with an additional computational cost, requiring better computational systems onboard (consider that the approach is extended to work with the traffic from both sides). Now consider wide roads in which the slower vehicles generally travel more on the left and the faster ones more on the right, but somehow the autonomous vehicle happens to lie on the left. On attempting overtake between two slow vehicles, it may not get cooperation and hence may have to follow the slower pool of vehicles, as ideally overtaking should take place on the right of the complete pack of vehicles. None of the previous approaches would be able to solve the problem either. Here, optimality would be of concern because one cannot expect the vehicle to complete the entire journey following slow vehicles.

Consider a vehicle intention analysis system is made which enables prediction of a vehicle's trajectory, using which the assumption of communication in GA, RRT and RRT-Connect algorithms can be dropped. RRT and RRT-Connect are noncooperative algorithms, and hence if a vehicle prefers to drive straight in the middle of the road on a two-way lane (assuming one-way), no other vehicle from behind can overtake it and would be forced to follow. Assuming the coordination has been made cooperative, these approaches solve all the above problems. GA may be better suited with an additional computational cost. A reverse limitation would of course exist. Even with the cooperation of the vehicle pool ahead, these approaches would make the vehicle overtake from wide off the pool.

It is common to see a large number of vehicles on a wide road, each travelling with some speed. On the roads with intermediate density, motor bikes may intervene in between the vehicles and cut their way with a large speed. Planning such motor bikes with these approaches would be difficult as the number of vehicles is very large. Any of the above approaches, especially the reactive ones, could have done so with the least computational cost. As the implementations are rather exploitative, on a wide road with a few vehicles, the vehicle may end up following the pack of vehicles ahead if the trajectory that overtakes them is rather odd.

In a scenario in which a large number of vehicles are autonomous, a common problem would be the traffic system giving options to choose one's lane when the lanes are divided by physical barriers. Using the above methods all the vehicles may initially choose to travel by the same lane, giving rise to problems later on. It was shown in the experimental results how multilevel planning solved the problem. Further, suppose vehicles are parked along one side of the road on a two-lane road, leaving only one lane active for part of the road (a common problem in English road travel). Now, vehicles coming from both sides of the road need to communicate with each other to decide whether the inbound vehicle would travel next or the outbound vehicle.

Such a problem cannot be solved by any of the above methods. In fact, humans use gestures and, sometimes, actual communication to solve the problem. The experimental results showed that planning using dynamic lanes can solve such a problem. These methods, however, cannot be used without communication between the vehicles. These methods cannot pass the motor bike scenario wherein they would make it follow the vehicle due to the computational constraints.

It is obvious that the scenarios mostly kept adding obstacles at places which may not be very practical, but the issue is what happens in such situations with no human aid. This is the reason why such cases are often not discussed in the literature. Further, mostly following a vehicle is considered a fair strategy. In fact, it is easy to build planning algorithms that simply follow the vehicle in front, or if either of the vehicles is too large or small, to wait for a reasonable separation to be made and move ahead. Due to its simplicity, this behaviour was not shown in the experiments presented throughout the book. The core focus of the research is to study situations when following a vehicle has a significant loss.

Based on the discussions it becomes even clearer that a perfect algorithm is not possible. The discussions tended to keep adding the loss of not deliberating and, hence, showcased the losses. In addition, of course, very deliberative algorithms cannot be used with a large number of vehicles. Even for this problem the trade-off between *being deliberative* and *being reactive* will remain. It will be best to have some *situation assessment* using which the algorithm can change forms. However, situational assessment is itself a big challenge. Computational availability is another constraining factor defining the choice. Autonomous vehicles can naturally store much computing infrastructure due to their large size; however,using means that are too deliberative with too many vehicles or obstacles may never be a good choice due to the exponential scalability issues.

16.2.3 Other Possibilities

A number of other approaches are possible which were not discussed. For simplicity, the only consideration given is whether a planning algorithm can result in enabling a close overtaking from happening, assuming a comfortable separation is available when the overtaking happens. The most obvious experiment is by assuming the road is divided by lanes, even though the other vehicles may or may not move within the marked lanes. This makes the planning simple and computationally inexpensive. However, clearly the overtaking will be called infeasible if half of the vehicle ahead is occupying the lane to be used for overtaking.

Similarly, *multiresolution planning techniques* (Chen et al., 1997; Kambhampati and Davis, 1986; Urdiales et al., 1998; Yahja et al., 1998; Kala et al., 2011) or any technique which makes a low-resolution map for coarser planning cannot be used. In such techniques, it is impossible to decide on a lower-resolution map whether the overtaking is possible. Lower-resolution means inaccurate information of the available separation for overtaking, which is important for deciding the feasibility of overtaking. The class of planning algorithms used for *structured environments* (Choset and Burdick, 1995; Fiorini and Shiller, 1998; Oommen et al., 1987;

Wesley and Lozano-Pérez, 1979) were also not used. These algorithms assume the structural information of the obstacle, like a polygon, to be precisely known. This was partly because of the additional assumption of the knowledge of the structure of the obstacles. Further, converting an obstacle into its structural equivalent may lead to a loss of width offered for overtaking and hence the overtaking may be judged incorrect.

The key issue is whether the *fusion of deliberative and reactive planning techniques* (Kala et al., 2010a,b) can take place to create a perfect hybrid algorithm. Certainly, using deliberative techniques on a coarser-level low-resolution map followed by a reactive planner is not possible as with the mobile robots, due to the previously highlighted reasons. The challenge here lies at the deliberative phase. Further, mostly mobile robots deal with deliberation for static obstacle avoidance, whereas the main challenge of trajectory planning of autonomous vehicles is enabling vehicles to intelligently avoid each other for which the dynamic obstacle avoidance strategy plans a major role. Currently, it seems difficult to think about the ways (other than the approaches discussed in the book) by which the information can be curtailed at the coarser level at the same time not affecting the overtaking decisions. This, of course, is an open question.

16.3 Intelligent Transportation Systems

The second part of the book was based upon the *intelligent management of the transportation system*. To fully understand the traffic system and motivate the design of solutions which intelligently manage the traffic system, traffic flow theory was used. The traffic flow theory leads to a fundamental relation that an increase in the traffic density reduces the mean speed and increases the traffic flow from free flow to a maximum of capacity flow; and thereafter, an increase of traffic density severely reduces the traffic flow and travel speeds leading to congestions. The model was extended for the cases of interruptions including traffic lights and mergers. This leads to the ability to simulate traffic given some road network graph and some traffic data.

The main takeaways for this part were not the algorithms to solve a particular problem, but the modelling of the problem using the various concepts. For this part of the book, the traffic was assumed to be moving within lanes. The various concepts are summarized in Table 16.3.

It should be possible to mix up the discussed concepts to create the transportation system of choice. The chief problem is *routing* which largely determines the travel efficiency. Apart from the map and the available road infrastructure, the assumption of recurrent or nonrecurrent traffic and the percentage of vehicles which are semiautonomous become important aspects. This determines how much it is possible to distribute the traffic on competing near-optimal routes, and how much in time bands, or simply by selling the road occupancy by time for money. The other elements of the transportation system can certainly aid the routing algorithm to further make the travel efficient.

Table 16.3 Summary of the Different Concepts for Intelligent Management of the Transportation System

S. No.	Concept	Features
1.	Routing objective/ considerations	Traffic density, congestion control, risk, traffic lights, expected travel time, best/worst travel time, time to reach destination, start time, reserved road (travel cost)
2.	Routing frequency	Frequent replanning, fixed plans, incomplete or complete plans
3.	Routing traffic assumptions	Recurrent, nonrecurrent, recurrent with some possibility of nonrecurrent trends
4.	Traffic lights	Cyclic, earliest vehicle first based, most late vehicles first based
5.	Lane change	Overtake based (extra lane primarily used for overtaking), cooperative to vehicles arriving more late, dynamic speed limit based, reserved lane (travel cost)
6.	Traffic	Entirely semiautonomous, mixed, manual

The type of traffic differs from place to place. Every place has a different degree of general congestion, different chance of getting congested, different degree of recurrence which may be used for prediction etc. This makes it difficult to state what would constitute a perfect transportation system. Consider typical British traffic. Traffic mostly seems clear at most places showing recurrent trends. However, for no clear reasons odd days see semihigh to high traffic jams on certain roads. As most roads are single lanes, much does not depend upon the mechanism of lane changes, although it may be profitable to let vehicles getting late pass through with a high priority on wider roads. A clear example of lane reservations is by keeping bus only lanes which provoke a natural question whether VIP vehicles, or other vehicles via a subscription fee, should also be allowed to use these lanes. It is already being questioned to raise the speed limits for autonomous vehicles, which means a strictly autonomous vehicle lane is a possibility which more generally may be a specific-speed capability-only vehicle lane.

Traffic lights hold a potential of regulating traffic to avoid the small congestions or near-congestions typically found. Currently, the greatest limitation is to detect the vehicles on all sides, only after which an intelligent scheduling framework may work. Such a framework can itself solve large congestion issues. A large queue of vehicles waiting for their turn, with the traffic light green for a road with no queue is clearly inefficient, which currently happens in the traffic systems. Such scenarios lead to small congestions which eventually get clear. Further, in most cities the traffic operates from the typical residential areas to the office areas in the mornings and vice versa in the evenings. This makes the distribution directional, causing congestions. Subsequently, preferring vehicles with the flow of traffic can lead to the traffic being static at high-density roads for lesser durations. This would further encourage the drivers to take a route with the general traffic flow instead of intercepting the traffic at intersection points. Currently, making some lanes one-ways partly does the same thing.

The most significant improvement can be made by *routing* which is clearly different for the recurrent and nonrecurrent cases. The problem is important because general traffic mostly shows recurrent trends with some temporary traffic jams which crop out of nowhere. Intuitively, one may say that the start times and the initial route can be computed with the algorithm discussed in chapter "Reaching Destination Before Deadline With Intelligent Transportation Systems," whereas postemergence the vehicle may actually be made to travel by the routing algorithm of chapter "Intelligent Transportation Systems With Diverse Vehicles." Although the former captures the recurrent trends to best place the vehicle in the domain of emergence times, the routing takes care of avoiding any nonrecurrent congestion that may happen. The approach was naturally not simulated as the former technique attempts to make the vehicle arrive within the specified time using competitive means, whereas the latter uses cooperative means to minimize the risk of traffic congestion, which makes the two very different objectives. Practically, both these objectives are important, and the mix would largely depend upon how much of what objective is preferred. Humans tend to be uncooperative, which is the reason for the First-Come-First-Served queues. Even if autonomy may not be a requirement, how the people would be made to sacrifice their personal best routes for overall more efficient traffic is a big question.

If the traffic demand is too high and an early departure is not acceptable, the only way out is to keep the road *reserved* for oneself. This is similar to the same road used for tram and vehicle traffic in certain cities, where the road may be called reserved by the tram. On the same lines, the natural question is, given that a road is under congestion, can some part of the traffic infrastructure be reserved for some special vehicles only, forcing the others to use alternate means (public transportation) or alternate times (too early start).

16.4 Limitations

The first part of the book was devoted to the trajectory planning of autonomous vehicles. The book leaves a pool of open questions, upon which the readers are encouraged to ponder. The first major assumption was the presence of vision, control, localization, and vehicle/obstacle-tracking algorithms which operate without lanes. It is important to model these algorithms accordingly. Further, the book largely considered only the problem of planning vehicles within a road segment. The planning for traffic jams, deadlocks, intersections, mergers, diversions, traffic signals, parking, take off from parking, sharp U-turns etc. for an unorganized operational scenario that needs to be carried out. The transfer of a vehicle from one segment to another or from one segment to any of these scenarios should be smooth. Deadlock recovery for an all autonomous vehicle scenarios is particularly challenging. Further, the algorithms assumed knowledge of the complete map, whereas in reality only part of the map may be known at any time.

Better strategies need to be made to understand vehicle intentions and hence to predict the future move. The algorithms need to be experimented on a physical vehicle or validated against real data sets. A wide variety of algorithms were presented and still no algorithm proved flawless. The task of understanding the situation and hence switching between the algorithms needs formal study. Although the former would only be made possible once the complete vehicle is engineered to operate without lanes, the latter would only be possible once a framework to track and record unorganized traffic data if designed and developed. Real-life traffic operates with social implications to various vehicles which further indicate which set of vehicles allow overtaking of which other set. This needs to be modelled. The ultimate goal is to enable the vehicles operate in traffic scenarios without lanes. The algorithms presented are a positive step towards this problem.

The other part of the book was devoted to the intelligent management of the transportation system at large. A limitation of the entire approach is that the diversities, number of vehicles getting late, number of recurrent and nonrecurrent traffic vehicles etc. are largely random. It would be good to take some real-life data, make realistic assumptions about anything absent in the data needed by the system, and to simulate the resultant approach. Testing on randomly created data gives good results. However, it is difficult to understand the physical interpretation of the results. A number of assumptions regarding the ability to track and inform the vehicles were made, which need to be better addressed. Better traffic prediction systems may be used. The experiments may be repeated for larger maps. It is particularly important to address how the vehicles with human drivers may be encouraged to cooperate with each other, or stopped from not cooperating. The current simulations are a long way from the simulations in the absence of lanes. The economic and social aspects were used to bias the system towards some vehicles. The manner in which these aspects are actually managed by the transportation authority is still not modelled.

16.5 Closing Remarks

Because long autonomous vehicle driving was seen as one of the unsolved and ultimate goals of Artificial Intelligence (AI). The current state of the art points towards feasibility of the technology, and it may be comfortably stated that the problem is partly solved. Although legal, ethical, nonideal scenarios and roads, difficult to manoeuvre roads, fear of unknown scenarios etc. may still create hurdles in the use of autonomous vehicles on the road (or on some roads), technology would eventually produce them.

The future of the traffic landscape as these autonomous vehicles replace the manually driven vehicles is still unclear. A number of things may currently be expected as the application specialists use the underlying autonomous vehicle technology for specific operations. A few things expected include vehicles operating in high-speed platoons, vehicles attaching to and detaching themselves from other vehicles, small vehicles going under other vehicles, vehicles fitting inside public vehicles, vehicles using both road and pedestrian areas as mobile robots, flying robots partly driving on the road in certain areas, vehicles which can deform as per the application requirements etc. As the future traffic landscape is still unclear, many of these problems cannot be modelled and studied within an AI framework. Each of these possible traffic implications poses difficult AI problems which would require new tools and algorithms in the future.

References

Chen, D.Z., Szcerba, R.J., Uhran, J.J., 1997. A framed-quadtree approach for determining Euclidean shortest paths in a 2-D environment. IEEE Transactions on Robotics and Automation 13 (5), 668−681.

Choset, H., Burdick, J., 1995. Sensor based planning. I. The generalized Voronoi graph. In: Proceedings of the 1995 IEEE International Conference on Robotics and Automation, pp. 1649−1655.

Fiorini, P., Shiller, Z., 1998. Motion planning in dynamic environments using velocity obstacles. The International Journal of Robotics Research 17 (7), 760−772.

Jin-ying, H., Hong-xia, P., Xi-wang, Y., Jing-da, L., 2008. Fuzzy controller design of autonomy overtaking system. In: Proceedings of the 12th IEEE International Conference on Intelligent Engineering Systems, Miami, Florida, pp. 281−285.

Kala, R., Shukla, A., Tiwari, R., 2010a. Fusion of probabilistic A* algorithm and fuzzy inference system for robotic path planning. Artificial Intelligence Review 33 (4), 275−306.

Kala, R., Shukla, A., Tiwari, R., 2010b. Dynamic environment robot path planning using hierarchical evolutionary algorithms. Cybernetics and Systems 41 (6), 435−454.

Kala, R., Shukla, A., Tiwari, R., 2011. Robotic path planning in static environment using hierarchical multi-neuron heuristic search and probability based fitness. Neurocomputing 74 (14−15), 2314−2335.

Kambhampati, S., Davis, L.S., 1986. Multiresolution path planning for mobile robots. IEEE Journal of Robotics and Automation 2 (3), 135−145.

Naranjo, J.E., González, C., García, R., de Pedro, T., 2008. Lane-change fuzzy control in autonomous vehicles for the overtaking maneuver. IEEE Transactions on Intelligent Transportation Systems 9 (3), 438−450.

Onieva, E., Naranjo, J.E., Milanés, V., Alonso, J., García, R., Pérez, J., 2011. Automatic lateral control for unmanned vehicles via genetic algorithms. Applied Soft Computing 11 (1), 1303−1309.

Oommen, B., Iyengar, S., Rao, N., Kashyap, R., 1987. Robot navigation in unknown terrains using learned visibility graphs Part I: the disjoint convex obstacle case. IEEE Journal of Robotics and Automation 3 (6), 672−681.

Tiwari, R., Shukla, A., Kala, R., 2013. Intelligent Planning for Mobile Robotics: Algorithmic Approaches. IGI Global Publishers, Hershey, PA.

Urdiales, C., Bantlera, A., Arrebola, F., Sandoval, F., 1998. Multi-level path planning algorithm for autonomous robots. IEEE Electronics Letters 34 (2), 223−224.

Wesley, M.A., Lozano-Pérez, T., 1979. An algorithm for planning collision-free paths among polyhedral obstacles. Communications of the ACM 22 (10), 560−570.

Yahja, A., Stentz, A., Singh, S., Brumitt, B.L., 1998. Framed-quadtree path planning for mobile robots operating in sparse environments. In: Proceedings of the IEEE International Conference on Robotics and Automation, pp. 650−655.

Index

Printed in the United States
By Bookmasters